READER'S DIGEST GUIDE TO
BRITAIN'S
WILDLIFE, PLANTS&FLOWERS

BRITAIN'S WILDLIFE PLANTS AND FLOWERS
was edited and designed by
The Reader's Digest Association Limited
London

Editor: Michael Wright
Art editor: Mavis Henley

First edition Copyright © 1987
The Reader's Digest Association Limited
11 Westferry Circus, Canary Wharf,
London E14 4HE
www.readersdigest.co.uk

We are committed to both the quality of our
products and the service we provide to our
customers. We value your comments, so please
feel free to contact us on 08705 113366,
or via our web site at www.readersdigest.co.uk
If you have any comments about the content of
our books, you can contact us at
gbeditorial@readersdigest.co.uk

Reprinted 2006

Copyright © 1987
Reader's Digest Association Far East Limited
Philippines Copyright 1987
Reader's Digest Association Far East Ltd

ISBN (10) 0 276 42635 5
ISBN (13) 978 0 276 42635 3
Book code 040-193-06
Oracle code 250005370H.00.24

Much of the material in this book was originally
published in the Reader's Digest Nature Lover's
Library series:
Animals of Britain
Birds of Britain
Butterflies and Other Insects of Britain
Trees and Shrubs of Britain
Water Life of Britain
Wild Flowers of Britain
Copyright © 1981, 1984
The Reader's Digest Association Ltd

® READER'S DIGEST, THE DIGEST and the Pegasus
logo are registered trademarks of
The Reader's Digest Association, Inc.
of Pleasantville, New York, U.S.A.

Printed and bound in Europe by Arvato Iberia

Print details for 2002 revision

Typesetting: Computape (Pickering) Limited
Page styling and make-up: Tony Rilett

Published by The Reader's Digest Association Limited
LONDON • NEW YORK • SYDNEY • MONTREAL

READER'S DIGEST GUIDE TO

BRITAIN'S
WILDLIFE,
PLANTS&FLOWERS

contributors

The publishers wish to express their gratitude to the following people for their major contributions to BRITAIN'S WILDLIFE, PLANTS AND FLOWERS

CONSULTANTS AND AUTHORS

Dr John Feltwell, FRES, FLS, MIBiol

Dr Peter Marren

Dr Peter Moore

Dr Pat Morris
Royal Holloway and Bedford New College, University of London

Tim Parmenter

John Parslow

Dr D. N. Pegler, FLS
Royal Botanic Gardens, Kew

Dr Franklyn Perring
Royal Society for Nature Conservation

Dr Keith Porter, FRES

Dr Anne Powell
Oxford Polytechnic

Phil Rye

Cyril A. Walker
British Museum (Natural History)

ARTISTS

Stephen Adams
Norman Arlott
Robin Armstrong
David Baird
Peter Barrett
Rachel Birkett
Richard Bonson
Leonora Box
Trevor Boyer
Wendy Bramall
Pierre Brochard
Josiane Campan
Jim Channell
Jeane Colville
Philippe Couté
Frankie Coventry

Helen Cowcher
Brian Delf
Colin Emberson
Maurice Espérance
Shirley Felts
Pat Flavel
Sarah Fox-Davies
John Francis
Ian Garrard
Robert Gillmor
Victoria Goaman
Marie-Claude Guyetand
Nicholas Hall
Stephanie Harrison
Tim Hayward

Rosalind Hewitt
Shirley Hooper
Roger Hughes
Delyth Jones
Brenda Katté
Norman Lacey
Stuart Lafford
Richard Lewington
Mick Loates
Erhard Ludwig
Line Mailhé
Helga Marxmüller
Guy Michel
Sean Milne
Robert Morton
Tricia Newell
Colin Newman
Marie-Claire Nivoix

Denys Ovenden
Liz Pepperell
Sandra Pond
Elizabeth Rice
John Rignall
Andrew Robinson
Eric Robson
Derek Rodgers
Jim Russell
David Salariya
Ann Savage
Marjory Saynor
Helen Senior
Sally Smith
Sue Stitt

Gill Tomblin
Libby Turner
François Vitalis
Barbara Walker
Phil Weare
Sue Wickison
Adrian Williams
Ann Winterbotham
Rosemary Wise
Ken Wood
Paul Wrigley
Peter Wrigley

A full list of the illustrations contributed by each artist appears at the end of the book.

contents

about this book

Densely populated, extensively built over and intensively farmed, the British Isles still manage to maintain an astonishing variety of wildlife. No matter where you live in these crowded islands, you can see part of nature's colourful living tapestry — even in your own back garden. Most people are close enough to coast or countryside where nature can expand, offering more species of flowers, trees, birds and animals for the knowing observer to enjoy.

Wildlife is literally all around you, to amuse, delight and reward your curiosity. But the reward is far greater if your curiosity is satisfied by knowledge — knowledge of what you are observing, why a creature behaves as it does, why a plant favours a particular habitat, what its place is in the intricate and almost infinite web of nature.

This book provides that knowledge. It describes more than 1600 species of plants and animals great and small, all illustrated in full colour. It includes the humblest of insects as well as all our birds and mammals, mighty trees as well as delicate wild flowers, ferns and fungi — everything you are likely to see out and about in the British countryside.

Look-alikes compared

Many of the illustrations have been selected from the bestselling Reader's Digest *Nature Lover's Library* series. But this book represents a significant extension of that work, including as it does additional species in many sections, plus an entirely new section on non-flowering plants. The salient identification points of each plant or animal are picked out and signposted in labels to the illustrations, and the book itself is organised to bring look-alikes together and allow the easiest possible comparison between similar species.

In the case of wild flowers, for example, the colour, number and arrangement of the petals are basic characteristics that help to distinguish one species from another. So this is the way in which the wild flower section is organised. If you were to find, for instance, a plant whose flowers have five blue petals arranged regularly, you should turn to pages 76—82 to discover what it is.

The other sections of the book, too, are arranged in a way that aids quick identification. Trees and shrubs are sorted by the shape and arrangement of their leaves, fungi by their overall form, fish by the number of fins on their back. With birds, butterflies and mammals, likeness tends to run in zoological families, so our system of organisation does likewise — for example, putting together all the ducks, all the hoofed mammals and all the white and yellow butterflies in self-contained groups.

Times and places

The main text for each animal or plant includes an indication of the habitat or area in which the species is most likely to be found. Where no specific places are mentioned, the information refers to the British Isles as a whole — both Britain (England, Wales and Scotland) and Ireland (the Republic and Northern Ireland).

Where appropriate, a calendar bar shows the months when you are most likely to see the plant or animal — the flowering season for wild flowers, the period of appearance above ground for non-flowering plants, the months when birds are usually present in some part of the British Isles, the times when insects and their like are to be seen in adult form, and the active period for mammals, amphibians and reptiles (some of which hibernate). Note that both the distribution and the seasonal information (which are both approximate) refer only to the main species named at the top of the column.

The size of each plant or animal is also given, together with its scientific name. Unless otherwise indicated, the size refers to the mature height of wild flowers, trees and other plants and the overall length of adult animals (including birds and fish). But, as with so many aspects of the natural world, there are myriad variations between individuals, so the figures are a guide only.

wild flowers

A rich tapestry of wild flowers is woven into the landscape of the British Isles, about 1700 species in all. Nearly 600 of them – including all those that you are most likely to encounter – are illustrated and described on the following pages. To aid identification they are divided into five major groups by flower colour. Within each category the flowers are subdivided according to the apparent number of petals (including sepals where these look like petals).

So, by observing the colour, counting the petals and looking at the shape of the flower that you are trying to identify, you can narrow down your search to a small number of pages. Within each section, flowers of similar appearance are placed side-by-side for comparison, with labels pinpointing the tell-tale differences. These commonly include the shape, or 'habit' of the growing plant, the shape of the leaves and how they are arranged, the shape of the seed-pod or other receptacle (known as a fruit, whether or not it is edible) and of the seeds themselves, and the way the flowers are arranged. The illustrations show the plants as they grow wild with common companions; only if there is risk of confusion is the companion plant labelled.

The search for wild flowers is a rewarding trail that can begin outside your own front door but can lead through our wildest and most beautiful countryside. The real nature lover takes pride in helping to preserve that beauty while observing and learning about it. By law, you must never uproot a wild plant without the landowner's express permission, and the rarest plants are protected from any disturbance – even picking the flowers or collecting seed. In any case you should not pick any flower without good reason, for a picked flower never produces seed. And try to disturb the wild flowers' environment as little as possible – remember that the bluebell, the glory of our woodlands in spring, is threatened as much by having its leaves trampled as by being picked or uprooted.

White or pale pink flowers

Cleavers

Galium aparine
6-48in (15-120cm)

Sweet woodruff

Galium odoratum
6-18in (15-45cm)

Cowberry

Vaccinium vitis-idaea
6-12in (15-30cm)

Common wintergreen

Pyrola minor
4-12in (10-30cm)

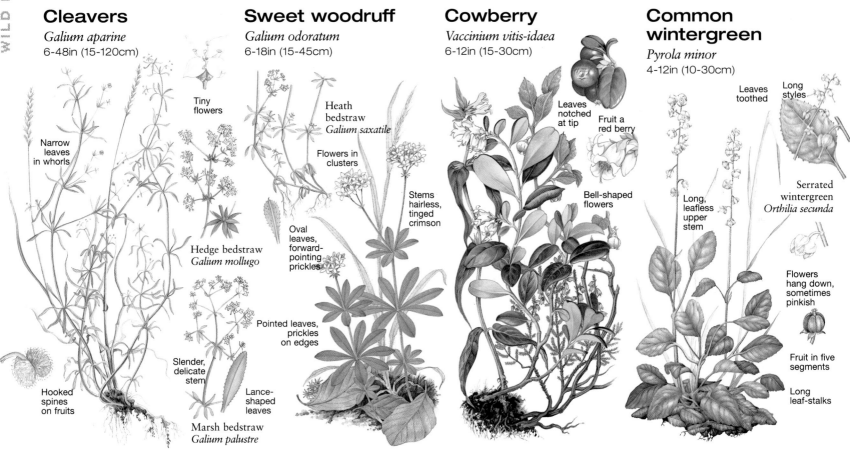

Tiny flowers

Narrow leaves in whorls

Hooked spines on fruits

Hedge bedstraw
Galium mollugo

Slender, delicate stem

Lance-shaped leaves

Marsh bedstraw
Galium palustre

Heath bedstraw
Galium saxatile

Flowers in clusters

Oval leaves, forward-pointing prickles

Pointed leaves, prickles on edges

Stems hairless, tinged crimson

Leaves notched at tip

Fruit a red berry

Bell-shaped flowers

Leaves toothed

Long styles

Serrated wintergreen
Orthilia secunda

Long, leafless upper stem

Flowers hang down, sometimes pinkish

Fruit in five segments

Long leaf-stalks

This tall, bright green plant is so common in our hedgerows that it is easily passed by without a second look. It is called cleavers because of the hooked hairs on its fruit which 'cleave' to clothes and animal hair. It is a favourite food of geese, and is also known as goosegrass.
Hedge bedstraw is a trailing plant with smoother stems, and marsh bedstraw a related species of wet places.

Hedges, ditches and wasteland.

This pretty woodland flower has no smell when fresh; but an hour or two after it is picked a magical scent – somewhere between vanilla and new-mown hay – emerges. The dried plant retains this smell for years, and used to be placed between sheets. Bedstraws are recognised by their slender, square stems, with whorls of 4-12 leaves and clusters of tiny flowers.
Heath bedstraw blackens when it is dried.

Woodlands throughout British Isles.

Like its close relative the bilberry, the cowberry belongs to the heather family and grows on dry, acid soils. While the bilberry's flowers are reddish, those of cowberry are much paler, appearing almost white against the glossy evergreen leaves. They are also more deeply lobed than those of bilberry. Unlike the sweet, black fruit of the bilberry, the cowberry's fruit is red and rather bitter.

Widespread on upland moors and heaths.

At a quick glance the flowers of the common wintergreen might easily be mistaken for lily of the valley. They hang like little bells on a long stem, keeping their distance from the round, glossy leaves. Wintergreen oil, pressed from the leaves, is used to flavour chewing-gum.
Serrated wintergreen is local in Wales and the north.

Locally in woods, moors and damp places in the north; less common in the south.

JFMAMJJASOND   JFMAMJJASOND   JFMAMJJASOND   JFMAMJJASOND

Three water plants and the snowdrop have flowers with only three petals, while the bedstraws have four joined petals, and square stems. Four petals arranged like a cross are typical of plants of the cabbage family, which differ from one another in the shape of their leaves and seed-pods.

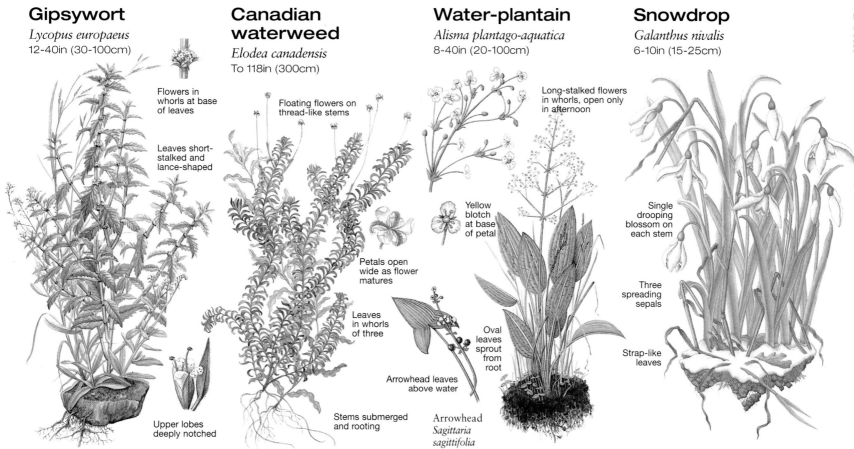

Gipsywort
Lycopus europaeus
12-40in (30-100cm)

Flowers in whorls at base of leaves

Leaves short-stalked and lance-shaped

Upper lobes deeply notched

Canadian waterweed
Elodea canadensis
To 118in (300cm)

Floating flowers on thread-like stems

Petals open wide as flower matures

Leaves in whorls of three

Arrowhead leaves above water

Stems submerged and rooting

Arrowhead
Sagittaria sagittifolia

Water-plantain
Alisma plantago-aquatica
8-40in (20-100cm)

Long-stalked flowers in whorls, open only in afternoon

Yellow blotch at base of petal

Oval leaves sprout from root

Snowdrop
Galanthus nivalis
6-10in (15-25cm)

Single drooping blossom on each stem

Three spreading sepals

Strap-like leaves

A common sight by a shady river bank in summer is the tight little whorls of purple-dotted white flowers decorating the long, stiff stems of the gipsywort. The plant belongs to the mint family, and though completely odourless it shares many characteristics of other mints. Its leaves are set in pairs all the way up the square stem, each pair being capped with flowers. Its lower leaves are deeply lobed.

Damp places, mainly in southern Britain.

The little flowers of this waterweed on the surface of a quiet country stream appear quite harmless. Yet not long ago the plant was viewed with alarm. After this stranger from Canada first appeared in Europe in 1836 it spread with startling rapidity, choking ponds, streams and ditches all over Britain. Then suddenly, in the late 1860s, it stopped its all-conquering expansion.

Widespread in slow-moving fresh water.

In the muddy shallows of ponds and rivers the water-plantain is immediately recognisable by its tall, slender stem rising high above a thicket of broad leaves. The stem culminates in a pyramid of branching spikes tipped with tiny flowers.
The name of arrowhead given to the related water plant found mainly in southern Britain refers only to the leaves growing above the water.

Widespread in mud by ponds and slow rivers.

Perhaps more than any other flower, the snowdrop signals that winter will soon be at an end. A small leaf-like sheath covers the top of the flower stem as it forces its way through the snow. Two bluish-green leaves appear first, followed by the flower, the inner petals of which are notched. The snowdrop was probably introduced from central Europe in medieval times.

Damp woods and river banks; not in Ireland.

| J | F | M | A | M | J | J | A | S | O | N | D |

| J | F | M | A | M | J | J | A | S | O | N | D |

| J | F | M | A | M | J | J | A | S | O | N | D |

| J | F | M | A | M | J | J | A | S | O | N | D |

White or pale pink flowers

Common whitlow-grass

Erophila verna
1-3in (25-75mm)

Deeply cleft petals

Long unbranching stems

Rounded seed-pod in two sections

Spear-shaped leaves in rosette at base

White whitlow-grass is very common on sandy ground throughout Britain, it is often overlooked. Historically it was believed to provide a cure for warts and whitlows. Hence the name, though strictly speaking it is not a grass but an annual herb of the cabbage family. It is a small plant, with deeply cleft petals at the end of long, leafless flower stems.

Widespread on open sandy ground and by walls, paths and rocks.

J F **M A M J** J A S O N D

Horse-radish

Armoracia rusticana
12-36in (30-90cm)

Petals twice as long as sepals

Flowers in dense leafy panicles

Long roots; spreads by underground stems

Lower leaves large, dark green and shiny

This large, distinctive plant was introduced in the 15th century and has spread far and wide beyond the vegetable garden, where it is cultivated for the sake of the pungent sauce made from its grated roots. The plant can be identified by its large, shiny leaves which at the base can grow to 24in (60cm). The many small flowers grow in dense, leafy panicles.

Wasteland, river banks and road verges, especially in England.

J F M A **M J J A** S O N D

Garlic mustard

Alliaria petiolata
8-48in (20-120cm)

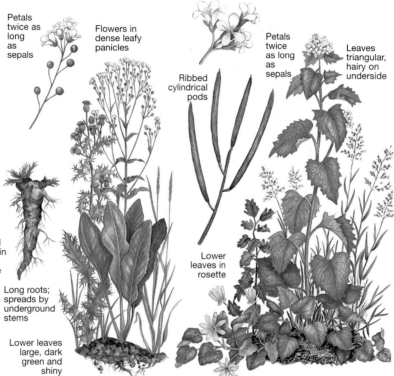

Petals twice as long as sepals

Ribbed cylindrical pods

Leaves triangular, hairy on underside

Lower leaves in rosette

Sometimes called hedge garlic, 'Jack-by-the-hedge' or 'poor man's mustard', this member of the wild cabbage family has small flowers and large, triangular leaves on a tall stem. The seeds are horn-shaped and almost black in colour, and the pods curve in at the base. The whole plant, but particularly the leaves when crushed, gives off a strong smell of garlic.

Wood margins and hedgerows, except in north-west.

J F **M A M J** J A S O N D

Field penny-cress

Thlaspi arvense
4-24in (10-60cm)

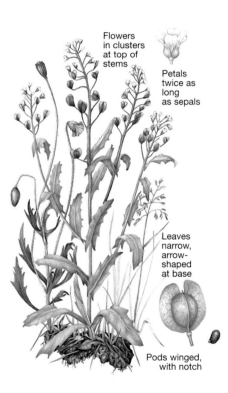

Flowers in clusters at top of stems

Petals twice as long as sepals

Leaves narrow, arrow-shaped at base

Pods winged, with notch

The ripe yellow fruit of field penny-cress brightens innumerable patches of wasteland throughout Britain. It is one of our four penny-cresses and is distinguished from its rarer relatives by the flat, almost circular, discs of its pods. The hairless plant has an upright, leafy stem, sometimes branched, that lengthens when it fruits. Apart from its fruit, the other distinctive feature of this plant is its unpleasant smell.

Wasteland throughout Britain.

J F M A **M J** J A S O N D

Hairy rock-cress

Arabis hirsuta
6-24in (15-60cm)

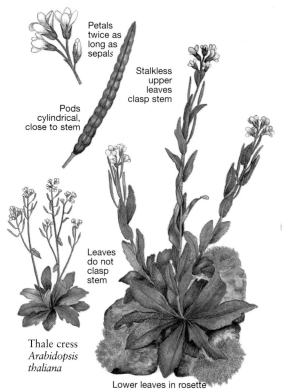

Petals twice as long as sepals

Stalkless upper leaves clasp stem

Pods cylindrical, close to stem

Leaves do not clasp stem

Thale cress
Arabidopsis thaliana

Lower leaves in rosette

Smith's pepperwort

Lepidium heterophyllum
6-36in (15-90cm)

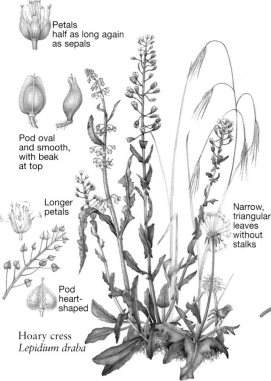

Petals half as long again as sepals

Pod oval and smooth, with beak at top

Longer petals

Narrow, triangular leaves without stalks

Pod heart-shaped

Hoary cress
Lepidium draba

Common scurvygrass

Cochlearia officinalis
4-20in (10-50cm)

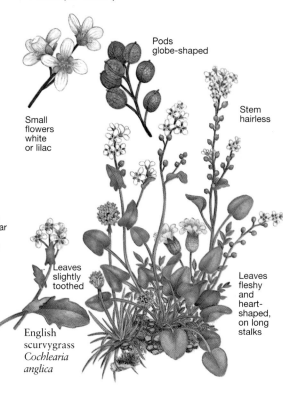

Pods globe-shaped

Small flowers white or lilac

Stem hairless

Leaves slightly toothed

English scurvygrass
Cochlearia anglica

Leaves fleshy and heart-shaped, on long stalks

Hirsuta means 'hairy', and the leaves and stem of this plant are unmistakably that. But it has its attractions nonetheless, particularly in the way its tall, erect seed-pods cluster round the upper stem. The reddish-brown seeds are slightly winged, and ripe ones burst open when touched. The stem is long, slender and usually unbranching, culminating in a long spike of small flowers whose petals are narrow and not notched. Thale cress is also hairy, but its seed-pods and leaves stand away from the stem.

Widespread on chalk slopes, dunes, hedgebanks, walls and rocks.

This weed of arable fields goes under a number of names, including 'Smith's cress' and 'hairy pepperwort'. The leaves, which are long, slender, pointed and toothed, grow straight from the stem, which is upright and often branched from the ground.
Hoary cress is another troublesome weed for the farmer. It has longer petals than Smith's pepperwort and its fruit is heart-shaped. The stem is branched in the upper part. The seeds of this plant were once ground for pepper.

Arable fields, dry banks and verges; commonest in western Britain and southern Ireland.

Spread like cream on top of a headland facing the sea, with plants such as thrift and sea campion, the flowers of common scurvygrass make a pretty sight. Scurvy, caused by prolonged deficiency of vitamin C, is remembered as a problem afflicting sailors in the days of long voyages in sailing ships. But the same dietary deficiencies were not uncommon on land, until gradually in the 17th century it became known that scurvygrass, taken in water, prevented the disease.
English scurvygrass has bigger flowers and toothed leaves.

Widespread on sea-cliffs and salt-marshes, but rarely inland.

J F M A M **J J A** S O N D   J F M A M **J J A** S O N D   J F M A M **J J A** S O N D   13

White or pale pink flowers
Up to four petals • Regular flowers

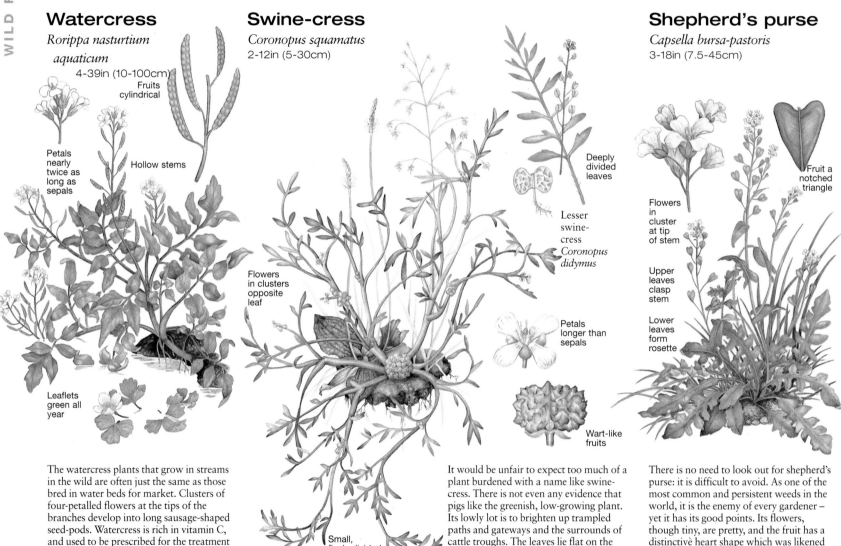

Watercress
Rorippa nasturtium aquaticum
4-39in (10-100cm)

Fruits cylindrical

Petals nearly twice as long as sepals

Hollow stems

Leaflets green all year

Swine-cress
Coronopus squamatus
2-12in (5-30cm)

Deeply divided leaves

Lesser swine-cress *Coronopus didymus*

Flowers in clusters opposite leaf

Petals longer than sepals

Wart-like fruits

Small, finely divided leaves

Shepherd's purse
Capsella bursa-pastoris
3-18in (7.5-45cm)

Fruit a notched triangle

Flowers in cluster at tip of stem

Upper leaves clasp stem

Lower leaves form rosette

The watercress plants that grow in streams in the wild are often just the same as those bred in water beds for market. Clusters of four-petalled flowers at the tips of the branches develop into long sausage-shaped seed-pods. Watercress is rich in vitamin C, and used to be prescribed for the treatment of scurvy. Plants should not be picked from stagnant water or streams flowing through pasture-land, where the eggs of liver fluke sometimes lie.

Throughout Britain in fast streams.

It would be unfair to expect too much of a plant burdened with a name like swine-cress. There is not even any evidence that pigs like the greenish, low-growing plant. Its lowly lot is to brighten up trampled paths and gateways and the surrounds of cattle troughs. The leaves lie flat on the ground, a fan-like background to the flowers, which grow in tight clusters. Lesser swine-cress has hairy stems, more deeply divided leaves and a notched fruit.

Widespread in the south and on coasts.

There is no need to look out for shepherd's purse: it is difficult to avoid. As one of the most common and persistent weeds in the world, it is the enemy of every gardener – yet it has its good points. Its flowers, though tiny, are pretty, and the fruit has a distinctive heart shape which was likened hundreds of years ago to a shepherd's purse. When the pods are ripe the seeds tumble out, as from a purse. The rustic associations add to the plant's appeal.

Widespread in wasteland, fields and gardens.

J F M A M J J A S O N D   J F M A M J J A S O N D   J F M A M J J A S O N D

White or pale pink flowers
Up to four petals
Irregular flowers

Hairy bittercress
Cardamine hirsuta
3-12in (7-30cm)

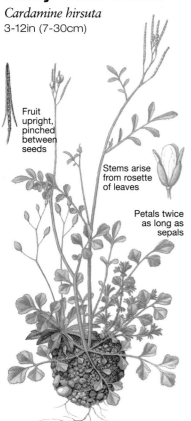

Fruit upright, pinched between seeds

Stems arise from rosette of leaves

Petals twice as long as sepals

This attractive common garden weed has small flowers surrounded by long, cylindrical, upright pods and a neat rosette of leaves at the base. Its stems and leaves have hairs, but they are not very conspicuous. When the seed-pods are ripe they burst with surprising force. Wavy bittercress (*Cardamine flexuosa*) differs from hairy bittercress in having wavy stems and six stamens instead of four

Widespread on bare ground, rocks, walls and dunes throughout British Isles.

Eyebright
Euphrasia nemorosa
1-12in (2-30cm)

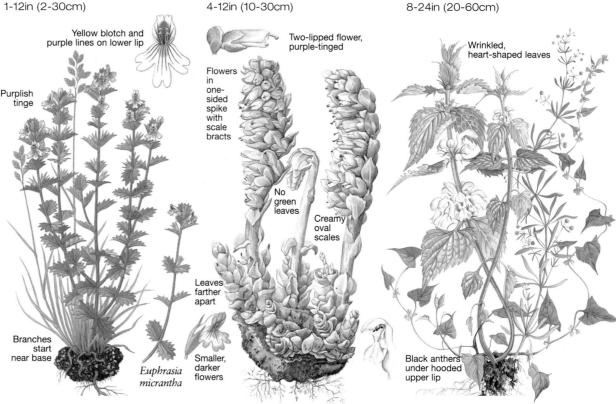

Yellow blotch and purple lines on lower lip

Purplish tinge

Branches start near base

Euphrasia micrantha

Smaller, darker flowers

With its striking, purple-streaked flowers, sturdy branched stem and many-pointed leaves, eyebright is one of the prettiest parasites in our countryside. It exists by attaching itself to the roots of plants such as clover and plantains. More than two dozen species of *Euphrasia* are found in Britain. *Euphrasia micrantha* has smaller, darker flowers and leaves which are farther apart on the stem.

Downs, pastures, heaths and woods, mainly in England and Wales.

Toothwort
Lathraea squamaria
4-12in (10-30cm)

Two-lipped flower, purple-tinged

Flowers in one-sided spike with scale bracts

No green leaves

Creamy oval scales

Leaves farther apart

The whole toothwort plant is cream or pale pink, with no green leaves at all. It is a parasite which attaches itself to the roots of trees, mainly hazel and elm. Its upright, unbranched stems are covered with two rows of flowers which are pollinated by bumble-bees. Instead of leaves it has broad, ivory-coloured scales or bracts which look like teeth, giving the plant its common name.

Damp woods and shady hedgerows, except in northern Scotland.

White dead-nettle
Lamium album
8-24in (20-60cm)

Wrinkled, heart-shaped leaves

Black anthers under hooded upper lip

The large, open-mouthed flowers of the white dead-nettle, a great attraction to nectar-seeking bumble-bees, are a common sight in hedgerows in summer. The flowers are set in a whorl round the square, upright, hairy stems, which often grow in large clumps. Its toothed, heart-shaped leaves are very similar to those of the true stinging nettle, but they are free of stinging hairs – hence the label 'dead'.

Widespread by roadsides, in hedgerows and wasteland.

J F M **A M J J A S** O N D

J F M A M J **J A S** O N D

J F M **A M** J J **A S** O N D

J F M **A M J J A S** O N D

White or pale pink flowers *Up to four petals • Irregular flowers*

White climbing fumitory

Ceratocapnos claviculata
8-32in (20-80cm)

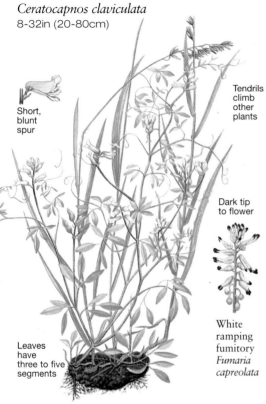

Short, blunt spur

Tendrils climb other plants

Dark tip to flower

White ramping fumitory *Fumaria capreolata*

Leaves have three to five segments

Goat's-rue

Galega officinalis
24-57in (60-145cm)

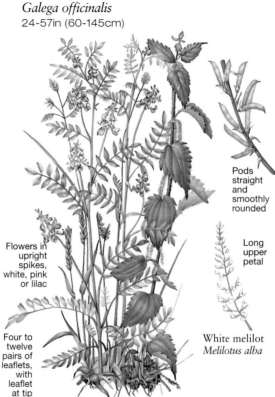

Pods straight and smoothly rounded

Flowers in upright spikes, white, pink or lilac

Long upper petal

Four to twelve pairs of leaflets, with leaflet at tip

White melilot *Melilotus alba*

Hare's-foot clover

Trifolium arvense
4-8in (10-20cm)

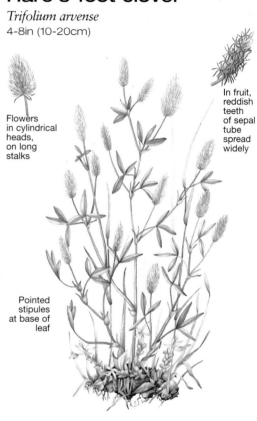

Flowers in cylindrical heads, on long stalks

In fruit, reddish teeth of sepal tube spread widely

Pointed stipules at base of leaf

On a walk through a shady wood in summer, the careful observer will often see the pale creamy-yellow flower of this plant nestling in a tangle of bracken. While the flowers are small, their shape is distinctive. Unlike the closely related fumitories, the leaves of this species end in branched tendrils which enable it to climb other plants, such as grass or bracken.
The rare white ramping fumitory is also a climber, but has smaller, creamy flowers tipped with dark pink.

Pyramids of flowers make goat's-rue an attractive ornamental plant, and it was in this role that it was introduced in the 16th century from southern and eastern Europe, where it is grown for fodder. It has now established itself on damp waste ground in the wild. A member of the pea family, goat's-rue is distinguished from the purple-flowered vetches by having a terminal leaflet, instead of a tendril, at the end of its leaves.
White melilot, also found on waste ground, has only three leaflets to its leaves and its fruits contain a single seed compared with several in each pod of goat's-rue.

Hold the soft, downy flower-head of this plant in your hand and its common name suddenly seems apt. From a distance the long, cylindrical flower-heads look just as distinctive, but call a different image to mind. They are like bottle brushes, waving on the ends of their long stalks. The flowers have a pinkish tinge caused by the reddish pointed teeth of the sepals, which are masked by a covering of long white hairs. The leaflets are narrow and slightly toothed.

Widespread in shady places, except far north and Ireland.

Rare. Naturalised in places on damp waste ground.

Dry, grassy places, commonest in south-east England and on coasts elsewhere.

Among the pea-like flowers in this group are several clovers and trefoils, easily recognised by having leaves that are 'trefoil', or divided into three. The non-stinging dead-nettle (on page 15) has large, toothed leaves while the broomrape (on page 46) is unusual in having no green leaves at all.

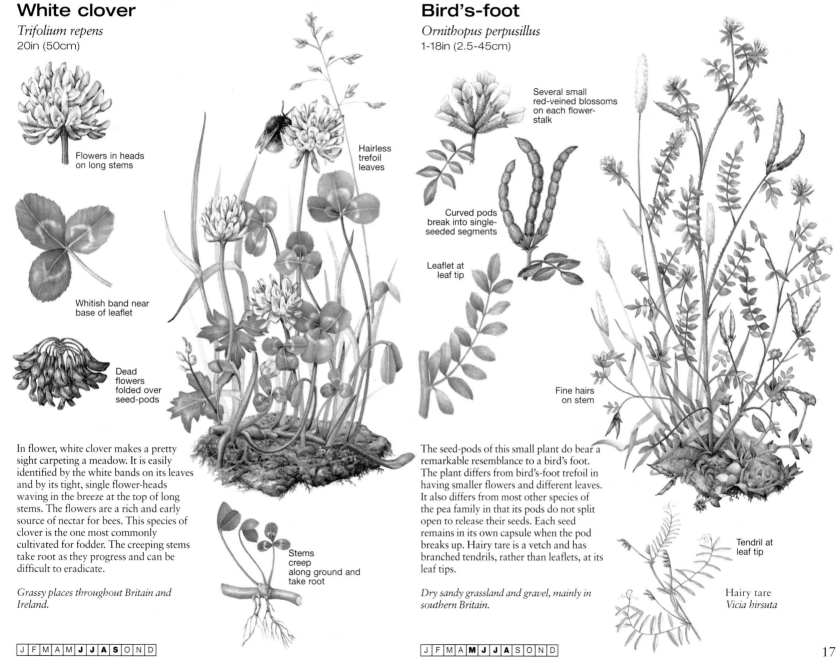

White clover

Trifolium repens
20in (50cm)

Flowers in heads on long stems

Whitish band near base of leaflet

Dead flowers folded over seed-pods

Hairless trefoil leaves

Stems creep along ground and take root

In flower, white clover makes a pretty sight carpeting a meadow. It is easily identified by the white bands on its leaves and by its tight, single flower-heads waving in the breeze at the top of long stems. The flowers are a rich and early source of nectar for bees. This species of clover is the one most commonly cultivated for fodder. The creeping stems take root as they progress and can be difficult to eradicate.

Grassy places throughout Britain and Ireland.

Bird's-foot

Ornithopus perpusillus
1-18in (2.5-45cm)

Several small red-veined blossoms on each flower-stalk

Curved pods break into single-seeded segments

Leaflet at leaf tip

Fine hairs on stem

Tendril at leaf tip

Hairy tare
Vicia hirsuta

The seed-pods of this small plant do bear a remarkable resemblance to a bird's foot. The plant differs from bird's-foot trefoil in having smaller flowers and different leaves. It also differs from most other species of the pea family in that its pods do not split open to release their seeds. Each seed remains in its own capsule when the pod breaks up. Hairy tare is a vetch and has branched tendrils, rather than leaflets, at its leaf tips.

Dry sandy grassland and gravel, mainly in southern Britain.

J F M A M **J** **J** **A** S O N D   J F M A M **J** **J** **A** S O N D

17

White or pale pink flowers *Five to seven petals • Regular flowers*

White campion

Silene latifolia
12-36in (30-90cm)

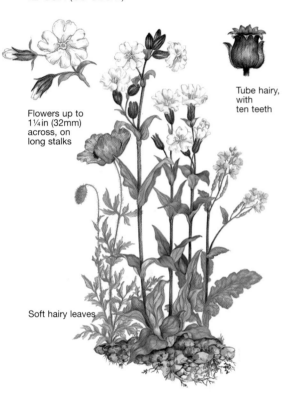

Flowers up to
1¼in (32mm)
across, on
long stalks

Tube hairy,
with
ten teeth

Soft hairy leaves

White campion grows mainly on roadside verges and arable land, unlike the closely related red campion which is found more often in shady hedgerows and woodlands. Where the two plants grow together they often cross-breed and produce pink-flowered plants. The stems are upright, with hairy leaves, and bear large flowers that are faintly scented at night and attract moths. A ring of ten teeth surrounds the opening of the narrow seed capsule. Wreaths of campions were once used to crown the 'champion' in public games.

Widespread in lowlands; rarer in north and west.

Bladder campion

Silene vulgaris
10-36in (25-90cm)

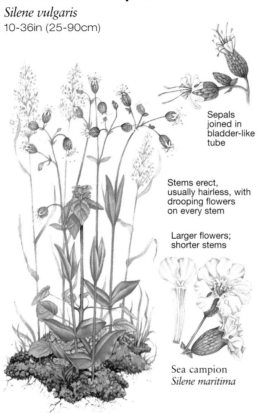

Sepals
joined in
bladder-like
tube

Stems erect,
usually hairless, with
drooping flowers
on every stem

Larger flowers;
shorter stems

Sea campion
Silene maritima

With its yellowish, sometimes purplish, veined sepal tube conspicuously distended, there is no mistaking how this particular campion got its English name. It has pointed, wavy-edged leaves, and the flowers, with their deeply cleft petals, look like frilly skirts at the end of the bulging sepal tubes. They are open day and night, but it is only as dusk falls that they exude their fragrant, clove-like scent, which is so attractive to bees and night-flying moths. Sea campion, found on rocky shores, has larger flowers, shorter stems and many non-flowering shoots.

Widespread on roadsides, open grassland and other disturbed ground in lowlands.

Thyme-leaved sandwort

Arenaria serpyllifolia ssp. *leptoclados*
1-10in (2.5-25cm)

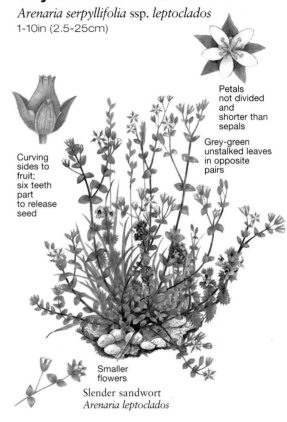

Petals
not divided
and
shorter than
sepals

Grey-green
unstalked leaves
in opposite
pairs

Curving
sides to
fruit;
six teeth
part
to release
seed

Smaller
flowers

Slender sandwort
Arenaria leptoclados

This sprawling plant, which seldom grows more than a few inches high, is easily overlooked. Thyme-leaved sandwort has delicate, grey-green stems, tipped with tiny flowers. These, once noticed, are worth a second look, with their petals set so elegantly against a background of sepals. The fruits also have an interesting shape, like miniature wine jars. The seeds are black and kidney-shaped.
Slender sandwort, or lesser thyme-leaved sandwort, is a more straggly plant with narrower leaves and even smaller flowers. Its fruit has straight sides.

Widespread on walls, cliff tops, chalk downs and arable land.

J F M A M J J A S O N D

J F M A M J J A S O N D

Two main leaf shapes subdivide this large group. Plants with simple leaves include campions, bindweeds and the tiny-flowered sandworts, stitchworts and chickweeds. Plants with leaves divided into leaflets include wild roses and parsleys, with their umbrella-shaped flower clusters.

Field mouse-ear

Cerastium arvense
To 12in (30cm)

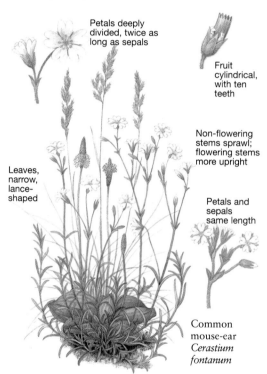

Petals deeply divided, twice as long as sepals

Fruit cylindrical, with ten teeth

Non-flowering stems sprawl; flowering stems more upright

Leaves, narrow, lance-shaped

Petals and sepals same length

Common mouse-ear
Cerastium fontanum

Sticky mouse-ear

Cerastium glomeratum
2-18in (5-45cm)

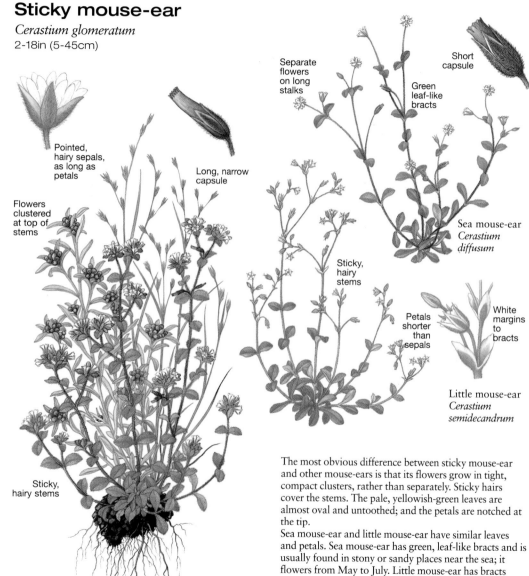

Pointed, hairy sepals, as long as petals

Flowers clustered at top of stems

Long, narrow capsule

Separate flowers on long stalks

Short capsule

Green leaf-like bracts

Sticky, hairy stems

Sea mouse-ear
Cerastium diffusum

Petals shorter than sepals

White margins to bracts

Little mouse-ear
Cerastium semidecandrum

Sticky, hairy stems

Summer grassland is often brightened by broad patches of field mouse-ear. The leaves are covered with short downy hairs, like the ears of a mouse. The 11 species of mouse-ear found in the British Isles are closely related to the stitchworts and chickweeds. Field mouse-ear is a low, hairy-stemmed plant with long, rooting shoots and erect flowering stems. Its leaves are paired at the top and form clusters at the base.
Common mouse-ear is a smaller but equally abundant species, with smaller leaves.

Widespread on dry, chalky grassland, banks and waysides, mainly in eastern Britain; very rare in Ireland.

The most obvious difference between sticky mouse-ear and other mouse-ears is that its flowers grow in tight, compact clusters, rather than separately. Sticky hairs cover the stems. The pale, yellowish-green leaves are almost oval and untoothed; and the petals are notched at the tip.
Sea mouse-ear and little mouse-ear have similar leaves and petals. Sea mouse-ear has green, leaf-like bracts and is usually found in stony or sandy places near the sea; it flowers from May to July. Little mouse-ear has bracts with broad white margins; it flowers from April to May.

Widespread in arable fields, paths, waste and grassland.

J F M A M J J A S O N D

J F M A M J J A S O N D

White or pale pink flowers

Five to seven petals • Regular flowers

Chickweed

Stellaria media
2-14in (5-36cm)

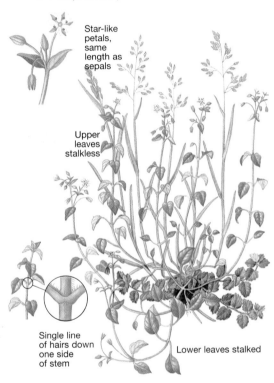

Star-like petals, same length as sepals

Upper leaves stalkless

Single line of hairs down one side of stem

Lower leaves stalked

Five deeply notched petals make the flower of chickweed look like the 'little star' of its scientific name. The reddish anthers on the stamens add a splash of colour. The stems have a line of water-absorbing hairs which changes sides at each pair of leaves. As dew settles on the stem, it runs along the line of hairs until checked by a pair of leaves, where some is absorbed through the hairs and the rest trickles onto the next pair of leaves. Although it looks weak and straggly, chickweed can grow and flower even in the depths of winter.

Widespread on waste and cultivated ground.

J F M A M J J A S O N D

Greater stitchwort

Stellaria holostea
6-24in (15-60cm)

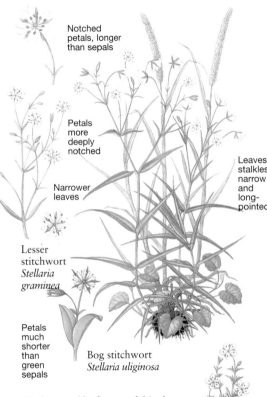

Notched petals, longer than sepals

Petals more deeply notched

Narrower leaves

Lesser stitchwort
Stellaria graminea

Petals much shorter than green sepals

Bog stitchwort
Stellaria uliginosa

Leaf clusters up stem

Petals twice as long as sepals

Knotted pearlwort
Sagina nodosa

The big, star-like flowers of this plant are conspicuous in hedgerows in spring. The petals are cleft halfway up and are longer than the sepals. The leaves are rough-edged and arranged in opposite pairs. Lesser stitchwort has smooth-edged leaves and petals cleft to the base. Bog stitchwort has long sepals forming a green star outside the petals. Knotted pearlwort has a similar flower to greater stitchwort, but with clusters of small leaves up the stem.

Widespread in woods and hedgerows.

J F M A M J J A S O N D

Three-nerved sandwort

Moehringia trinervia
4-16in (10-40cm)

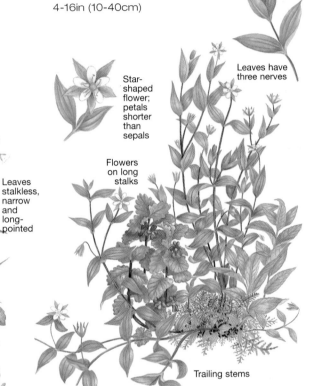

Leaves have three nerves

Star-shaped flower; petals shorter than sepals

Flowers on long stalks

Leaves stalkless, narrow and long-pointed

Trailing stems

The undivided petals distinguish this plant from the superficially similar chickweeds and stitchworts, whose petals are forked. Its leaves have three conspicuous veins or nerves on the underside – another distinctive feature. This is the biggest of the sandworts, with many delicate trailing branches, and the only one that grows in woodland. Its tiny seeds have an oily appendage that attracts ants, which help to distribute them.

Widespread throughout British Isles in dry woods; not western and northern isles of Scotland.

J F M A M J J A S O N D

Corn spurrey

Spergula arvensis
3-16in (7.5-40cm)

Fairy flax

Linum catharticum
2-8in (5-20cm)

Spring beauty

Claytonia perfoliata
4-12in (10-30cm)

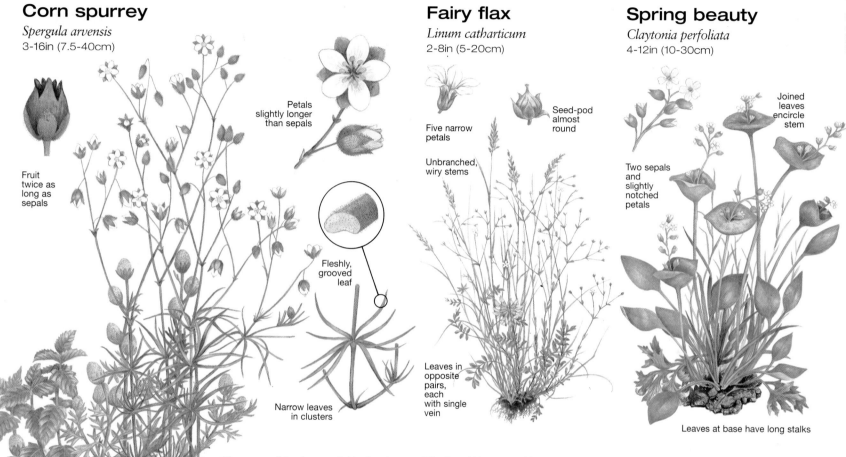

Petals slightly longer than sepals

Fruit twice as long as sepals

Fleshly, grooved leaf

Narrow leaves in clusters

Stems branch at base, then bend upwards

Five narrow petals

Unbranched, wiry stems

Seed-pod almost round

Leaves in opposite pairs, each with single vein

Joined leaves encircle stem

Two sepals and slightly notched petals

Leaves at base have long stalks

The most striking feature of this plant is not the flower, which opens prettily for a few hours from midday and emits an unpleasant smell, but its leaves. They are long and narrow and arranged in whorl-like clusters at the swollen joints of the stems. When the fruits are ripe the flower-stalks turn downwards and the tops of the capsules open to release the seeds, which have a winged edge. Corn spurrey often grows with pineapple weed.

Widespread on waste and arable ground.

The thread-like stems of fairy flax, tipped with their dainty flowers, are often seen on moors and in grassland. The plant, shown here growing with horseshoe vetch, lacks the big blue flower which is the glory of pale flax and perennial flax. Although small and white, however, its flowers do develop into the typical five-celled seed-pod of the flax family. Unlike the other flaxes, the leaves of fairy flax are in opposite pairs. The flowers are in loose clusters.

Widespread on grassland, heaths and fens.

An explosion of tiny flowers from a horn of leaves trumpets the presence of spring beauty. The flowers themselves are unexceptional, but they are presented in green cups, which later spread into curved plates, made by pairs of fused leaves surrounding the stems. The plant was introduced to Britain from western North America in the 18th century. Unlike the chickweeds and spurreys, spring beauty has two sepals instead of five.

Sandy soils in south-east Britain.

J F M A M **J J A** S O N D

J F M A M **J J A S** O N D

J F M A **M J J A** S O N D

White or pale pink flowers

Five to seven petals · Regular flowers

Starry saxifrage
Saxifraga stellaris
To 10in (25cm)

Red anthers; two yellow spots at base of each petal

Sepals bent back

Upper stems leafless

Rosette of toothed leaves at base

Meadow saxifrage
Saxifraga granulata
4-20in (10-50cm)

Single stem, straight and hairy

Narrow petals with green veins

Stalked leaves hairy and toothed

Mossy saxifrage
Saxifraga hypnoides
2-8in (5-20cm)

Reddish tinge

Sepals remain upright

Three-fingered leaves

Rue-leaved saxifrage
Saxifraga tridactylites

Little foliage on flower stems

Leafy lower shoots form tangled mat

White stonecrop
Sedum album
6-10in (15-25cm)

Ring of tiny scales round base of fruit

Star-like flowers with long stamens

Fleshy leaves, flushed with red

Upper leaves flushed with crimson

English stonecrop
Sedum anglicum

This attractive little plant thrives on wet, rocky ledges on high mountain-sides. But it can be seen at lower altitudes, provided the ground is rocky and well supplied with water from above. For all its toughness, it is a slender, graceful plant, sprouting clusters of star-like flowers with distinctive crimson anthers. The name saxifrage means 'stonebreaker', reflecting the old belief that the plant must somehow have caused the cracks in which it roots.

Widespread in mountainous areas.

The tallest of the native saxifrages is an increasingly rare plant of old meadows. The straight, single stems are almost leafless and stickily hairy, and carry up to 12 flowers. But the most striking feature is the tiny brown, nut-like stem bulbs. The tiny rue-leaved saxifrage has a reddish tinge, and stems which are sticky and hairy. It grows on the top of old walls and other dry, lime-rich places.

Widespread in lowlands, but rare in north and west.

Mossy saxifrage is to be found most commonly on mountain ledges and screes where the soil is not acidic, and on open, grassy hillsides. It differs from the other saxifrages in having narrower and more pointed lobes to the leaves and prostrate, non-flowering shoots. But like the meadow saxifrage it often has a cluster of bulbils between the stem and the stalks of lower leaves.

Widespread on high rocky ground or hilly grassland.

Walls, rocks and cliff faces are often adorned with white stonecrop. Its fleshy, cylindrical leaves grow out of the stems in spirals, storing water to tide the plant over the frequent dry spells which are a natural hazard of living on brick and rock. The thick stems rise like little palm tree trunks topped by a broad head of small flowers. English stonecrop is smaller, with fewer branches and stubby red-tinged leaves.

Introduced; widespread on rocks, walls, dunes and shingle.

J F M A M J J A S O N D   J F M A M J J A S O N D   J F M A M J J A S O N D   J F M A M J J A S O N D

Grass-of-Parnassus

Parnassia palustris
4-12in (10-30cm)

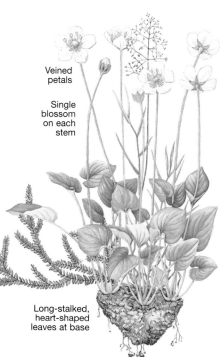

Veined petals

Single blossom on each stem

Long-stalked, heart-shaped leaves at base

Round-leaved sundew

Drosera rotundifolia
2½-10in (6-25cm)

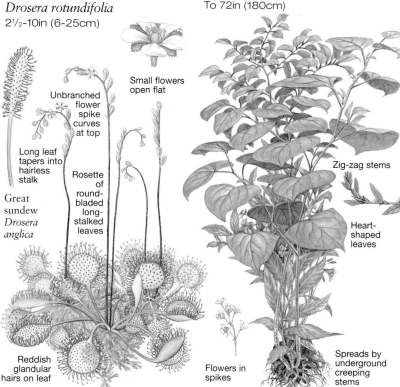

Small flowers open flat

Unbranched flower spike curves at top

Long leaf tapers into hairless stalk

Great sundew
Drosera anglica

Rosette of round-bladed long-stalked leaves

Reddish glandular hairs on leaf

Japanese knotweed

Fallopia japonica
To 72in (180cm)

Zig-zag stems

Heart-shaped leaves

Flowers in spikes

Spreads by underground creeping stems

Knotgrass

Polygonum aviculare
To 40in (100cm)

Flowers in clusters at base of upper leaf-stalks

Oval leaves

The plant's name comes from the Greek mountain where it was seen and written about as early as the 1st century AD. However, it is not a grass and does not look like one. The honey-scented flower has five modified stamens fringed with glands, distinguishing it from the saxifrages with which it is sometimes confused. The loose rosette of heart-shaped leaves on long stalks is the other main feature.

Widespread but local; not southern England and most of Wales.

The 'dew' of this plant is really a glue for catching insects. They are attracted by the shiny drops on the tips of the red hairs, mistaking them for water in which to lay eggs. They are held fast while the leaf curls to enclose them. After a few days the leaf opens and the trap is set for the next meal. One plant can catch as many as 2000 insects in a summer, and these make up for the lack of nutrients in the moorland soil. Great sundew has larger, narrower leaves.

Widespread on wet heaths or boggy moorland.

This plant was introduced in 1825 and became a favourite with Victorian gardeners. It is now a persistent weed, often forming tall, dense thickets. It is fast-growing, with thick, pinkish, zig-zag stems. The flowers are borne in branching spikes, growing from the base of heart-shaped leaves almost 6in (15cm) across. It has extensive underground creeping stems.

Widespread on roadsides, wasteland and river banks throughout British Isles.

One of the most common weeds of the British Isles, knotgrass is a member of the dock family. When found among corn it has upright stems, but on wasteland and seashores its stems are long and straggly and it lies flat on the ground. There is a marked difference in the size of the leaves, which are small on the flowering stems and much bigger on the main branches. Its flowers are small; birds eat the seeds.

Widespread on waste ground, arable fields and seashores.

J F M A M J **J A S O** N D J F M A M J **J J A S** O N D J F M A M J **J A S** O N D J F M A M J **J A S O** N D   23

White or pale pink flowers

Five to seven petals • Regular flowers

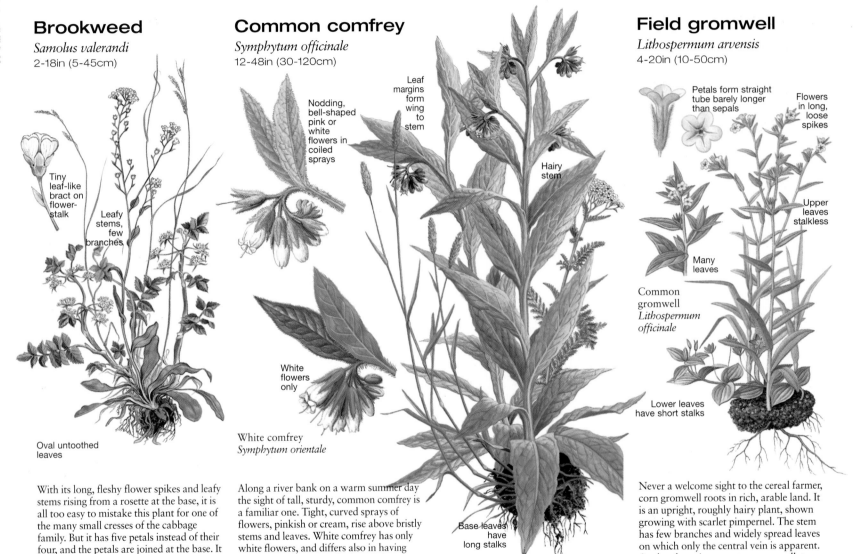

Brookweed
Samolus valerandi
2-18in (5-45cm)

Tiny leaf-like bract on flower-stalk

Leafy stems, few branches

Oval untoothed leaves

Common comfrey
Symphytum officinale
12-48in (30-120cm)

Nodding, bell-shaped pink or white flowers in coiled sprays

Leaf margins form wing to stem

Hairy stem

White flowers only

White comfrey
Symphytum orientale

Base leaves have long stalks

Field gromwell
Lithospermum arvensis
4-20in (10-50cm)

Petals form straight tube barely longer than sepals

Flowers in long, loose spikes

Upper leaves stalkless

Many leaves

Common gromwell
Lithospermum officinale

Lower leaves have short stalks

With its long, fleshy flower spikes and leafy stems rising from a rosette at the base, it is all too easy to mistake this plant for one of the many small cresses of the cabbage family. But it has five petals instead of their four, and the petals are joined at the base. It is usually found on or near the coast where a small stream runs out to the sea, and on damp cliffs above a beach.

Widespread in wet places, especially near sea.

Along a river bank on a warm summer day the sight of tall, sturdy, common comfrey is a familiar one. Tight, curved sprays of flowers, pinkish or cream, rise above bristly stems and leaves. White comfrey has only white flowers, and differs also in having heart-shaped leaves at the base. It is an introduced plant of grassy places, mainly in eastern England.

Damp places; scarce in north and west Scotland and in Ireland.

Never a welcome sight to the cereal farmer, corn gromwell roots in rich, arable land. It is an upright, roughly hairy plant, shown growing with scarlet pimpernel. The stem has few branches and widely spread leaves on which only the central vein is apparent. A related species, common gromwell, grows in long grass or woodland margins. It is leafier, and the leaves have lateral veins as well as a central vein.

Arable land, mainly in south-east England.

J F M A M J J A S O N D

J F M A M J J A S O N D

Hedge bindweed

Calystegia sepium
To 10ft (3m)

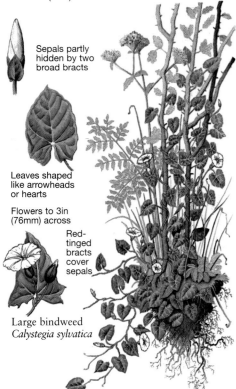

Sepals partly hidden by two broad bracts

Leaves shaped like arrowheads or hearts

Flowers to 3in (76mm) across

Red-tinged bracts cover sepals

Large bindweed
Calystegia sylvatica

Given a free run, hedge bindweed can dress a hedge as if for a wedding with its brilliant trumpet flowers. Indeed, the blooms are so attractive it is easy to forget that they are the product of a very troublesome weed. Although lacking in scent, the flowers are rich in nectar and particularly attractive to the convolvulus hawk-moth. Common or field bindweed (p. 106) has a smaller pink candy-striped flower. Large bindweed bears flowers which are among the largest of all British wild flowers.

Hedges, fen and woodland throughout British Isles; sparser in northern England and Scotland.

J F M A M J **J A S** O N D

Red valerian

Centranthus ruber
12-32in (30-80cm)

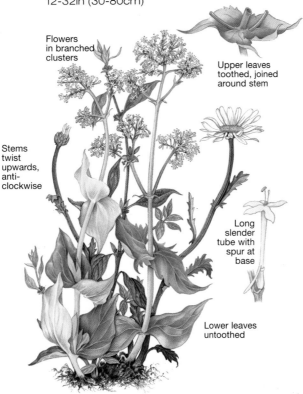

Flowers in branched clusters

Upper leaves toothed, joined around stem

Stems twist upwards, anti-clockwise

Long slender tube with spur at base

Lower leaves untoothed

Dense clusters of little flowers bursting from the end of long slender tubes makes this fragrant herb easy to spot. The flowers can be red or pink, as well as white: the plant is called red valerian after the species name *ruber*, Latin for 'red'. The bluish-green leaves are untoothed at the base, but the upper leaves have curved, tooth-like indentations. The sepals unroll to form a feathery parachute, carrying the seed away on the wind. Sprouting vigorously from old walls and dry banks, valerian shows its Mediterranean origins.

Widespread on wasteland, especially in south-west England.

J F M A M J **J A S** O N D

Black nightshade

Solanum nigrum
To 24in (60cm)

Small flowers in drooping flower-heads

Single, bluish-green leaf blade

Green fruits ripen to dull black

In autumn, when the green berries of this nightshade ripen to black, they look very much like blackcurrants. But they contain the poisonous alkaloid solanine, and should not be eaten. The plant has dark, often blackish, stems and pointed, sometimes toothed, leaves. At a quick glance the little flowers look rather like those of its relative the potato, and the plant is often found growing between potato rows. But it is fond of all well-nourished soil.

Wasteland in England, rarer in Wales and absent from most of Scotland and Ireland.

J F M A M J **J A S** O N D

White or pale pink flowers *Five to seven petals · Regular flowers*

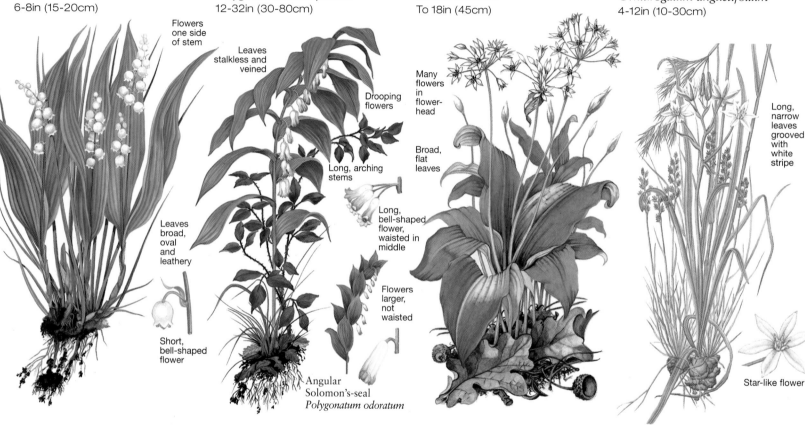

Lily-of-the-valley
Convallaria majalis
6-8in (15-20cm)

Flowers one side of stem

Leaves broad, oval and leathery

Short, bell-shaped flower

Solomon's-seal
Polygonatum multiflorum
12-32in (30-80cm)

Leaves stalkless and veined

Drooping flowers

Long, arching stems

Long, bell-shaped flower, waisted in middle

Flowers larger, not waisted

Angular Solomon's-seal
Polygonatum odoratum

Ramsons
Allium ursinum
To 18in (45cm)

Many flowers in flower-head

Broad, flat leaves

Star of Bethlehem
Ornithogalum angustifolium
4-12in (10-30cm)

Long, narrow leaves grooved with white stripe

Star-like flower

Among the delights of a woodland walk is the sight of the short, bell-shaped flowers of lilies-of-the-valley; their sweet scent adds to the charm. The stems have two shoots, one ending in dark green leaves, the other in one-sided spikes of white flowers which look as if they might tinkle if shaken. All parts of the plant are poisonous.

Widespread in dry woods, England and parts of Wales and Scotland.

Though not so conspicuously pretty as lily-of-the-valley, Solomon's-seal is attractive in a more subdued way. The arching stem carries broad, drooping leaves beneath which hang the slender-waisted flowers. The use of its powdered roots as a cure for bruises may have had the approval of King Solomon.
The stem of angular Solomon's-seal is ribbed, with a single flower to each leaf.

Woods, mainly southern England and Wales.

Crush the stems or leaves of ramsons and the small of garlic can be overpowering. But the plant has other attractions. Its flower-heads can carry up to 25 little star-like flowers which rise above broad, bright green leaves with winged, triangular stalks. Ramsons is a vigorous plant, tending to dominate its habitat to the exclusion of everything else.

Widespread in woods and shady places.

As well as growing in profusion in Palestine, Star of Bethlehem also grows readily in parts of Britain, and may be a native plant. It opens only in the brightest part of the day, from mid-morning to early afternoon: hence its other common name, 'eleven o'clock lady'. Its petals have a broad green stripe down the back which can clearly be seen until the flower is fully open.

On wasteland and in woods; not in Ireland.

J F M A **M J** J A S O N D

J F M A **M J** J A S O N D

J F M A **M J** J A S O N D

J F M A **M J** J A S O N D

Wood-sorrel

Oxalis acetosella
2-6in (5-15cm)

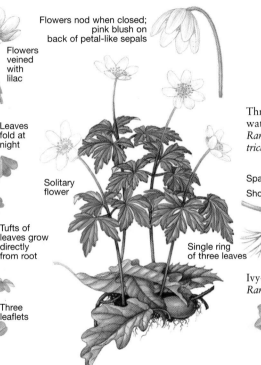

Flowers veined with lilac

Leaves fold at night

Tufts of leaves grow directly from root

Three leaflets

The leaves of wood-sorrel look very much like clover, though the plants are not related. To compound the confusion, the leaves of both fold down for the night. Wood-sorrel produces two sorts of flowers. One is the familiar, cup-shaped, lilac-veined flower which blooms in abundance on woodland floors, but produces very little seed. The other, found on short stalks close to the ground, is self-pollinating and fertile.

Common in woods, hedges and shady spots.

Wood anemone

Anemone nemorosa
2-12in (5-30cm)

Flowers nod when closed; pink blush on back of petal-like sepals

Solitary flower

Single ring of three leaves

One of the delights of a spring walk in deciduous woodland is to find a carpet of wood anemones. The big, demurely drooping flowers respond immediately to the sun and raise their heads, petals wide open, to take it in, only to close again as soon as cloud or evening comes. But they wither quickly if picked. After the plant has flowered, one or two long-stalked leaves grow direct from the underground stem.

Widespread in deciduous woods and coppices.

Common water-crowfoot

Ranunculus aquatilis
Stems to 48in (120cm)

Flowers held just above water

Thread-leaved water-crowfoot
Ranunculus trichophyllus

Spaced petals

Shorter leaves

Lobed floating leaves

Feathery submerged leaves

Globe-shaped fruits hang downwards into water

Ivy-leaved crowfoot
Ranunculus hederaceus
Small flowers

Ivy-shaped leaves

River water-crowfoot
Ranunculus fluitans

Petals overlap

All leaves submerged, like long tassels

A floating bed of deep green leaves supports a multitude of little white and yellow flowers, which just keep their heads above the water. Most of the plant lies submerged, with finely divided frond-like leaves. Ivy-leaved crowfoot has shiny, ivy-shaped leaves, and flowers only ¼in (6mm) across. Neither river water-crowfoot, a plant of fast waters, nor thread-leaved water-crowfoot has floating leaves.

Widespread in ponds, streams and ditches.

J F M A **M J J** A S O N D   J F M A **M J** J A S O N D   J F M A **M J** J A S O N D

27

White or pale pink flowers
Five to seven petals • Regular flowers

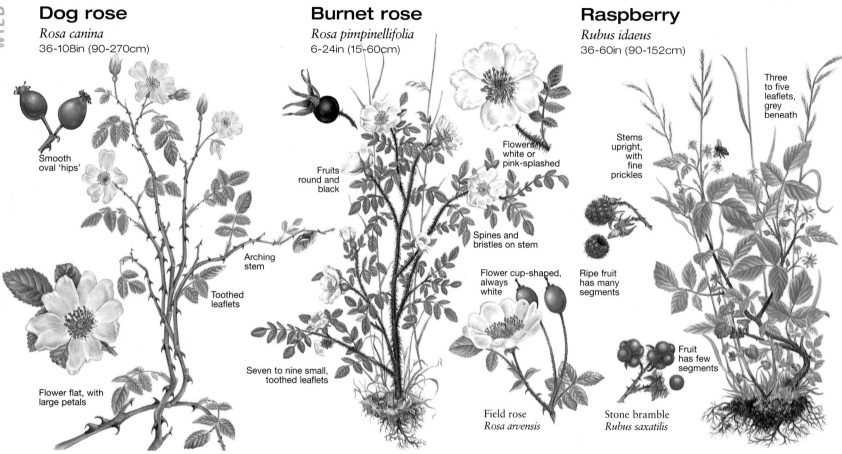

Dog rose
Rosa canina
36-108in (90-270cm)

Smooth oval 'hips'

Arching stem

Toothed leaflets

Flower flat, with large petals

Burnet rose
Rosa pimpinellifolia
6-24in (15-60cm)

Fruits round and black

Seven to nine small, toothed leaflets

Flowers white or pink-splashed

Spines and bristles on stem

Flower cup-shaped, always white

Field rose
Rosa arvensis

Raspberry
Rubus idaeus
36-60in (90-152cm)

Three to five leaflets, grey beneath

Stems upright, with fine prickles

Ripe fruit has many segments

Fruit has few segments

Stone bramble
Rubus saxatilis

Symbol of the Tudor monarchs and since then of England itself, the dog rose is the commonest of our wild roses. It has the biggest flowers, blooming at the end of tall, sturdy stems which are liberally supplied with hooked thorns. The flowers, which have numerous yellow stamens, are sometimes flushed with pink and the fruit is the familiar red rose hip, from which the syrup is made. The dog rose is cultivated by nurserymen to provide strong rootstocks onto which the more delicate garden roses are grafted.

Hedgerows and scrubland throughout England and Wales. Rare in Scotland.

This is a wild rose which has won considerable popularity with gardeners. It is low-growing and, spreading by suckers, can form big patches filled with its small, fragrant flowers. It is unique among wild roses in having rounded, purple-black hips instead of the red or scarlet fruits of other wild roses. Both the common and scientific names derive from the close resemblance of its leaves to those of the burnet-saxifrage.
The field rose has trailing stems and curved thorns.

Downland, heath and dunes throughout Britain and Ireland, especially near sea.

The wild fruit is not as big as the cultivated varieties of raspberry, but in the hilly regions of the north and east the taste is at least as good. The tall, woody plant sends up shoots which bear fruit the following year. Each fruit is made up of many globe-shaped segments which are really separate fruits each growing round its own pip.
The related stone bramble is a smaller, low-growing plant found in hilly districts throughout Britain. Its runners spread above ground.

Heathland and wooded upland all over Britain and Ireland.

J F M A M J J A S O N D   J F M A M J J A S O N D   J F M A M J J A S O N D

Wild strawberry

Fragaria vesca
2-12in (5-30cm)

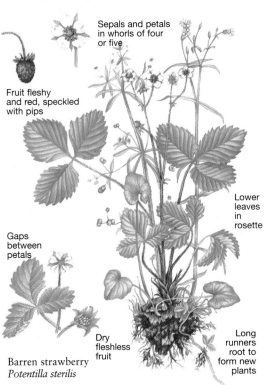

Sepals and petals in whorls of four or five

Fruit fleshy and red, speckled with pips

Gaps between petals

Barren strawberry
Potentilla sterilis

Dry fleshless fruit

Lower leaves in rosette

Long runners root to form new plants

Small but superbly flavoured, the fruit of the wild strawberry is well worth seeking out when in season. It also has a curious feature. Its seeds or pips are scattered over the surface of the fruit, while in cultivated strawberries the seeds are embedded in the fruit. The plant has clusters of bright green trefoil leaves, hairy on the underside. The flowers of wild strawberry are on upright stalks and have petals which touch or overlap. By contrast the petals of the barren strawberry do not touch, and the fruits are dry and fleshless.

Woods, scrub and grassland throughout Britain and Ireland.

Bramble

Rubus fruticosus
To 36in (90cm)

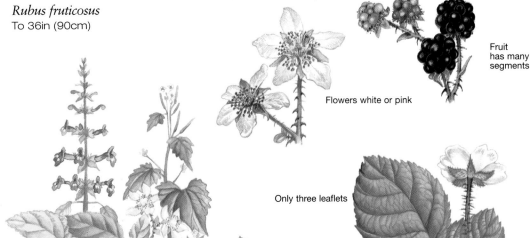

Fruit has many segments

Flowers white or pink

Only three leaflets

Short, weak prickles

Fruit has few segments

Dewberry
Rubus caesius

Leaves have three to five leaflets

Sprawling, prickly stems root to form new plants

Rubus fruticosus is often used to refer to the scientific grouping of several hundred similar types, known as microspecies, each of which has its own pattern of thorns, leaf shape and colour varying from deep pink to white. In autumn, woods, hedges and heathland are filled with the familiar fruit, but not all varieties are equally good to eat. They are all popular with birds, which scatter the seed far and wide.

The dewberry's fruit has fewer and larger segments, and is covered with a bluish bloom. It is found in similar areas.

Woods, hedges and scrub in most of England, Wales and Ireland; less common in northern Scotland.

White or pale pink flowers
Five to seven petals • Regular flowers

Dropwort
Filipendula vulgaris
6-18in (15-45cm)

Sweet cicely
Myrrhis odorata
24-71in (60-180cm)

Rough chervil
Chaerophyllum temulum
12-36in (30-90cm)

Sanicle
Sanicula europaea
8-24in (20-60cm)

Dropwort is similar to meadowsweet (p. 61), but its flowers are creamy-white, with a less yellowish tinge. Unlike the meadowsweet, the dropwort's sprays of flowers are hardly scented, and the blooms fewer but larger. Its leaves are divided into numerous, heavily toothed dark green leaflets. The common name refers to the plant's roots, which are swollen in places into pea-sized tubers – the 'drops'.

Widespread on dry grassland, especially on chalk downs in England; rare elsewhere.

The crushed leaves of sweet cicely give off a strong smell of aniseed, for which it was once cultivated as a flavouring for food. Its flowers are densely packed together in tight, many-branched heads, while its leaves are large, dark and much divided, often with small white flecks. The stalks of the stem leaves form a sheath round the hairy stem. The fruits are long and slender with points and ridges.

Widespread, particularly in north, in shady and grassy places.

Roadsides and hedgebanks are dominated by the 'big four' of the parsley family during the summer months. First on the scene is cow parsley, followed by hogweed (p. 37), rough chervil and upright hedge-parsley. Rough chervil differs from the others in having a purple-spotted stem, which is also coarsely hairy. The other main difference is the conspicuous swelling where the leaf-stalks join the stem. They all have big umbrella-shaped heads of flowers.

Hedgerows mainly in England and Wales.

Most people would recognise this plant as a member of the parsley family; but many would find it hard to explain the name. It may be a contraction of St Nicholas, who was, like the plant, credited with exceptional curative powers. Sanicle is happiest in beech and oak woods, where it is often found in extensive patches. The lobed leaves are shiny and deeply toothed. The fruit is covered with hooked bristles.

Dry deciduous woodland throughout British Isles.

Upright hedge-parsley

Torilis japonica
To 48in (120cm)

Many stalked flowers on each main branch of flower-head

Egg-shaped fruit covered with curved spines

Straight bristles on one half of fruit only

Knotted hedge-parsley
Torilis nodosa

Solid stems, with bent-back hairs

A fruit which looks like some unpleasant little insect has earned upright hedge-parsley the local name of 'Devil's nightcap'. The hooked spines ensure that the fruit is carried away on the back of any passing animal to seed elsewhere, most commonly by the roadside. Of the three main roadside varieties of parsley, much alike in appearance, upright hedge-parsley flowers last (from July to August), after cow parsley (April to June) and rough chervil (June to July).
Knotted hedge-parsley is a less elegant, sprawling plant, often found behind sea walls with a sunny aspect.

Widespread on woodland edges, hedgebanks and roadsides.

Bur chervil

Anthriscus caucalis
12-36in (30-90cm)

Flowers very small

Flower-heads grow opposite leaves

Stems sprawling and hairless, with narrow ridges

Bur chervil is much less common than the three main hedgebank parsleys – upright hedge-parsley, rough chervil and cow parsley – and tends to confine itself to certain areas, particularly East Anglia. It is a distinctive plant, smaller and more delicate than cow parsley, with much more finely cut leaves and hairless stems. Like most other British species of parsley, the stems are hollow. The flowers are small, umbrella-like clusters on stalks opposite the leaves, which are hairy on the underside. The fruits are covered with very short, hooked bristles.

Wasteland and sandy banks, mainly in eastern England, especially near sea.

Cow parsley

Anthriscus sylvestris
24-48in (60-120cm)

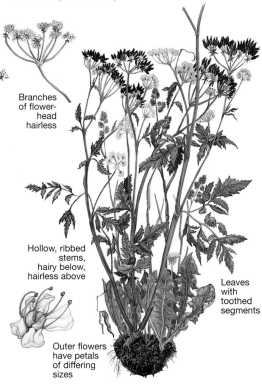

Branches of flower-head hairless

Hollow, ribbed stems, hairy below, hairless above

Outer flowers have petals of differing sizes

Leaves with toothed segments

The massed flowers by the roadside seem to foam like the crest of a breaking wave. This frothy appearance long ago inspired a comparison which gave cow parsley its alternative name of Queen Anne's lace. It is the first of the main roadside species of parsley to flower and does so in abundance in spring. Cow parsley differs from the later flowering rough chervil in having much less hairy leaves and unspotted stems. It differs from the still later upright hedge-parsley in its broader, more dissected leaves, and stems which are not downy above. Its fruit is long, smooth and black.

Widespread in hedges, wood edges, wasteland and roadsides.

31

White or pale pink flowers *Five to seven petals • Regular flowers*

Burnet-saxifrage

Pimpinella saxifraga
12-36in (30-90cm)

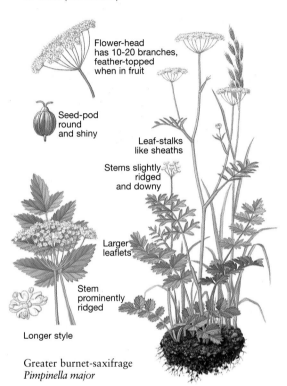

Flower-head
has 10-20 branches,
feather-topped
when in fruit

Seed-pod
round
and shiny

Leaf-stalks
like sheaths

Stems slightly
ridged
and downy

Larger
leaflets

Stem
prominently
ridged

Longer style

Greater burnet-saxifrage
Pimpinella major

Despite its name, this is neither a burnet nor a saxifrage but a member of the parsley family, with its characteristic branched 'umbrella' of flowers. It is an upright, elegant plant with tough, almost solid stems which are rounded, slightly ridged and covered in downy hairs. All the leaves are divided into leaflets, but those on the stem are much more finely cut than those growing from the base. Greater burnet-saxifrage is darker green and much taller, growing to 48in (120cm). It is a more robust plant, with larger leaflets and longer styles.

Scattered in dry grassy places throughout Britain and Ireland; less common in north.

Pignut

Conopodium majus
8-20in (20-50cm)

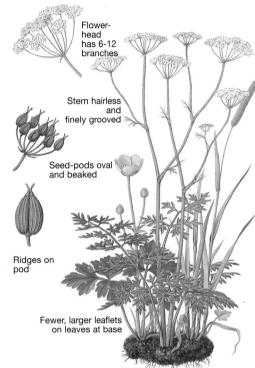

Flower-
head
has 6-12
branches

Stem hairless
and
finely grooved

Seed-pods oval
and beaked

Ridges on
pod

Fewer, larger leaflets
on leaves at base

This relative of the carrot is best known for its edible, swollen brown tubers, which have a pleasantly nutty flavour eaten raw or cooked. Pigs are certainly fond of them, hence the name. But they were also popular among country children in the days when sweets were scarce. Slender, hollow, hairless stems rise from the tubers and end in little umbrellas of flowers, with fewer branches than those of burnet-saxifrage. The leaves are much divided, those growing from the base withering before the plant flowers. Another difference from burnet-saxifrage is the shape of the seed-pod.

Meadows, hedgerows and woods throughout British Isles.

Ground elder

Aegopodium podagraria
12-24in (30-60cm)

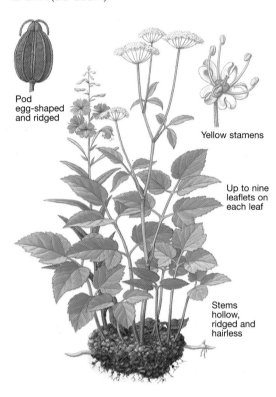

Pod
egg-shaped
and ridged

Yellow stamens

Up to nine
leaflets on
each leaf

Stems
hollow,
ridged and
hairless

Gardeners have their reasons for disliking this plant, which in Northern Ireland is known, appropriately, as garden plague. Its most striking feature is the large leaves which form a canopy shutting out all light from anything attempting to grow below, though taller plants like rosebay willowherb often grow alongside it. Ground elder also has long, white, underground runners, enabling it to spread safely, secretly and rapidly. Its flowers are minute, in dense heads which can have up to 20 slender umbrella-like branches. The prominent stamens give a yellowish tinge to the flower-head.

Common throughout British Isles.

J F M A M **J J A** S O N D

J F M A M **J J** A S O N D

J F M A M **J J** A S O N D

Hemlock

Conium maculatum
36-84in (90-210cm)

Flower-
heads
on side
branches
or at end
of stem

Stems greyish
and hairless, with
purple spots

Leaves
finely
divided

Fruits round,
with
wavy ridges

One of the tallest members of the parsley family, with hairless, purple-blotched stems and fern-like leaves, hemlock is fortunately an easy plant to identify. Every part of the plant is poisonous, but the seeds contain the biggest concentration of alkaloids, particularly coniine, which paralyses the respiratory nerves. Socrates was poisoned by hemlock, and the plant formed part of the witches' brew in Shakespeare's *Macbeth*. The flowers grow in dense little umbrellas, and the upper bracts sprout on the outer side only.

By roads, streams and on waste ground throughout Britain and Ireland; commonest in south and south-east.

Fool's watercress

Apium nodiflorum
12-36in (30-90cm)

Flower-head
short-stalked,
opposite leaf

Fruit oval,
with equal
ridges

Leaves
shiny, stalkless,
with toothed
edges

Sprawling stems finely furrowed

The leaves of this plant are similar to those of watercress (hence its common name), though the leaflets of fool's watercress have shallow teeth. Although less tasty than true watercress it can safely be eaten. The plant could also be mistaken for lesser water-parsnip though, unlike that plant, it has no leaf-like bracts below the flower-head. The main distinguishing feature is the way the short stalks bearing the flower-heads sprout from the stem opposite a leaf.

Marshy places, ditches and slow streams; rare in Scotland.

Lesser marshwort

Apium inundatum
4-20in (10-50cm)

Flower-head
on long stem,
opposite leaf,
with two to four
branches

Stalkless
upper
leaflets

Fruit
narrow and
oval, with
ridges

Grooved,
upright
stems

Lower
leaflets
on
separate
stalks

Rooting
stems
floating or
submerged

Hairlike
submerged
leaves

Wild celery
Apium graveolens

Seen floating in a pond, lesser marshwort looks at first glance more like water-crowfoot than a member of the parsley family. But its flowers, though they are small and few, grow in the typical umbrella-shaped head of parsleys. It is a small plant, growing in and under water, rooting as it creeps. The upper leaves are divided into narrow leaflets; those near the water have deeply divided lobes; and those under water are thin and hairlike.
Wild celery, the ancestor of cultivated celery, is a tough and wiry plant of salty grassland and has the characteristic celery smell.

Lakes, ponds and ditches, mainly in southern Britain.

J F M A M **J J** A S O N D   J F M A M J **J A** S O N D   J F M A M J **J A** S O N D   33

White or pale pink flowers *Five to seven petals • Regular flowers*

Lesser water-parsnip
Berula erecta
12-36in (30-90cm)

Hemlock water-dropwort
Oenanthe crocata
12-36in (30-90cm)

Parsley water-dropwort
Oenanthe lachenalii
12-36in (30-90cm)

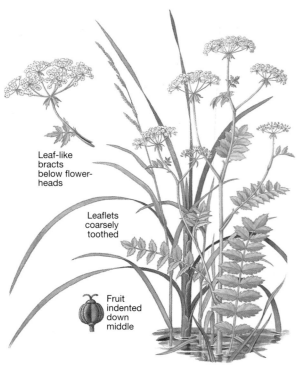

Leaf-like bracts below flower-heads

Leaflets coarsely toothed

Fruit indented down middle

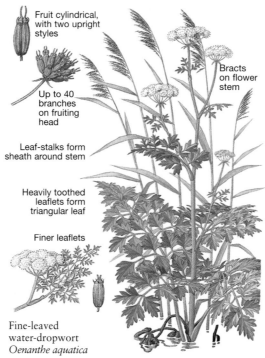

Fruit cylindrical, with two upright styles

Up to 40 branches on fruiting head

Leaf-stalks form sheath around stem

Heavily toothed leaflets form triangular leaf

Finer leaflets

Fine-leaved water-dropwort
Oenanthe aquatica

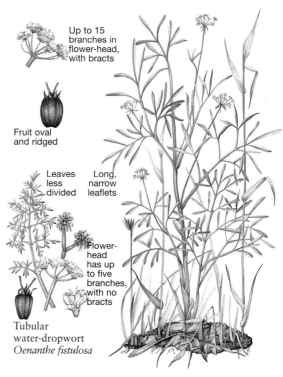

Up to 15 branches in flower-head, with bracts

Fruit oval and ridged

Bracts on flower stem

Leaves less divided

Long, narrow leaflets

Flower-head has up to five branches, with no bracts

Tubular water-dropwort
Oenanthe fistulosa

Though often mistaken for fool's watercress, lesser water-parsnip has leaves that are more coarsely toothed and bluish-green, with a distinct purple ring towards the bottom of the leaf-stalk. Unlike fool's watercress, there are leaf-like bracts below the lower flower-stalks. The umbrella-shaped flower-heads grow on stalks opposite the leaves. The lower leaves have long stalks with five to nine pairs of leaflets. The plant spreads by means of stolons – sprawling stems which bend to the ground or below the water and take root. The roots were once used to make a poultice to cleanse sores and disperse swellings.

Fens, ditches and ponds; rare in northern Britain.

It smells of parsley and has the characteristic umbrella-shaped flower-head of the parsley family. But hemlock water-dropwort is also one of Britain's most poisonous plants, dangerous to humans and animals. It is often seen in broad clumps in or on the edge of fresh or brackish water. A stout, hairless plant, it has large flower-heads and deep green leaves made up of broad, triangular, heavily toothed leaflets.
Fine-leaved water-dropwort has leaves that are paler as well as finer. It is found most commonly in eastern England in slow, still or stagnant water.

By fresh or brackish water, mainly in south and west.

Although a close relative of the hemlock water-dropwort and bearing some similarity to it, this plant is not poisonous. The main difference in appearance between the two plants is in the leaves – parsley water-dropwort has long, narrow leaflets, greyish in colour, compared with the triangular, deeper green leaves of hemlock water-dropwort. It also has a vine-like smell – unlike hemlock water-dropwort which smells of parsley.
Tubular water-dropwort is a small plant, often hard to detect among tall vegetation. The flower-heads are flat-topped and very dense, and the fruits angular.

Fresh water and brackish marsh, especially near coast.

| J F M A M J J A S O N D |

| J F M A M J J A S O N D |

Fool's parsley

Aethusa cynapium
2-50in (5-127cm)

Bracts below flower-heads

Fruit deeply ridged

Finely lined, greyish-green stems

Leaves triangular, finely divided

Fool's parsley contains an alkaloid called coniine, the main active ingredient of hemlock, which is extremely poisonous. A hairless plant, its most striking feature is the presence of long, streamer-like little leaves called bracts, hanging beneath the umbrella-shaped flower-heads. The flower-heads themselves are always at the end of a branch, opposite a leaf. The fruit is egg-shaped and deeply ridged.

Widespread, but less common in north.

Wild angelica

Angelica sylvestris
12-98in (30-250cm)

Flattened oval fruit

Broad sheath where upper leaves join stem

Large leaves grow from base

A mass of pink-tinged blossom at the top of tall, stout and purplish stems makes wild angelica easy to identify. It is also notable for the broad sheathing of the stalks on the upper leaves, which are much smaller than the lower leaves. The 17th-century herbalist Nicholas Culpeper recommended angelica against 'all epidemical diseases' and also as a candy, for which the stalks of the related garden angelica are still crystallised today.

Wet woods, fens and damp meadows.

Yarrow

Achillea millefolium
6-18in (15-45cm)

Flowers in dense, flat heads

Each flower has many florets

Stems furrowed and woody

Long, feathery leaves

The Greek warrior-hero Achilles was said to have used yarrow to purge and heal wounds made by iron weapons – hence the plant's generic name *Achillea*. Its ancient role now forgotten, yarrow still thrives as a roadside plant, thanks to its deep, water-gathering roots. The broad, flat flower-heads and numerous dark green feathery leaves – the *millefolium* of the plant's species name – make it an easy plant to identify.

Grassy places, hedges and roadsides.

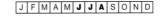

Bogbean

Menyanthes trifoliata
4-59in (10-150cm)

Three large leaflets

White hairs fringe petals

Upright stems and leaves above water

Base of leaves forms sheath around stem

Fruit has single style

Spikes of feathery, pinkish-white flowers set against a background of large, three-lobed leaves make this a far prettier plant than its name would suggest, and one of the most attractive plants of bogs and fens. Creeping underwater stems enable the plant to colonise large stretches of marshland. The leaves were once used in the north of England in place of hops to impart a bitter flavour to beer.

Ponds, lake edges, bogs and fens.

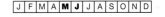

35

White or pale pink flowers
Five to seven petals • Irregular flowers

Greater butterfly orchid
Platanthera chlorantha
8-24in (20-60cm)

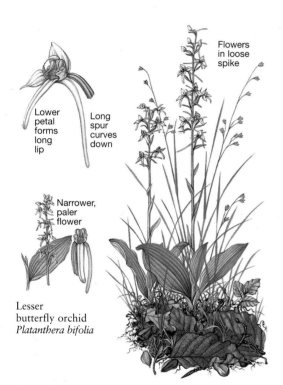

Flowers in loose spike

Lower petal forms long lip

Long spur curves down

Narrower, paler flower

Lesser butterfly orchid *Platanthera bifolia*

Autumn lady's tresses
Spiranthes spiralis
3-6in (7.5-15cm)

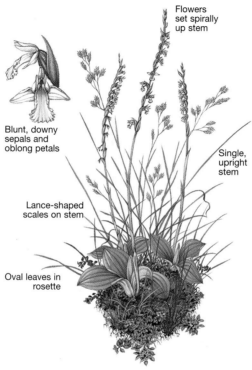

Flowers set spirally up stem

Blunt, downy sepals and oblong petals

Single, upright stem

Lance-shaped scales on stem

Oval leaves in rosette

Common cottongrass
Eriophorum angustifolium
8-24in (20-60cm)

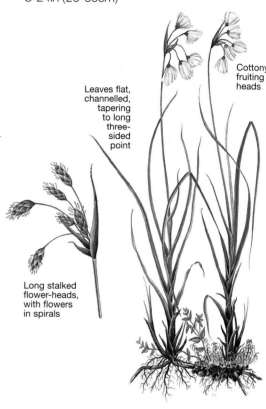

Leaves flat, channelled, tapering to long three-sided point

Cottony fruiting heads

Long stalked flower-heads, with flowers in spirals

The scent of vanilla in woodland or pasture indicates the presence of this attractive wild orchid. The greenish-white flowers grow in a loose, pyramid-shaped spike; each flower has a long, narrow, hanging lip in the front, and a spur so lengthy that only long-tongued butterflies and moths can reach the bottom of it. Below the flower spike, on the same single stem, are a few small unstalked leaves which contrast with the broad leaves at the base. The lesser butterfly orchid is a smaller plant but much more widespread, seen most often on acid moorland soils.

Widespread in woods and pastures on lime-rich soils.

This wild orchid is often plentiful on downland and dry pastureland, where its fragrance scents the evening air. But it is a small plant with less showy flowers than most orchids, and so is easily overlooked among the abundance of tall grasses. The rosette of leaves at the base of the plant dies back in June, when a flowering spike starts to grow. Unusually among orchids, the buds sprout spirally in a single row up the stem, which has a few, scale-like leaves. Autumn lady's tresses sometimes disappears for several years then suddenly reappears.

Dry grassy places, especially on chalk in southern areas.

The dun colours of peat bogs are often broken with startling splashes of white as the flowers of the common cottongrass ripen into fruit. The flower-heads suddenly sprout what at a distance look like tiny balls of cottonwool. They are the long white hairs which grow round each fruit as it ripens, ready to carry the seed away on the wind. The flowers themselves are brownish-green with yellow anthers and can easily pass unnoticed in the surrounding dull green vegetation. Then, with the ripening fruit, the plant is transformed.

Widespread in bogs and swampy ground.

J F M A **M** J J A S O N D

J F M A M J J **A S** O N D

J F M A **M J** J A S O N D

Petals unequal in size are characteristic of the orchid family, three members of which are predominantly white in colour. In the parsley family, however, equal-sized petals are more common and three species which have irregular flowers stand out as exceptions.

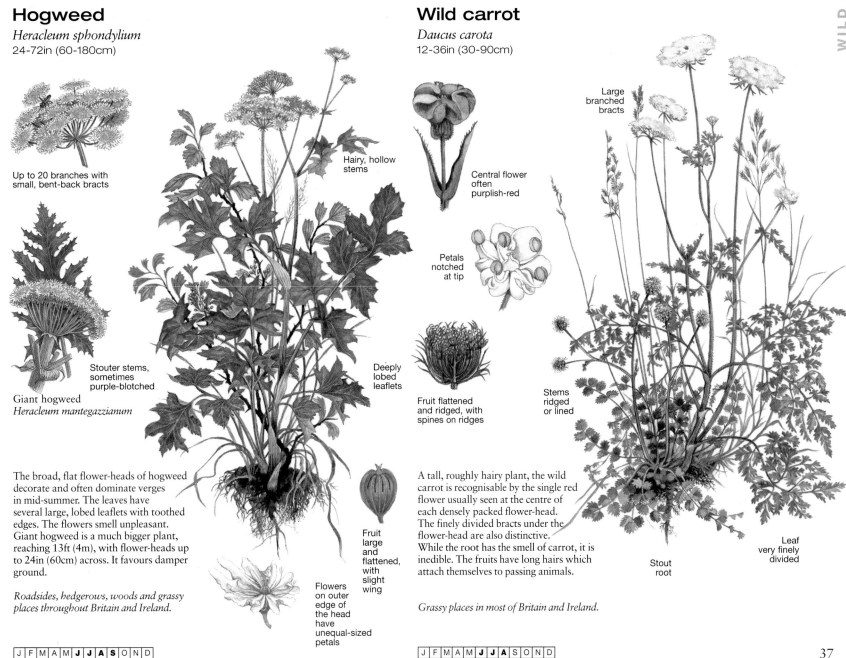

Hogweed
Heracleum sphondylium
24-72in (60-180cm)

Up to 20 branches with small, bent-back bracts

Stouter stems, sometimes purple-blotched

Giant hogweed
Heracleum mantegazzianum

Hairy, hollow stems

Deeply lobed leaflets

Fruit large and flattened, with slight wing

Flowers on outer edge of the head have unequal-sized petals

The broad, flat flower-heads of hogweed decorate and often dominate verges in mid-summer. The leaves have several large, lobed leaflets with toothed edges. The flowers smell unpleasant. Giant hogweed is a much bigger plant, reaching 13ft (4m), with flower-heads up to 24in (60cm) across. It favours damper ground.

Roadsides, hedgerows, woods and grassy places throughout Britain and Ireland.

J F M A M J J A S O N D

Wild carrot
Daucus carota
12-36in (30-90cm)

Large branched bracts

Central flower often purplish-red

Petals notched at tip

Fruit flattened and ridged, with spines on ridges

Stems ridged or lined

Stout root

Leaf very finely divided

A tall, roughly hairy plant, the wild carrot is recognisable by the single red flower usually seen at the centre of each densely packed flower-head. The finely divided bracts under the flower-head are also distinctive. While the root has the smell of carrot, it is inedible. The fruits have long hairs which attach themselves to passing animals.

Grassy places in most of Britain and Ireland.

J F M A M J J A S O N D

37

White or pale pink flowers *Eight or more petals*

White water-lily

Nymphaea alba
Stems to 9ft (2.75m)

Globe-shaped fruit

Sepals green on back, white inside

Many petals; four shorter sepals

Leaves circular, green above, reddish beneath

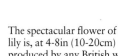

Stout underground stems anchor plant

The spectacular flower of the white water-lily is, at 4-8in (10-20cm) across, the largest produced by any British wild plant. Typical of its many local names is that used in Cheshire, where it is known as the 'lady of the lake'. But it needs a sunny day to catch its full beauty: the flowers which float on the surface of the water, at the end of the stems up to 9ft (2.75m) long, only open fully in sunshine and start to close as soon as the sun begins to wane.

Lakes and ponds throughout British Isles.

Chickweed wintergreen

Trientalis europaea
4-10in (10-25cm)

Ripe fruit splits into five parts

Starry flower, 5-9 petals

Single unbranched stem

Whorl of large leaves below flowers

This pretty little flower is best seen in its favourite setting, beneath tall Scots pines in a remote glen in Scotland. Occasionally it can be found in northern England on moist, mossy ground. One or two delightful flowers rise on tall, thin stems from the middle of a whorl of broad, pale green leaves. The leaves in turn surmount a single, unbranched stem from which a few much smaller leaves grow below.

Common in pine woods and on moors in Scotland, and locally in northern England.

Mountain avens

Dryas octopetala
1-19in (2.5-50cm)

Long green sepals

Yellow stamens

Fruit has long feathery styles

Single flower on stem

Leaves dark green above, silvery beneath

This beautiful evergreen, often seen in garden rockeries, is much less common in the wild, where it favours inaccessible rocky ledges. Each attractive flower rises on a separate stem from a fan of leaves. It has a mass of golden stamens and is clasped by long, green sepals. The leaves are distinctive, too, being dark green and very oak-like, with a silvery underside.

Rare in England. Local on rock ledges and steep slopes in Scotland, Wales and north-west Ireland.

| J | F | M | A | M | J | J | A | S | O | N | D |

| J | F | M | A | M | J | J | A | S | O | N | D |

| J | F | M | A | M | J | J | A | S | O | N | D |

The dense flower-heads of the daisy family are made up of as many as 200 individual florets, some forming a central disc and others surrounding it like rays. Mayweeds and chamomiles, very similar to each other in appearance, are often most easily distinguished from one another by their scent.

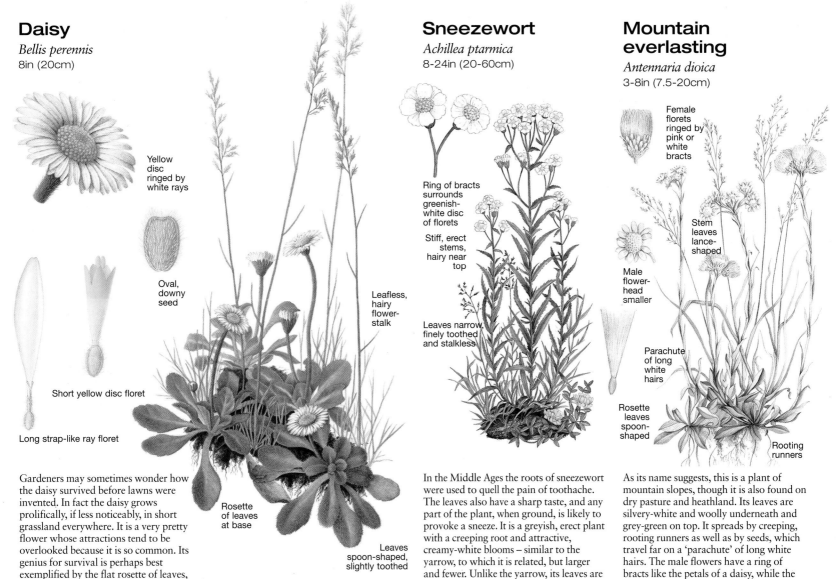

Daisy
Bellis perennis
8in (20cm)

Yellow disc ringed by white rays

Oval, downy seed

Short yellow disc floret

Long strap-like ray floret

Rosette of leaves at base

Leafless, hairy flower-stalk

Leaves spoon-shaped, slightly toothed

Gardeners may sometimes wonder how the daisy survived before lawns were invented. In fact the daisy grows prolifically, if less noticeably, in short grassland everywhere. It is a very pretty flower whose attractions tend to be overlooked because it is so common. Its genius for survival is perhaps best exemplified by the flat rosette of leaves, which spreads too close to the ground for animals to bite or for mowers to cut.

Common everywhere, especially on lawns.

Sneezewort
Achillea ptarmica
8-24in (20-60cm)

Ring of bracts surrounds greenish-white disc of florets

Stiff, erect stems, hairy near top

Leaves narrow, finely toothed and stalkless

In the Middle Ages the roots of sneezewort were used to quell the pain of toothache. The leaves also have a sharp taste, and any part of the plant, when ground, is likely to provoke a sneeze. It is a greyish, erect plant with a creeping root and attractive, creamy-white blooms – similar to the yarrow, to which it is related, but larger and fewer. Unlike the yarrow, its leaves are long, narrow and undivided, with a saw-toothed edge.

Damp grassland throughout British Isles.

Mountain everlasting
Antennaria dioica
3-8in (7.5-20cm)

Female florets ringed by pink or white bracts

Stem leaves lance-shaped

Male flower-head smaller

Parachute of long white hairs

Rosette leaves spoon-shaped

Rooting runners

As its name suggests, this is a plant of mountain slopes, though it is also found on dry pasture and heathland. Its leaves are silvery-white and woolly underneath and grey-green on top. It spreads by creeping, rooting runners as well as by seeds, which travel far on a 'parachute' of long white hairs. The male flowers have a ring of bracts like the petals of a daisy, while the female flowers (on a separate plant) are bigger, with long pink or white bracts.

Dry uplands, mainly in north.

White or pale pink flowers *Eight or more petals*

Sea mayweed
Tripleurospermum maritimum
6-24in (15-60cm)

Stinking chamomile
Anthemis cotula
8-24in (20-60cm)

Ox-eye daisy
Leucanthemum vulgare
8-24in (20-60cm)

Feverfew
Tanacetum parthenium
10-24in (25-60cm)

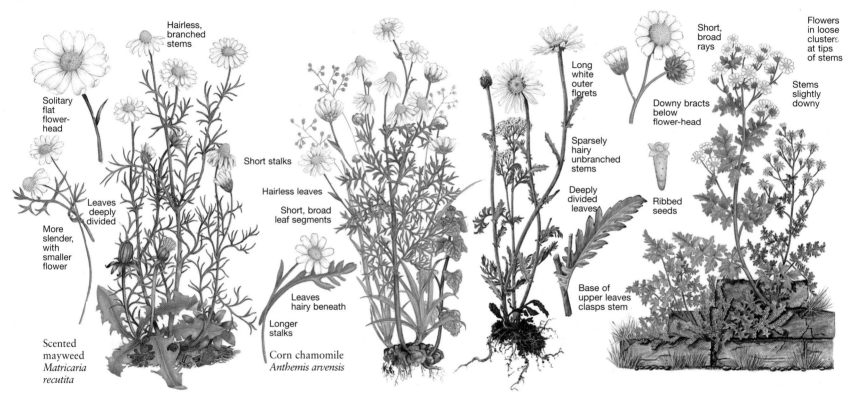

Hairless, branched stems

Solitary flat flower-head

Leaves deeply divided

More slender, with smaller flower

Scented mayweed
Matricaria recutita

Short stalks

Hairless leaves

Short, broad leaf segments

Leaves hairy beneath

Longer stalks

Corn chamomile
Anthemis arvensis

Long white outer florets

Sparsely hairy unbranched stems

Deeply divided leaves

Base of upper leaves clasps stem

Short, broad rays

Downy bracts below flower-head

Ribbed seeds

Flowers in loose clusters at tips of stems

Stems slightly downy

The big, daisy-like flowers of the sea mayweed make a happy sight by roadsides and on wasteland in summer. It has the biggest flowers of all the mayweeds, up to 1½in (38mm) across. It is also a common weed of arable fields.
Scented mayweed is more slender and upright, with smaller flowers; it has a pleasant, chamomile smell. It is most common in the Midlands and south-east.

Waste and cultivated ground, sand, shingle, rocks and walls throughout British Isles.

Compared with sea mayweed, stinking chamomile has shorter, broader leaf segments on slightly hairy stems and downy-white bracts with a narrow, green midrib. It also has a strong, sickly-sweet and unpleasant smell. In fruit the flat yellow centre of the flower becomes a solid cone of seeds.
Corn chamomile is slightly scented, with grey, downy stems and leaves downy-white underneath.

Arable and wasteland, especially in south.

Countless young lovers have pulled off the petals of ox-eye daisy one by one, wondering whether 'he (or she) loves me, he (or she) loves me not'. Far from imperilling its existence, this treatment seems to have encouraged the ox-eye daisy to thrive, since it decorates grassy wasteland everywhere. The solitary flowers are certainly eye-catching, like giant daisies up to 2in (50mm) across.

Widespread on most good soil; less common in northern Scotland.

Seen in a mass, densely covered with daisy-like flowers, there are few plants prettier than feverfew. But it was not for its appearance that it was brought to Britain by medieval herbalists. They grew it for a drug it contains, which was used to reduce fevers and relieve headaches. It has pungently aromatic yellowish-green leaves and flowers with white rays almost as broad as they are long.

Waste ground, hedgebanks and walls throughout most of British Isles.

J F M A M J **J A S** O N D   J F M A M J **J A S** O N D   J F M A M J **J A S** O N D   J F M A M J **J A S** O N D

Yellow flowers *Up to four petals • Regular flowers*

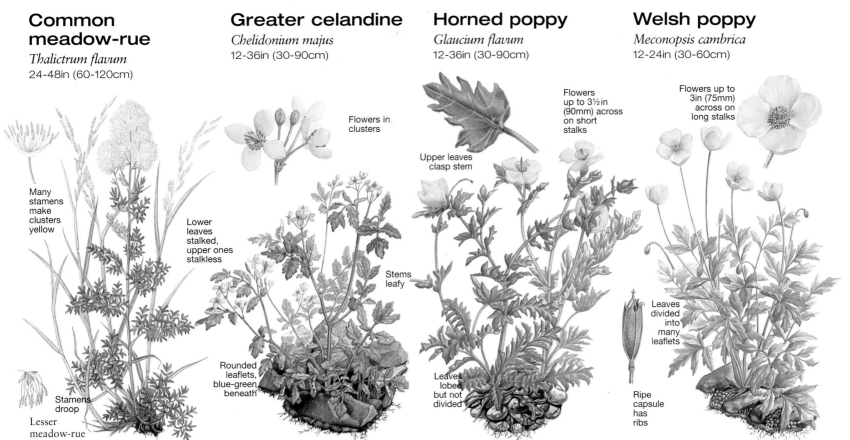

Common meadow-rue
Thalictrum flavum
24-48in (60-120cm)

Many stamens make clusters yellow

Lower leaves stalked, upper ones stalkless

Stamens droop

Lesser meadow-rue
Thalictrum minus

Greater celandine
Chelidonium majus
12-36in (30-90cm)

Flowers in clusters

Rounded leaflets, blue-green beneath

Horned poppy
Glaucium flavum
12-36in (30-90cm)

Flowers up to 3½in (90mm) across on short stalks

Upper leaves clasp stem

Stems leafy

Leaves lobed but not divided

Welsh poppy
Meconopsis cambrica
12-24in (30-60cm)

Flowers up to 3in (75mm) across on long stalks

Leaves divided into many leaflets

Ripe capsule has ribs

The flowers of this perennial plant, a relative of the anemones, have four small sepals and many long, rigid, pollen-bearing stamens which give the head a fluffy look. The stem is grooved and hairless and usually has no branches. The leaves are dark green above and paler below, and divided into numerous segments. Lesser meadow-rue is more slender and has drooping stamens. It grows in grassland or on dunes.

Moist ground, mainly in south-east England.

Ants feed on the oil glands of the plant's black seeds and carry off the seeds stuck to their bodies. The flowers, less than 1in (25mm) across, grow in clusters on leafy stems. The leaves, which are blue-green underneath, are almost hairless and the terminal leaf is usually three-lobed. The caustic orange sap of greater celandine has been used in Asia for burning away warts and corns since the early days of Chinese civilisation.

Hedgerows and banks, except in far north.

The horned poppy is the most colourful of Britain's seashore plants. Its flowers are about 3in (75mm) across. These are pollinated by small flies of the shingle and sand-dunes, and are followed in autumn by long, narrow seed-pods. These grow up to 12in (30 cm) long and are more like the pods of some members of the cabbage family than those of cornfield poppies. The orange sap is foul-smelling, and all parts of the plant are poisonous.

Shingle banks, except in far north.

The Welsh poppy still grows in the wild in damp upland woods and rocky places, but is now more commonly cultivated as a garden flower. It has many branches on its hairy stem. The leaves at the base have long stalks while the upper ones are much shorter. Each stem is crowned by a single flower which lacks the dark blotches of other poppies. The ripe seed capsule opens by flaps.

Damp, shady soil Wales, Ireland and south-west England.

JFMAMJJASOND   JFMAMJJASOND   JFMAMJJASOND   JFMAMJJASOND   41

Yellow flowers

Up to four petals • Regular flowers

Lady's bedstraw

Galium verum
6-40 in (15-100cm)

Hairy stems, often tinged light brown

Small flowers in clusters

Fruits green at first, turning black

Yellow-green hairy stem

Crosswort
Cruciata laevipes

Large-flowered evening primrose

Oenothera glazioviana
20-71in (50-180cm)

Large cup-shaped flowers in spike

Sepals red-striped

Thick pithy stem

Narrow-toothed leaves

Weld

Reseda luteola
20-60in (50-152cm)

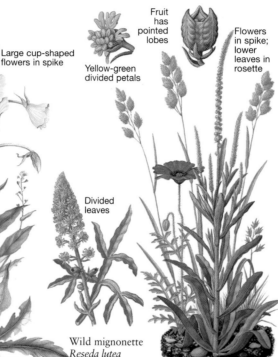

Fruit has pointed lobes

Yellow-green divided petals

Flowers in spike; lower leaves in rosette

Divided leaves

Wild mignonette
Reseda lutea

Yellow bird's-nest

Monotropa hypopitys
3-12in (7.5-30cm)

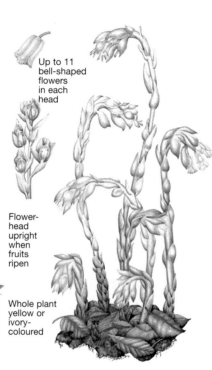

Up to 11 bell-shaped flowers in each head

Flower-head upright when fruits ripen

Whole plant yellow or ivory-coloured

Legend says the Virgin Mary lay on a bed of this plant in Bethlehem, hence the name and the belief that it led to safe and easy childbirth. The stems, upright or sprawling, are often tinged light brown, with short hairs. The leaves are single- veined, also with short hairs. The related crosswort has similar flowers, but its sprawling stem is yellow-green and its very hairy leaves have three veins.

Widespread on grassland, hedgebanks and sand-dunes throughout British Isles.

The flowers of evening primrose open at dusk and emit a delicate fragrance which attracts night-flying moths. The flowers are often more than 2in (50mm) across and arranged in spikes on a thick, pithy stem. The leaves are set spirally round the stem. Evening primrose is related more closely to willowherb than to the primrose family. It was introduced from North America as a garden flower, but escaped.

Dunes and waste ground in England, Wales and southern Scotland.

The flowers of weld turn on their stalks during the day to follow the sun. The upright, hairless stem has few, if any, branches. The lower leaves are narrow, and the upper leaves wavy-edged and oblong. A brilliant yellow dye produced from weld, or dyer's rocket, has been used since the Stone Age.
Wild mignonette is a shorter plant, growing to only 24in (60cm), and differs from weld also in having divided leaves.

Open ground in lowland; rarer in north.

This waxy-looking plant grows in the darkest parts of woods, usually among pine and beech. It does not need sunlight and feeds on the decaying vegetable matter that covers the woodland floor. The leaves are scale-like and clasp the simple, unbranched stem. The plant's common name comes from the tangle of short, fleshy branched roots which look something like a bird's nest.

Very local in woodlands, mainly in southern England.

J F M A M J J A S O N D

J F M A M J J A S O N D

J F M A M J J A S O N D

J F M A M J J A S O N D

Oilseed rape is the most conspicuous of the many yellow-flowered plants of the cabbage family, which all have petals arranged in the shape of a cross and long seed-pods. Outside the cabbage family are several bedstraws and poppies, the handsome evening primrose and the unusual semi-parasite yellow bird's-nest.

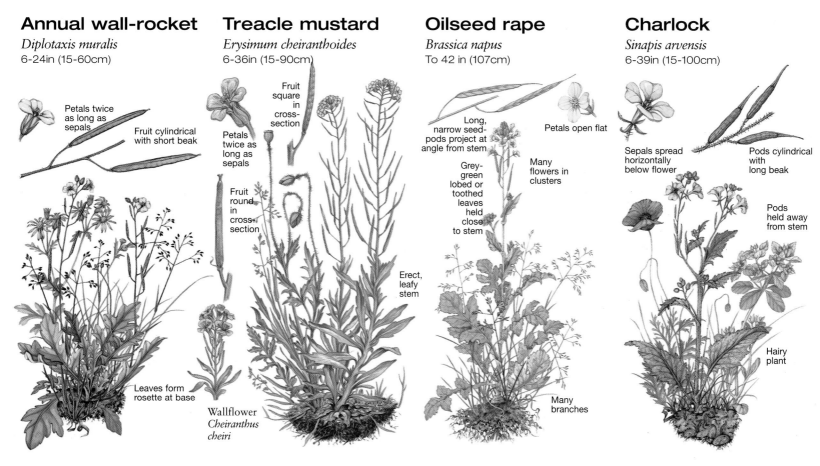

Annual wall-rocket

Diplotaxis muralis
6-24in (15-60cm)

Petals twice as long as sepals

Fruit cylindrical with short beak

Leaves form rosette at base

Treacle mustard

Erysimum cheiranthoides
6-36in (15-90cm)

Fruit square in cross-section

Petals twice as long as sepals

Fruit round in cross-section

Erect, leafy stem

Wallflower
Cheiranthus cheiri

Oilseed rape

Brassica napus
To 42 in (107cm)

Long, narrow seed-pods project at angle from stem

Petals open flat

Grey-green lobed or toothed leaves held close to stem

Many flowers in clusters

Many branches

Charlock

Sinapis arvensis
6-39in (15-100cm)

Sepals spread horizontally below flower

Pods cylindrical with long beak

Pods held away from stem

Hairy plant

A common alternative name for annual wall-rocket is stinkweed, because its stem contains sulphuretted hydrogen, the 'rotten eggs' chemical beloved by schoolboys for making stink-bombs. Its unbranched stem carries upright seed-pods, each of which contains two rows of yellow-brown seeds. The petals are twice as long as the sepals, which spread slightly. Despite its name, it lives for two years.

Common on walls, and waste and sandy ground, mainly in southern England.

Herbalists used treacle mustard to counteract animal bites and poison in the 16th century, and its seeds were given to children to drive out intestinal worms. The leaves may be smooth-edged or have shallow teeth; the lower ones have stalks but the upper ones do not. The young flower-head is flat-topped and the seed-pods are downy. The related wallflower is easily identified by its hairy fruits, which are round in cross-section.

Cultivated and waste ground; rarer in north.

This plant of the mustard family has become a familiar sight in the countryside, its multi-flowered heads turning fields into carpets of bright yellow and largely replacing the black mustard which is similar in appearance. It often spreads to form clumps on road verges. Oilseed rape is cultivated for its seeds, which yield an oil used in lubricants or refined to produce cooking oil.

Widespread on farmland and roadsides throughout Britain; rare in Ireland.

A choking weed which has been in the past a serious menace to arable land, charlock today is becoming rarer as a result of the use of weed-killers. The pods hold two rows of dark red-brown seeds. These seeds can persist in the soil for up to 50 years and burst forth when old pastures are ploughed. The sepals spread out horizontally below the flowers. Charlock was once sold as a vegetable in Ireland.

Field margins, roadsides and wasteland all over lowland British Isles.

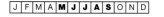
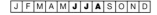
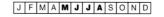
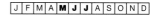

Yellow flowers *Up to four petals • Regular flowers*

Hedge mustard

Sisymbrium officinale
12-36in (30-90cm)

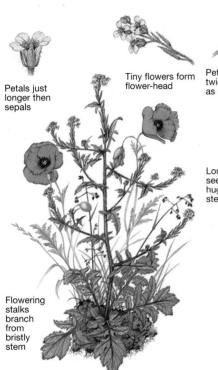

Petals just longer then sepals

Tiny flowers form flower-head

Flowering stalks branch from bristly stem

Black mustard

Brassica nigra
24-80in (60-200cm)

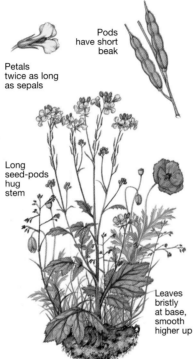

Pods have short beak

Petals twice as long as sepals

Long seed-pods hug stem

Leaves bristly at base, smooth higher up

March yellow-cress

Rorippa palustris
10-24in (25-60cm)

Petals same length as sepals

Lower leaves stalked, upper ones stalkless

Fruit oblong and squat

Wild radish

Raphanus raphanistrum
8-24in (20-60cm)

Many leaflets

Dark veins on petals

'Waisted' and jointed pod

Sea radish *Raphanus raphanistrum* ssp. *maritimus*

Each leaf has up to four pairs of leaflets

The branches of this wiry wasteland plant, seen growing with poppies, protrude almost at right angles from the rigid stem, and its jagged leaves have their points curled inwards towards the stem. The small flowers produce ribbed and hairy seed-pods; each has a short beak. In 16th and 17th-century France, actors, singers and politicians used infusions from this plant as a gargle. In Britain its sap was mixed with honey as a cure for asthma.

Widespread on waste and arable land.

The flower-heads of black mustard used to paint splashes of yellow across the countryside much as oilseed rape does today. It was grown for its black-brown seeds, which are still crushed to be used in mustard baths and poultices. Today, however, rape is a more profitable crop, and fields of black mustard are much rarer. The plant persists as a weed, however, long after fields have been cleared. The long seed-pods grow upright and hug the stem.

Waysides and cliffs in southern Britain.

This sprawling plant likes a site which is under water in winter but which dries out in the summer. The stem is hollow and its leaves are strongly lobed; the base of each leaf-stalk half-clasps the stem. The petals are the same length as the sepals, and the small brown seeds are held in a pointed pod which is about the same length as its stalk. Though a relative of salad cresses, marsh yellow-cress is itself inedible.

Widespread on pond sides and river banks.

Wild radish is an irksome weed of arable land. The colour of its flowers varies according to its location. Usually they are yellow or lilac in the north and west. The stems are rough and hairy. The fruit bears up to eight seeds in a pod about 2in (50mm) long.
The rarer seaside sea radish has leaves divided into many overlapping and crowded leaflets. Its fruit contains no more than three seeds.

Arable land, rarer in Scotland.

J F M A **M J J A S** O N D   J F M A **M J J A S** O N D   J F M A **M J J A S** O N D   J F M A **M J J A S** O N D

Yellow flowers
Up to four petals
Irregular flowers

Common winter-cress

Barbarea vulgaris
12-36in (30-90cm)

Petals twice as long as sepals

Flower-head lengthens as fruits develop

Hairless plant with shiny leaves

Short beak on seed-pod

Tormentil

Potentilla erecta
2-20in (5-50cm)

Buttercup-like flowers

Small solid fruits

Stalkless leaves with three leaflets and two leaf-like stipules

Stems grow from rosette of leaves

Honeysuckle

Lonicera periclymenum
20ft (6m)

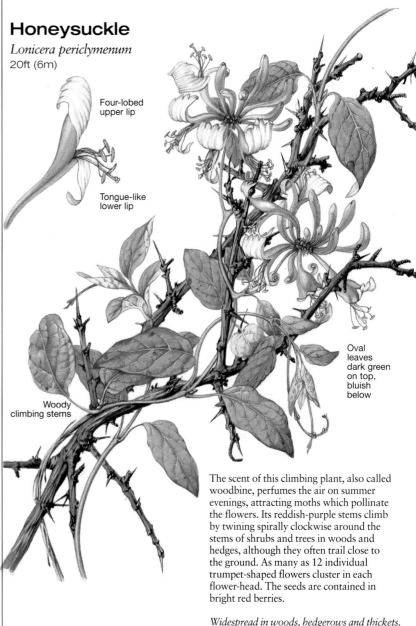

Four-lobed upper lip

Tongue-like lower lip

Woody climbing stems

Oval leaves dark green on top, bluish below

The wallflower-like dense flower spikes of common winter-cress brighten the banks of ditches, streams and ponds in early summer. The hairless plant has a branched stem and the lower leaves are deeply lobed, while the upper ones clasp the stem. The plant is a rich source of Vitamin C. Its botanical name comes from St Barbara, the patron saint of gunners, quarrymen and miners, because the leaves were once used to cover wounds caused by explosives.

Hedges and damp places, rarer in Scotland.

The four-petalled flower of tormentil distinguishes it from the five-petalled creeping tormentil (p. 56). Its upper leaves are stalkless and have three leaflets and two leaf-like appendages called stipules. The flowering stems grow from a rosette of leaves that often wither before the flowers appear. Tormentil's astringent roots were the source of a herbal remedy which was used in the 17th century to relieve the 'torment' of toothache.

Widespread in fens, bogs and heathlands.

The scent of this climbing plant, also called woodbine, perfumes the air on summer evenings, attracting moths which pollinate the flowers. Its reddish-purple stems climb by twining spirally clockwise around the stems of shrubs and trees in woods and hedges, although they often trail close to the ground. As many as 12 individual trumpet-shaped flowers cluster in each flower-head. The seeds are contained in bright red berries.

Widespread in woods, hedgerows and thickets.

J F M A **M J J A S O** N D   J F M A **M J J A S O** N D   J F M A **M J J A S O** N D

Yellow flowers *Up to four petals • Irregular flowers*

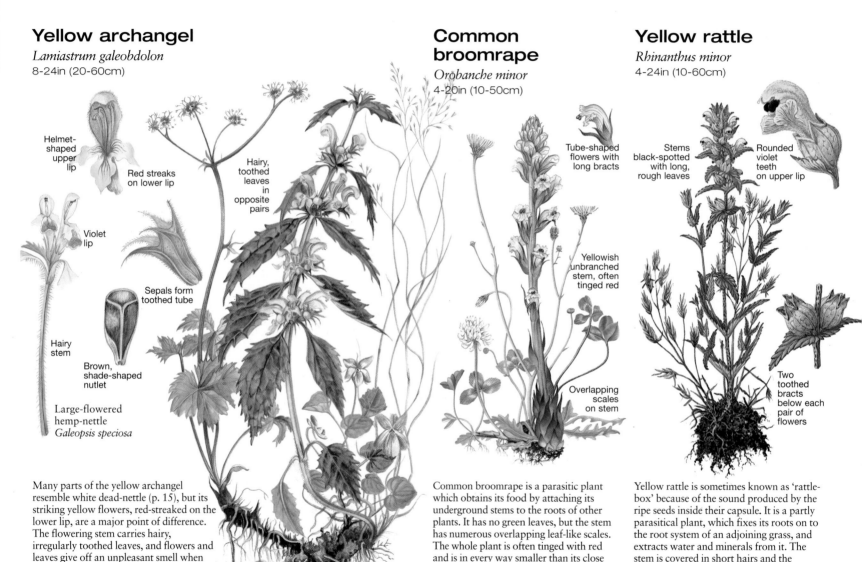

Yellow archangel
Lamiastrum galeobdolon
8-24in (20-60cm)

Helmet-shaped upper lip

Red streaks on lower lip

Hairy, toothed leaves in opposite pairs

Violet lip

Sepals form toothed tube

Hairy stem

Brown, shade-shaped nutlet

Large-flowered hemp-nettle
Galeopsis speciosa

Many parts of the yellow archangel resemble white dead-nettle (p. 15), but its striking yellow flowers, red-streaked on the lower lip, are a major point of difference. The flowering stem carries hairy, irregularly toothed leaves, and flowers and leaves give off an unpleasant smell when they are bruised. The related large-flowered hemp-nettle has a hairier stem and violet-lipped flowers.

Woodland in England and Wales; very rare in Scotland and Ireland.

Common broomrape
Orobanche minor
4-20in (10-50cm)

Tube-shaped flowers with long bracts

Yellowish unbranched stem, often tinged red

Overlapping scales on stem

Common broomrape is a parasitic plant which obtains its food by attaching its underground stems to the roots of other plants. It has no green leaves, but the stem has numerous overlapping leaf-like scales. The whole plant is often tinged with red and is in every way smaller than its close relative, greater broomrape. The two lips of its petal tube have toothed edges. The plant often damages clover crops.

Clover fields and roadsides, mainly in south and east England.

Yellow rattle
Rhinanthus minor
4-24in (10-60cm)

Stems black-spotted with long, rough leaves

Rounded violet teeth on upper lip

Two toothed bracts below each pair of flowers

Yellow rattle is sometimes known as 'rattle-box' because of the sound produced by the ripe seeds inside their capsule. It is a partly parasitical plant, which fixes its roots on to the root system of an adjoining grass, and extracts water and minerals from it. The stem is covered in short hairs and the narrow leaves are regularly toothed. The flower is two-lipped, with two rounded, usually violet teeth on the upper lip. The fruit is round and has a short beak.

Grassy places throughout British Isles.

J F M A **M J J** A S O N D   J F M A **M J J** A S O N D   J F M A **M J J** A S O N D

Leaf shape distinguishes two groups of plants with pea-like flowers. Trefoils and medicks have leaves divided into three, while vetches have rows of leaflets with, usually, a tendril at the end of the leaf. Some members of the mint and figwort families also have pea-like flowers.

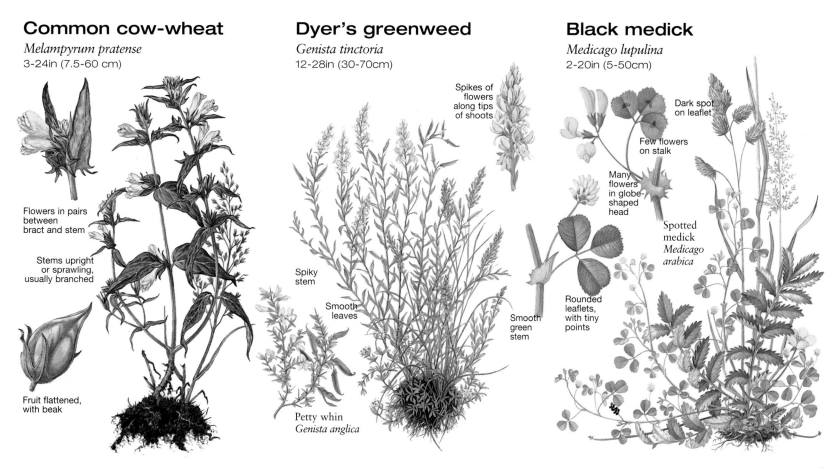

Common cow-wheat

Melampyrum pratense
3-24in (7.5-60 cm)

Flowers in pairs between bract and stem

Stems upright or sprawling, usually branched

Fruit flattened, with beak

Dyer's greenweed

Genista tinctoria
12-28in (30-70cm)

Spikes of flowers along tips of shoots

Spiky stem

Smooth leaves

Petty whin
Genista anglica

Black medick

Medicago lupulina
2-20in (5-50cm)

Dark spot on leaflet

Few flowers on stalk

Many flowers in globe-shaped head

Spotted medick
Medicago arabica

Smooth green stem

Rounded leaflets, with tiny points

It was once believed that cows which fed on this plant would produce the finest and yellowest butter, and that pregnant women would bear male children if they ate a flour prepared from the blackish, wheat-like seeds. The stem, either hairy or sparsely hairy, carries untoothed leaves which vary from oval to lance-shaped. Flowers grow in pairs in the angle between a leaf-like bract and the stem. The petal tube is two-lipped, often with the mouth closed and the lower lip pointing forwards.

Woods, heaths and grassy places throughout Britain.

The wild shrub of today, resembling a dwarf broom, was used in the 14th century by Flemish immigrants for the yellow dye it produced. Cloth dipped in the dye turns bright yellow, then green when dipped again in a solution of blue woad. The dye was known as Kendal green, after the Cumbrian town where it was developed. The plant's smooth stems carry stalkless, undivided leaves and flowers in spikes at their tips.
The less common petty whin has much smaller leaves and a spiny stem. It flowers in May and June on moors and heaths scattered over most of mainland Britain.

Clay or chalk grassland in England and Wales.

The name of black medick, here seen growing with cinquefoil, has nothing to do with medicine: it means plant of the Medes, a people of ancient Persia. It has trefoil leaves, and is distinguished from the otherwise similar lesser trefoil by a tiny pointed tip in the rounded end of each leaflet. The flower-head contains between 10 and 20 separate flowers, and the black pods are kidney-shaped, unlike the straight pods of clover.
Spotted medick has spots on the leaflets, no more than five flowers in each head and a seed-pod that is spirally coiled and spiny.

Grassy lowlands in Britain and Ireland.

J	F	M	A	**M**	J	**J**	**A**	**S**	O	N	D

J	F	M	A	M	J	**J**	**A**	**S**	O	N	D

J	F	M	A	**M**	J	**J**	**A**	**S**	O	N	D

Yellow flowers
Up to four petals • Irregular flowers

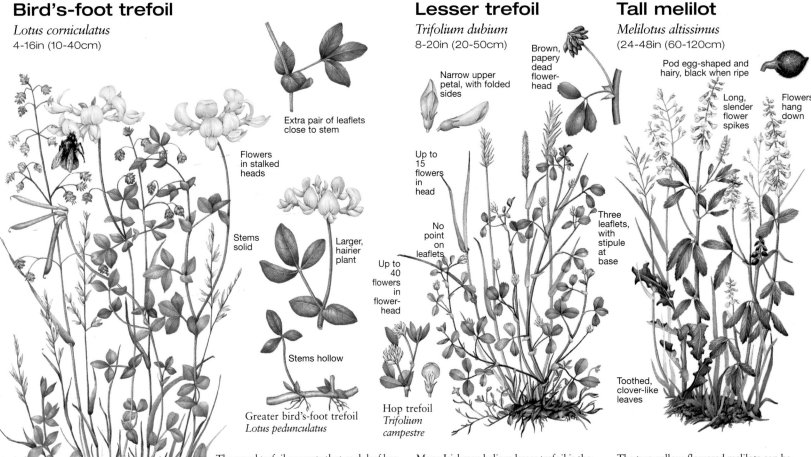

Bird's-foot trefoil
Lotus corniculatus
4–16in (10–40cm)

Extra pair of leaflets close to stem

Flowers in stalked heads

Stems solid

Larger, hairier plant

Stems hollow

Greater bird's-foot trefoil
Lotus pedunculatus

Lesser trefoil
Trifolium dubium
8–20in (20–50cm)

Narrow upper petal, with folded sides

Brown, papery dead flower-head

Up to 15 flowers in head

No point on leaflets

Up to 40 flowers in flower-head

Three leaflets, with stipule at base

Hop trefoil
Trifolium campestre

Tall melilot
Melilotus altissimus
(24–48in (60–120cm)

Pod egg-shaped and hairy, black when ripe

Long, slender flower spikes

Flowers hang down

Toothed, clover-like leaves

The word trefoil suggests that each leaf has only three leaflets. In bird's-foot trefoil, however, each leaf has an extra pair of leaflets close to the stem. The red streaks sometimes seen on the flowers give the plant the alternative common name of 'bacon-and-eggs'. The seed-pods are arranged like the toes of a bird's foot. Greater bird's-foot trefoil has a hollow and more upright stem. It prefers damper ground.

Roadsides and grassland everywhere.

Many Irishmen believe lesser trefoil is the shamrock used by St Patrick to explain the Holy Trinity to the people of Ireland, and this is one of several clover-like plants they wear on St Patrick's Day. In flower, lesser trefoil can be confused with black medick, but it lacks the small point at the tip of the leaflet.
Hop trefoil has bigger flower-heads, containing up to 40 flowers and resembling hop cones when brown.

Common in grassy places, rarer in far north.

The two yellow-flowered melilots can be most easily distinguished by the shape of their flowers. In that of tall melilot the lower petal, or keel, is the same length as the wing petals on each side and the petal above; in ribbed melilot the lower petal is shorter than the others. The flowers of both species are a rich source of nectar for honey-bees. The seed-pods of tall melilot are egg-shaped, hairy and black.

Roadsides and wood margins, rare in far north and in Ireland.

J F M A M **J J A** S O N D

J F M A M **J J A** S O N D

J F M A M **J J A** S O N D

Ribbed melilot

Melilotus officinalis
24-48in (60-120cm)

Flowers in long
tapering spikes

Seed-pod
wrinkled and
hairless, brown
when ripe

Trefoil
leaves

Short
lower petal

Although the leaves of ribbed melilot
resemble those of clover, the flowers are
quite different, being borne in long spikes.
They develop into brown, hairless pods
with the wrinkles that give the species its
name. The lower petal or 'keel' is shorter
than that of tall melilot. Melilots were
introduced from Europe in the 16th
century and were used to make poultices.
The dried plant smells of new-mown hay.

*Common on roadsides and embankments,
especially in the south and east.*

Horseshoe vetch

Hippocrepis comosa
4-16in (10-40cm)

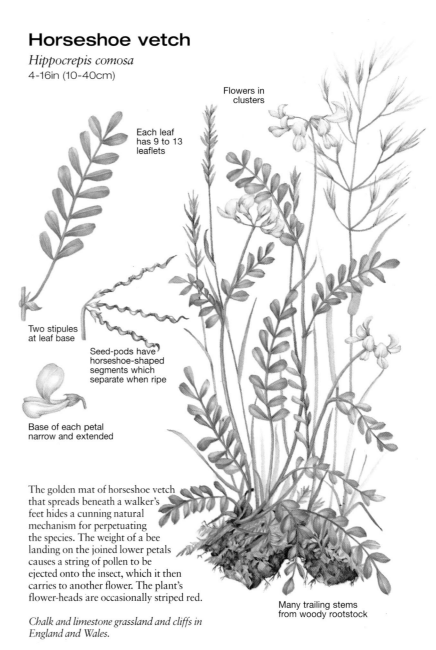

Flowers in
clusters

Each leaf
has 9 to 13
leaflets

Two stipules
at leaf base

Seed-pods have
horseshoe-shaped
segments which
separate when ripe

Base of each petal
narrow and extended

The golden mat of horseshoe vetch
that spreads beneath a walker's
feet hides a cunning natural
mechanism for perpetuating
the species. The weight of a bee
landing on the joined lower petals
causes a string of pollen to be
ejected onto the insect, which it then
carries to another flower. The plant's
flower-heads are occasionally striped red.

*Chalk and limestone grassland and cliffs in
England and Wales.*

Many trailing stems
from woody rootstock

Yellow flowers *Up to four petals • Irregular flowers*

Kidney vetch
Anthyllis vulneraria
To 24in (60cm)

Meadow vetchling
Lathyrus pratensis
12-48in (30-120cm)

Greater bladderwort
Utricularia vulgaris
6-18in (15-45cm)

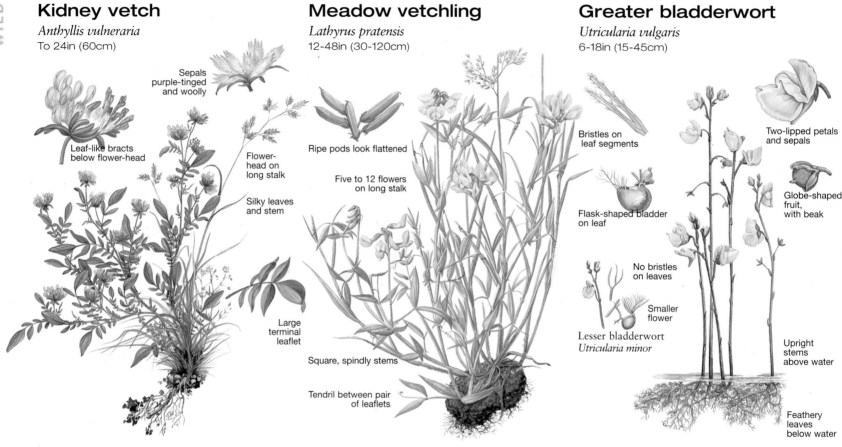

Sepals
purple-tinged
and woolly

Leaf-like bracts
below flower-head

Flower-
head on
long stalk

Silky leaves
and stem

Large
terminal
leaflet

Ripe pods look flattened

Five to 12 flowers
on long stalk

Square, spindly stems

Tendril between pair
of leaflets

Bristles on
leaf segments

Flask-shaped bladder
on leaf

No bristles
on leaves

Smaller
flower

Lesser bladderwort
Utricularia minor

Two-lipped petals
and sepals

Globe-shaped
fruit,
with beak

Upright
stems
above water

Feathery
leaves
below water

The old idea that kidney vetch could cure kidney diseases is now known to be false. There is more aptness in the plant's alternative name of lady's finger, a reference to the silky, finger-like bracts which appear just below the flower-heads. The seeds are borne in pods enclosed by the woolly, purple-tipped sepals. The leaf carries pairs of leaflets with a terminal leaflet much larger than the others. Kidney vetch is a rich source of nectar, but only large insects such as the bumble bee can force its petals apart to reach it.

Widespread in dry grassland, especially by sea.

Farmers once encouraged this slender, scrambling plant to grow in their meadows because the nodules of its roots draw nitrogen from the air and thus increase the richness of the soil. The plant, particularly its seeds, is also rich in protein so it adds to the food value of pasturage and hay. Although a forked tendril for climbing springs from each pair of leaflets, meadow vetchling tends not to be a great climber. Instead, the angular stems rise from the rootstock in profusion, relying on each other and surrounding plants for support. Five to 12 flowers are borne on a stalk that is longer than the leaves.

Grassy places throughout British Isles.

In summer the greater bladderwort speckles the surface of deep still waters with its flowers; but beneath this innocent display lurks a sinister world where tiny bladders wait to trap insects. The bladders, borne on leaves, have hairs at one end. When a water insect brushes these hairs, a trap-door is triggered and the insect is drawn into the bladder on the inrush of water. The trap closes and the insect is left to die. The plant absorbs the mineral salts left by its decomposed victim.
Lesser bladderwort has smaller flowers and leaves which may be colourless and buried in the mud.

Ponds and deep ditches throughout Britain and Ireland.

Yellow flowers *Five to seven petals • Regular flowers*

Bog asphodel

Narthecium ossifragum
2-16in (5-40cm)

Head turns deep orange after flowering

Stamens woolly, with orange tips

Flowers in spike at top of stem

Fruit has brown seeds with tail

Short leaves sheathe stem

The bog asphodel brings welcome splashes of colour to many sombre moors throughout Britain. In fact, it offers more than one colour, depending on the season. When its flowers are out, growing straight from the stem on short stalks, they are a brilliant orange-yellow. Then in the autumn, after flowering, the petals, sepals, ovary and flower-stalk take on a peachy tint before turning deep orange. Its curved leaves are like those of the iris, and its seeds have long tails at each end which help them to float during periods of flooding.

Bogs, moors and mountains over most of Britain and Ireland; rarer in central and eastern England.

Wild daffodil

Narcissus pseudonarcissus
To 14 in (36cm)

Whorl of petals and sepals around darker yellow trumpet

Papery spathe protects bud

Single flower on stem

Fruit splits into three parts when ripe

Bluish-green leaves

The poet William Wordsworth made the wild daffodil very much his own, and it is in fact very much an English plant. It is rare in Wales and almost totally absent from Scotland and Ireland. Even in England it has suffered from the depredations of drainage schemes and over-keen gardeners anxious to add to their stocks. But its bright, nodding flowers can still be seen in abundance in damp woods in southern and south-western England. They are smaller than most of the 10,000 cultivated forms, with a darker yellow 'trumpet' as long as the outer ray petals.

Widespread in damp open woodland, heathland, common and meadow in southern and western England.

Yellow iris

Iris pseudacorus
16-60in (40-152cm)

Outer petals hang down

Two crested lobes on style

Fine, parallel veins on leaf

Creeping underground stem

Large seed capsules in groups of two or three

Leaves long, flat and strap-shaped

This handsome flower is also called flag, possibly recalling Clovis, the 5th-century king of the Franks, who first wore the iris as a heraldic device. Its leaves are appropriately sharp-edged and sword-shaped, and can cut if not handled carefully. But it is the flower which makes this such a striking plant, especially when seen in large clumps, usually on the edge of pools and streams. Each stem bears two or three flower buds, wrapped in a papery spathe. The huge green seed capsules are almost as conspicuous as the flowers, and last well into the autumn.

Widespread in wet ground or shallow water throughout Britain and Ireland.

Yellow flowers
Five to seven petals • Regular flowers

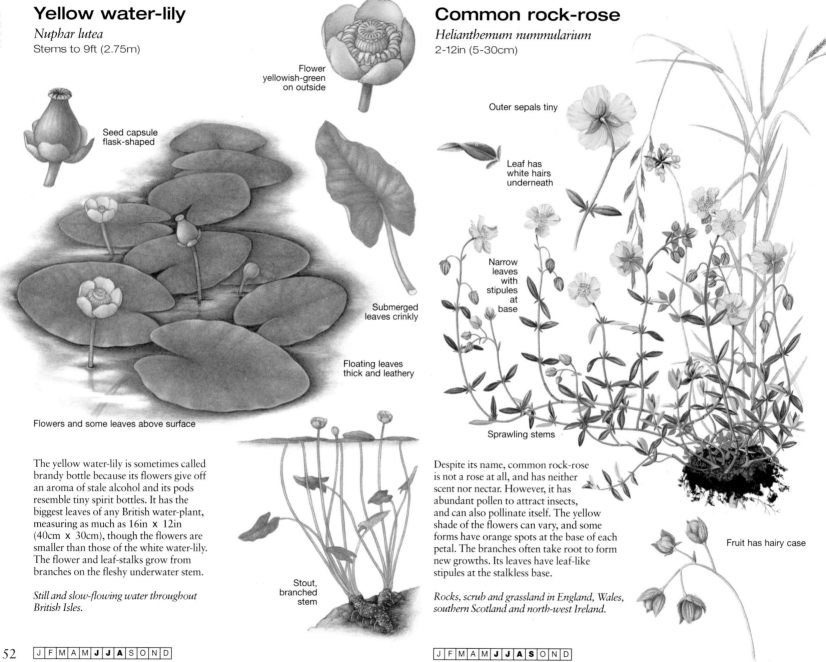

Yellow water-lily
Nuphar lutea
Stems to 9ft (2.75m)

Flower yellowish-green on outside

Seed capsule flask-shaped

Submerged leaves crinkly

Floating leaves thick and leathery

Flowers and some leaves above surface

The yellow water-lily is sometimes called brandy bottle because its flowers give off an aroma of stale alcohol and its pods resemble tiny spirit bottles. It has the biggest leaves of any British water-plant, measuring as much as 16in x 12in (40cm x 30cm), though the flowers are smaller than those of the white water-lily. The flower and leaf-stalks grow from branches on the fleshy underwater stem.

Still and slow-flowing water throughout British Isles.

Stout, branched stem

Common rock-rose
Helianthemum nummularium
2-12in (5-30cm)

Outer sepals tiny

Leaf has white hairs underneath

Narrow leaves with stipules at base

Sprawling stems

Despite its name, common rock-rose is not a rose at all, and has neither scent nor nectar. However, it has abundant pollen to attract insects, and can also pollinate itself. The yellow shade of the flowers can vary, and some forms have orange spots at the base of each petal. The branches often take root to form new growths. Its leaves have leaf-like stipules at the stalkless base.

Rocks, scrub and grassland in England, Wales, southern Scotland and north-west Ireland.

Fruit has hairy case

The large buttercup family includes plants with feathery leaves, such as the meadow buttercup itself, and others with heart-shaped leaves like the lesser celandine. Primroses and cowslips are easily recognised, and the group also includes members of the St John's-wort and rose families.

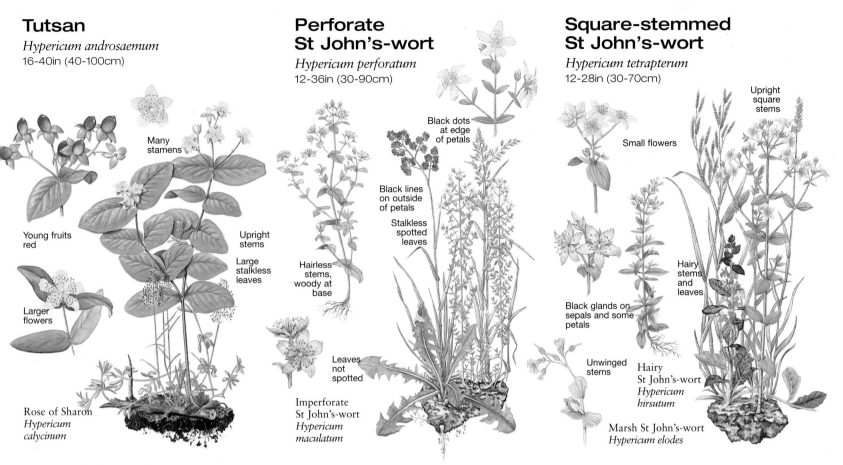

Tutsan
Hypericum androsaemum
16-40in (40-100cm)

Many stamens

Young fruits red

Upright stems

Large stalkless leaves

Larger flowers

Rose of Sharon
Hypericum calycinum

Perforate St John's-wort
Hypericum perforatum
12-36in (30-90cm)

Black dots at edge of petals

Black lines on outside of petals

Stalkless spotted leaves

Hairless stems, woody at base

Leaves not spotted

Imperforate St John's-wort
Hypericum maculatum

Square-stemmed St John's-wort
Hypericum tetrapterum
12-28in (30-70cm)

Upright square stems

Small flowers

Hairy stems and leaves

Black glands on sepals and some petals

Unwinged stems

Hairy St John's-wort
Hypericum hirsutum

Marsh St John's-wort
Hypericum elodes

Stamens bristle like pins in a pin-cushion in the flowers of tutsan, which rise in small clusters above large stalkless leaves. Medieval herbalists laid these leaves across flesh wounds to help them heal, because of the plant's genuine antiseptic properties; its name is derived from the French *toute-saine*, 'all healthy'. When dried, the leaves give off a pleasant smell and they were once used as bookmarks, particularly for Bibles. The related rose of Sharon is low-growing and has very large solitary flowers which, like tutsan, have a pin-cushion of stamens.

Damp places of woods and hedges, especially southern and western Britain and Ireland.

Translucent glands on the leaves, which look like perforations when held up to the light, give this plant its common name. It differs from other similar St John's-worts in having two narrow wings along its hairless stem. Its flowers are about 1in (25mm) across, with black dots at the edge of the petals and with the stamens grouped in three bundles. During the crusades, the Knights of St John of Jerusalem used the leaves of this plant to heal wounds received in battle. Imperforate St John's-wort lacks the leaf spots but has winged stems and petals with black lines and dots.

Grassland, hedges and open woods throughout Britain.

As its name implies, the shape of the stem is the main distinguishing feature of this species; the squared appearance comes from four narrow wings which run along it. The flowers are similar to those of perforate St John's-wort but only half as large. The petals have a very few black glands at the edges, while the leaves are stalkless and partly clasp the stem.
Marsh St John's-wort has unwinged stems and is covered in white down. Hairy St John's-wort has hairy stems and leaves, and stalked flowers with black glands.

Marshy and damp grassy places throughout British Isles, except northern Scotland.

| J | F | M | A | M | J | J | A | S | O | N | D |

| J | F | M | A | M | J | J | A | S | O | N | D |

| J | F | M | A | M | J | J | A | S | O | N | D |

Yellow flowers
Five to seven petals • Regular flowers

Slender St John's-wort
Hypericum pulchrum
6-20in (15-50cm)

Rounded
sepals with
black glands

Heart-shaped
leaves; bases
clasp stem

Smooth,
unridged
stems

Sprawling
stems

Smaller
flowers

Trailing St John's-wort
Hypericum humifusum

Reflexed stonecrop
Sedum rupestre
6-12in (15-30cm)

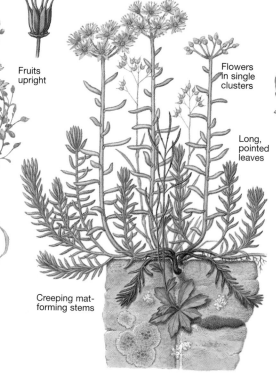

Fruits
upright

Flowers
in single
clusters

Long,
pointed
leaves

Creeping mat-
forming stems

Biting stonecrop
Sedum acre
1-4in (2.5-10cm)

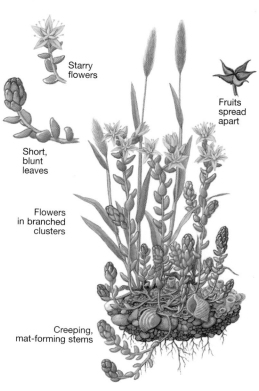

Starry
flowers

Fruits
spread
apart

Short,
blunt
leaves

Flowers
in branched
clusters

Creeping,
mat-forming stems

Unlike perforate St John's-wort, slender St John's-wort dislikes lime-rich soils and grows mainly on acid heaths and commons. Its heart-shaped, hairless leaves grow in pairs, their bases clasping the slender, smooth stems. The sepals are rounded and fringed with black glands, and the golden petals have dark red spots at their edges. Trailing St John's-wort, commonest in the south and west, has sprawling stems and much smaller flowers. Its pale green leaves are oval and the upper ones have transparent dots.

Woodland fringes and rough grassland throughout British Isles.

A dense head of flowers called a crop and a habit of growing on stone walls give the family of creeping plants known as stonecrops their name. Reflexed stonecrop was introduced from the Continent and is often cultivated in gardens, from where it has escaped and become naturalised extensively. The stems produce short, spreading non-flowering shoots and long, erect flowering shoots. A single cluster of flowers grows at the top of each stem. The leaves are fleshy and pointed, and sometimes curved back, or 'reflexed'; at one time they were eaten as a spring salad, and used by herbalists to stop bleeding.

Rocks and walls, mainly in southern Britain.

Biting stone crop is known sometimes as wall-pepper because of the sharp taste of its yellowish-green leaves, best avoided because they are poisonous in quantity. It is the smallest of the British stonecrops. Its short stalks are very numerous, some bearing star-like flowers, others with overlapping and succulent leaves. The flowers have broad-based golden petals. The young leaves are tipped with crimson and the young fruits are yellowish, turning brown when they spread apart. Superstition says that if the stonecrop is planted on the roof of a house it will ward off thunderstorms.

Grassland, shingle, dunes and walls throughout British Isles.

J F M A M J J A S O N D

J F M A M J J A S O N D

Primrose

Primula vulgaris
To 6in (15cm)

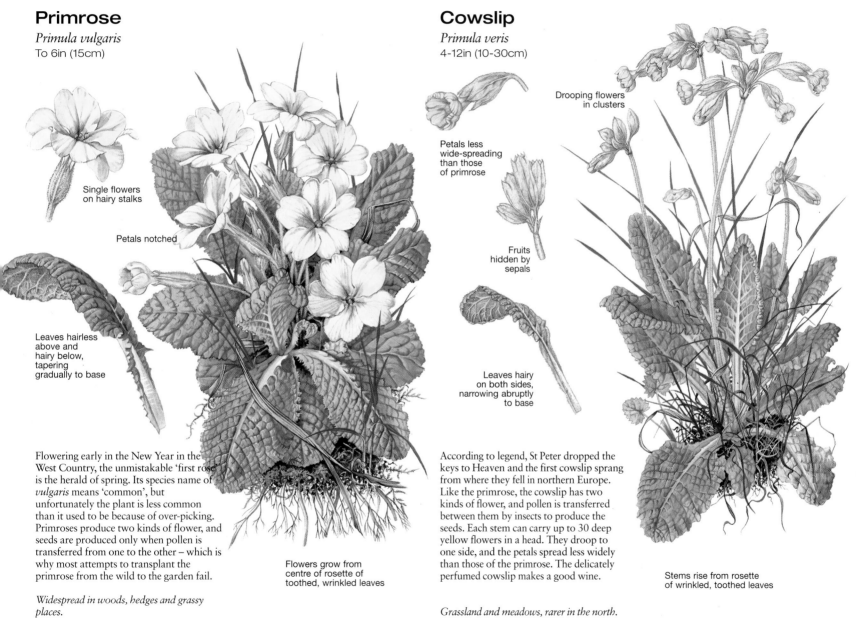

Single flowers
on hairy stalks

Petals notched

Leaves hairless
above and
hairy below,
tapering
gradually to base

Flowers grow from
centre of rosette of
toothed, wrinkled leaves

Flowering early in the New Year in the West Country, the unmistakable 'first rose' is the herald of spring. Its species name of *vulgaris* means 'common', but unfortunately the plant is less common than it used to be because of over-picking. Primroses produce two kinds of flower, and seeds are produced only when pollen is transferred from one to the other – which is why most attempts to transplant the primrose from the wild to the garden fail.

Widespread in woods, hedges and grassy places.

Cowslip

Primula veris
4-12in (10-30cm)

Drooping flowers
in clusters

Petals less
wide-spreading
than those
of primrose

Fruits
hidden by
sepals

Leaves hairy
on both sides,
narrowing abruptly
to base

According to legend, St Peter dropped the keys to Heaven and the first cowslip sprang from where they fell in northern Europe. Like the primrose, the cowslip has two kinds of flower, and pollen is transferred between them by insects to produce the seeds. Each stem can carry up to 30 deep yellow flowers in a head. They droop to one side, and the petals spread less widely than those of the primrose. The delicately perfumed cowslip makes a good wine.

Grassland and meadows, rarer in the north.

Stems rise from rosette
of wrinkled, toothed leaves

J F **M A M** J J A S O N D   J F **M A M** J J A S O N D

Yellow flowers
Five to seven petals • Regular flowers

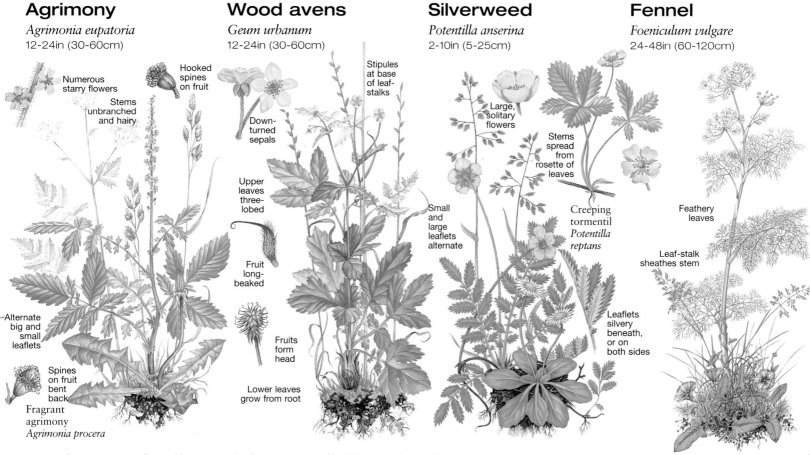

Agrimony
Agrimonia eupatoria
12-24in (30-60cm)

Numerous starry flowers

Stems unbranched and hairy

Hooked spines on fruit

Alternate big and small leaflets

Spines on fruit bent back

Fragrant agrimony
Agrimonia procera

Wood avens
Geum urbanum
12-24in (30-60cm)

Stipules at base of leaf-stalks

Down-turned sepals

Upper leaves three-lobed

Fruit long-beaked

Fruits form head

Lower leaves grow from root

Silverweed
Potentilla anserina
2-10in (5-25cm)

Large, solitary flowers

Stems spread from rosette of leaves

Small and large leaflets alternate

Creeping tormentil *Potentilla reptans*

Leaflets silvery beneath, or on both sides

Fennel
Foeniculum vulgare
24-48in (60-120cm)

Feathery leaves

Leaf-stalk sheathes stem

A scent of apricots attracts flies and bees to agrimony's slender, crowded spike of little starry flowers. In pagan times the plant was thought to have magical properties; nowadays its leaves, deeply divided into coarsely toothed leaflets, are still used as a stimulating alternative to tea.
Fragrant agrimony is a bigger plant, with more branches and fragrant leaves which have sticky hairs underneath.

Field margins, road verges and hedgerows over most of lowland British Isles.

The alternative name of herb bennet derives from the flower's association with St Benedict. In the 15th century it was widely used to ward off evil spirits. It is a short plant, with downy stems and deeply divided leaves which have a large lobed leaflet at the end. Little leaf-like stipules grow in pairs at the junction of the stem and leaf-stalk. The flowers are small and widely spread, with a long, hooked hairy style which is retained on the fruit.

Widespread in damp shady places.

Silverweed is easily identified by its long-stalked solitary flowers and silvery silky-hairy leaves of many leaflets. The plant spreads by overground runners, often forming large patches. It was once cultivated in western Scotland for its edible roots, which taste like turnips.
Creeping tormentil is a related but bigger species, often found in hedgerows. It flowers from June to September.

Widespread in damp, grassy and waste places, and by roadsides throughout British Isles.

With its large, dense umbrellas of little mustard-coloured flowers and its distinctive, aniseed smell, fennel is different from every other British wild flower. It is a tall plant, with feathery sprays of bright green leaves, like asparagus fern, with thread-like lobes and stalks that sheathe the stems. The Romans probably brought it to Britain as a medicine – particularly for indigestion – and to flavour their food.

Hedgebanks, cliffs and waste places near sea in England, Wales and eastern Ireland.

| J | F | M | A | M | **J** | **J** | **A** | S | O | N | D |

| J | F | M | A | M | **J** | **J** | **A** | S | O | N | D |

| J | F | M | A | **M** | **J** | **J** | **A** | **S** | O | N | D |

| J | F | M | A | M | **J** | **J** | **A** | **S** | O | N | D |

Garden parsley

Petroselinum crispum
8-18in (20-45cm)

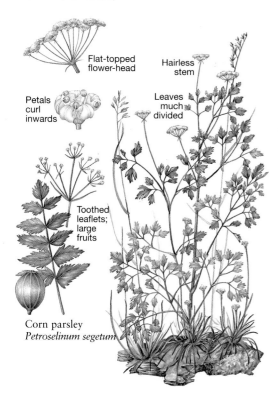

Flat-topped flower-head

Hairless stem

Petals curl inwards

Leaves much divided

Toothed leaflets; large fruits

Corn parsley
Petroselinum segetum

Any plant of this species, growing wild in Britain, will have escaped from domestic stock, which was brought into Britain from the Mediterranean in the Middle Ages. It differs from the present-day cultivated form in not having the deeply curled or frilled leaves which give a delicate feathery quality to the kitchen garden plant. It has, however, the characteristic umbrella-shaped flower-head of the parsley family, with up to 15 branches. The leaf-stalks clasp the stem at their base like a sheath. Corn parsley has smaller flower-heads, but bigger, toothed leaflets in rows on either side of the stem.

Occasional in grassy places by rocks, especially near sea.

J F M A M **J J A** S O N D

Pepper-saxifrage

Silaum silaus
12-36in (30-90cm)

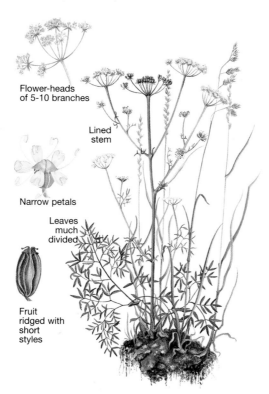

Flower-heads of 5-10 branches

Lined stem

Narrow petals

Leaves much divided

Fruit ridged with short styles

Although it can be found in damp meadows throughout Britain, pepper-saxifrage is typical of areas of heavy clay such as are found in the Weald of Sussex and Kent. Its name is inappropriate, since it is neither peppery nor a saxifrage. It is, in fact, a stiff, hairless perennial, with leaves divided into a spread of finely toothed lobes which mostly surround the base of the plant. Another identifying feature is the way tiny leaf-like bracts sprout above the junctions on the flower stems, but not below them. It is a food plant for the caterpillar of the swallowtail butterfly.

Common in damp ancient lowland meadows.

J F M A M **J J A** S O N D

Wild parsnip

Pastinaca sativa
12-48in (30-120cm)

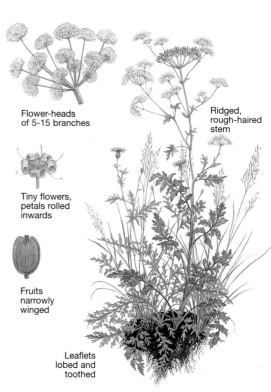

Flower-heads of 5-15 branches

Tiny flowers, petals rolled inwards

Fruits narrowly winged

Ridged, rough-haired stem

Leaflets lobed and toothed

Like the carrot, the domesticated parsnip has been developed by long and careful cultivation from its wild and fairly inedible ancestor. The 16th-century herbalist John Gerard described the root of the wild parsnip as 'small, hard, woodie, and not fit to be eaten'. Apart from the root, however, the wild parsnip looks much like its domesticated relation. It is a tall, hairy, strong-smelling plant, with leaves divided into toothed and lobed leaflets. The flowers are clustered at the top of long, branching stems, in large, umbrella-shaped flower-heads.

Widespread on dry, grassy and waste ground.

J F M A M **J** **J A S** O N D

Yellow flowers

Five to seven petals • Regular flowers

Yellow pimpernel

Lysimachia nemorum
To 15in (38cm)

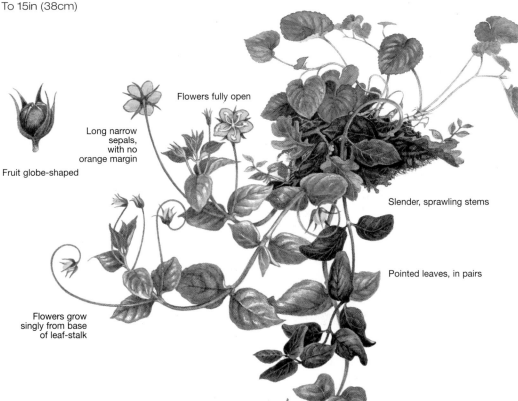

Fruit globe-shaped

Long narrow sepals, with no orange margin

Flowers fully open

Slender, sprawling stems

Pointed leaves, in pairs

Flowers grow singly from base of leaf-stalk

Sepals broad

Flowers only half open

Creeping Jenny
Lysimachia nummularia

Rounded leaves

A creeping plant which is a member of the primrose family, yellow pimpernel has oval leaves, with very short stalks, in opposite pairs. The flowers grow singly from the base of each leaf-stalk. The sepals are long and thin but lack the orange margin found on yellow loosestrife. The seed capsules protrude from the surrounding sepals. Creeping Jenny thrives in similar conditions to yellow pimpernel and is also grown in rock gardens and hanging baskets. Its leaves are rounder and its flowers only half open, with broader sepals.

Woods and shady hedges, rare in drier districts.

Yellow loosestrife

Lysimachia vulgaris
24-60in (60-152cm)

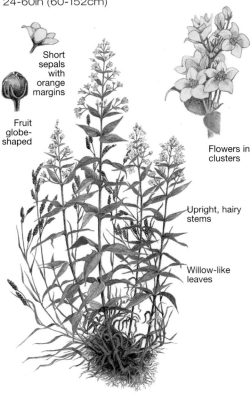

Short sepals with orange margins

Fruit globe-shaped

Flowers in clusters

Upright, hairy stems

Willow-like leaves

Willow-like leaves make this tall member of the primrose family easy to identify. They grow either in pairs or in whorls of three to four, their upper surface dotted with minute orange or black glands. The tops are bright green and the undersides bluish-green. The leaves were once burned in homes to drive away troublesome flies and gnats. The flowers are scentless and contain no nectar, but the plant is pollinated by wasps and one species of bee. Yellow loosestrife often grows in large colonies, spread by strong underground stems.

Rivers, lake margins and fens, except in far north.

J F M A M J J A S O N D

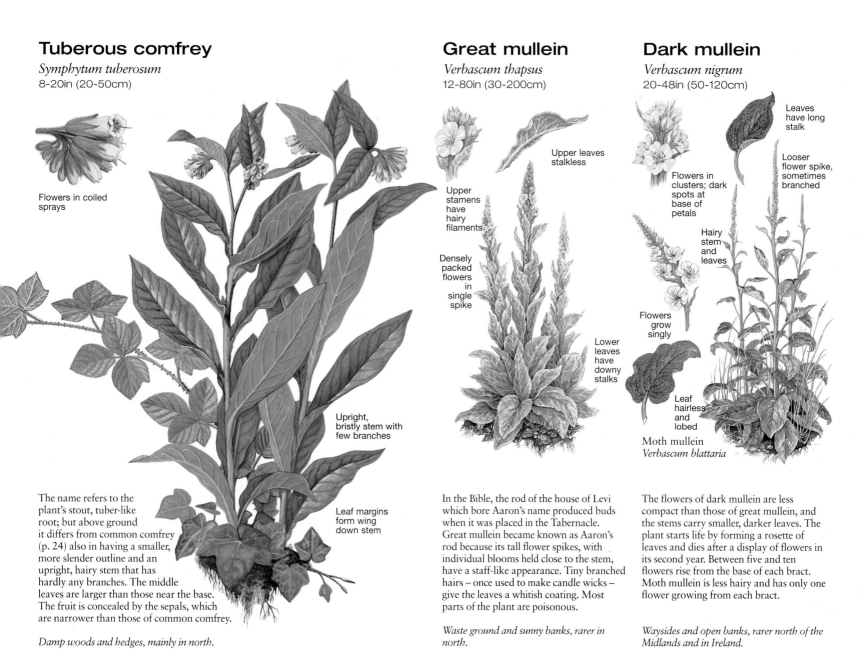

Tuberous comfrey

Symphytum tuberosum
8-20in (20-50cm)

Flowers in coiled sprays

Upright, bristly stem with few branches

Leaf margins form wing down stem

The name refers to the plant's stout, tuber-like root; but above ground it differs from common comfrey (p. 24) also in having a smaller, more slender outline and an upright, hairy stem that has hardly any branches. The middle leaves are larger than those near the base. The fruit is concealed by the sepals, which are narrower than those of common comfrey.

Damp woods and hedges, mainly in north.

Great mullein

Verbascum thapsus
12-80in (30-200cm)

Upper leaves stalkless

Upper stamens have hairy filaments

Densely packed flowers in single spike

Lower leaves have downy stalks

In the Bible, the rod of the house of Levi which bore Aaron's name produced buds when it was placed in the Tabernacle. Great mullein became known as Aaron's rod because its tall flower spikes, with individual blooms held close to the stem, have a staff-like appearance. Tiny branched hairs – once used to make candle wicks – give the leaves a whitish coating. Most parts of the plant are poisonous.

Waste ground and sunny banks, rarer in north.

Dark mullein

Verbascum nigrum
20-48in (50-120cm)

Leaves have long stalk

Looser flower spike, sometimes branched

Flowers in clusters; dark spots at base of petals

Hairy stem and leaves

Flowers grow singly

Leaf hairless and lobed

Moth mullein
Verbascum blattaria

The flowers of dark mullein are less compact than those of great mullein, and the stems carry smaller, darker leaves. The plant starts life by forming a rosette of leaves and dies after a display of flowers in its second year. Between five and ten flowers rise from the base of each bract. Moth mullein is less hairy and has only one flower growing from each bract.

Waysides and open banks, rarer north of the Midlands and in Ireland.

J F M A M J J A S O N D   J F M A M J J A S O N D   J F M A M J J A S O N D   59

Yellow flowers

Five to seven petals • Regular flowers

Goldilocks buttercup

Ranunculus auricomus
8-16in (20-40cm)

Perfect flower has five petals

One or more petals often missing

Sepals tinged purple

Stem leaves deeply divided, often stalkless

Leaves hairless

Lower leaves on long stalks, vary from deeply lobed to roundish or kidney-shaped

A striking feature of this buttercup is the flawed appearance of many of its flowers, which give the impression of having been pecked by birds. One flower may have up to five petals while others have none at all, and many of the petals are small and malformed. Goldilocks has deeply divided upper leaves, but the lower ones vary greatly in shape. Overall, it looks like a smaller and less hairy meadow buttercup. It is the main British woodland species.

Widespread in woods; rare in north and west.

Marsh-marigold

Caltha palustris
12-24in (30-60cm)

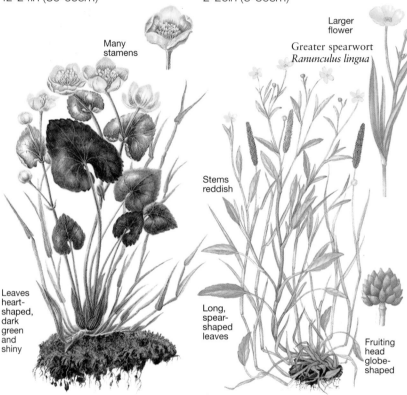

Many stamens

Leaves heart-shaped, dark green and shiny

Marsh-marigolds light up damp places with their brilliant flowers and glossy leaves as early as March, when there is still snow on the ground. The flowers are sometimes as much as 2in (50mm) across, with as many as 100 stamens. The heart-shaped leaves have long stalks and grow from the base of the hairless plant. The plant's alternative name of kingcup is derived from the Old English *cop*, 'button'.

Wet woodland, ditches and fens all over Britain and Ireland.

Lesser spearwort

Ranunculus flammula
2-20in (5-50cm)

Larger flower

Greater spearwort
Ranunculus lingua

Stems reddish

Long, spear-shaped leaves

Fruiting head globe-shaped

Lesser spearwort has killed cattle and sheep that have eaten it because it has the poisonous sap common to all buttercups; the name *flammula*, Latin for 'little flame', refers to the plant's burning taste. The creeping stems often root at intervals. The leaves are usually spear-shaped, the lower ones stalked, the upper ones stalkless. Greater spearwort is much taller, reaching 48in (120cm), and rarer.

Wet or marshy places throughout Britain and Ireland.

| J F M A M J J A S O N D |

| J F M A M J J A S O N D |

Meadow buttercup

Ranunculus acris
12-36in (30-90cm)

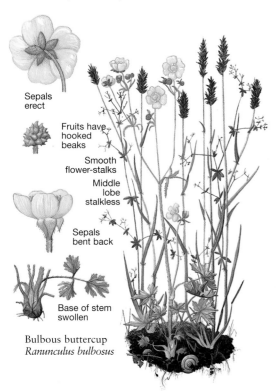

Sepals erect

Fruits have hooked beaks

Smooth flower-stalks

Middle lobe stalkless

Sepals bent back

Base of stem swollen

Bulbous buttercup
Ranunculus bulbosus

Fields bright with meadow buttercups are a less common sight than they once were, as traditional meadows are ploughed and resown with grass and clover. But where they still occur they are one of the most delightful sights of early summer. The meadow buttercup is the tallest and most graceful of several related species. The leaves are hairy and deeply cut, while the flowers are on smooth stalks on a branching stem.

The bulbous buttercup flowers earlier than the meadow species, from March onwards, and prefers drier grassland.

Damp meadows throughout Britain and Ireland.

Creeping buttercup

Ranunculus repens
2-20in (5-50cm)

Flowers single or in clusters

Stem and leaves hairy

Fruits in globe-shaped head

Three leaflets, middle one stalked

Smaller flowers

Long fruit

Celery-leaved buttercup
Ranunculus sceleratus

This might well be called the problem buttercup, for its creeping overground runners can spread rapidly, in time taking over whole fields. In pastureland this can be serious, for like all buttercups it is poisonous. Cattle avoid it, but as they eat the grass around the buttercup, so they make room for it to spread, until the whole field is covered. Ploughing only exacerbates the problem, chopping the runners up into little pieces and thereby creating thousands of potential new plants.

Celery-leaved buttercup has smaller flowers and hairless lower leaves without a stalked middle lobe.

Widespread on disturbed land, damp pasture and roadsides.

Meadowsweet

Filipendula ulmaria
24-48in (60-120cm)

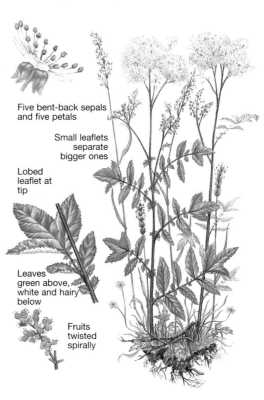

Five bent-back sepals and five petals

Small leaflets separate bigger ones

Lobed leaflet at tip

Leaves green above, white and hairy below

Fruits twisted spirally

A flower with such an attractive name cannot fail to win friends. It smells good and, seen in great drifts by a river bank in summer, it looks superb. In medieval and Tudor times the flowers were often strewn among the rushes on the floors of houses to sweeten the domestic air. Queen Elizabeth much approved of this practice. But the flowers are best seen in a natural setting, where the dense clusters of blossom sprout like creamy-yellow foam at the end of tall, hairless, branching stems. Meadowsweet is in fact a corruption of an older name, mede-sweete, given to it by the Anglo-Saxons who used it to flavour their mead.

Widespread in wet woodland, meadows and fens.

| J | F | M | A | **M** | **J** | **J** | **A** | **S** | O | N | D |

| J | F | M | A | **M** | **J** | **J** | **A** | **S** | O | N | D |

| J | F | M | A | **M** | **J** | **J** | **A** | **S** | O | N | D |

Yellow flowers

Five to seven petals
Irregular flowers

Pansies differ from violets in having side petals which are held horizontally, instead of drooping towards the lower petal. This group includes three members of the figwort family.

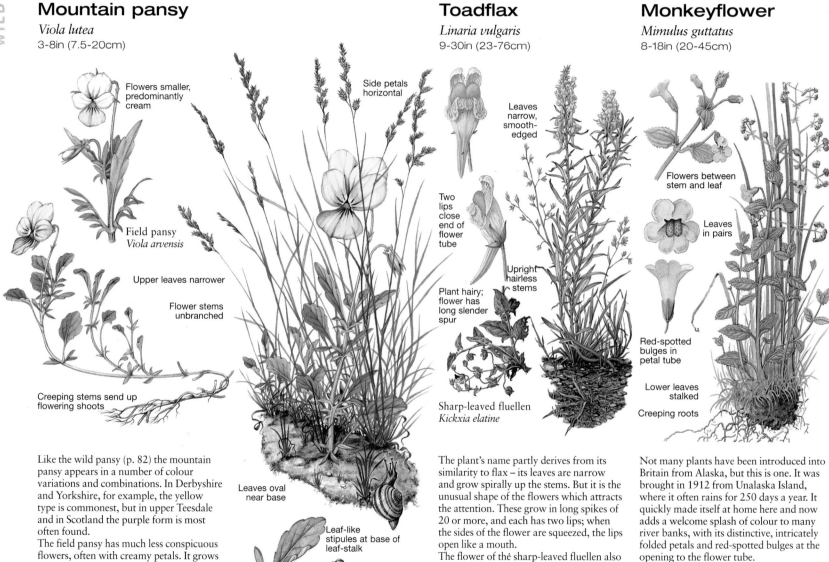

Mountain pansy

Viola lutea
3-8in (7.5-20cm)

Flowers smaller, predominantly cream

Field pansy
Viola arvensis

Upper leaves narrower

Flower stems unbranched

Creeping stems send up flowering shoots

Side petals horizontal

Leaves oval near base

Leaf-like stipules at base of leaf-stalk

Fruit capsules split into three

Toadflax

Linaria vulgaris
9-30in (23-76cm)

Leaves narrow, smooth-edged

Two lips close end of flower tube

Upright hairless stems

Plant hairy; flower has long slender spur

Sharp-leaved fluellen
Kickxia elatine

Monkeyflower

Mimulus guttatus
8-18in (20-45cm)

Flowers between stem and leaf

Leaves in pairs

Red-spotted bulges in petal tube

Lower leaves stalked

Creeping roots

Like the wild pansy (p. 82) the mountain pansy appears in a number of colour variations and combinations. In Derbyshire and Yorkshire, for example, the yellow type is commonest, but in upper Teesdale and in Scotland the purple form is most often found.
The field pansy has much less conspicuous flowers, often with creamy petals. It grows on waste ground and in arable fields.

Widespread in grassy, hilly areas in northern England, Wales, Scotland and Ireland.

The plant's name partly derives from its similarity to flax – its leaves are narrow and grow spirally up the stems. But it is the unusual shape of the flowers which attracts the attention. These grow in long spikes of 20 or more, and each has two lips; when the sides of the flower are squeezed, the lips open like a mouth.
The flower of thé sharp-leaved fluellen also has a long, slender spur.

Meadows, cultivated fields and waste ground; commonest in south.

Not many plants have been introduced into Britain from Alaska, but this is one. It was brought in 1912 from Unalaska Island, where it often rains for 250 days a year. It quickly made itself at home here and now adds a welcome splash of colour to many river banks, with its distinctive, intricately folded petals and red-spotted bulges at the opening to the flower tube.

By streams and in marshy meadows throughout Britain and most of Ireland.

J F M A M J J A S O N D

J F M A M J J A S O N D

J F M A M J J A S O N D

Yellow flowers *Eight or more petals*

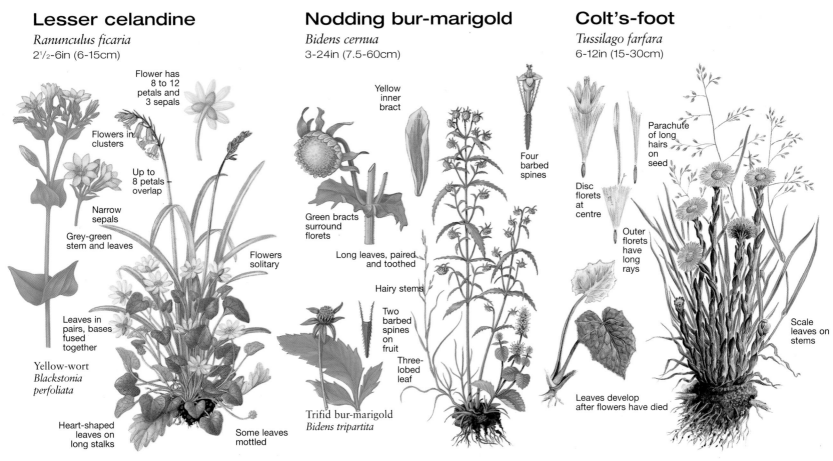

Lesser celandine
Ranunculus ficaria
2½-6in (6-15cm)

Flower has 8 to 12 petals and 3 sepals

Flowers in clusters

Up to 8 petals – overlap

Narrow sepals

Grey-green stem and leaves

Flowers solitary

Leaves in pairs, bases fused together

Yellow-wort
Blackstonia perfoliata

Heart-shaped leaves on long stalks

Some leaves mottled

Nodding bur-marigold
Bidens cernua
3-24in (7.5-60cm)

Yellow inner bract

Four barbed spines

Green bracts surround florets

Long leaves, paired and toothed

Hairy stems

Two barbed spines on fruit

Three-lobed leaf

Trifid bur-marigold
Bidens tripartita

Colt's-foot
Tussilago farfara
6-12in (15-30cm)

Parachute of long hairs on seed

Disc florets at centre

Outer florets have long rays

Leaves develop after flowers have died

Scale leaves on stems

Lesser celandine is one of the first wild flowers to appear, often carpeting woodland in early spring. Each of the solitary flowers can have up to 12 yellow petals which sometimes fade to white. The leaves often have light or dark markings. William Wordsworth wrote a poem praising the flower as shining 'bright as the sun himself'. Yellow-wort, a chalkland annual growing to 18in (45cm), flowers all summer. It is easily identified by the fused pairs of stem leaves; it has a rosette of larger leaves at the base. The butter-yellow flowers grow in clusters.

Damp grassland, shady places and gardens all over Britain and Ireland.

The tightly packed flower-heads of this plant really do nod on their slender stalks, giving a certain charm to an otherwise undistinguished wild flower. A member of the daisy family, it has a compact flower-head which looks like a single flower, but is in fact made up of many tiny individual flowers, surrounded by a ring of green leaf-like bracts. The flowers are unscented and the leaves unstalked and undivided.
Trifid bur-marigold, which grows in similar places, has leaves with three spear-shaped lobes, the middle one being the longest.

Locally common by still water, rarer in north.

As winter subsides, colt's-foot is so eager to burst into flower that it does not wait for its leaves to form: they appear later. The flowers comprise both disc and ray florets, the latter often tinged with orange. Then there are the many overlapping, fleshy, purple scales which cover the stems right up to the flower-heads. Finally, there are the large, toothed leaves, which appear after the flowers; their hoof-like shape gives the plant its name. The 'parachutes' by which the seeds are spread need only the slightest of draughts to keep them airborne.

Widespread on bare and thinly covered ground throughout Britain and Ireland.

Yellow flowers *Eight or more petals*

Common fleabane
Pulicaria dysenterica
8-24in (20-60cm)

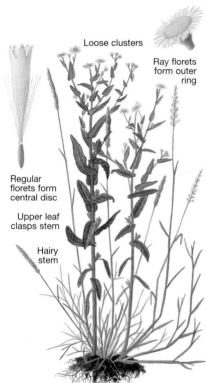

Loose clusters

Ray florets form outer ring

Regular florets form central disc

Upper leaf clasps stem

Hairy stem

The best way to identify this plant is by the leaf rather than the flower. The upper leaves are stalkless and notched at the base, so that they half-clasp the stem. All the leaves are wrinkled, wavy-edged and woolly, and the whole plant is greyish in colour. The flower is daisy-like, with up to 600 separate florets packed into each flower-head. When the plant was burnt, the smoke was said to drive away fleas.

Widespread on marshy ground and meadows, except in northern England and Scotland.

J F M A M J J A S O N D

Common cudweed
Filago vulgaris
2-12in (5-30cm)

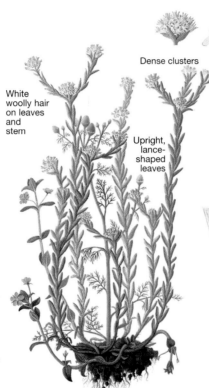

White woolly hair on leaves and stem

Dense clusters

Upright, lance-shaped leaves

The little flowers of the common cudweed are easily overlooked, half-hidden behind a thicket of silver-grey woolly hairs and spreading side branches. The flowers are gathered in clusters of 20-40 in small, ball-like heads without rays, at the tips of tough, wiry stems. Tapered leaves grow spirally upwards round, and close to, the stems. Medieval farmers gave the plant to cattle that had stopped ruminating, to replace their 'cud'.

Widespread on heaths and dry pastures.

J F M A M J J A S O N D

Goldenrod
Solidago virgaurea
2-24in (5-60cm)

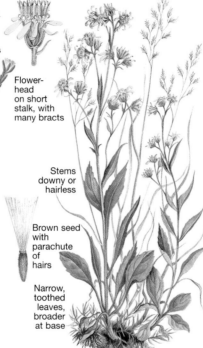

Flower-head on short stalk, with many bracts

Stems downy or hairless

Brown seed with parachute of hairs

Narrow, toothed leaves, broader at base

The Elizabethans failed to notice that this bright and brilliant plant was native to this country, and instead spent large sums of money importing it for use in the treatment of wounds. As soon as its presence was noticed on Hampstead Heath its market price crashed. The flower-heads have short rays and grow on branched spikes. The narrow, surrounding bracts look like sepals and are greenish-yellow.

Widespread on mountains and in woods, hedgerows and dunes.

J F M A M J J A S O N D

Canadian fleabane
Conyza canadensis
3-36in (7.5-90cm)

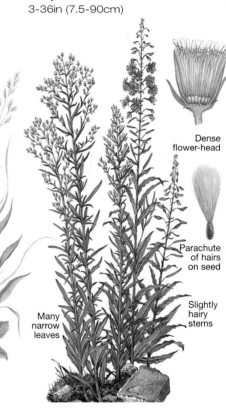

Dense flower-head

Parachute of hairs on seed

Slightly hairy stems

Many narrow leaves

This New World immigrant, here seen growing with rosebay willowherb, grows in pavement cracks and on building sites as readily as by country roads. It came to Britain about 200 years ago, and its spread was helped by the development of the railways. Its seeds have parachute hairs and the draughts caused by passing trains aided their dispersal. The small flower-heads grow in open, branched spikes.

Common on roadsides and embankments, particularly in south and south-east England.

J F M A M J J A S O N D

In such a large family as the daisies, many members are superficially alike and few are as distinctive as the dandelion. In all of them, each flower-head consists of hundreds of individual florets. The species can be told apart by the arrangement of the flower-heads and the pattern of the leaves.

Corn marigold

Chrysanthemum segetum
6-18in (15-45cm)

Hawkweed

Hieracium umbellatum
10-48in (25-120cm)

Prickly sow-thistle

Sonchus asper
8-60in (20-152cm)

Smooth sow-thistle

Sonchus oleraceus
8-60in (20-152cm)

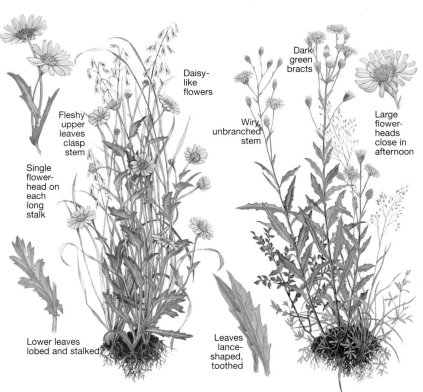

Daisy-like flowers

Fleshy upper leaves clasp stem

Single flower-head on each long stalk

Lower leaves lobed and stalked

Dark green bracts

Wiry, unbranched stem

Large flower-heads close in afternoon

Leaves lance-shaped, toothed

Spiny, toothed leaves clasp stem

Green triangular bracts

Flower-heads in clusters

Rounded lobes clasp stem

Perennial sow-thistle *Sonchus arvensis*

Leaves toothed but spineless, with points at base, extending beyond stem

Now that modern herbicides have reduced the threat that this plant once posed to corn harvests, it can be admired where it survives on waste ground for its finer points, particularly its rich, golden-yellow flower-heads. These are too big to miss, usually about 2in (50mm) across, growing singly on stalks thickened at the top, with long rays. The leaves are narrow, fleshy and jaggedly toothed. They are notched at the base and partly clasp the stem.

Light, sandy soils throughout lowland areas.

Hawkweeds are not the easiest of plants to identify. In most cases the seed is produced without fertilisation, leading to the perpetuation of minor differences and to the creation in time of what amounts to a new species. Botanists recognise over 250 hawkweeds in the British Isles alone. One of the commonest species is this tall, erect plant with numerous alternate leaves diminishing in length up the stem.

Lowland areas, roadsides, hedgerows and open woods in Britain; rare in Ireland.

The white, milky fluid contained in the stem and roots of this plant is attractive to animals – including sows, whose milk flow it is thought to improve. As a defence, the plant has armed itself with prickly, glossy green leaves which are also its most interesting feature. They are rounded at the base and wrap themselves around the stem. The white, fluffy fruiting head resembles a small dandelion 'clock'.

Widespread near walls, on dunes and disturbed ground throughout British Isles.

It is not only the absence of prickles that differentiates this plant from the prickly sow-thistle. The leaves are dull, pale green and divided, with the terminal lobe much the largest. They are pointed at the base, loosely clasping the stem. The two plants' flower-heads are, however, very similar. Perennial sow-thistle has bigger flower-heads which, like its stems, are thickly covered in glandular hairs.

Common on waste and cultivated land and roadsides, often by walls.

J F M A M J J A S O N D   J F M A M J J A S O N D   J F M A M J J A S O N D   J F M A M J J A S O N D

Yellow flowers *Eight or more petals*

Bristly ox-tongue

Picris echioides
12-36in (30-90cm)

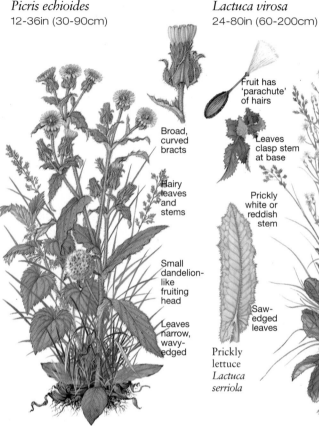

Broad, curved bracts

Hairy leaves and stems

Small dandelion-like fruiting head

Leaves narrow, wavy-edged

With its leaves covered in what at first glance look like white blisters, and the whole bristling with stiff hairs, bristly ox-tongue can hardly be described as a pretty plant. But it can fairly claim to be one of the most distinctive of the dandelion-like flowers. It is a close relative of the sow-thistle, but less common. The pale yellow flower-heads grow in irregular clusters.

Locally common on roadsides and waste ground in lowland England and Wales, and eastern Ireland.

Great lettuce

Lactuca virosa
24-80in (60-200cm)

Fruit has 'parachute' of hairs

Leaves clasp stem at base

Prickly white or reddish stem

Saw-edged leaves

Prickly lettuce *Lactuca serriola*

Leaves have bristles below

There is nothing in the appearance of this wild lettuce to link it with the garden variety. It is blue-green, bristly and very tall, sometimes reaching a height of over 6ft (1.8m). The leaves are gathered mainly on the lower half of the stem, below clusters of flower-heads.
Prickly lettuce is a 'compass' plant; when fully exposed to the sun, its saw-edged leaves lie in a north-south plane.

Roadside verges and disturbed chalky ground, mainly in south and eastern England.

Goat's beard

Tragopogon pratensis
12-28in (30-70cm)

Solitary flower-heads

Fruiting head a downy 'clock'

Long grassy leaves

Leaves sheathe stem

Mouse-ear hawkweed *Hieracium pilosella*

Also called Jack-go-to-bed-at-noon, this plant attracts attention on sunny mornings when its large, dandelion-like flowers are out. They quickly disappear about midday, hidden within long, pointed finger-like bracts. In mid-afternoon, with no flowers in sight, the long, narrow leaves look much like grass and can easily go unnoticed. They twist and sheathe the stem in an unusual manner for dandelion-like plants.

Widespread by roadsides, waste ground, dunes, meadows and grassland.

Cat's ear

Hypochoeris radicata
8-24in (20-60cm)

Bristly bracts

Wiry stems

Toothless leaves

Broad teeth

Hairy leaves in rosette

Small, dark, scale-like bracts grow spirally up the flower stems of this plant, like miniature cat's ears. The brilliant yellow flower does much to brighten grassy places – including, without invitation, garden lawns. The leaves are broadly toothed. Mouse-ear hawkweed has creeping, rooting stems and woolly white hairs on the underside of its leaves, which form a rosette.

Pastures, grassy dunes and waysides throughout British Isles.

J F M A M J J A S O N D   J F M A M J J A S O N D   J F M A M J J A S O N D   J F M A M J J A S O N D

Autumn hawkbit
Leontodon autumnalis
2-24in (5-60cm)

Many florets in flower-head, often reddish beneath

Scale-like bracts

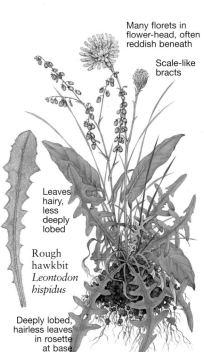

Leaves hairy, less deeply lobed

Rough hawkbit
Leontodon hispidus

Deeply lobed, hairless leaves in rosette at base

The autumn hawkbit does for roadsides in August and September what the dandelion does in May, turning the verge into a glowing border of golden-yellow flowers. Its leaves are almost identical to the dandelion's – long, slender, and deeply lobed. The stems are branched, the flower-heads surrounded by scale-like bracts. Rough hawkbit is very hairy, with much less deeply lobed leaves.

Roadsides and grassy places throughout British Isles.

Nipplewort
Lapsana communis
8-36in (20-90cm)

Many small flower-heads

Lance-shaped bracts

Lower leaves have large terminal lobe

Nipplewort is readily identified by its clusters of flower-heads, resembling tiny dandelions, on long, slender, many-branched stalks. They close in the early afternoon, and do not open at all in bad weather. Each flower-head is enclosed in a ring of lance-shaped bracts. The lower leaves are even more distinctive in having a large, toothed, terminal lobe with, below it, smaller, wing-shaped lobes.

Roadsides, hedges, waste ground and wood margins throughout British Isles.

Smooth hawk's-beard
Crepis capillaris
8-36in (20-90cm)

Two rows of bracts

Stem-leaves narrow

Tiny flower-heads

Arrow-shaped lobes

Lower leaves in rosette

This plant has the smallest flowers of all the hawk's-beards – ½in (13mm) across. In bud they are often reddish in appearance, the outer florets being tinted underneath. They grow in branched clusters, each flower-head surrounded by two rows of bracts. The leaves are dandelion-shaped, forming a rosette at the base. They are shiny and divided into arrow-shaped lobes, with winged stalks.

Common in pastures, heaths and waste places throughout British Isles.

Dandelion
Taraxacum officinale
2-12in (5-30cm)

White-haired fruits form dandelion 'clock'

Long, hollow stalk

Leaves in rosette

Away from the garden, where its reputation is decidedly low, the dandelion can be appreciated as a flower full of character, filling roadsides and fields with a golden blaze of colour in the summer months. The large flower-heads are made up of some 200 ray florets, which close at night or in dull weather. Its leaves give the dandelion its name – a corruption of the French *dent de lion*, 'lion's tooth'.

Widespread in grassland and on banks and wasteland.

J F M A M **J J A S O** N D | J F M A M **J J A S** O N D | J F M A M **J J A S** O N D | J F **M A M J J A S O** N D

Yellow flowers *Eight or more petals*

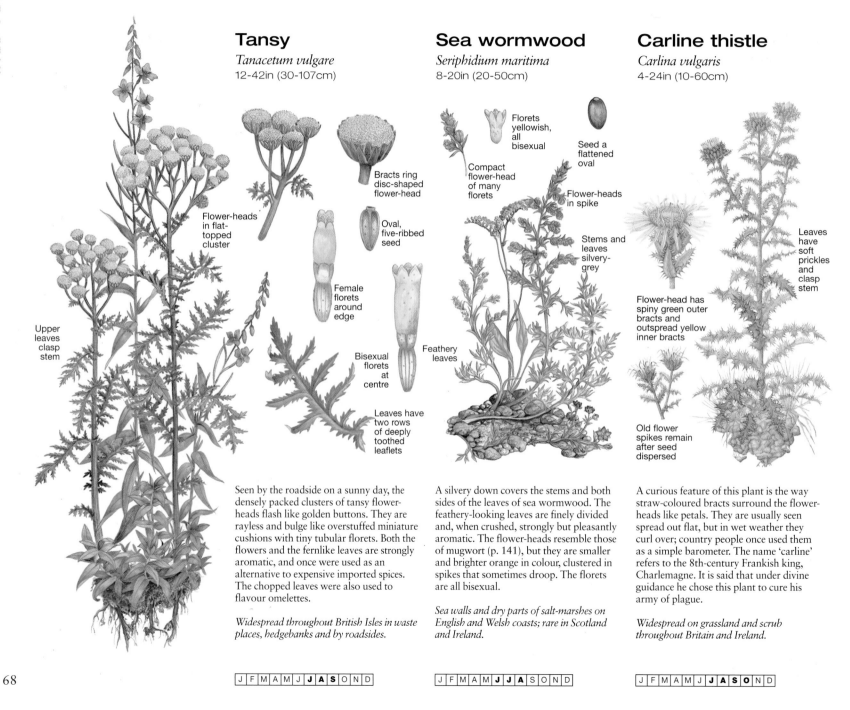

Tansy
Tanacetum vulgare
12-42in (30-107cm)

Bracts ring disc-shaped flower-head

Flower-heads in flat-topped cluster

Oval, five-ribbed seed

Female florets around edge

Bisexual florets at centre

Upper leaves clasp stem

Leaves have two rows of deeply toothed leaflets

Sea wormwood
Seriphidium maritima
8-20in (20-50cm)

Florets yellowish, all bisexual

Seed a flattened oval

Compact flower-head of many florets

Flower-heads in spike

Stems and leaves silvery-grey

Feathery leaves

Carline thistle
Carlina vulgaris
4-24in (10-60cm)

Leaves have soft prickles and clasp stem

Flower-head has spiny green outer bracts and outspread yellow inner bracts

Old flower spikes remain after seed dispersed

Seen by the roadside on a sunny day, the densely packed clusters of tansy flower-heads flash like golden buttons. They are rayless and bulge like overstuffed miniature cushions with tiny tubular florets. Both the flowers and the fernlike leaves are strongly aromatic, and once were used as an alternative to expensive imported spices. The chopped leaves were also used to flavour omelettes.

Widespread throughout British Isles in waste places, hedgebanks and by roadsides.

A silvery down covers the stems and both sides of the leaves of sea wormwood. The feathery-looking leaves are finely divided and, when crushed, strongly but pleasantly aromatic. The flower-heads resemble those of mugwort (p. 141), but they are smaller and brighter orange in colour, clustered in spikes that sometimes droop. The florets are all bisexual.

Sea walls and dry parts of salt-marshes on English and Welsh coasts; rare in Scotland and Ireland.

A curious feature of this plant is the way straw-coloured bracts surround the flower-heads like petals. They are usually seen spread out flat, but in wet weather they curl over; country people once used them as a simple barometer. The name 'carline' refers to the 8th-century Frankish king, Charlemagne. It is said that under divine guidance he chose this plant to cure his army of plague.

Widespread on grassland and scrub throughout Britain and Ireland.

J F M A M J **J A S** O N D

J F M A M **J J A S** O N D

J F M A M J **J A S** O N D

Common groundsel

Senecio vulgaris
3-12in (7.5-30cm)

White fruiting heads

Black-pointed bracts

Leaves have toothed lobes

Leaves more deeply toothed

Leaves widely spaced

Heath groundsel *Senecio sylvaticus*

Florets have tongue

After flowering, the tiny, yellow cylindrical flower-heads turn into fluffy balls of white hair. This feature is the origin of the plant's scientific name, from the Latin *senex*, 'old man'. Until this transformation occurs the flower-heads, which grow in loose clusters, are inconspicuous apart from the bracts, which have black, pointed tips. The plant's common name is an accurate assessment of its reputation as a garden weed: it comes from the Anglo-Saxon *grondeswyle*, meaning 'ground glutton'. Heath groundsel has more deeply toothed leaves. The outer florets have a tongue-like projection.

Widespread on cultivated ground and in waste places.

Common ragwort

Senecio jacobaea
12-48in (30-120cm)

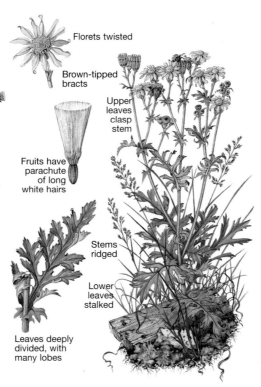

Florets twisted

Brown-tipped bracts

Fruits have parachute of long white hairs

Upper leaves clasp stem

Stems ridged

Lower leaves stalked

Leaves deeply divided, with many lobes

The deeply divided leaves give the plant a ragged look, which accounts for its name, but this is nevertheless a handsome plant, with fine daisy-like flowers. The outer florets spread out from the centre disc like the rays of a symbolic sun, twisting and curling at the ends. Growing in large, dense, flat-topped clusters, the flower-heads make an attractive sight. The inner bracts surrounding the flower-heads are brown-tipped. The leaves are mildly poisonous to cattle, but are the favourite food of the black and yellow striped caterpillar of the cinnabar moth.

Widespread on wasteland, neglected pastureland and by roadsides.

Oxford ragwort

Senecio squalidus
8-12in (20-30cm)

Black-tipped bracts

Upper leaves clasp stem

Leaves divided into a few narrow lobes

Fruit

Hairless stem

Green bracts

Greyish hairs on stem

Sticky groundsel *Senecio viscosus*

Lower leaves stalked

The 19th-century naturalist C.A. Johns claimed that Oxford ragwort was the prettiest of the British ragwort species. From its sheltered beginnings in this country in Oxford's Botanic Gardens, after being introduced from Italy in the 18th century, it is rapidly becoming one of the most common ragworts, especially around towns. Part of its success is no doubt due to its prolific flowering; one plant may produce up to 10,000 seeds.
Sticky groundsel is covered in sticky glandular hairs, which give it a grey-hued appearance.

Common on walls, roadsides and railway banks, and on waste and arable land; rare in northern Britain and most of Ireland.

Blue flowers

Up to four petals
Regular flowers

This small group includes three meadow plants, unlike in other respects, and one plant of the shoreline. Its regular petals distinguish devil's-bit scabious from two close relatives.

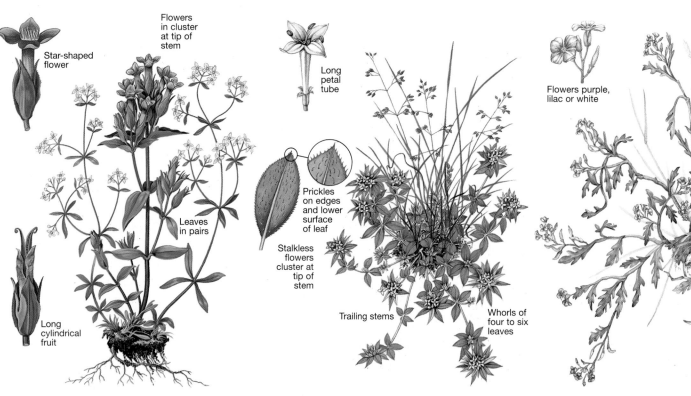

Field gentian
Gentianella campestris
4-12in (10-30cm)

Star-shaped flower

Flowers in cluster at tip of stem

Leaves in pairs

Long cylindrical fruit

Field madder
Sherardia arvensis
2-16in (5-40cm)

Long petal tube

Prickles on edges and lower surface of leaf

Stalkless flowers cluster at tip of stem

Trailing stems

Whorls of four to six leaves

Sea rocket
Cakile maritima
6-12in (15-30cm)

Flowers purple, lilac or white

Fleshy leaves

Bushy, straggling stems

Counting the petals is the best way to distinguish between field gentian, seen growing with marsh bedstraw, and the closely related autumn gentian (p. 107). Whereas the star-shaped flower of field gentian has four points, that of autumn gentian has five points. In addition, field gentian is a duller purple in colour, and much rarer in southern England. The plant's name is said to be derived from Gentius, a king of Illyria in the Balkans who lived in the 2nd century BC and is said to have been the first to discover the medicinal properties of the powdered roots of field gentian.

Grassland and sand-dunes, mainly in northern Britain.

Long tangled stems of field madder sprawl over waste ground, unnoticed until sprays of tiny lilac flowers burst from the tips of the shoots. As if to compensate for their small size, the flowers keep on appearing through most of the summer and autumn. A red dye extracted from the roots of field madder was once used to colour cloth, while another species of madder yielded an even brighter dye used as long ago as Anglo-Saxon times. But these vegetable dyes have been largely superseded by synthetic products.

Widespread on arable and waste ground; rarer in Scotland.

Amid the line of driftwood and litter at the top of a sand or shingle beach, the delicate purple, lilac or white flowers of sea rocket provide a contrasting splash of colour. It also grows higher up the shore, where it can help to stabilise young sand-dunes, though it is an annual plant and relies on the tides to disperse its buoyant seed-pods. The fleshy leaves conserve every drop of fresh water that the roots can find, and help the plant to survive salt spray and even burial by sand.

Strand line of beaches around Britain and Ireland.

J F M A M J J A S O N D

J F M A M J J A S O N D

J F M A M J J A S O N D

Blue flowers
Up to four petals • Irregular flowers

Devil's-bit scabious

Succisa pratensis
6-40in (15-100cm)

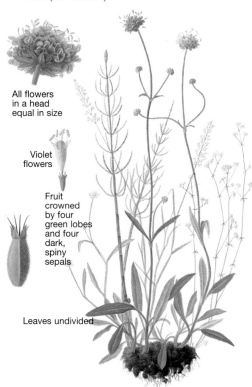

All flowers in a head equal in size

Violet flowers

Fruit crowned by four green lobes and four dark, spiny sepals

Leaves undivided

Field scabious

Knautia arvensis
10-40in (25-100cm)

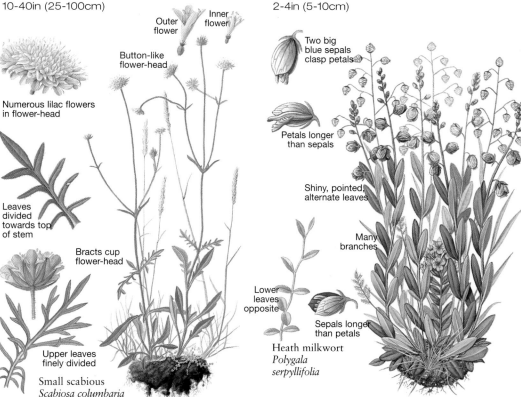

Outer flower

Inner flower

Button-like flower-head

Numerous lilac flowers in flower-head

Leaves divided towards top of stem

Bracts cup flower-head

Upper leaves finely divided

Small scabious
Scabiosa columbaria

Common milkwort

Polygala vulgaris
2-4in (5-10cm)

Two big blue sepals clasp petals

Petals longer than sepals

Shiny, pointed, alternate leaves

Many branches

Lower leaves opposite

Sepals longer than petals

Heath milkwort
Polygala serpyllifolia

The button-like heads of Britain's three scabious species consist of as many as 50 individual flowers. Devil's-bit scabious has violet-blue flowers, and undivided leaves all the way up the stem. Its flower-heads have equal-sized individual flowers, unlike those of field scabious. Devil's-bit scabious is particularly at home in wet places such as marshes and damp woodlands, though it may grow also on the same drier grasslands favoured by field scabious. The plant's name is associated with its short root: legend has it that the Devil, jealous of its reputed skin-healing power, bit off part of the root.

Grassy places throughout British Isles.

Field scabious, tallest of the three British species, has hairy leaves which are undivided at the base of the plant but become progressively more divided towards the top. The flower-heads are lilac-blue, and the outer flowers are larger than the inner ones. The name scabious derives from the reputed power these plants had for curing skin diseases such as scabies. Caterpillars of several butterflies and moths feed on the dull green leaves.
Small scabious is lower-growing and has much more finely segmented leaves on the upper part of its stem.

Dry fields throughout British Isles, but rare in northern Scotland and northern Ireland.

A flower's sepals are normally green and insignificant, but the two inner sepals of the milkworts are much larger than the three outer ones and brightly coloured – usually blue, but sometimes pink or white. These two inner sepals, called wings, all but conceal the flower's actual petals, which are joined together in a tube extending just beyond the sepals.
Heath milkwort is a smaller plant, different also in having some pairs of opposite leaves at the base of the stem. Milkworts were supposed to increase the milk yield of cows that fed on them.

Dry grassland and sand-dunes throughout British Isles.

| J | F | M | A | M | **J** | **J** | **A** | **S** | O | **N** | D |

| J | F | M | A | M | **J** | **J** | **A** | **S** | O | N | D |

| J | F | M | A | **M** | **J** | **J** | **A** | **S** | O | N | D |

71

Blue flowers
Up to four petals • Irregular flowers

Heath speedwell
Veronica officinalis
4-12in (10-30cm)

Large upper petals

Flowers in long spikes at tip of stem

Heart-shaped fruit

Hairy, greyish leaves

The creeping roots of heath speedwell spread to form a vivid blue carpet on the dry soils of heaths or woodland regions. The flowers are grouped in long, loose spikes. The individual flowers are on very short stalks, and their upper petals are larger than the lower ones. The greyish leaves are hairy on both sides. The name of speedwell may refer to the plant's supposed curative powers in respect of a wide range of ailments.

Dry grassland, heaths and open woodlands throughout British Isles.

Brooklime
Veronica beccabunga
8-24in (20-60cm)

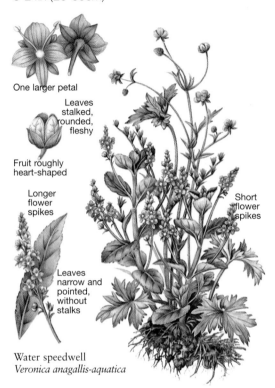

One larger petal

Leaves stalked, rounded, fleshy

Fruit roughly heart-shaped

Longer flower spikes

Leaves narrow and pointed, without stalks

Short flower spikes

Water speedwell
Veronica anagallis-aquatica

The banks of streams and ponds in summer are often brightened by the vivid blue splashes of brooklime and water speedwell. The 'lime' of brooklime's name comes from the Latin *limus*, 'mud'. The flowers of both species grow in spikes; the individual flowers of water speedwell are smaller than those of brooklime, but they are grouped in longer spikes. The round, fleshy, stalked leaves of brooklime are quite different from the leaves of water speedwell, which are lance-shaped and stalkless. Water speedwell is commonest in southern and eastern England, but is scattered elsewhere.

Wet places throughout British Isles, except far north.

Germander speedwell
Veronica chamaedrys
4-12in (10-30cm)

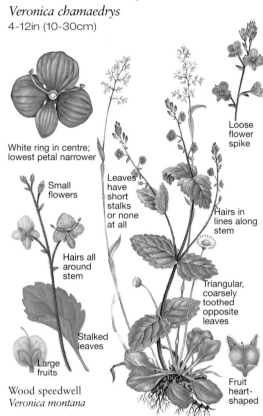

White ring in centre; lowest petal narrower

Small flowers

Loose flower spike

Leaves have short stalks or none at all

Hairs in lines along stem

Hairs all around stem

Triangular, coarsely toothed opposite leaves

Stalked leaves

Large fruits

Fruit heart-shaped

Wood speedwell
Veronica montana

This is the familiar speedwell of gardens, hedges and roadsides. Its large bright blue flowers, seen peeping out of the foliage on a summer's day, give it a special charm, and the eye-like white ring at the centre of each flower has given the plant the alternative name of bird's-eye speedwell. Though many speedwells are superficially similar, germander speedwell has one unique feature: a sharply defined line of hairs down either side of the stem. Wood speedwell has hairs all around the stem and smaller flowers than germander speedwell. It favours damp woodlands in the south and west.

Hedges, woods and grassy places throughout British Isles.

J F M A **M J J A S** O N D

J F M A **M J J A S** O N D

J F M A **M J J** A S O N D

The large family of speedwells fall into two main groups – those bearing solitary flowers at the base of a small leaf, and those with flowers in a spike. Plants of the equally large mint family have square stems, two-lipped flowers and hairs which exude the characteristic minty fragrance.

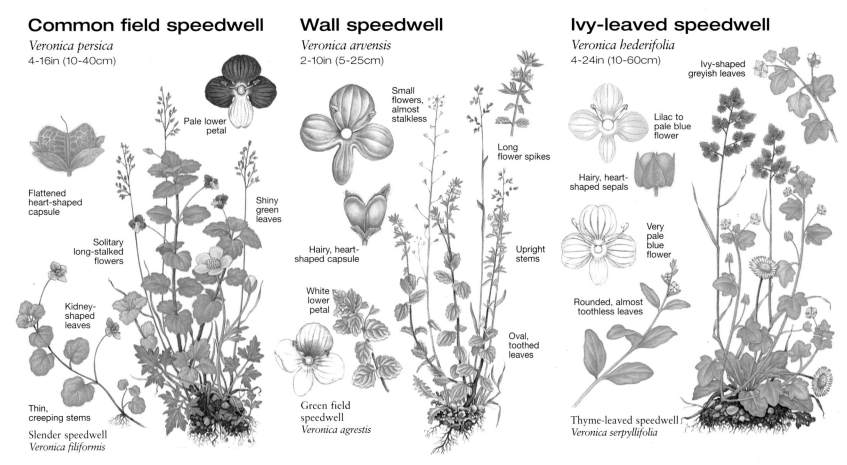

Common field speedwell

Veronica persica
4-16in (10-40cm)

Pale lower petal

Flattened heart-shaped capsule

Shiny green leaves

Solitary long-stalked flowers

Kidney-shaped leaves

Thin, creeping stems

Slender speedwell
Veronica filiformis

Wall speedwell

Veronica arvensis
2-10in (5-25cm)

Small flowers, almost stalkless

Long flower spikes

Hairy, heart-shaped capsule

Upright stems

White lower petal

Oval, toothed leaves

Green field speedwell
Veronica agrestis

Ivy-leaved speedwell

Veronica hederifolia
4-24in (10-60cm)

Ivy-shaped greyish leaves

Lilac to pale blue flower

Hairy, heart-shaped sepals

Very pale blue flower

Rounded, almost toothless leaves

Thyme-leaved speedwell
Veronica serpyllifolia

Introduced from western Asia early in the 19th century, the common field speedwell, or Buxbaum's speedwell, has become one of the commonest speedwells on cultivated ground all over Britain. It flowers all the year round. The flower is borne on a long stalk arising from the angle of leaf and stem; it is up to 1/2in (13mm) across, and has a pale lower lip.
Slender speedwell has thread-like stems which quickly take root and form a mat over riverbanks, roadside verges – and garden lawns.

Cultivated ground all over British Isles, except north and west Scotland and northern Ireland.

The location in which it grows is one way of identifying wall speedwell. As its name suggests, it is a plant that can find a foothold in seemingly inhospitable crevices in rocks and garden walls and on scree slopes, from which it sends up a tall straight stem bearing a long spike of bright blue flowers. But wall speedwell is equally at home in cultivated fields and dry grassland. Here the very short flower-stalk, or total absence of stalk, identifies it.
Green field speedwell has a single flower on a stalk rather than a spike of flowers. In this case the flower has a white lower petal. It is commoner in the south than the north.

Dry, bare places throughout British Isles.

The name indicates this speedwell's main distinguishing feature. As well as having from three to seven lobes or teeth, the leaves are greyish in colour. The small flowers, borne singly on long stalks, are lilac or a pale Cambridge blue – or sometimes almost white – and the sepals hairy and heart-shaped. Ivy-leaved speedwell flowers early in the year and dies down in summer.
Thyme-leaved speedwell has rounded, almost toothless leaves, without hairs. It grows in damp grassland, and bears pale blue flowers from March to October.

Arable fields, gardens and bare ground throughout British Isles; rare in north and west.

J F M A M J J A S O N D

J F M A M J J A S O N D

J F M A M J J A S O N D

73

Blue flowers · *Up to four petals* · *Irregular flowers*

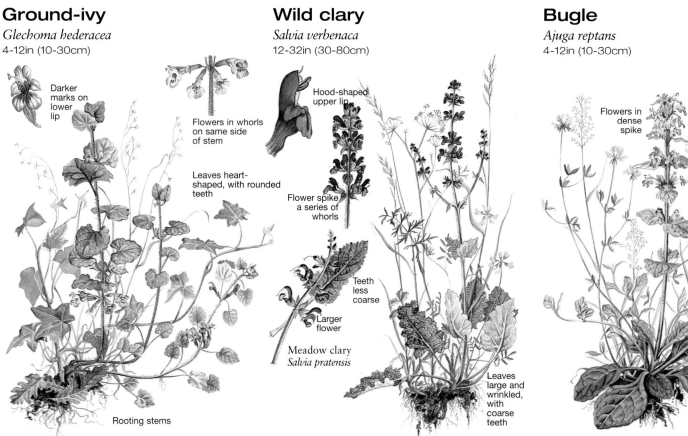

Ground-ivy
Glechoma hederacea
4-12in (10-30cm)

Darker marks on lower lip

Wild clary
Salvia verbenaca
12-32in (30-80cm)

Hood-shaped upper lip

Flowers in whorls on same side of stem

Leaves heart-shaped, with rounded teeth

Flower spike a series of whorls

Teeth less coarse

Larger flower

Meadow clary
Salvia pratensis

Leaves large and wrinkled, with coarse teeth

Bugle
Ajuga reptans
4-12in (10-30cm)

Shorter upper lip

Flowers in dense spike

Leaves glossy green

Leafy rooting stems

Rosette of leaves at base

Rooting stems

Despite its name, ground-ivy is neither related to ivy, nor does it look like it. Although remaining green in winter and spreading by long, creeping stems, it is more likely to be mistaken for the violet, for its leaves have the same heart shape and its bluish flowers bloom in clusters in spring. The name of 'alehoof' given to the plant in the West Country and parts of Yorkshire is more accurate, since before the introduction of hops the leaves of ground-ivy were used to flavour and clarify beer.

Open woods, hedges and grassland all over Britain and Ireland; rarer in far north and west.

Two sizes of flower often appear on the same plant of wild clary. The smaller ones never fully open and are self-pollinating. The more conspicuous ones, more than ¹/₂in (13mm) long, are in whorls making up a long spike characteristic of the mint family to which it belongs. Wild clary can claim garden sage as a relative, though it has itself no culinary use.
Meadow clary, a rare plant of limestone grassland, has less coarsely toothed leaves and bigger flowers with curved upper lips.

Roadsides and dry grassland; commonest in southern England.

Small spikes of blue flowers on short, stiff stems colour damp woods and meadows all over Britain in summer. Bugle spreads quickly by long runners and often forms large patches. It tolerates shade, and is often found deep in oak woods. The opposite pairs of leaves typical of the mint family form a rosette at the base, and are often so glossy and dark as to appear almost purple. These lower leaves have stalks, whereas the smaller leaves higher up the stem are stalkless.

Widespread in woods and damp meadows.

Selfheal

Prunella vulgaris
2-12in (5-30cm)

Basil thyme

Clinopodium acinos
4-8in (10-20cm)

Skullcap

Scutellaria galericulata
6-20in (15-50cm)

Tufted vetch

Vicia cracca
24-80in (60-200cm)

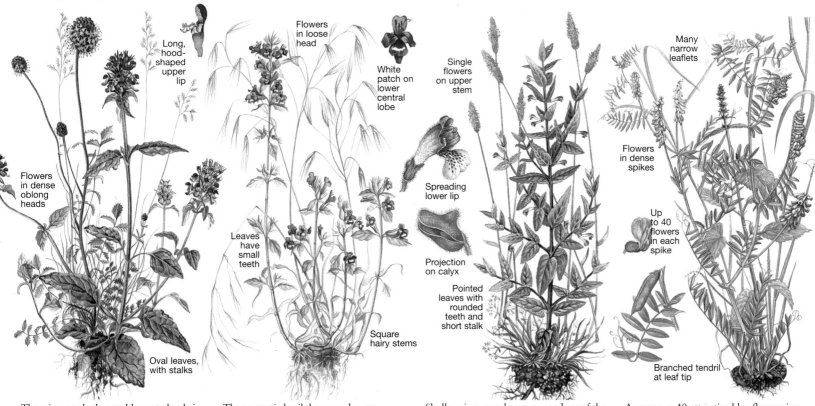

Long, hood-shaped upper lip

Flowers in dense oblong heads

Oval leaves, with stalks

Flowers in loose head

White patch on lower central lobe

Leaves have small teeth

Square hairy stems

Single flowers on upper stem

Spreading lower lip

Projection on calyx

Pointed leaves with rounded teeth and short stalk

Many narrow leaflets

Flowers in dense spikes

Up to 40 flowers in each spike

Branched tendril at leaf tip

There is a marked resemblance to bugle in the way the closely packed purple flower-heads of selfheal rise above the short grass of meadows and woodland clearings in summer. Like bugle, selfheal also spreads quickly by runners to form a purple carpet along dry, grassy banks. But while the flowers of bugle have a very short upper lip, those of selfheal have a long hood-shaped upper lip.

Grasslands and waste ground everywhere.

The aromatic basil thyme makes an attractive show in hedges and on banks in limestone areas of southern England, with its violet, two-lipped flower showing a white patch on the lower lip. The white mark guides bees to the nectar inside the bloom. Although both are members of the mint family, basil thyme is a different plant from the less aromatic wild basil (p. 94).

Cultivated land and open grasslands, mainly in lowland Britain.

Skullcap is unusual among members of the mint family in that most of its flowers grow in pairs facing the same way. Each pair sprouts from the angle between the stem and the leaves. The plant's common name is derived from the calyx or sheath from which the flower springs; a projection on the top of this was thought to give it the appearance of a *galerum*, the leather skull-helmet worn by Roman soldiers.

Riverbanks, marshes and pond margins throughout Britain; scarcer in Ireland.

As many as 40 attractive blue flowers in a single spike make tufted vetch one of the most distinctive plants that scramble over grasslands and hedgerows. The pea-like flowers grow on only one side of the long stalk, and emerge continuously through the summer. The leaves consist of up to 30 greyish leaflets. From the tip of the leaf emerges a branched tendril which winds around neighbouring plants and enables tufted vetch to climb 6ft or more.

Grasslands and bushy places everywhere.

J F M A M **J J A S** O N D   J F M A M J **J A S** O N D   J F M A M J **J A S** O N D   J F M A M J **J A S** O N D

Blue flowers
Five to seven petals • Regular flowers

Lesser periwinkle
Vinca minor
12-24in (30-60cm)

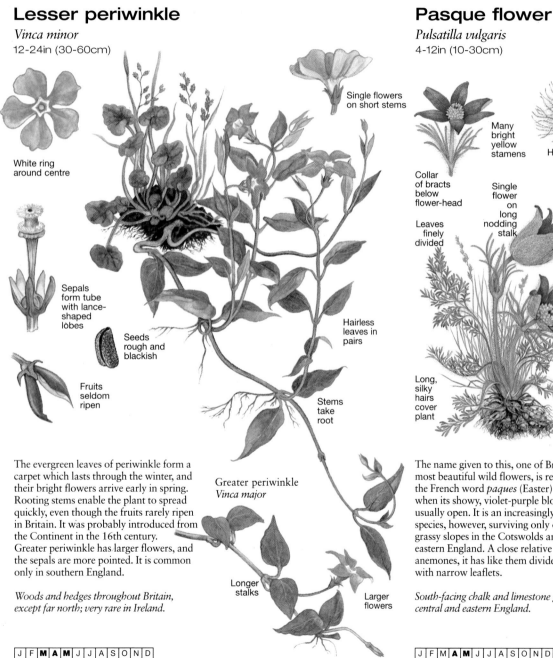

White ring around centre

Sepals form tube with lance-shaped lobes

Seeds rough and blackish

Fruits seldom ripen

Single flowers on short stems

Hairless leaves in pairs

Stems take root

Greater periwinkle
Vinca major

Longer stalks

Larger flowers

The evergreen leaves of periwinkle form a carpet which lasts through the winter, and their bright flowers arrive early in spring. Rooting stems enable the plant to spread quickly, even though the fruits rarely ripen in Britain. It was probably introduced from the Continent in the 16th century. Greater periwinkle has larger flowers, and the sepals are more pointed. It is common only in southern England.

Woods and hedges throughout Britain, except far north; very rare in Ireland.

Pasque flower
Pulsatilla vulgaris
4-12in (10-30cm)

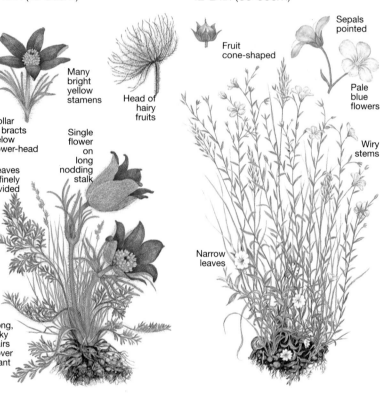

Many bright yellow stamens

Head of hairy fruits

Collar of bracts below flower-head

Single flower on long nodding stalk

Leaves finely divided

Long, silky hairs cover plant

The name given to this, one of Britain's most beautiful wild flowers, is related to the French word *paques* (Easter) – the time when its showy, violet-purple blooms usually open. It is an increasingly rare species, however, surviving only on certain grassy slopes in the Cotswolds and in eastern England. A close relative of the anemones, it has like them divided leaves with narrow leaflets.

South-facing chalk and limestone grasslands, central and eastern England.

Pale flax
Linum bienne
12-24in (30-60cm)

Sepals pointed

Fruit cone-shaped

Pale blue flowers

Wiry stems

Narrow leaves

Since the coming of mass-produced cotton in the 19th century, the flax once widely cultivated for making linen has become rare in Britain. But its close relative pale flax survives in dry grassland in the south and west. It has wiry stems and narrow leaves, and as the name implies its flowers are paler and also smaller than those of cultivated flax. Some linen flax is still grown in Northern Ireland.

Grassy places, especially near sea, mainly in West Country.

J F **M** A **M** J J A S O N D

J F M A **M** J J A S O N D

J F M A **M** J **J** **A** S O N D

Most members of the forget-me-not family have flowers in curved sprays; different species are identified by the shape of the leaf and the way the flower is formed. Bellflowers are easier to tell apart. This group also includes individual members of large families such as crane's-bills.

Meadow crane's-bill

Geranium pratense
12-30in (30-76cm)

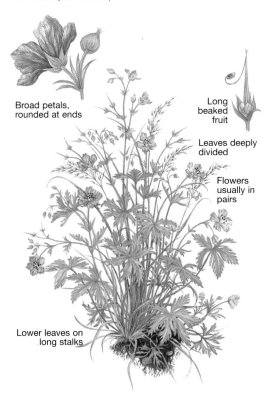

Broad petals, rounded at ends

Long beaked fruit

Leaves deeply divided

Flowers usually in pairs

Lower leaves on long stalks

The 'bill' of the plant's title is the long central beak of the fruit. Round the base of this beak are clustered five seeds, which are ejected as the outer wall of the lobes containing them springs upwards. There are several forms of crane's-bills, most of which are pink-flowered. Meadow crane's-bill is distinguished by its large violet-blue flowers, with radiating crimson veins on its petals to guide bees to the nectar. In bud the hairy flower-stalks are upright, but they droop after flowering and rise to the vertical again when the fruit ripens. Meadow crane's-bill is often planted in gardens.

Locally common in meadows and on roadsides in Britain.

Sea holly

Eryngium maritimum
12-36in (30-90cm)

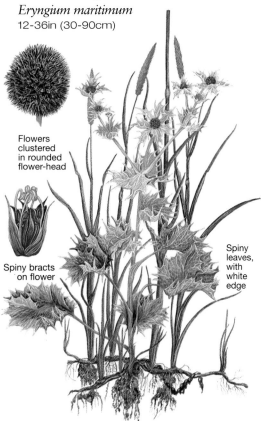

Flowers clustered in rounded flower-head

Spiny bracts on flower

Spiny leaves, with white edge

The thistly leaves of sea holly have inflicted painful stabs on the bare feet of many a bather, and because of this it has been eradicated from many beaches. However, it is now cultivated in gardens for its showy display of metallic-blue flowers and waxy leaves. The thistle-like heads are dried and used in flower arrangements. The thick skin on stalks and leaves represents an adaptation to seashore life, diminishing loss of moisture and protecting the plant against salt spray. For centuries the fleshy roots of sea holly were candied and sold as a popular delicacy, under the name of 'eringoes'.

Sand-dunes, except in north-east.

Bittersweet

Solanum dulcamara
12-80in (30-200cm)

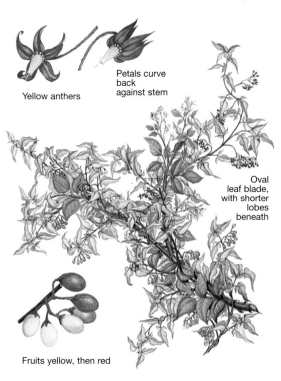

Yellow anthers

Petals curve back against stem

Oval leaf blade, with shorter lobes beneath

Fruits yellow, then red

The slender stems of bittersweet, twisting around vegetation such as brambles, drape a colourful curtain over hedgerows in summer. The display starts with the blue-and-yellow flowers of midsummer and continues with the berries, green at first then ripening to yellow and finally deep red. (The related, but much less common, deadly nightshade has black berries.) Bittersweet is not as poisonous as deadly nightshade, but all parts of the flower can cause sickness. An alkaloid in the plant is said to give its berries a bitter taste at first, turning sweet, giving the plant its name.

Woods, hedges, beaches throughout Britain, except far north.

Blue flowers *Five to seven petals • Regular flowers*

Oysterplant

Mertensia maritima
To 24in (60cm)

Flowers pink, then blue

Blue-grey leaves

Bell-shaped flower

Fleshy stems and leaves

Although a member of the forget-me-not family, oysterplant has distinctive fleshy leaves which conserve water, an essential requirement at the seashore where plants lose moisture to the keen winds. Its flowers are bell-shaped rather than star-like, but like some other forget-me-nots they are pinkish when they open and only turn blue with age.

The sprawling stems creep to form blue-grey patches on shingle in northern areas. The plant's name is derived from the tangy flavour of its leaves, which were once eaten raw or cooked as greens.

North-western coasts, especially Scotland.

Green alkanet

Pentaglottis sempervirens
12-40in (30-100cm)

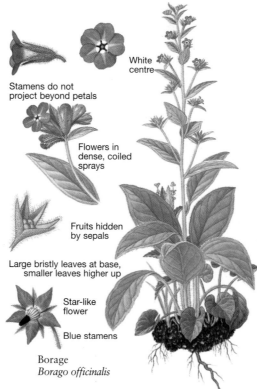

White centre

Stamens do not project beyond petals

Flowers in dense, coiled sprays

Fruits hidden by sepals

Large bristly leaves at base, smaller leaves higher up

Star-like flower

Blue stamens

Borage
Borago officinalis

The bright blue flowers with a white centre of green alkanet resemble those of the garden forget-me-not, which belongs to the same family. It also has the same curved sprays of flowers, which rise from the tip of a long stalk growing from the angle between the stem and an upper leaf. Alkanet was brought to Britain in the Middle Ages for the red dye which its roots yield.

Borage has narrower petals, giving the flower a star-like appearance. Its leaves taste of cucumber and can be used in salad or infused to make a refreshing drink. Cultivated herbs have escaped to waste ground in the south.

Hedges and wood margins; widely scattered.

Bugloss

Anchusa arvensis
6-20in (15-50cm)

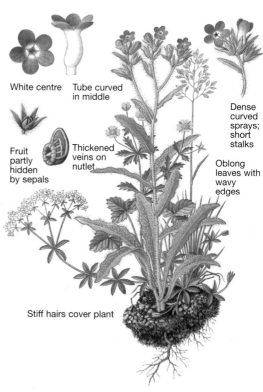

White centre

Tube curved in middle

Dense curved sprays; short stalks

Fruit partly hidden by sepals

Thickened veins on nutlet

Oblong leaves with wavy edges

Stiff hairs cover plant

The bright blue flowers of bugloss which dot cultivated fields throughout Britain in summer are characteristic of the forget-me-not family to which it belongs, though the plant stands taller than many of its relatives. However, the modern use of weed-killers means that the plant is rarer than it used to be. It is a very hairy and bristly plant, the stiff hairs having distinct swollen bases. The bristly leaf surface that is another trademark is reflected in the plant's common name, which is derived from two Greek words meaning 'ox-tongue'.

Arable fields and sandy heaths throughout Britain and north and east Ireland, especially near sea.

J F M A M **J** **J** A S O N D

J F M A M **M** **J** A S O N D

J F M A M **J** **J** **A** S O N D

Field forget-me-not

Myosotis arvensis
6-12in (15-30cm)

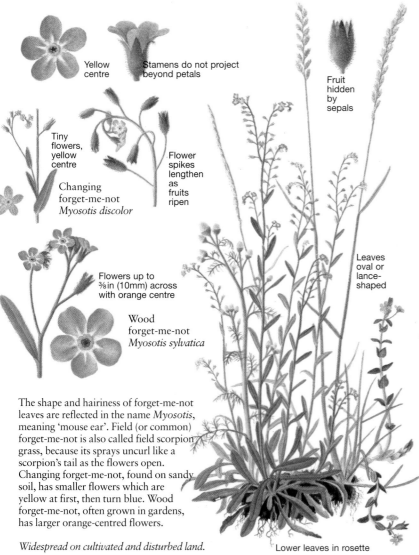

Yellow centre

Stamens do not project beyond petals

Tiny flowers, yellow centre

Changing forget-me-not *Myosotis discolor*

Flower spikes lengthen as fruits ripen

Fruit hidden by sepals

Flowers up to ⅜in (10mm) across with orange centre

Wood forget-me-not *Myosotis sylvatica*

Leaves oval or lance-shaped

Lower leaves in rosette

The shape and hairiness of forget-me-not leaves are reflected in the name *Myosotis*, meaning 'mouse ear'. Field (or common) forget-me-not is also called field scorpion grass, because its sprays uncurl like a scorpion's tail as the flowers open. Changing forget-me-not, found on sandy soil, has smaller flowers which are yellow at first, then turn blue. Wood forget-me-not, often grown in gardens, has larger orange-centred flowers.

Widespread on cultivated and disturbed land.

Tufted forget-me-not

Myosotis laxa ssp. *caespitosa*
8-16in (20-40cm)

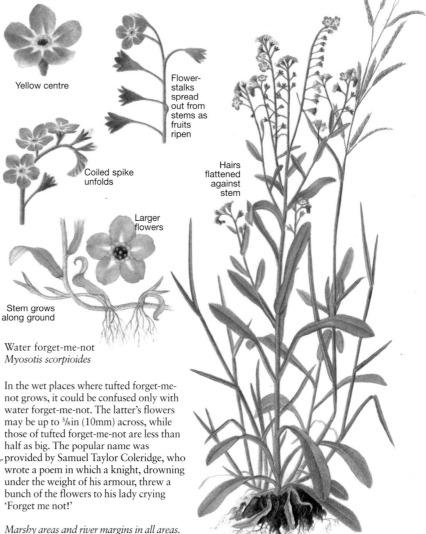

Yellow centre

Flower-stalks spread out from stems as fruits ripen

Coiled spike unfolds

Larger flowers

Hairs flattened against stem

Stem grows along ground

Water forget-me-not *Myosotis scorpioides*

In the wet places where tufted forget-me-not grows, it could be confused only with water forget-me-not. The latter's flowers may be up to ⅜in (10mm) across, while those of tufted forget-me-not are less than half as big. The popular name was provided by Samuel Taylor Coleridge, who wrote a poem in which a knight, drowning under the weight of his armour, threw a bunch of the flowers to his lady crying 'Forget me not!'

Marshy areas and river margins in all areas.

Blue flowers

Five to seven petals • Regular flowers

Viper's-bugloss

Echium vulgare
12-36in (30-90cm)

Funnel-shaped flowers, opening from base of stalk

Tall, hairy stem

Rough, tongue-shaped leaves with whitish hairs

Bristly leaves and short curved sprays of flowers are features which viper's-bugloss shares with other members of the forget-me-not family. But it is a taller plant, bearing its flowers on a hairy spike, and the long leaves come to a sharp point. The pink buds open to bright blue flowers with pink stamens, making viper's-bugloss a decorative plant, especially noticeable on cliffs and sand-dunes. Bees and butterflies are attracted by its sugary nectar.

Widespread on dry, sandy soils.

Sheep's-bit

Jasione montana
2-20in (5-50cm)

Single flower-head at tip of stem

Somewhat spreading stems, upper part leafless

The button-like flower-heads of sheep's-bit look more like those of daisies and scabious than those of the bellflower family to which it belongs. The heads are formed from a cluster of many separate tiny flowers, surrounded by a ring of green scales, giving the alternative country names of 'blue bonnets' and 'blue buttons'. Sheep's-bit is cropped, or 'bit', by sheep in the rough pastures where it frequently grows.

Pastures and cliffs, especially in the west.

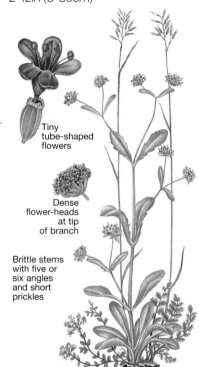

Button-like head with many flowers

Long strap-like petals

Narrow oblong leaves spiral up lower stem

Valves at top of capsule open to release seeds

Common cornsalad

Valerianella locusta
2-12in (5-30cm)

Tiny tube-shaped flowers

Dense flower-heads at tip of branch

Brittle stems with five or six angles and short prickles

Large, green, leaf-like bracts almost enclose the tiny lilac flower-heads of common cornsalad, making it an easy plant to overlook. This small member of the valerian family soon dies after it has flowered. The brittle stem is angled and covered in bristles. The leaves are at their greenest and crispest at the end of winter, when lambs are born and, being rich in vitamins, form an important ingredient of their diet – giving the plant its alternative name of lamb's lettuce.

Widespread on rocks, banks and dunes.

Nettle-leaved bellflower

Campanula trachelium
20-40in (50-100cm)

Stalked bell-shaped flowers in twos or threes

Broad, toothed leaves

Clustered bellflower

Campanula glomerata
1-8in (2.5-20cm)

Flowers open from top of spike

Flowers in tight cluster at top of stem

Upper leaves stalkless, clasping stem

Stalkless violet flowers

Narrow long-stalked leaves at base

Harebell

Campanula rotundifolia
6-16in (15-40cm)

Upper leaves narrow and stalkless

Light blue style in centre of flower

Bell-shaped flower hangs down

Capsule releases seed through pores at base

Rooting stems produce long-stalked, heart-shaped leaves

Few nodding flowers on each stem

Slender unbranched stems

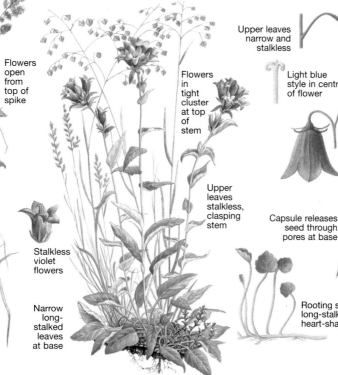

Toothed, heart-shaped leaves like those of stinging nettles give this bellflower its name. The lower leaves are on long stalks, while the upper leaves are smaller and have short stalks. The flowers open from the top of the spike downwards, unlike those of the related giant bellflower, *Campanula latifolia*, which open from the base upwards. This woodland plant of northern England and southern Scotland has paler but larger flowers.

Woodland clearings, mainly in south.

A much smaller plant than nettle-leaved and giant bellflowers, clustered bellflower has one other distinctive feature, signalled by its name. While the flowers of its relatives grow on stalks down the spike, those of clustered bellflower are stalkless and open in tight clusters at the tips of tall stems, making the stems, short as they are, appear top-heavy. The hairy leaves have only shallow teeth.

Limestone grassland, shell-rich dunes in England and eastern Scotland.

The flowers of harebell – or bluebell to the Scots – hang on fine stems and nod in the slightest breeze, their delicate blue bells often half-veiled by the grasses around them. Occasionally the flowers may be white or pinkish. When not in flower, harebell can be recognised by its narrow, stalkless stem leaves. The rounded leaves at the base of the plant have often withered by the time the plant is in flower.

Dry grassy places throughout Britain and Ireland.

J F M A M J **J A S** O N D   J F M A M **J J A S** O N D   J F M A M J **J A S** O N D   81

Blue flowers

Five to seven petals • Regular flowers

Five to seven petals
Irregular flowers

Bluebell

Hyacinthoides non-scripta
8-20in (20-50cm)

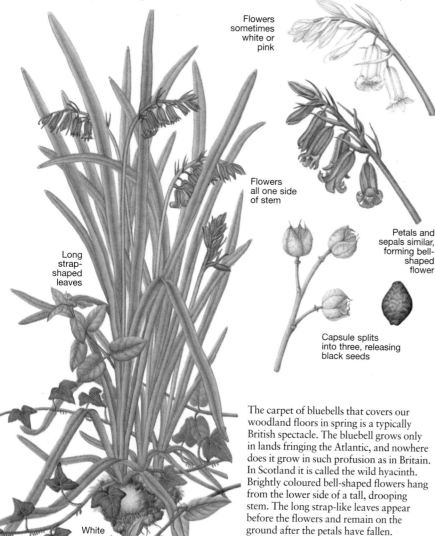

Flowers
sometimes
white or
pink

Flowers
all one side
of stem

Long
strap-
shaped
leaves

Petals and
sepals similar,
forming bell-
shaped
flower

Capsule splits
into three, releasing
black seeds

White
bulb

Spring squill

Scilla verna
2-8in (5-20cm)

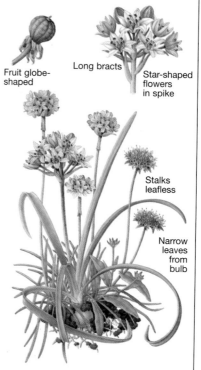

Fruit globe-
shaped

Long bracts

Star-shaped
flowers
in spike

Stalks
leafless

Narrow
leaves
from
bulb

Wild pansy

Viola tricolor
To 12in (30cm)

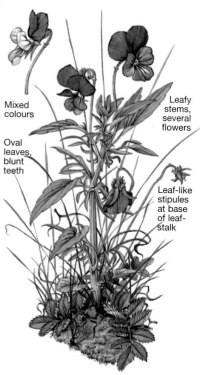

Mixed
colours

Oval
leaves,
blunt
teeth

Leafy
stems,
several
flowers

Leaf-like
stipules
at base
of leaf-
stalk

The carpet of bluebells that covers our woodland floors in spring is a typically British spectacle. The bluebell grows only in lands fringing the Atlantic, and nowhere does it grow in such profusion as in Britain. In Scotland it is called the wild hyacinth. Brightly coloured bell-shaped flowers hang from the lower side of a tall, drooping stem. The long strap-like leaves appear before the flowers and remain on the ground after the petals have fallen.

Woods throughout British Isles.

Like the bluebell a member of the lily family, spring squill is a much smaller plant, with star-shaped flowers on either side of its short stem, above pairs of leafy bracts. Its long, narrow, dark green leaves appear before the flowers, like those of bluebells, and survive after them. Flowers seen in August and September will be of the autumn squill, a similar plant confined almost entirely to the coasts of Devon and Cornwall.

Grassy places near sea in west and north.

The *tricolor* of the plant's scientific name refers to the fact that the large flowers may be violet, yellow or white or a combination of these three colours in a single blossom. This gives the plant something of the showiness of the garden pansy which has been bred from it. Pansies differ from violets in having lobed stipules, looking like extra leaves, at the base of each leaf-stalk and in having more erect side petals.

Widespread on cultivated or waste ground, grassland or dunes throughout British Isles.

| J | F | M | A | M | J | J | A | S | O | N | D |

| J | F | M | A | M | J | J | A | S | O | N | D |

| J | F | M | A | M | J | J | A | S | O | N | D |

The distinctive flower of violets and pansies has a lower petal developed into a spur extending backwards beyond the sepals. The size and precise colouring of the petals differs from species to species. The two-lipped flower of toadflaxes is shared by the insect-eating common butterwort.

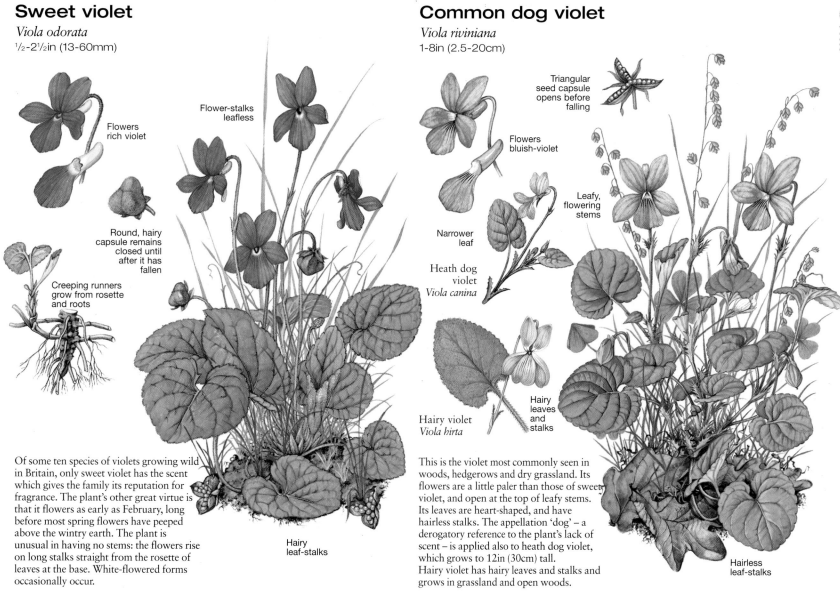

Sweet violet

Viola odorata
½-2½in (13-60mm)

Flowers rich violet

Flower-stalks leafless

Round, hairy capsule remains closed until after it has fallen

Creeping runners grow from rosette and roots

Hairy leaf-stalks

Of some ten species of violets growing wild in Britain, only sweet violet has the scent which gives the family its reputation for fragrance. The plant's other great virtue is that it flowers as early as February, long before most spring flowers have peeped above the wintry earth. The plant is unusual in having no stems: the flowers rise on long stalks straight from the rosette of leaves at the base. White-flowered forms occasionally occur.

Woods and hedges, mostly in England.

Common dog violet

Viola riviniana
1-8in (2.5-20cm)

Triangular seed capsule opens before falling

Flowers bluish-violet

Narrower leaf

Leafy, flowering stems

Heath dog violet
Viola canina

Hairy violet
Viola hirta

Hairy leaves and stalks

Hairless leaf-stalks

This is the violet most commonly seen in woods, hedgerows and dry grassland. Its flowers are a little paler than those of sweet violet, and open at the top of leafy stems. Its leaves are heart-shaped, and have hairless stalks. The appellation 'dog' – a derogatory reference to the plant's lack of scent – is applied also to heath dog violet, which grows to 12in (30cm) tall.
Hairy violet has hairy leaves and stalks and grows in grassland and open woods.

Woods, hedges, heaths throughout British Isles.

Blue flowers
Five to seven petals • Irregular flowers

Ivy-leaved toadflax
Cymbalaria muralis
4-30in (10-76cm)

Glossy, ivy-shaped leaves

Fruit opens by several teeth

Seed has thick, wavy ridges

Stems root at intervals

Two-lipped flower, like snapdragon, with spur at rear end

Loose flower-head

Narrow leaves

Pale toadflax
Linaria repens

Common butterwort
Pinguicula vulgaris
2-4in (5-10cm)

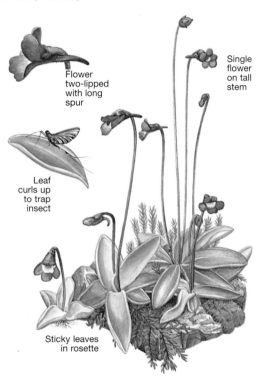

Flower two-lipped with long spur

Single flower on tall stem

Leaf curls up to trap insect

Sticky leaves in rosette

Old walls are often draped with this attractive trailing plant, introduced from the Mediterranean three centuries ago and now naturalised throughout Britain. Though its leaves resemble those of ivy, its flowers could hardly be more different, being shaped like those of snapdragon and coloured lilac with a yellow centre. The seed capsules are carried on long stalks which curve away from the light and shed the seeds in dark crevices. Once established, the plant spreads by long runners.
Pale toadflax has similar-shaped flowers, but they are borne in a long, loose head on upright flowering stems.

Widespread on old walls, except in northern Scotland.

An insect landing on the sticky leaf of common butterwort quickly finds itself trapped, as the leaf curls round it like a clutching hand. The plant engulfs its prey and extracts mineral salts from the body, before the leaf reopens and the dried remains of the insect are blown away. The deadly leaves spread out in a rosette at the base of the stems like a yellow-green starfish. Above, the stems are tall and leafless, bearing at their tops delicate two-lipped flowers. Butterwort is also known as bog violet from its favoured habitat of wet heaths and moors.

Wet ground, mainly in north and west. Rare in south.

J F M A **M J J** A S O N D   J F M A **M J J** A S O N D

Blue flowers

Eight or more petals

Four members of the daisy family have blue flowers, each of which is made up – as in all daisies – from a large number of separate florets. They include the rare alpine blue-sow-thistle, found only in north-east Scotland.

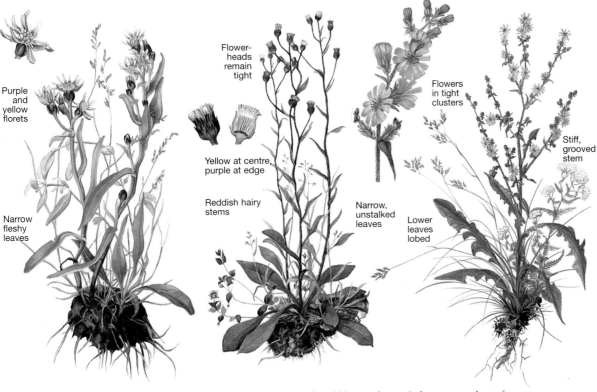

Sea aster

Aster tripolium
6-24in (15-60cm)

Purple and yellow florets

Narrow fleshy leaves

Blue fleabane

Erigeron acer
4-16in (10-40cm)

Flower-heads remain tight

Yellow at centre, purple at edge

Reddish hairy stems

Narrow, unstalked leaves

Chicory

Cichorium intybus
12-48in (30-120cm)

Flowers in tight clusters

Stiff, grooved stem

Lower leaves lobed

Alpine blue-sow-thistle

Cicerbita alpina
22-80in (55-200cm)

Sharp-toothed leaves

Flower-heads in dense spike

Tall upright plant with stout stem

Fleshy leaves enable the sea aster to retain as much fresh water as possible on the salt-marshes and sea-cliffs where it grows. In September whole areas of salt-marsh take on a bluish tinge as the flower-heads open. Each is a tight cluster of small florets, yellow at the centre and bluish-purple around the edge. Sometimes in southern England the outer florets are lacking. Sea aster was a popular garden plant in Elizabethan times.

Salt-marshes all around coast.

Because its outer rings of petal-like purple florets remain upright rather than spreading, blue fleabane looks perpetually in bud instead of opening into full flower. It is a plant of chalk downs, dunes and gravel pits. The purple florets enclose an inner core of pale yellow, tube-shaped florets. White hairs on the flower-head act as parachutes to disperse the seeds, which can lie dormant for years until the soil is turned.

Dry grassy places, mainly in southern Britain; rare in Ireland.

In late autumn, long after most summer flowers have faded, chicory still blooms in fields and on road verges. The bright blue flowers open with the sun in the morning and close about midday; in dull weather they stayed closed. The lower leaves are long and thistle-like, while the upper leaves are stalkless. The leaves of this native of the Orient have long been eaten as a vegetable and used as animal fodder.

Lime-rich grassland; commonest in south-east and rare in Scotland and Ireland.

This attractive member of the daisy family is one of Britain's rarest wild flowers. It is confined to a few colonies in the Scottish Highlands, where it grows on moist rock ledges well out of reach of man and grazing animals. Even if the plant is safely inaccessible, its powdery blue flower-heads may be glimpsed from a distance. They appear in a dense spike at the top of stout stems that rise to more than 72in (180cm), well above any surrounding vegetation.

Rare, north-east Scotland only.

J F M A M J **J A S O** N D   J F M A M J **J A S O** N D   J F M A M J **J A S O** N D   J F M A M J **J A S O** N D

Red flowers *Up to four petals • Regular flowers*

Heather

Calluna vulgaris
To 24in (60cm)

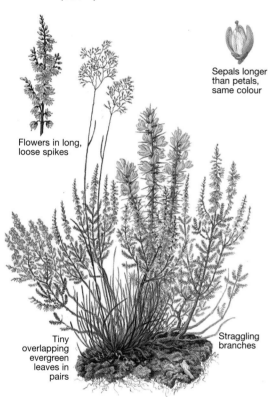

Sepals longer than petals, same colour

Flowers in long, loose spikes

Tiny overlapping evergreen leaves in pairs

Straggling branches

The commonest of several evergreen members of the heath family that brighten moors with their purple blooms in autumn is true heather, or ling. Its flowers and leaves differ from those of bell heather and cross-leaved heath. The petals are shorter than the sepals, similar in colour and texture, which surround them; and the leaves are in opposite pairs, overlapping one another. Though the flowers are usually purple, some white blooms are found. The young shoots of heather are the main food of grouse, and heather has been used for thatching, fencing and making brooms and baskets.

Heaths or moors, most parts of Britain and Ireland.

Bell heather

Erica cinerea
To 24in (60cm)

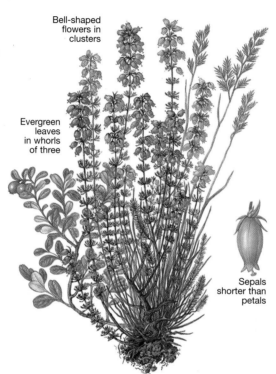

Bell-shaped flowers in clusters

Evergreen leaves in whorls of three

Sepals shorter than petals

The plant's name highlights its most easily recognised feature – the elegantly shaped, deep crimson-purple bells. These are larger than the flowers of true heather, and hang in clusters near the top of the stem. When the plant is not in bloom it can be recognised by the evergreen leaves which grow on short stalks in whorls, usually of three. The short, leafy side branches look like bunches of leaves. The habitat of bell heather helps to distinguish it from the superficially similar cross-leaved heath. Bell heather grows on the driest tussocks of a moor, while cross-leaved heath grows in the wetter parts.

Dry heaths and moors throughout Britain and Ireland.

Cross-leaved heath

Erica tetralix
To 24in (60cm)

Sepals shorter than petals

Evergreen leaves in whorls of four

Drooping flowers in rounded heads at top of stem

Whorls of four thin leaves form a cross when looked at from above, giving this plant its name. The leaves are very hairy and have long bristles, each tipped with a round gland. Although the bell-shaped flowers resemble those of bell heather, they are lighter in colour and grouped in more compact, rounded heads at the top of the stem. The bells droop until the fruits ripen, when the dying flowers become upright. The bushes of all heathers begin to sprawl with age. As a bush spreads outwards its branches begin to root and spread the bush even farther.

Bogs and wet heaths throughout Britain and Ireland.

J F M A M J J A S O N D

J F M A M J J A S O N D

J F M A M J J A S O N D

Large, bright petals make poppies conspicuous wherever they grow, and the tall stems of willowherb tower over many other wayside flowers. The heather family includes several fruiting shrubs, while many mints have an aroma that makes them useful as herbs.

Cranberry

Vaccinium oxycoccos
12in (30cm)

Petals curl back

Flowers on long, thin stalks

Evergreen oval leaves

Globular or pear-shaped berries

Bilberry

Vaccinium myrtillus
24in (60cm)

Drooping flowers, single or in pairs

Leaves finely toothed, fall in winter

Blue-grey bloom on berries

Green, upright stems

Crowberry

Empetrum nigrum
6-18in (15-45cm)

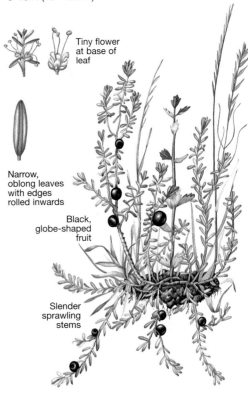

Tiny flower at base of leaf

Narrow, oblong leaves with edges rolled inwards

Black, globe-shaped fruit

Slender sprawling stems

While the flowers of bilberry hide modestly among the leaves, those of cranberry rise proudly above the leaves on tall, thin stalks. Perhaps the similarity of the shape of the stalk to that of a crane's neck gave the shrub its name. The flowers of cranberry are also quite different in shape from those of other heathers in having petals which curl back from the stamens. The evergreen shrub is a plant of wet areas, especially bogs, and it spreads outwards by means of slender rooting stems. Its sharp red berries make tasty jam and sauce and a filling for pies.

Bogs and wet heaths in widely scattered sites.

The dainty bell-shaped flowers of bilberry mark the shrub as a member of the heather family, but unlike other heathers it sheds its leaves in winter. The flowers are often half concealed among the bright green leaves, which have conspicuous veins. Bilberry thrives on poor, acid soil and grows abundantly in high places such as mountain forests and moors. The berries turn blue-grey from July onwards and have an excellent flavour eaten raw or made into pies and jellies. They are also known as whortleberries or blaeberries.

Woods, moors throughout Britain and Ireland; less common in south and east.

The evergreen leaves of crowberry are specially adapted to life on the dry rocky moors where it thrives. They curl inwards, almost meeting down the centre, to form a narrow tube that reduces the loss of moisture by evaporation through the pores on the leaf surface. Male and female flowers are borne on separate plants, the male flowers having three long projecting stamens. The slender sprawling stems, often reddish when young, hug the ground and spread rapidly. The black fruits, acid to the taste, are eaten by grouse, ptarmigan and other birds.

Moors and mountains; in Scotland, northern England, Wales and Ireland.

J F M A M J J A S O N D

J F M A M J J A S O N D

J F M A M J J A S O N D

Red flowers *Up to four petals • Regular flowers*

Water mint
Mentha aquatica
6-36in (15-90cm)

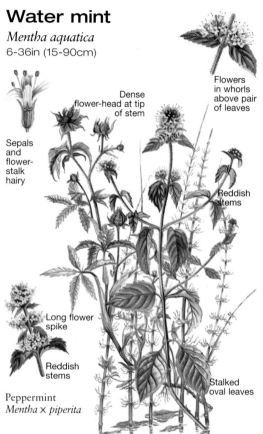

Flowers in whorls above pair of leaves

Dense flower-head at tip of stem

Sepals and flower-stalk hairy

Reddish stems

Long flower spike

Reddish stems

Stalked oval leaves

Peppermint
Mentha × piperita

Spearmint
Mentha spicata
To 24in (60cm)

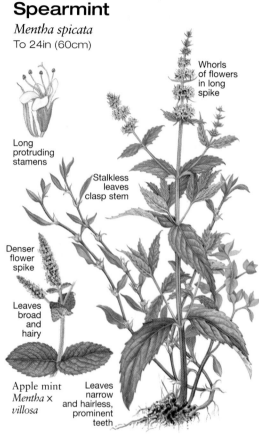

Whorls of flowers in long spike

Long protruding stamens

Stalkless leaves clasp stem

Denser flower spike

Leaves broad and hairy

Apple mint
Mentha × villosa

Leaves narrow and hairless, prominent teeth

Corn mint
Mentha arvensis
4-24in (10-60cm)

Flower whorls spaced out along stem

Tuft of leaves at tip of stem

Sepals and flower-stalk hairy

Stems green and hairy

Leaves oval and stalked

This wild form of mint found growing in wet places could well have been the first mint to be cultivated as a kitchen herb by the Romans 2000 years ago. Though its place in the kitchen garden has now been taken by other species, water mint is today the commonest of all wild mints, and the pungent smell of its leaves crushed underfoot perfumes many a walk beside marshes and streams. Peppermint is a hybrid, with the reddish stem of water mint and the flower spike of its other parent, spearmint. It is cultivated for the oil in its leaves, used to flavour sweets and medicines, but also grows wild in damp places.

Marshes and fens throughout British Isles.

The mint cultivated in gardens for the pungent smell of its leaves is in fact spearmint, introduced from central Europe as a kitchen herb. It has escaped and become established widely on roadsides and other waste places. The stems are smooth and the leaves hairless and shiny. Flowering stems grow in the angle between unstalked leaf and main stem, and end in a long spike of pale lilac flowers in tight whorls.
Another mint cultivated as a herb is apple mint, a cross between spearmint and the rarer round-leaved mint. It has a more delicate flavour than spearmint.

Widespread, but very scarce in Ireland.

One mint which lacks the normal minty smell is the wild corn mint. Instead it has a strong acrid smell like overripe cheese. Corn mint reproduces by seeds and grows widely in woodland clearings and on arable fields, where it can become a pest. It is usually found in drier places than the other common wild species, water mint. Corn mint's whorls of flowers are less conspicuous than those of most mints, being half hidden at the base of the broad hairy leaves. Corn mint is also unusual among mints in having no known medicinal or culinary properties.

Widespread, except northern Scotland.

Wild thyme

Thymus polytrichus ssp. *britannicus*
To 3in (75mm)

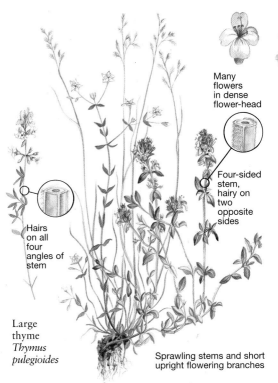

Many flowers in dense flower-head

Four-sided stem, hairy on two opposite sides

Hairs on all four angles of stem

Large thyme *Thymus pulegioides*

Sprawling stems and short upright flowering branches

The sprawling, low-growing stems of wild thyme are in sharp contrast with the upright spikes of the familiar garden herb, which comes from the Mediterranean and does not grow wild in Britain. Wild thyme forms a dense carpet on grassland and sand-dunes, often excluding other plants. Its short stems bear dense flower-heads which attract insects, especially honey-bees.
Large thyme, found only in southern England, is a more upright plant, differentiated from its more widespread relative by having hairs on the angles of its squarish stem instead of on the opposite faces.

Widespread on dry grassland, heaths and dunes.

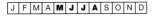

Flowering-rush

Butomus umbellatus
To 60in (152cm)

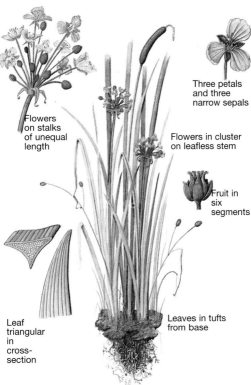

Three petals and three narrow sepals

Flowers on stalks of unequal length

Flowers in cluster on leafless stem

Fruit in six segments

Leaf triangular in cross-section

Leaves in tufts from base

Though it grows alongside rushes beside rivers and ponds, flowering-rush is not a true rush but a much more ornamental plant whose graceful stance and brightly coloured flowers have earned it a place in many gardens. The greyish-green leaves grow in a rosette from the base, and each leafless unbranched stem supports a single cluster of flowers. The plant spreads by underground creeping stems called rhizomes.

Water edges; scattered in British Isles, north to central Scotland where it has often been introduced.

Hoary plantain

Plantago media
12in (30cm)

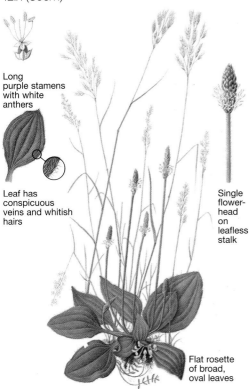

Long purple stamens with white anthers

Leaf has conspicuous veins and whitish hairs

Single flower-head on leafless stalk

Flat rosette of broad, oval leaves

The hoary plantain's most obvious feature is its tall leafless spike rising from a flat rosette of leaves at the base. Purple stamens give the flower-head a reddish appearance, unlike the green or brownish flowers of other plantains. These flowers are scented and attract insect pollinators; in this the hoary plantain differs from other plantains which are pollinated by the wind and do not need the attention of insects. The description 'hoary' refers to the whitish hairs covering the leaves, which also bear conspicuous veins.

Chalk grassland in southern England.

Red flowers *Up to four petals • Regular flowers*

Broad-leaved willowherb
Epilobium montanum
20-24in (50-60cm)

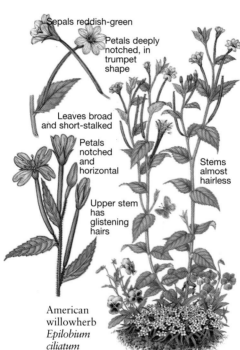

Sepals reddish-green

Petals deeply notched, in trumpet shape

Leaves broad and short-stalked

Petals notched and horizontal

Upper stem has glistening hairs

Stems almost hairless

American willowherb
Epilobium ciliatum

Great willowherb
Epilobium hirsutum
32-60in (80-152cm)

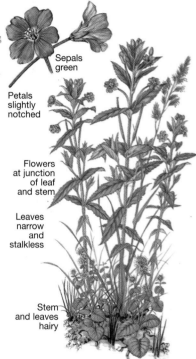

Sepals green

Petals slightly notched

Flowers at junction of leaf and stem

Leaves narrow and stalkless

Stem and leaves hairy

Enchanter's nightshade
Circaea lutetiana
8-28in (20-70cm)

Petals deeply notched

Fruits droop down

Slightly toothed leaves

Flowers in long leafless spikes

Opium poppy
Papaver somniferum
12-36in (30-90cm)

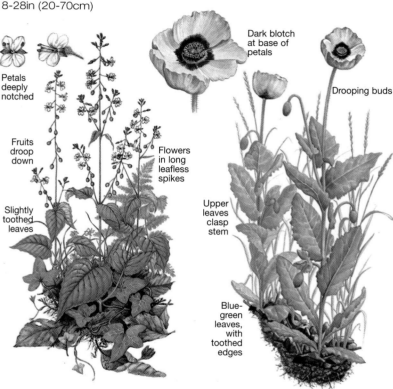

Dark blotch at base of petals

Drooping buds

Upper leaves clasp stem

Blue-green leaves, with toothed edges

Willowherbs are named after their narrow willow-like leaves, but the leaves of this species are slightly broader than those of its relatives. They are in opposite pairs, growing out horizontally from the stem on short stalks. The flowers droop in bud, then straighten up as they open. American willowherb, first recorded in Britain in 1891, is now common on old railway lines and waste places. Its upper stem is covered in glistening hairs.

Widespread in shady places, hedgerows.

The height of great willowherb is as distinctive as the hairiness of its stem and leaves which gives the species its scientific name of *hirsutum*. The upright stems bear solitary rosy flowers, which form bright patches of colour by the edge of rivers or in field-side ditches. The plant's seeds are plumed to ensure wide dispersal, but the plant also spreads by fleshy stems just below ground. These give rise to dense clumps, which exclude other plants.

Widespread by streams and in marshes.

Despite its name, enchanter's nightshade is not a nightshade but a willowherb. It bears its flowers in tall spikes, its leaves are pointed and lightly toothed, and the hairy leaf-stalk has a channel along the top edge. The leaves are in pairs set at right angles to each other up the hairy stem. Unlike other willowherbs, the seeds of enchanter's nightshade are not dispersed by the wind. Instead, each egg-shaped fruit has bristles which catch onto clothing or animal fur.

Woods and shady places, except in far north.

The flowers of the opium poppy are much larger than those of other poppies, up to 7in (18cm) across, and its lilac colouring is distinctive. It is a relative of the plant cultivated in the Far East as a source of opium and, itself, used to be widely grown in Britain to provide a sedative known as syrup of poppies, extracted from the seed capsules in their immature form. The ripe seeds are not narcotic, and are baked into bread and cakes and crushed for oil.

Waste ground, widely scattered.

J F M A M J J A S O N D

J F M A M J J A S O N D

J F M A M J J A S O N D

J F M A M J J A S O N D

Common poppy

Papaver rhoeas
8-24in (20-60cm)

Prickly poppy

Papaver argemone
6-20in (15-50cm)

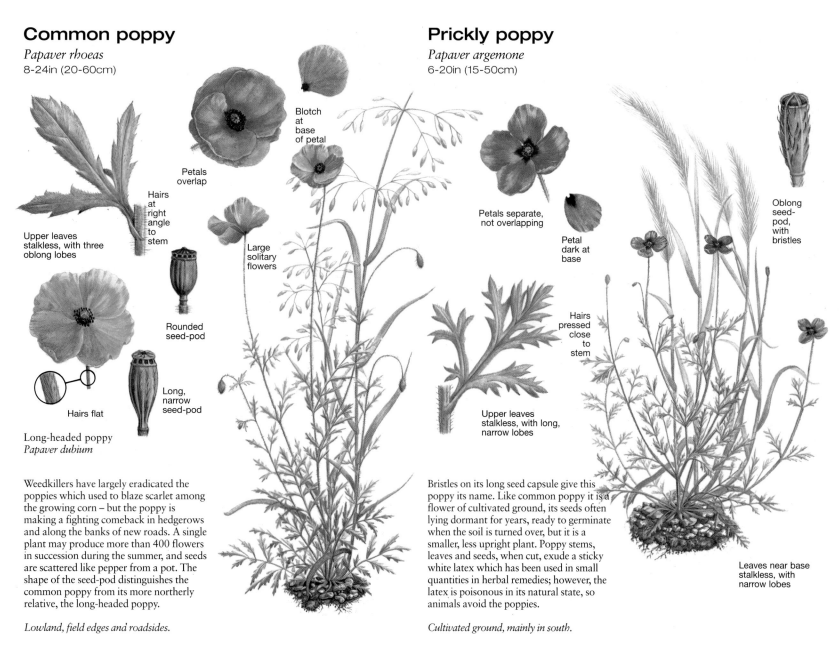

Upper leaves stalkless, with three oblong lobes

Hairs at right angle to stem

Petals overlap

Blotch at base of petal

Large solitary flowers

Rounded seed-pod

Hairs flat

Long, narrow seed-pod

Long-headed poppy
Papaver dubium

Petals separate, not overlapping

Petal dark at base

Hairs pressed close to stem

Upper leaves stalkless, with long, narrow lobes

Oblong seed-pod, with bristles

Leaves near base stalkless, with narrow lobes

Weedkillers have largely eradicated the poppies which used to blaze scarlet among the growing corn – but the poppy is making a fighting comeback in hedgerows and along the banks of new roads. A single plant may produce more than 400 flowers in succession during the summer, and seeds are scattered like pepper from a pot. The shape of the seed-pod distinguishes the common poppy from its more northerly relative, the long-headed poppy.

Lowland, field edges and roadsides.

Bristles on its long seed capsule give this poppy its name. Like common poppy it is a flower of cultivated ground, its seeds often lying dormant for years, ready to germinate when the soil is turned over, but it is a smaller, less upright plant. Poppy stems, leaves and seeds, when cut, exude a sticky white latex which has been used in small quantities in herbal remedies; however, the latex is poisonous in its natural state, so animals avoid the poppies.

Cultivated ground, mainly in south.

J F M A **M J J A S** O N D

J F M A **M J J A S** O N D

Red flowers

Up to four petals
Regular flowers

Great burnet

Sanguisorba officinalis
12-36in (30-90cm)

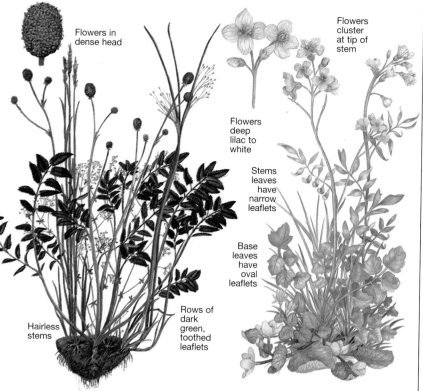

Flowers in dense head

Hairless stems

Rows of dark green, toothed leaflets

Cuckoo flower

Cardamine pratensis
6-24in (15-60cm)

Flowers cluster at tip of stem

Flowers deep lilac to white

Stems leaves have narrow leaflets

Base leaves have oval leaflets

Damp grassland and fens provide a haven for the great burnet, which in many areas has lost ground to intensive farming. It is an upright perennial, with many hairless branches. The leaves are composed of paired rows of toothed leaflets, and the tiny crimson flowers are packed in an oblong head about ½in (13mm) long. The related salad burnet (p. 124) is a smaller, green-flowered plant.

Damp grassland, central England and Wales.

The large lilac petals of cuckoo flower often peep above the grass on fragile, criss-crossing stems in unmown pastures in spring. Its lower leaves look like those of watercress. The flowering time of cuckoo flower coincides with the arrival of the cuckoo, while its alternative name of lady's smock may derive from the associations of the white-flowered form with milkmaids' smocks.

Widespread in damp pastures.

Red flowers

Up to four petals
Irregular flowers

Rosebay willowherb

Chamerion angustifolium
To 48in (120cm)

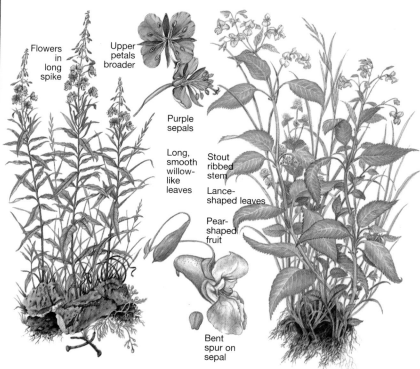

Flowers in long spike

Upper petals broader

Purple sepals

Long, smooth willow-like leaves

Stout ribbed stem

Lance-shaped leaves

Pear-shaped fruit

Bent spur on sepal

Indian balsam

Impatiens glandulifera
4-80in (100-200cm)

Tall stems bright with purple flowers in summer and white with fluffy seed heads in autumn make this one of the most easily recognised flowers of waste ground. It thrives especially on bonfire sites in cleared woodland – hence its alternative name of 'fireweed'. The Victorians often grew rosebay willowherb in gardens. Left to itself in such surroundings, however, it would quickly drive out other flowers.

Widespread on waste ground.

Touched at the base, the ripe pods of the Indian balsam shower the passer-by in autumn with a fusillade of seeds, released by the spring action of the opening sides of the pod. The plant was brought to Britain in 1839 and grown in greenhouses before it escaped into the wild, where damp river banks are its commonest habitat. Bees are attracted by nectar in the lower sepal, which closes round the insect.

Widespread on river banks and waste places.

J F M A M J J A S O N D

J F M A M J J A S O N D

J F M A M J J A S O N D

J F M A M J J A S O N D

Tubular flowers with two lips are characteristic of the foxglove and its relatives the louseworts, and also of the mints and dead-nettles, which have square stems and opposite leaves. Trefoil leaves characterise the clovers, while vetches have pea-like flowers and a climbing tendril at the end of a much-divided leaf.

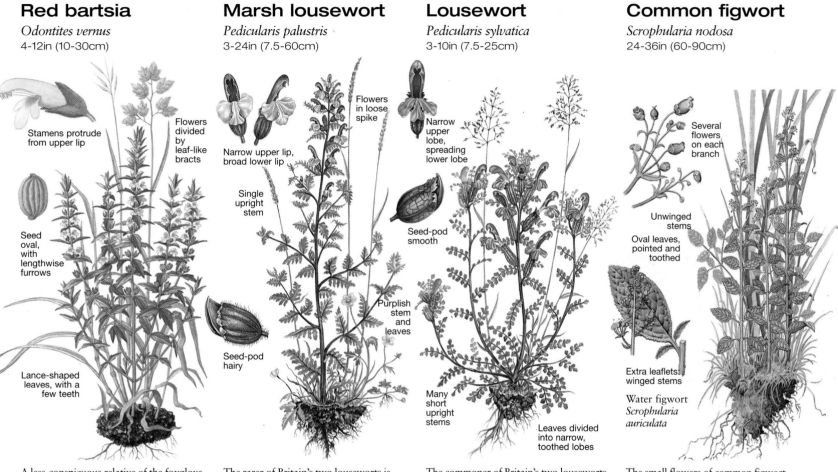

Red bartsia
Odontites vernus
4-12in (10-30cm)

Stamens protrude from upper lip

Flowers divided by leaf-like bracts

Seed oval, with lengthwise furrows

Lance-shaped leaves, with a few teeth

Marsh lousewort
Pedicularis palustris
3-24in (7.5-60cm)

Flowers in loose spike

Narrow upper lip, broad lower lip

Single upright stem

Seed-pod hairy

Purplish stem and leaves

Lousewort
Pedicularis sylvatica
3-10in (7.5-25cm)

Narrow upper lobe, spreading lower lobe

Seed-pod smooth

Many short upright stems

Leaves divided into narrow, toothed lobes

Common figwort
Scrophularia nodosa
24-36in (60-90cm)

Several flowers on each branch

Unwinged stems

Oval leaves, pointed and toothed

Extra leaflets: winged stems

Water figwort
Scrophularia auriculata

A less-conspicuous relative of the foxglove, red bartsia has small pinkish flowers which are almost lost among the leaf-like bracts below them. In the north of Britain the flowers are shorter than the bracts, and the branches are upright; in the south, the flowers are longer than the bracts, and the branches make almost a right angle with the stem. Hairs growing all over the plant give it a dusty appearance.

Widespread in fields, wasteland and roadsides.

The rarer of Britain's two louseworts is also called red rattle because of the pinkish-purple colouring of its stem and leaves and the noise made when the ripe seeds are shaken inside the pods. Marsh lousewort lives partly off other plants, attaching its roots to those of neighbouring grasses to tap their supply of water and mineral salts, without harming the host plant.

Wet, grassy places; rarer in south-east.

The commoner of Britain's two louseworts is a creeping plant with many short upright stems. The leaves of both species are so deeply divided that they look like two rows of leaflets. Lousewort is, however, a paler plant than red rattle, or marsh lousewort. The seed-pods also differ, those of lousewort being hairless. Lousewort's reputation for spreading lice to grazing sheep is unproven.

Damp heaths, except in English Midlands.

The small flowers of common figwort attract insects not by their bright colours but by their strong smell, unpleasant to humans. The insects, mostly wasps, pollinate the flowers as they feed on the nectar at the base of the petal tube. The plant spreads by underground stems (rhizomes), which send up tall stems. Water figwort grows beside rivers and ponds, and has winged leaf stems.

Damp places throughout Britain.

J F M A M J J A S O N D   J F M A M J J A S O N D   J F M A M J J A S O N D   J F M A M J J A S O N D

Red flowers

Up to four petals • Irregular flowers

Wild marjoram

Origanum vulgare
12-24in (30-60cm)

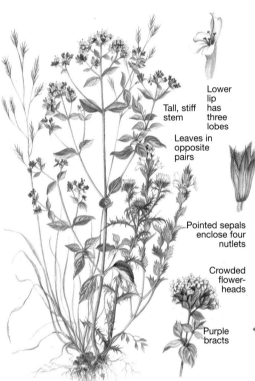

Lower lip has three lobes

Tall, stiff stem

Leaves in opposite pairs

Pointed sepals enclose four nutlets

Crowded flower-heads

Purple bracts

The cultivated marjoram used as a herb in the kitchen comes from the Mediterranean, but even the wild variety growing on chalky grassland in Britain contains an oil once sold as oil of thyme and used as a painkiller. The leaves, which exude the scent, can be used to make a herb tea and herb sachets. Like those of other mints, wild marjoram's flowers are tube-shaped, with two lips. The actual flowers are pale pink, but the leaf-shaped bracts beneath them give a purplish tinge to the flower-heads. These are carried at the top of the tall, upright stem, which has leaves in opposite pairs.

Dry grassland on chalky soils, rarer in north.

Wild basil

Clinopodium vulgare
12-32in (30-80cm)

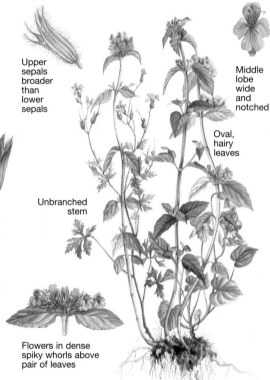

Upper sepals broader than lower sepals

Middle lobe wide and notched

Oval, hairy leaves

Unbranched stem

Flowers in dense spiky whorls above pair of leaves

Though less strongly scented than thyme or marjoram, wild basil is another member of the mint family which has been used in medicine and cookery over the centuries. Its name, derived from the Greek *basilikon*, 'kingly', reflects the esteem in which it was once held. Its leaves smell faintly like those of thyme when crushed. The little two-lipped pink flowers of wild basil are grouped in dense, spiky-looking whorls, each whorl being immediately above a pair of leaves. The stems are tallish and often rather straggly, with few side branches, and hairs covering the leaves make them appear dull.

Hedges and wood edges; scarcer in Scotland and Ireland.

Common hemp-nettle

Galeopsis tetrahit
4-40in (10-100cm)

Flower whorls in loose spike

Dark markings on lower lip

Stems swollen below leaves

Fruit has four nutlets

Red-tipped hairs on stem

Long sharp teeth on sepal tube

The most easily detected difference between common hemp-nettle and the similar-looking red dead-nettle and black horehound is the way in which the stems of hemp-nettle are swollen below each pair of leaves and covered in downward-pointing, red-tipped hairs. The name 'hemp' was applied to the plant because of the apparent similarity of its leaves to those of Indian hemp – but there is no cannabis to be extracted from our innocent British hemp-nettle. Indeed, unlike many other members of the mint family, it has no herbal or medicinal value, though it shares their square stems and tube-shaped flowers.

Widespread in cultivated land and woodland clearings.

Marsh woundwort

Stachys palustris
16-40in (40-100cm)

White markings on lower lip

Narrow leaves with fine teeth, on short stalks

Whorls of flowers in loose spike

Heart-shaped leaves

Field woundwort
Stachys arvensis

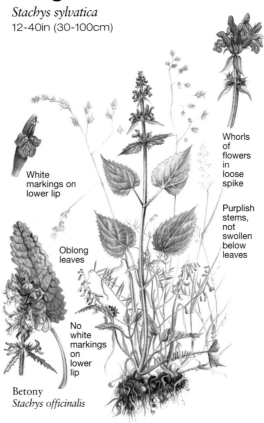

Even if woundworts are no longer used in medicine, modern experiments have shown that the oil that some of them yield does have antiseptic qualities. Like other woundworts, marsh woundwort has the two-lipped flowers arranged in a spike which gives the genus its name of *Stachys*, Greek for 'spike'. The main difference between this damp-loving plant and its drier-soil relative, hedge woundwort, is its narrow, short-stalked leaves. The much smaller field woundwort is a low-growing farmland weed that became rarer with the burning of corn stubble and earlier ploughing.

Widespread in ditches, swamps and fens.

J F M A M J J A S O N D

Hedge woundwort

Stachys sylvatica
12-40in (30-100cm)

White markings on lower lip

Oblong leaves

No white markings on lower lip

Betony
Stachys officinalis

Whorls of flowers in loose spike

Purplish stems, not swollen below leaves

The offensive smell of its crushed leaves belies the usefulness of hedge woundwort to man since the days of the ancient Greeks, as a poultice to treat wounds and to stem bleeding. It is easily recognised by its hairy, heart-shaped leaves and by its whorls of long, claret-coloured flowers, borne in loose spikes on the top of purplish stems. Betony has also been used as a remedy for many ills, including lowering fevers and curing digestive ailments. It can be distinguished from hedge woundwort by its oblong leaves, round at the tip and crinkly at the edges rather than coarsely toothed.

Widespread in woods, hedges and shady places.

J F M A M J J A S O N D

Red dead-nettle

Lamium purpureum
4-18in (10-45cm)

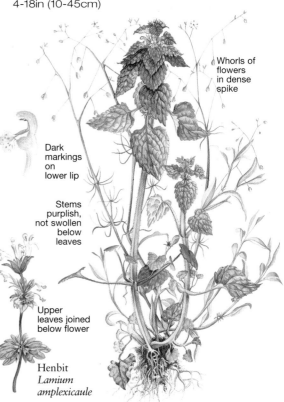

Whorls of flowers in dense spike

Dark markings on lower lip

Stems purplish, not swollen below leaves

Upper leaves joined below flower

Henbit
Lamium amplexicaule

When its flowers are out, nobody should mistake the dead-nettle, which does not sting, from the unpopular stinging nettle. Red dead-nettle has the characteristic two-lipped flowers, brightly coloured and with a hooded upper lip, of the mint family; stinging nettle, on the other hand, has the inconspicuous green flowers of the true nettle. When not in flower, the plants are easier to confuse. However, the stems of red dead-nettle, though hairy, lack the bristles which give true nettle its sting. Henbit, another dead-nettle, has stalkless upper leaves fused together to form a disc below each whorl of flowers.

Widespread on waste and cultivated ground.

J F M A M J J A S O N D

95

Red flowers
Up to four petals • Irregular flowers

Black horehound
Ballota nigra ssp *meridionalis*
16-40in (40-100cm)

Coarse leaves

Hairy stems, not swollen below leaves

Whorls of flowers on upper stem

White markings on lower lip of flower

Cattle avoid black horehound because of its offensive smell, noticeable to humans when the leaves are crushed between the fingers. As a result the plant flourishes unchecked on waste ground, its tall leafy stems ringed towards the top by whorls of dull purple flowers. These are distinct from those of related members of the mint family in being funnel-shaped, with white markings. Black horehound is a larger, coarser plant than red dead-nettle (p. 95).

Hedgerows in England and Wales.

Lesser skullcap
Scutellaria minor
8-16in (20-40cm)

Flowers usually in pairs

Helmet-shaped sepal tube

Leaves almost toothless

Unusually among members of the mint family, skullcap has flowers in pairs, each pair facing the same way and growing from the base of the dark green leaves. What gives the plant its name, however, is the shape of its sepal tube, said to resemble the leather skull helmet worn by Roman soldiers. The common skullcap (p. 75) has blue flowers; the lesser is a smaller, bushier plant, usually only 8in (20cm) tall.

Wet heaths in southern and western Britain and southern Ireland.

Wild teasel
Dipsacus fullonum
18-78in (45-195cm)

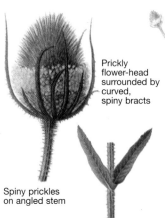

Prickly flower-head surrounded by curved, spiny bracts

Spiny prickles on angled stem

Long pointed leaves, joined at base to form cup

Bristly sepals remain on top of ripe fruit.

The spiny oval heads of teasel tower above surrounding vegetation – and win a commanding place in many a dried-flower arrangement. The spikiness of teasel is not confined to the heads. The rest of the plant bristles with discouragements to the human hand – toothed bracts below the heads, prickles along the stem and pointed, saw-toothed leaves.

Rough pastures and roadsides in England, Wales and southern Scotland; rare elsewhere.

Common fumitory

Fumaria officinalis
20in (50cm)

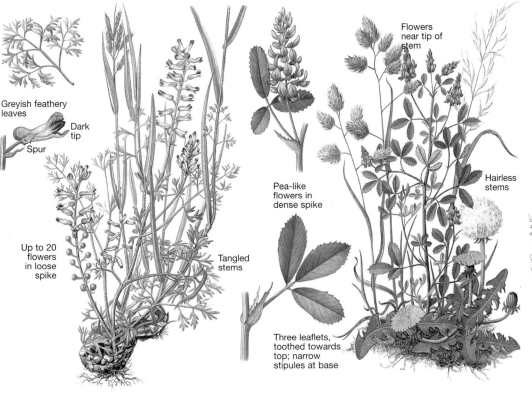

Greyish feathery leaves

Dark tip

Spur

Up to 20 flowers in loose spike

Tangled stems

Seen in a mass across a distant field, the common fumitory is a blue, smoky haze. The name fumitory derives from Latin words meaning 'smoke of the earth', and the plant's smoky-grey, feathery leaves certainly give that impression. In addition, when the plant is pulled up the roots smell of smoke. In summer the colour of the leaves is offset by the soft, dusky pink of the flowers which darken to a blackish-red at the tip. The colour of the flower distinguishes common fumitory from the related white climbing fumitory and white ramping fumitory (p. 16).

Widespread on cultivated ground and by roads.

Lucerne

Medicago sativa
12-36in (30-90cm)

Flowers near tip of stem

Pea-like flowers in dense spike

Hairless stems

Three leaflets, toothed towards top; narrow stipules at base

This is an invaluable plant, nourishing the soil with nitrogen taken from the air and providing, in its stems and leaves, fodder rich in vitamins and proteins for cattle and sheep. Nor is that all. Originally a Mediterranean plant, it has deep roots which enable it to resist drought. It is also prolific, providing farmers with four or five cuts a year. And it is far from dull. The seed-pods are delightful little spiral houses, and seedlings grown from the seeds make a nutty-tasting garnish for salads. It is also known as alfalfa.

Grown as field crop; also roadsides and waste ground, especially in central and eastern England.

Restharrow

Ononis repens
12-24in (30-60cm)

Pea-like flower

Pod shorter than joined sepals

Three leaflets with two toothed stipules at base

Stems hairy

Spiny stems

Spiny restharrow
Ononis spinosa

Trailing stems root at base

The deep, tough roots and matted stems of restharrow used to be the bane of farmers in the days of horse-drawn ploughs. Now the plant can be admired for its more positive qualities, particularly its very pretty flowers. These are an unusual shape, like pink tongues reaching out from the plant's hairy stems. The leaves are also unusual in having, in addition to toothed leaflets, leaf-like stipules at the base which half-clasp the stem.
Spiny restharrow has more upright spiny stems, which do not root like those of common restharrow. It is limited to well-drained soils in south and central England.

Rough, grassy places and sand-dunes.

J F M A **M J J A S** O N D

J F M A M **J J A S** O N D

J F M A M **J J A S** O N D

97

Red flowers   *Up to four petals • Irregular flowers*

Red clover

Trifolium pratense
To 24in (60cm)

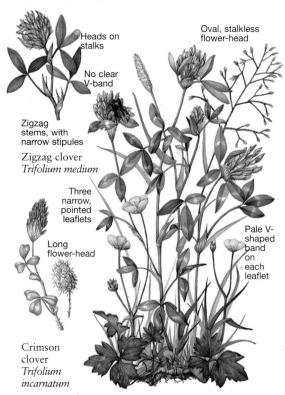

Heads on stalks

Oval, stalkless flower-head

No clear V-band

Zigzag stems, with narrow stipules

Zigzag clover
Trifolium medium

Three narrow, pointed leaflets

Long flower-head

Pale V-shaped band on each leaflet

Crimson clover
Trifolium incarnatum

The flowers of the red clover look like Olympic torches, held aloft on hairy stems. The trefoil leaves are distinctive, too, with their pale, crescent-shaped bar in the middle of each leaflet. Occasionally a four-leaf clover is found – in folklore a supreme piece of good fortune. Red clover is a precious plant, nourishing the soil with nitrogen, providing fodder for animals and, being entirely pollinated by bees, providing superb honey.
Zigzag clover has narrow, papery stipules at the base of the leaf. Crimson clover has furry-looking crimson heads up to 2¹/₂in (60mm) long.

Grassy places throughout Britain and Ireland.

Sainfoin

Onobrychis viciifolia
4-32in (10-80cm)

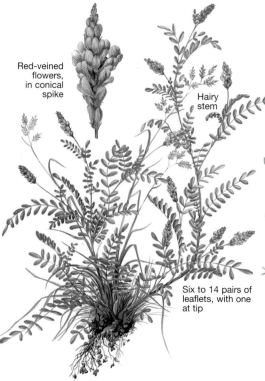

Red-veined flowers, in conical spike

Hairy stem

Six to 14 pairs of leaflets, with one at tip

Spikes of magenta-pink flowers up to 32in (80cm) high make sainfoin one of our more spectacular grassland plants. It began to be widely grown only in the 1930s to provide fodder for cattle, and this use is reflected in its common name, which comes from French words for 'wholesome hay'. From farmlands it has spread to chalk downs and roadside verges. The pea-like flowers are borne in a cone-shaped spike on the top of long stems. Below them is a delicate tracery of feathery, grey-green leaves consisting of numerous pairs of leaflets.

Chalk and limestone grassland in southern England.

Purple milk-vetch

Astragalus danicus
2-14in (5-36cm)

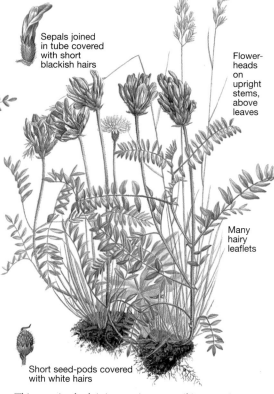

Sepals joined in tube covered with short blackish hairs

Flower-heads on upright stems, above leaves

Many hairy leaflets

Short seed-pods covered with white hairs

This creeping herb is inconspicuous until it comes into flower in May and produces heads of pea-like flowers on long stems. The flowers are very like those of clover, but the leaves are totally different, being divided into numerous leaflets covered with soft whitish hairs. The seed-pods, too, are covered with white hairs. Vast sweeps of purple milk-vetch bring rich colour to sandy hills in the Breckland region of Norfolk and other eastern areas; it is uncommon in the west. The plant's name derives from a belief that it increased the milk yield of goats.

Chalk and limestone grassland, mainly in eastern Britain.

Bush vetch

Vicia sepium
12-40in (30-100cm)

Up to
six blooms in
each flower-head

Stems climb by
branched tendril
at tip of leaf

Five to
nine pairs
of small
leaflets

Larger
flowers and
leaflets

Wood vetch
Vicia sylvatica

Common vetch

Vicia sativa ssp. *sativa*
6-48in (15-120cm)

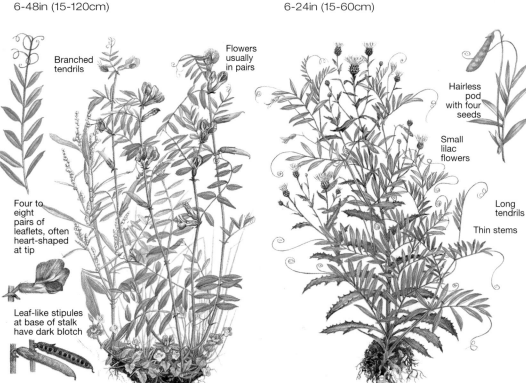

Branched
tendrils

Flowers
usually
in pairs

Four to
eight pairs of
leaflets, often
heart-shaped
at tip

Leaf-like stipules
at base of stalk
have dark blotch

Pods slightly hairy,
with long beak

Smooth tare

Vicia tetrasperma
6-24in (15-60cm)

Hairless
pod
with four
seeds

Small
lilac
flowers

Long
tendrils

Thin stems

Climbing by tendrils which grow from the tips of the leaves, bush vetch scrambles high over other plants, draping them with its long feathery leaves and short spikes of pea-like flowers. It is one of the commonest British vetches, and its flowers appear throughout the summer. Bumble bees are particularly attracted to the nectar, and force their way to the tightly enclosed flower bases to get at it.

Wood vetch has larger and more accessible flowers, often pollinated by wasps. It is most prolific in woods, but grows also on cliffs and shingle by the sea.

Rough, grassy places and hedges throughout British Isles.

To distinguish common vetch from other British vetches, look first at the base of the leaf. Here, common vetch alone has a pair of flowers and two black blotches on the leaf-like stipules. Despite its name, common vetch is only really common in south-east England, where it was introduced from the Continent for cattle food and has since spread onto waste ground. Like other vetches it climbs by branched tendrils sprouting from the tip of the leaf. At the end of the summer the flowers develop into seed-pods 2-3in (50-75mm) long, which are smooth or slightly hairy, with depressions between the seeds.

Grassy places, mainly in south-east England.

The name tare, originally applied to vetch seed, is today used for several small vetch species. In the Biblical parable, tare was the weed sown by the farmer's enemies to adulterate his wheat harvest. In 17th-century England, too, tare used to twine its way through crops of wheat, barley and oats; but this was mainly the white-flowered hairy tare (p. 17). Smooth tare – with deep lilac flowers and smooth four-seeded pods rather than the hairy two-seeded pods of its relative – is a plant of waste places, found among long grass and low bushes. It has similar tendrilled leaves to other vetches.

Grassy places, mainly on clay soils in lowland Britain.

J F M A **M J J A S** O N D   J F M A **M J J A S** O N D   J F M A **M J J A S** O N D

Red flowers
Up to four petals • Irregular flowers

Bitter vetch

Lathyrus linifolius
6-16in (15-40cm)

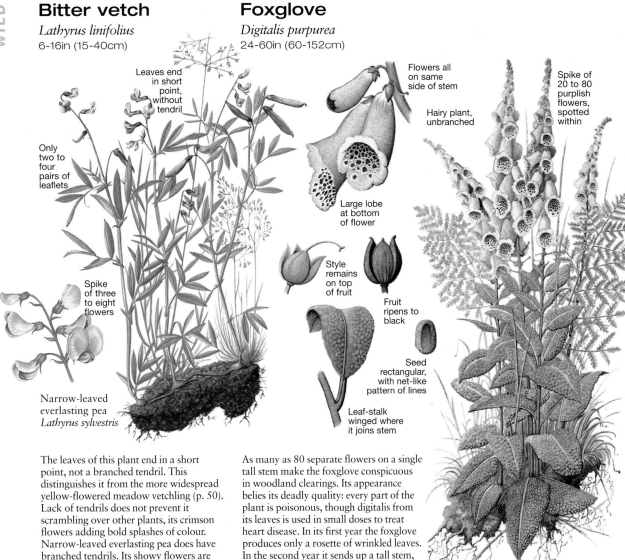

Leaves end in short point, without tendril

Only two to four pairs of leaflets

Spike of three to eight flowers

Narrow-leaved everlasting pea
Lathyrus sylvestris

The leaves of this plant end in a short point, not a branched tendril. This distinguishes it from the more widespread yellow-flowered meadow vetchling (p. 50). Lack of tendrils does not prevent it scrambling over other plants, its crimson flowers adding bold splashes of colour. Narrow-leaved everlasting pea does have branched tendrils. Its showy flowers are among the largest of the pea family.

Woods, thickets and shady hedgerows, except in some east and central areas.

J F M A M J J A S O N D

Foxglove

Digitalis purpurea
24-60in (60-152cm)

Flowers all on same side of stem

Hairy plant, unbranched

Spike of 20 to 80 purplish flowers, spotted within

Large lobe at bottom of flower

Style remains on top of fruit

Fruit ripens to black

Seed rectangular, with net-like pattern of lines

Leaf-stalk winged where it joins stem

Oval or lance-shaped wrinkled leaves, forming rosette at base

As many as 80 separate flowers on a single tall stem make the foxglove conspicuous in woodland clearings. Its appearance belies its deadly quality: every part of the plant is poisonous, though digitalis from its leaves is used in small doses to treat heart disease. In its first year the foxglove produces only a rosette of wrinkled leaves. In the second year it sends up a tall stem, and clusters of flowers open from the bottom upwards, on one side only.

Woods, heaths and rocks in all parts.

J F M A M J J A S O N D

Pink water speedwell

Veronica catenata
To 12in (30cm)

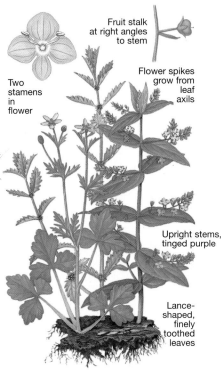

Fruit stalk at right angles to stem

Flower spikes grow from leaf axils

Two stamens in flower

Upright stems, tinged purple

Lance-shaped, finely toothed leaves

Once a farmer has cleared his drainage ditches, the flower-heads of water speedwell are among the first to make an appearance. This pink-flowered species often grows together with blue water speedwell (p. 72), with which it will hybridise. The two plants differ from most other speedwells in having narrow, pointed leaves in opposite pairs. The leaves have tiny teeth and no stalks.

Wet places, especially ponds and streams, mainly in southern and eastern England.

J F M A M J J A S O N D

Red flowers

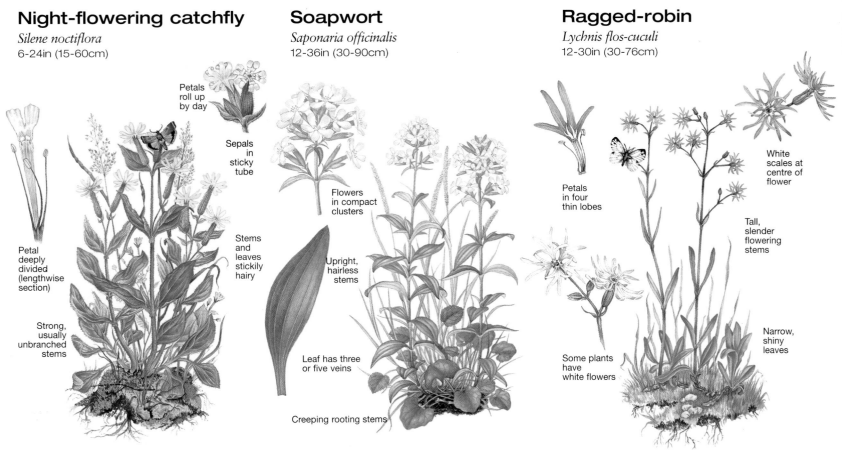

Night-flowering catchfly

Silene noctiflora
6-24in (15-60cm)

Petals roll up by day

Sepals in sticky tube

Petal deeply divided (lengthwise section)

Strong, usually unbranched stems

Soapwort

Saponaria officinalis
12-36in (30-90cm)

Flowers in compact clusters

Upright, hairless stems

Stems and leaves stickily hairy

Leaf has three or five veins

Creeping rooting stems

Ragged-robin

Lychnis flos-cuculi
12-30in (30-76cm)

White scales at centre of flower

Petals in four thin lobes

Tall, slender flowering stems

Narrow, shiny leaves

Some plants have white flowers

This is one of very few plants in Britain that opens its flowers at dusk. During the day the petals of its deeply cleft, rose-tinged flowers curl into the centre of the flower, so that only their yellow undersides are on view. At dusk the flowers open, filling the air with a heavy, sweet-smelling scent and drawing insects, particularly moths, towards them. Fine sticky hairs on the plant's leaves and stem temporarily trap the insects and ensure that by the time they reach the flower its pollen will adhere to them.

The soapwort looks something like the garden sweet william, but it has taller and thicker flowering stems, and the petals are untoothed. Its flowers grow in pretty clusters which make it an attractive plant in its own right that is sometimes grown in borders. The flowers are fragrant, with undivided petals which have two slender scales at the base. The green parts of soapwort were once boiled to produce a lathery liquid used for washing wool, and much of the wild soapwort seen today is growing near the fields in which it was once grown commercially.

The name suggests a somewhat unkempt flower, but this is far from the case. It is a delightful, exceptionally pretty plant, immediately recognisable by the long, finger-lobed petals from which it gets its common name. The flowers are more delicate than ragged, though the leaves are rough. The seed capsule has five short teeth which bend back to open. Despite the appeal of the flowers, it has traditionally been considered unlucky to pick them and take them indoors.

Widespread locally on arable land, mostly in south and east.

Scattered wasteland, and hedgerows.

Widespread in marshes, wet meadows and damp woodland.

J F M A M **J** **J** **A** S O N D   J F M A M J J **A** **S** **O** N D   J F M A M **M** **J** J A S O N D

Red flowers *Five to seven petals • Regular flowers*

Red campion

Silene dioica
12-36in (30-90cm)

Petals deeply divided

Inner ring of white flaps on petals

Upright flowering stems

Upper leaves unstalked

Hairy stems and leaves

Lower leaves have long winged stalks

Maiden pink

Dianthus deltoides
6-18in (15-45cm)

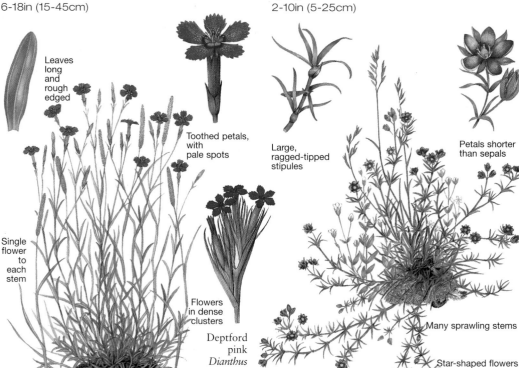

Leaves long and rough edged

Toothed petals, with pale spots

Single flower to each stem

Flowers in dense clusters

Deptford pink *Dianthus armeria*

Sand spurrey

Spergularia rubra
2-10in (5-25cm)

Large, ragged-tipped stipules

Petals shorter than sepals

Many sprawling stems

Star-shaped flowers

Unlike white campion (p. 18) which has a faint scent, red campion is scentless. Both are hedgerow plants, but the red variety is more shade-loving and can be found in woods throughout the British Isles. It is a tall, handsome plant, with masses of red flowers which would look well in any garden. Male and female flowers grow on separate plants; when red and white campions grow together the two often hybridise, producing plants with flowers in a wide range of pinks. The seed-pods have short, curled teeth.

Hedgerows and woods on rich soils in most parts of Britain and Ireland.

Maiden pink is a wild flower which is well-known to gardeners as a rockery plant. While it lacks scent, it is nevertheless an attractive plant, with its long, branching shoots which suddenly burst into deep pink flower. The flowers are freckled with spots, paler or darker than the rest of the petals. They close in dull weather. The leaves are long, slender and rough-edged.
Maiden pink has a mixture of flowering and non-flowering shoots. By contrast, all the shoots of Deptford pink bear flowers. Deptford pink is increasingly rare, and confined to parts of southern England and Wales.

Occasional, on scattered dry banks and hill pastures.

Of the six varieties of spurrey which grow in the British Isles, sand spurrey is the only one which grows inland; the rest are found near the sea. The five petals of its star-shaped flower are separated by five green sepals, slightly longer than the petals. But perhaps the most noticeable feature of the plant is the narrow, pointed leaves and, in particular, the distinctive triangular stipules growing at their base. The stipules of the sand spurrey are unusually long, in some cases measuring as much as one-third of the length of the leaf.

On lime-free ground in most of Britain, but rare in northern Scotland and in Ireland.

J F **M A M J J A S O** N D

J F M A M **J J A S** O N D

J F M A **M J J A S O** N D

This large group ranges from the bright-flowered campion and common mallow of the roadside to the thrift and sea lavender of the coast; and from tall loosestrife to low-growing pimpernels. Crane's-bills get their name from their pointed fruits.

Orpine
Sedum telephium
8-24in (20-60cm)

Flowers on long stalks in globe-shaped heads

Purple fruit

Leaves pale green, fleshy and toothed

Common mallow
Malva sylvestris
12-36in (30-90cm)

Round fruit has many nutlets

Narrow, dark-veined petals

Stem leaves ivy-shaped

Basal leaves rounded

Stem leaves deeply divided

Stems upright and unbranched, often reddish

Musk mallow
Malva moschata

Basal leaves rounded and lobed

Pink purslane
Claytonia sibirica
6-15in (15-38cm)

Petals long and deeply notched; two sepals

Stem leaves unstalked but not joined

Capsule oval, as long as sepals

Long-stalked leaves from base

There is no mistaking this rose-red flower, even in the half light of shady woodland, where it is most often found. It has two striking features: the big, broad heads of little flowers, with their spreading petals; and the large, fleshy, oblong leaves which grow all the way up the unbranching stems. The leaves and stems store water, enabling the plant to survive prolonged drought and to remain fresh for some time after being picked. It has a carrot-like underground root tuber which also stores water.

Locally common in woods and hedges in most parts of Britain.

The plant may be widespread, but there is nothing common about its flower. It could fairly be called the English hibiscus, with its rich heliotrope-pink petals lined with darker veins. In fact it is related to the hibiscus, as well as to the hollyhock. With flowers as eye-catching as this, there are few problems in identification. Two other clues are the crinkly leaves, which usually have a dark spot where the leaf joins the stalk, and the fruit – a round disc of nutlets which are edible and taste like peanuts. Musk mallow has lighter, rose-pink flowers even bigger than the common mallow's; they grow in spikes.

Roadsides and waste ground throughout lowland areas.

At first sight pink purslane seems so unlike its close relation spring beauty (p. 21) that there would appear to be no connection. The purslane's flowers are pink rather than white, and much bigger, while the leaves are separate and not joined. The connection is revealed on the back of the flowers; unlike the rather similar chickweeds and spurreys, spring beauty and pink purslane have two sepals instead of five. The leaves of pink purslane grow in unstalked pairs on the stems. The basal leaves are long-stalked.

Widespread but local in damp, shady places, in woods and along banks of streams.

J F M A M J **J A S** O N D   J F M A M J **J J A S** O N D   J F M A M **J** J **A** S O N D

Red flowers   *Five to seven petals • Regular flowers*

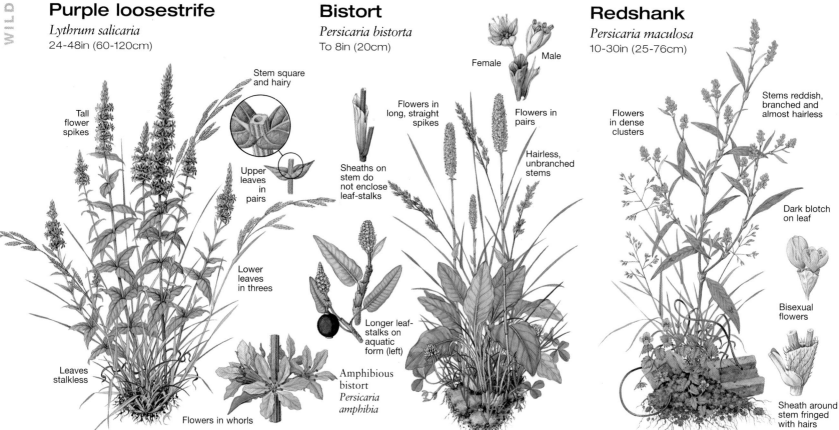

Purple loosestrife
Lythrum salicaria
24-48in (60-120cm)

Tall flower spikes

Stem square and hairy

Upper leaves in pairs

Lower leaves in threes

Leaves stalkless

Flowers in whorls

Bistort
Persicaria bistorta
To 8in (20cm)

Female   Male

Flowers in long, straight spikes

Sheaths on stem do not enclose leaf-stalks

Longer leaf-stalks on aquatic form (left)

Amphibious bistort
Persicaria amphibia

Flowers in pairs

Hairless, unbranched stems

Redshank
Persicaria maculosa
10-30in (25-76cm)

Stems reddish, branched and almost hairless

Flowers in dense clusters

Dark blotch on leaf

Bisexual flowers

Sheath around stem fringed with hairs

This tall, beautiful plant brings a touch of the tropics to the edge of still, inland waters in Britain, with its long spikes of vivid purple-red flowers. They cluster on the stout, square stems in whorls, and sometimes grow in such abundance by lakesides that the water, seen from a distance, seems edged with a purple mist. The leaves are untoothed and grow in opposite pairs on the upper parts of the stems and in whorls of three below. The upper leaves are usually much shorter.

Common in marshes and by lakes and slow-flowing rivers throughout British Isles, except in northern Scotland.

In the Lake District the young leaves of bistort are an essential ingredient of Easter-ledge pudding – a dish whose name is almost as odd as that of the plant itself. This comes from the Latin *bistorta*, meaning 'twice twisted', and refers to the contorted shape of the roots. The plant often grows in large patches in which the long straight spikes of small pink flowers make a fine display. Amphibious bistort comes in two forms. In water it has long, narrow hairless leaves which float on the surface. On land it has hairy leaves with shorter stalks.

Widespread but local in damp meadows and grassy places throughout British Isles.

Redshank can quickly be identified not only by its reddish stems, which have given the plant its common name, but even more certainly by the large, black blotches on its leaves. Why this discoloration should be there and what purpose it serves is a mystery, but it is certainly a help to recognition. Another unusual feature of the redshank is the fringed sheath which surrounds the flower stem just above each leaf-stalk. The plant can be a troublesome weed in gardens and fields, but the sense spikes of little flowers are undeniably pretty.

Cultivated ground, waste places and beside ponds throughout the British Isles.

Thrift

Armeria maritima
To 6in (15cm)

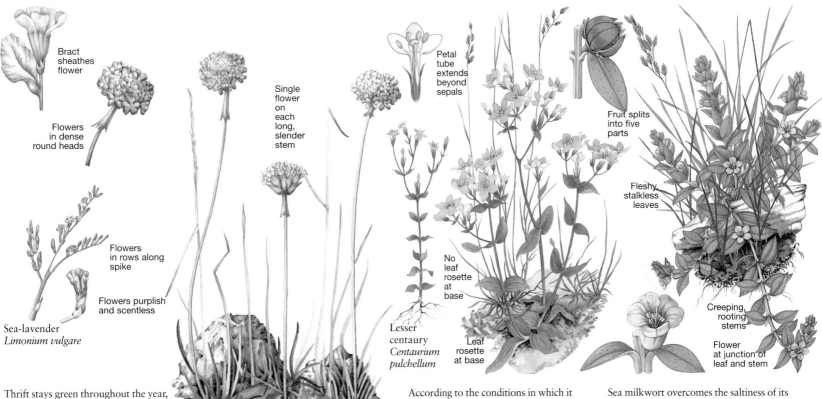

Bract sheathes flower

Flowers in dense round heads

Single flower on each long, slender stem

Flowers in rows along spike

Flowers purplish and scentless

Sea-lavender
Limonium vulgare

Narrow blue-green leaves in rosettes

Common centaury

Centaurium erythraea
2-12in (5-30cm)

Petal tube extends beyond sepals

No leaf rosette at base

Lesser centaury
Centaurium pulchellum

Leaf rosette at base

Sea milkwort

Glaux maritima
4-12in (10-30cm)

Fruit splits into five parts

Fleshy, stalkless leaves

Creeping, rooting stems

Flower at junction of leaf and stem

Thrift stays green throughout the year, and when dried it is virtually everlasting. It has been a garden favourite for more than 400 years, but to appreciate its true beauty it needs to be seen in a great mass, preferably on top of a Cornish cliff on a sunny day.
Sea-lavender is a plant of salt-marshes, colouring whole stretches purple-pink when in flower.

Sea-cliffs and salt-marshes on all coasts.

According to the conditions in which it grows, common centaury can be a single stem only 2in (50mm) high – or a plant with several stems and many branches 12in (30cm) or more tall. One helpful clue is the rosette of leaves, with three to seven prominent veins, at the base of the stem. The flowers are on short stalks and grow in dense clusters.
Lesser centaury has no rosette of leaves and fewer flowers, on longer stalks.

Dunes and dry grassy places in England.

Sea milkwort overcomes the saltiness of its environment by storing the limited fresh water it can obtain in its fleshy leaves. It is a pale green, hairless plant which usually grows in a thick, spreading mat, so reducing water loss by evaporation. The small, oblong leaves grow in profusion along the creeping stems. The apparent petals of the pretty little flowers are, in fact, sepals.

Rocks, cliffs and salt-marshes round coast; occasionally inland.

Red flowers
Five to seven petals • Regular flowers

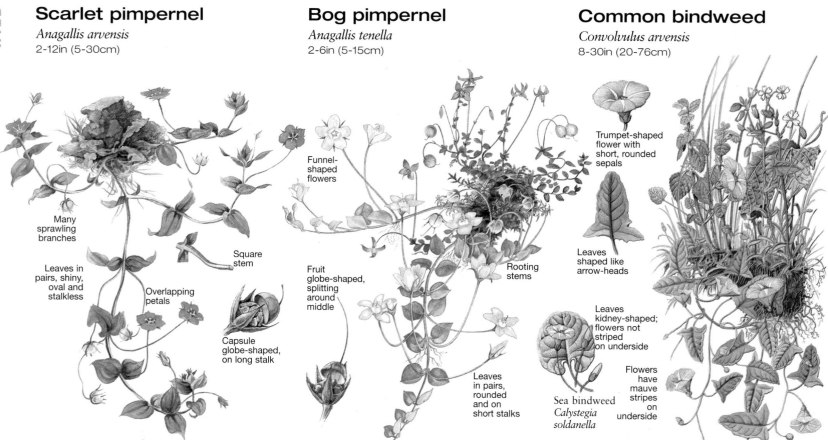

Scarlet pimpernel
Anagallis arvensis
2-12in (5-30cm)

Many sprawling branches

Leaves in pairs, shiny, oval and stalkless

Square stem

Overlapping petals

Capsule globe-shaped, on long stalk

Bog pimpernel
Anagallis tenella
2-6in (5-15cm)

Funnel-shaped flowers

Fruit globe-shaped, splitting around middle

Rooting stems

Leaves in pairs, rounded and on short stalks

Common bindweed
Convolvulus arvensis
8-30in (20-76cm)

Trumpet-shaped flower with short, rounded sepals

Leaves shaped like arrow-heads

Leaves kidney-shaped; flowers not striped on underside

Flowers have mauve stripes on underside

Sea bindweed
Calystegia soldanella

Although the bright, starry flowers of this pimpernel are usually scarlet, they can also be pink, white, lilac or blue. It has even been known for red and blue flowers to be found on the same plant. Nor is it so 'damned elusive' as the Pimpernel hero of Baroness Orczy's famous novel – it is found all over Europe. It is elusive in another sense, though, in that its flowers close in mid-afternoon and are always shut during dull or wet weather. The flowers grow on long, drooping stalks and have petals tipped with glandular hairs.

Widespread on cultivated and waste land and dunes.

This delicately coloured relative of the scarlet pimpernel has flowers of pastel-pink and tiny, pale green leaves. The combination creates an extremely pretty effect. Unlike the scarlet pimpernel, bog pimpernel is a creeping plant with rooting stems, often growing so densely that it forms a mat. The leaves are more rounded than those of scarlet pimpernel, and very small, growing in opposite pairs on short stalks. The elegant, funnel-shaped flowers grow on long, slender, erect stalks. They open only when the sun is out. Bog pimpernel is illustrated growing with the deeper pink cranberry (p. 87).

Locally common in bogs and damp grasslands in most areas.

The speed at which bindweed wraps itself round the nearest convenient support has been timed. The ends of the thin but strong, thread-like stems twine themselves round anything they touch and complete a full circle, travelling anti-clockwise, in under two hours. Not only does bindweed strangle other plants, but it has a deep and extensive root system that exhausts the soil. The trumpet-shaped flowers smell of almonds.
Sea bindweed has sprawling stems which do not climb. It is confined to the coast, and absent from the far north.

Cultivated land and open grassland, especially road verges in lowlands.

J F M A M J J A S O N D

J F M A M J J A S O N D

J F M A M J J A S O N D

Common dodder
Cuscuta epithymum

Hound's-tongue
Cynoglossum officinale
12-36in (30-90cm)

Autumn gentian
Gentianella amarella
2-12in (5-30cm)

Wild onion
Allium vineale
To 33in (84cm)

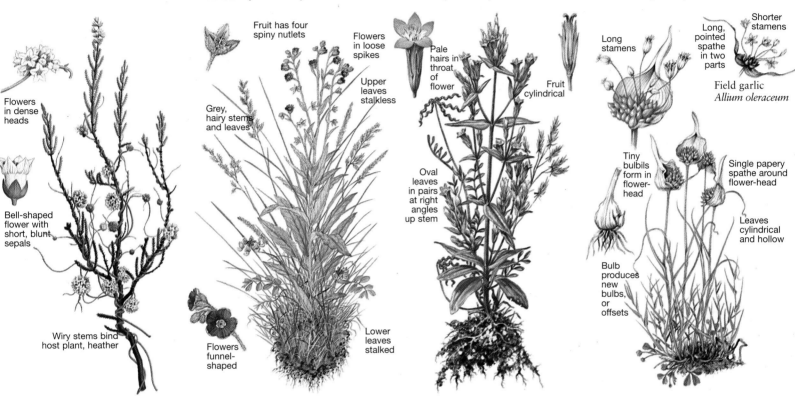

Flowers in dense heads

Bell-shaped flower with short, blunt sepals

Wiry stems bind host plant, heather

Fruit has four spiny nutlets

Flowers in loose spikes

Upper leaves stalkless

Grey, hairy stems and leaves

Flowers funnel-shaped

Lower leaves stalked

Pale hairs in throat of flower

Fruit cylindrical

Oval leaves in pairs at right angles up stem

Long stamens

Long, pointed spathe in two parts

Shorter stamens

Field garlic
Allium oleraceum

Tiny bulbils form in flower-head

Single papery spathe around flower-head

Leaves cylindrical and hollow

Bulb produces new bulbs, or offsets

This leafless parasite grows by twining its red, wiry stem anti-clockwise round its victim – often gorse or heather – and sending out suckers which penetrate the stem of the host. Once established, the root of the dodder then withers and dies and all its food requirements are drawn through the suckers from the host plant, which is gradually weakened. The dodder, however, has dense clusters of pretty pink flowers.

Widespread in scattered localities throughout England and Wales.

In the 17th century, a leaf of hound's-tongue placed under the big toe was said to prevent the annoyance of barking dogs. The plant's effectiveness in silencing canines was perhaps connected with its pungent, unpleasant smell. The plant does have one attractive feature, however; its flowers are small, maroon-coloured and very pretty, with velvety scales at the centre, closing the mouth.

Widespread on grassland, at wood edges and on sand-dunes.

Flowering later than its four-petalled relative, field gentian (p. 70), autumn gentian is also reddish-purple rather than bluish-purple in colour. Its opposite pairs of leaves are set at right angles to each other up the stem. Gentian roots yield a bitter liquid used in medicinal tonics. Early gentian is a smaller plant which flowers in May and June on some chalk grasslands in southern England.

Widespread on grassland and dunes in England; rarer elsewhere.

Known also as crow garlic, wild onion was once a common weed in meadows; eaten by cattle, it gave milk and butter an unpleasant flavour. Modern methods of cultivation have, however, almost eliminated it from arable land. It is a strange-looking plant, with its large, papery spathes shielding each flower-head like a monk's cowl. The related field garlic is much less widespread.

Widespread on dry grassy dunes and commons, especially in south.

J F M A M J J A S O N D   J F M A M J J A S O N D   J F M A M J J A S O N D   J F M A M J J A S O N D   107

Red flowers

Five to seven petals • Regular flowers

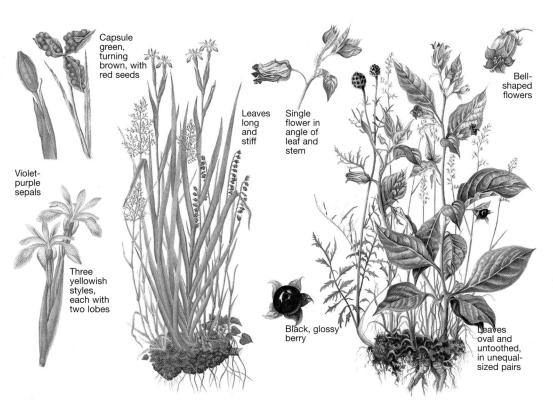

Stinking iris
Iris foetidissima
12-32in (30-80cm)

Capsule green, turning brown, with red seeds

Leaves long and stiff

Violet-purple sepals

Three yellowish styles, each with two lobes

Deadly nightshade
Atropa belladonna
To 60in (152cm)

Single flower in angle of leaf and stem

Bell-shaped flowers

Black, glossy berry

Leaves oval and untoothed, in unequal-sized pairs

Henbane
Hyoscyamus niger
To 32in (80cm)

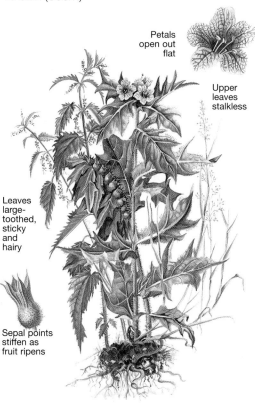

Petals open out flat

Upper leaves stalkless

Leaves large-toothed, sticky and hairy

Sepal points stiffen as fruit ripens

When the dagger-sharp leaves of this plant are crushed they give off a smell vaguely reminiscent of roast beef. It is shorter than the other species of wild iris native to Britain, the yellow iris (p. 51), and has flowers of an unusual purple-violet colour, with darker veins. The three styles, or female parts, are yellow and petal-like and are situated above the petals and sepals. The flowers are short-lived, but the bright red seeds cling to the split capsule for some time.

Open woods, hedgebanks and sea-cliffs in south of England and on Welsh coast.

Deadly nightshade has a unique notoriety among poisonous plants which is perhaps due to the fact that it looks so benign. It is hard to believe that the black, shiny berries, tempting as cherries, could have such violent effects. Pheasants eat them with no ill results, but two or three berries are enough to kill a child. Unlike henbane, it is neither unpleasantly hairy, sticky nor evil smelling. It is a tall plant with dull green, pointed, oval leaves – looking large for the size of the plant – and dull reddish-purple flowers which grow singly in the angle of leaf and stem.

Woodland, hedges and thickets in the south; rare elsewhere.

More deadly even than deadly nightshade, henbane is an evil-looking plant which nobody would want to linger over unless, like Dr Crippen, murder was in mind. All parts of the plant are packed with poison, including the chemical hyoscine, which was used by Crippen to murder his wife in 1910. Sticky white hairs cover the plant, which also emits a disagreeable smell. The flowers, yellow with purple veins, grow in a leafy one-sided spike. The fruit is a capsule packed with seeds, which opens by a cap at the top. Although once common on waste ground, henbane is today quite rare.

Wasteland in England, seaside in Wales and eastern Scotland.

J F M A M **J J** A S O N D   J F M A M **J J A** S O N D   J F M A M **J J** A S O N D

Hedgerow crane's-bill

Geranium pyrenaicum
9-24in (23-60cm)

Downy, upright stems

Rounded leaves, lobed halfway to midrib

Flowers in pairs, with deeply notched petals

Dove's-foot crane's-bill

Geranium molle
4-16in (10-40cm)

Dark flowers

Rounded petals and sepals

Petals slightly notched

Leaves deeply divided

Dusky crane's-bill
Geranium phaeum

Cut-leaved crane's-bill
Geranium dissectum

Petals not notched

Stems hairless, leaves glossy

Sepals pointed and hairy

Shining crane's-bill
Geranium lucidum

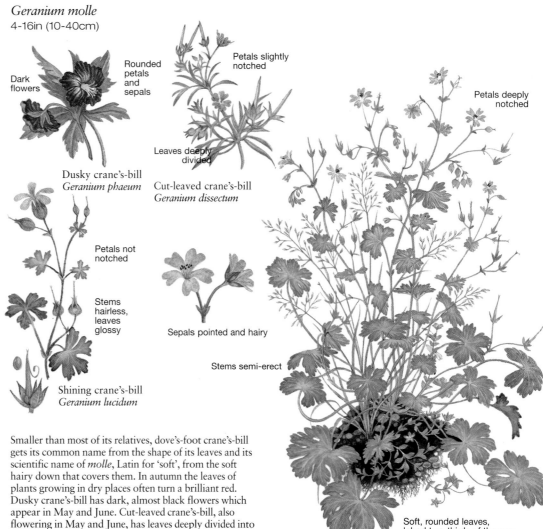

Petals deeply notched

Stems semi-erect

Soft, rounded leaves, lobed two-thirds of the way to midrib

The crane's-bills get their name from the shape of their fruits, which end in a long pointed beak resembling that of a crane. In spite of their generic name they are quite distinct from garden geraniums, which belong to the genus *Pelargonium*. The various species of crane's-bills differ from each other in their colour, and in the way their petals are notched and their leaves divided. The small, pinkish flowers of hedgerow, or mountain, crane's-bill make a delicate contrast with the tangle of leaves and stalks. A native of southern Europe, it was not recorded in this country until 1762.

Hedgerows and waste ground, mainly in south and east.

Smaller than most of its relatives, dove's-foot crane's-bill gets its common name from the shape of its leaves and its scientific name of *molle*, Latin for 'soft', from the soft hairy down that covers them. In autumn the leaves of plants growing in dry places often turn a brilliant red. Dusky crane's-bill has dark, almost black flowers which appear in May and June. Cut-leaved crane's-bill, also flowering in May and June, has leaves deeply divided into five to seven lobes. Shining crane's-bill has hairless, glossy leaves and is common on limestone walls and banks, flowering from May to August.

Widespread in fields and waste places.

J F M A M J J A S O N D

J F M A M J J A S O N D

Red flowers
Five to seven petals • Regular flowers

Common stork's-bill
Erodium cicutarium
To 20in (50cm)

Fruit splits from base, each segment carrying corkscrew-tailed seed

Petals longer than sepals

Beak shorter

Leaves undivided

Leaves in two rows of leaflets

Sea stork's-bill
Erodium maritimum

Stems erect or prostrate, usually hairy

As its common name suggests, this sand-loving plant, related to the crane's-bill, has fruits of beak-like shape. Each seed has a corkscrew-like 'tail', which changes shape with fluctuations in humidity and causes the seed to burrow into the ground. The flowers vary greatly in colour, from rose-purple to white; after opening early in the morning, they have often lost their petals by midday. They are pollinated by a wide variety of insects, or can be self-pollinating. The petals often have a black spot at the base.

Heaths, dry grassland and sandy seaside areas throughout British Isles.

Herb-robert
Geranium robertianum
4-20in (10-50cm)

Rounded petals, not notched

Stems branch from base

Leaves have three to five lobes

In autumn, the deeply divided fern-like leaves and hairy stems of this crane's-bill turn a fiery red, which may account for the second element of its Latin name, possibly derived from the Latin *ruber*, meaning 'red'. Alternatively, it may get its name from Robert, an early Duke of Normandy. The leaves, when crushed, give off a pungent smell, similar to that of the garden geranium, which has led to the plant's nickname of 'Stinking Bob'.

Widespread in shade on hedge-banks and walls or in woods.

Marsh cinquefoil
Potentilla palustris
6-18in (15-45cm)

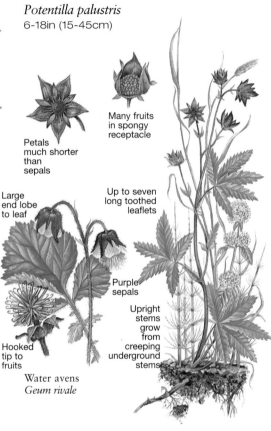

Petals much shorter than sepals

Many fruits in spongy receptacle

Large end lobe to leaf

Up to seven long toothed leaflets

Purple sepals

Upright stems grow from creeping underground stems

Hooked tip to fruits

Water avens
Geum rivale

The leaves of this marsh-loving plant are usually found in groups of five – hence the name cinquefoil, from the French for 'five leaves'. It can be found growing up to an altitude of 3000ft, spreading by means of creeping underground stems, or rhizomes. The flowers are reddish, as are the fruits, which are carried in spongy receptacles. In the Isle of Man the plant is called 'marsh strawberry'. Similar swampy conditions suit the water avens, which has nodding orange-pink flowers and hooked, bur-like fruit heads.

Fens and bogs in northern and upland Britain and Ireland; rarer in Midlands and south.

J F M A M J J A S O N D

J F M A M J J A S O N D

J F M A M J J A S O N D

Common valerian

Valeriana officinalis
8-60in (20-150cm)

Harsh downy rose

Rosa tomentosa
36-72in (90-180cm)

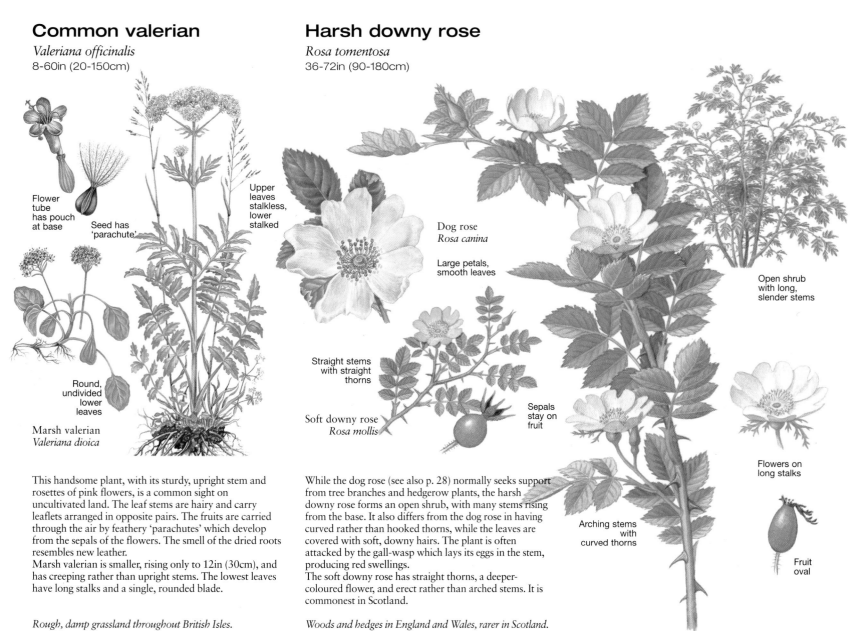

Flower tube has pouch at base

Seed has 'parachute'

Round, undivided lower leaves

Marsh valerian
Valeriana dioica

Upper leaves stalkless, lower stalked

Dog rose
Rosa canina

Large petals, smooth leaves

Straight stems with straight thorns

Soft downy rose
Rosa mollis

Sepals stay on fruit

Open shrub with long, slender stems

Flowers on long stalks

Arching stems with curved thorns

Fruit oval

This handsome plant, with its sturdy, upright stem and rosettes of pink flowers, is a common sight on uncultivated land. The leaf stems are hairy and carry leaflets arranged in opposite pairs. The fruits are carried through the air by feathery 'parachutes' which develop from the sepals of the flowers. The smell of the dried roots resembles new leather.

Marsh valerian is smaller, rising only to 12in (30cm), and has creeping rather than upright stems. The lowest leaves have long stalks and a single, rounded blade.

Rough, damp grassland throughout British Isles.

While the dog rose (see also p. 28) normally seeks support from tree branches and hedgerow plants, the harsh downy rose forms an open shrub, with many stems rising from the base. It also differs from the dog rose in having curved rather than hooked thorns, while the leaves are covered with soft, downy hairs. The plant is often attacked by the gall-wasp which lays its eggs in the stem, producing red swellings.

The soft downy rose has straight thorns, a deeper-coloured flower, and erect rather than arched stems. It is commonest in Scotland.

Woods and hedges in England and Wales, rarer in Scotland.

Red flowers

Five to seven petals
Regular flowers

Water violet

Hottonia palustris
To 16in (40cm) total height

This graceful, aquatic member of the primrose family lives suspended in the water. Its leafless stems, crowned with delicate pinkish-purple flowers with deep yellow throats, rise from the surface; below the water are the green whorls of its finely divided leaves, and its long silvery roots.

Clear water and ditches mainly in eastern and south-east England.

| J | F | M | A | **M** | **J** | J | A | S | O | N | D |

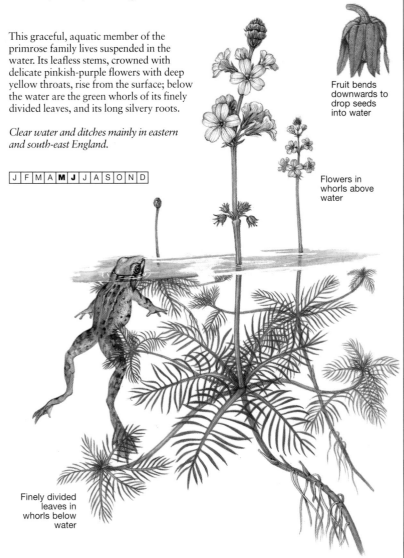

Fruit bends downwards to drop seeds into water

Flowers in whorls above water

Finely divided leaves in whorls below water

Red flowers

Five to seven petals
Irregular flowers

Small toadflax

Chaenorhinum minus
3-10in (7.5-25cm)

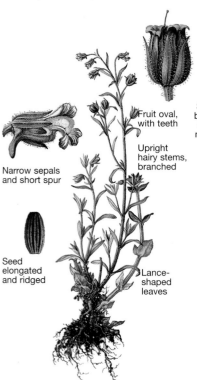

Fruit oval, with teeth

Upright hairy stems, branched

Narrow sepals and short spur

Seed elongated and ridged

Lance-shaped leaves

This small purple-flowered annual has spread widely since the growth of the railways in the 19th century, as it has adapted itself to growing in the ballast along railway tracks. Related to the snapdragon, its flower tube is partially closed by the 'palate' or fold of the lower lip, and has a short protruding spur at the lower end. The narrow leaves mostly grow on alternate sides up the stem.

Waste places and arable land throughout British Isles except the north.

Vervain

Verbena officinalis
12-24in (30-60cm)

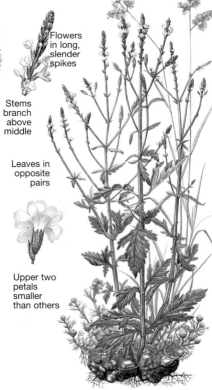

Flowers in long, slender spikes

Stems branch above middle

Leaves in opposite pairs

Upper two petals smaller than others

The tiny pink flowers of this common wayside plant are carried in spikes on tough, upright stems. They are pollinated by a variety of insects, including butterflies, small bees and hoverflies, and are also self-pollinating. The hairy leaves are often divided into lobes with curved teeth. Among the plant's close relatives are lemon verbena, whose scented leaves are used to make perfume and tea.

Roadsides and waste places, locally in England, Wales and southern Ireland.

| J | F | M | A | **M** | **J** | **J** | **A** | **S** | **O** | N | D |

| J | F | M | A | **M** | J | **J** | **A** | **S** | O | N | D |

Helleborines and other orchids add an exotic touch to Britain's woods and chalk downlands. Their colourful flower spikes consist of numerous individual flowers in bizarre shapes that vary according to species; the flower of one orchid even mimics the bee that pollinates it.

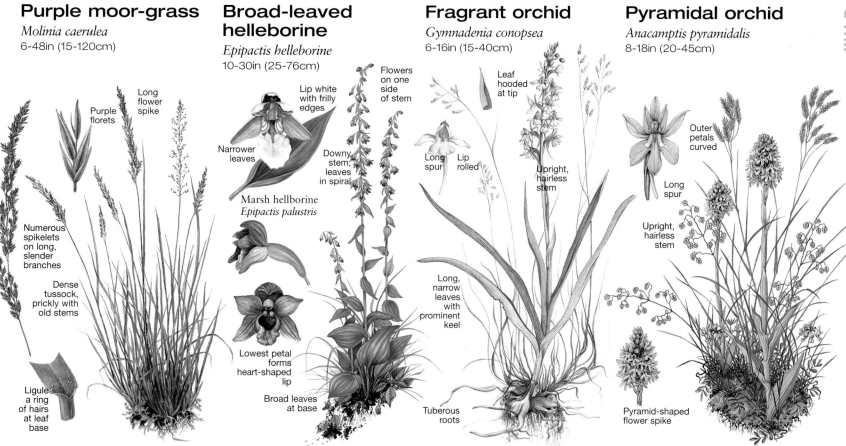

Purple moor-grass

Molinia caerulea
6-48in (15-120cm)

Long flower spike

Purple florets

Numerous spikelets on long, slender branches

Dense tussock, prickly with old stems

Ligule a ring of hairs at leaf base

Broad-leaved helleborine

Epipactis helleborine
10-30in (25-76cm)

Lip white with frilly edges

Narrower leaves

Marsh hellborine
Epipactis palustris

Flowers on one side of stem

Downy stem; leaves in spiral

Lowest petal forms heart-shaped lip

Broad leaves at base

Fragrant orchid

Gymnadenia conopsea
6-16in (15-40cm)

Leaf hooded at tip

Long spur

Lip rolled

Upright, hairless stem

Long, narrow leaves with prominent keel

Tuberous roots

Pyramidal orchid

Anacamptis pyramidalis
8-18in (20-45cm)

Outer petals curved

Long spur

Upright, hairless stem

Pyramid-shaped flower spike

Walkers on soggy moorland will often come across clumps of this tall grass, which has a remarkable botanical feature. As with other grasses, each leaf joins the stem at a node; the leaf is wrapped tightly round the stem, and held at the join by a watertight seal or 'ligule'. On nearly all grasses, this ligule consists of a flap of tissue, usually whitish in colour. But on the moor-grass it takes the form of a ring of tiny hairs at the base of each leaf blade.

Marshes and moors throughout British Isles.

This woodland orchid bears an amazing number of little flowers, sometimes as many as a hundred, down one side of its downy stem. The flowers are a subtle combination of greens and purples, with the unmistakable hanging lower lip of the orchid a deep purplish-red. The ribbed leaves are much larger at the base. The marsh helleborine is a smaller plant, and the lip of its flower is white with red veins and frilly edges. It grows in fenland.

Locally common in woodland in most areas.

A plant mainly of chalk and limestone grassland, this pinkish-flowered orchid can fluctuate enormously in numbers from year to year. It gets its name from its clove-like scent, which is irresistibly attractive to moths and other long-tongued insects. The individual pollen grains are bound into tiny clusters or 'pollinia' which are carried by the insect from one flower to the next. The flowers form long spikes.

Grassland, fens and marshes, scattered throughout British Isles.

A striking plant of the chalk downland, the pyramidal orchid gets its name from its triangular clusters of pink flowers. This shape distinguishes it from the fragrant orchid, with which it might otherwise be confused. Its foxy scent is also different from the pleasant, clove-like scent of fragrant orchid, and is highly attractive to moths and butterflies. The nectar-filled flower spurs are unusually long.

Chalk and limestone pasture in most areas of British Isles; coastal in Wales and Scotland.

J F M A M J **J A S O** N D   J F M A M J **J A S O** N D   J F M A M J **J J A S** O N D   J F M A M J **J J A S** O N D   113

Red flowers   *Five to seven petals • Irregular flowers*

Early purple orchid

Orchis mascula
6-24in (15-60cm)

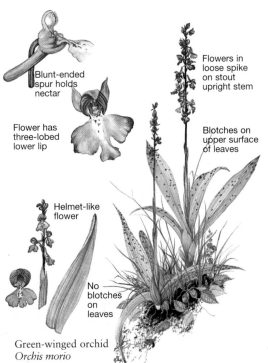

Blunt-ended spur holds nectar

Flower has three-lobed lower lip

Helmet-like flower

No blotches on leaves

Flowers in loose spike on stout upright stem

Blotches on upper surface of leaves

Green-winged orchid
Orchis morio

When Ophelia drowned in Shakespeare's *Hamlet*, among the flowers draped over her body were 'long purples' – the blooms of this orchid, which brighten woodland in early spring. The flower has a three-lobed lip, on which insects land, and a blunt-ended spur which holds nectar. Even before the flowers appear the plant can be recognised by the rosettes of lance-shaped leaves with round, purplish blotches on the upper surface.
The green-winged orchid is smaller, has no blotches on the leaves, and grows on grassland.

Woodland throughout British Isles.

Bee orchid

Ophrys apifera
6-24in (15-60cm)

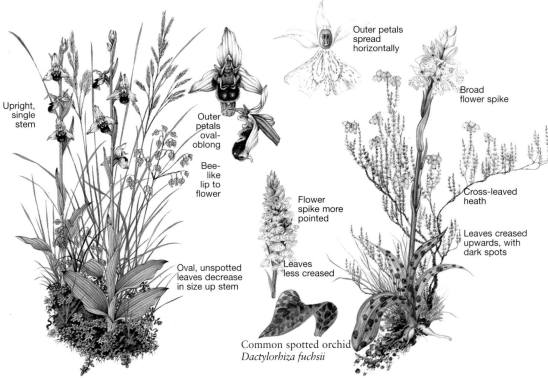

Upright, single stem

Outer petals oval-oblong

Bee-like lip to flower

Oval, unspotted leaves decrease in size up stem

This fascinating and beautiful plant lives up to its name: each flower looks as though it has a bee resting on its outer rim. The purpose of the deception is to lure a real bee to mate with the false one; while this activity is taking place, pollen attaches itself to the bee's head, to be transferred to the stigma of the next orchid it visits. In fact, the plants are usually self-pollinating. The leaves are oval and unspotted, and decrease in size up the single stem of the plant. Like the fragrant orchid (p. 113), bee orchid numbers vary enormously from year to year in any particular location.

Turf and dunes throughout Britain, except northern Scotland.

Heath spotted orchid

Dactylorhiza maculata ssp. *ericetorum*
6-18in (15-45cm)

Outer petals spread horizontally

Broad flower spike

Cross-leaved heath

Leaves creased upwards, with dark spots

Flower spike more pointed

Leaves less creased

Common spotted orchid
Dactylorhiza fuchsii

Unlike many other orchids, which flourish on chalky soils, the heath spotted orchid likes acid conditions and can be found growing on bogs on top of sphagnum moss. The leaves, marked with dark circular spots, vary from oblong to lance-shaped, folding upwards to a sharp tip. The large flowers, carried in a dense cluster, may be purple, pink or white, with small crimson blotches. The outer petals of the lip are larger than the middle one. The closely related common spotted orchid prefers alkaline soils, and is often found in damp meadows and woodland clearings.

Damp acid soils throughout Britain.

J F M A M J J A S O N D   J F M A M J J A S O N D

Red flowers *Eight or more petals*

Hottentot fig
Carpobrotus edulis
3-4in (7.5-10cm)

Five lobes around fleshy fruit

Curving, fleshy leaves

Trailing stems

The many-petalled magenta or sometimes yellow flowers of Hottentot fig hang in festoons over the cliffs in the south west, in summer. The plant was introduced to Britain from South Africa in about 1690 and has become fully naturalised. During the last century it has been widely planted with marram grass to stabilise sand-dunes. The trailing stems bear narrow, fleshy, upward-curving leaves, triangular in cross-section and reddish towards the tip. The plant gets its name from the fact that its fleshy fruits were eaten by the nomadic Hottentot tribesmen of Africa.

Near sea in West Country, Scilly, Channel Islands and other scattered localities.

Butterbur
Petasites hybridus
6-30in (15-76cm)

Long-stalked female flower-head produces plumed seeds

Male flower-head has shorter stalk

Dense flower-heads of many florets

Many bracts on stem

Leaves up to 36in (90cm) across, felted beneath

This creeping plant, with rhubarb-like leaves up to 36in (90cm) across, is often found in damp woods and by streams and ditches. The leaves, once used for wrapping butter, develop fully during the summer, after the flowers have died. Male and female flowers are carried on separate plants which are often visited by bees. The cylindrical seed is borne on the wind by a 'parachute' of whitish hairs, but the butterbur spreads to cover large areas by means of its underground roots. In the Middle Ages, these roots were used to remove skin blemishes.

Damp ground throughout British Isles except far north.

Lesser burdock
Arctium minus
24-48in (60-120cm)

Hooked bracts around flower-head

Furrowed stems

Oval, pointed leaves, cottony on underside

Larger flower-heads; longer stalks

Leaves arranged spirally up stem

Rounded, blunter leaves

Greater burdock
Arctium lappa

The hooked, thistle-like flower-heads or burs of this common roadside plant often have to be picked off clothing or extracted from dogs' coats after a country walk. The burs carry the seed far and wide, enabling lesser burdock quickly to colonise any patch of waste ground. The furrowed stems are reddish and woolly, with leaves arranged spirally up them. At one time the juice was used to soothe burns and sores, and the young stalks were eaten in salads. The flowers are often purple-tinted. Greater burdock is a larger and sturdier plant, with leaves that are blunter-tipped than those of lesser burdock.

Roadsides and waste ground throughout British Isles.

Red flowers *Eight or more petals*

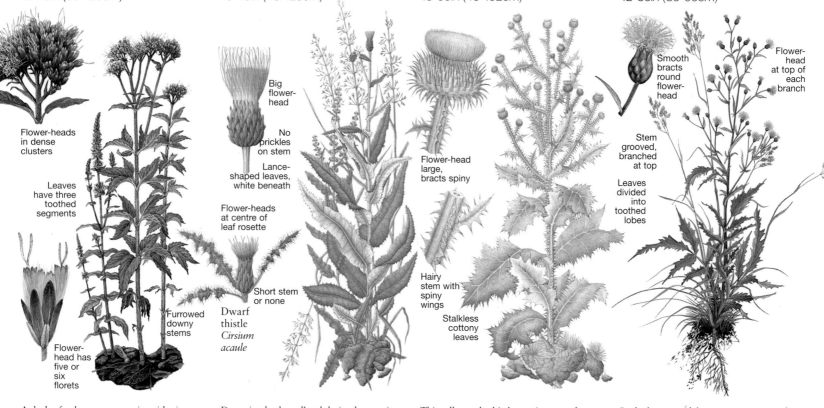

Hemp-agrimony
Eupatorium cannabinum
12-48in (30-120cm)

Flower-heads in dense clusters

Leaves have three toothed segments

Furrowed downy stems

Flower-head has five or six florets

Melancholy thistle
Cirsium heterophyllum
18-48in (45-120cm)

Big flower-head

No prickles on stem

Lance-shaped leaves, white beneath

Flower-heads at centre of leaf rosette

Short stem or none

Dwarf thistle *Cirsium acaule*

Cotton thistle
Onopordum acanthium
18-60in (45-152cm)

Flower-head large, bracts spiny

Hairy stem with spiny wings

Stalkless cottony leaves

Saw-wort
Serratula tinctoria
12-36in (30-90cm)

Smooth bracts round flower-head

Flower-head at top of each branch

Stem grooved, branched at top

Leaves divided into toothed lobes

A dash of colour on many riversides is provided by hemp-agrimony's clusters of reddish and white flowers on tall, hairy stems. The flowers are known in Dorset as 'raspberries and cream'; each flower-head consists of five or six florets, surrounded by purple-tipped bracts. In spite of its name, hemp-agrimony is related neither to hemp nor to agrimony. The downy leaves were once used to make a purge and an emetic.

Fens and damp woods throughout Britain and Ireland; rarer and mainly coastal in Scotland.

Drooping buds and undulating leaves give this thistle a somewhat downcast appearance. The large flower-heads, up to 1½in (40mm) across, usually grow singly, though more may be found in a cluster. Stemless thistle is the most unobtrusive of all thistles. Its low rosettes of leaves, with up to four flower-heads at the centre of each rosette, grow on close-cropped turf on dry, shallow, lime-rich soils.

Hill pastures in northern England and mid-Wales.

This tall, sturdy thistle gets its name from the white cottony hairs that cover its stem and strongly spined leaves. Though popularly known as the Scottish thistle, this is a misnomer, as it is a rare plant in Scotland and the thistle of Scottish heraldry is in fact the spear thistle (p. 117). The flower-head is large and usually solitary, and the florets are purple or sometimes white.

Roadsides and field edges in England and Wales; commonest in eastern England.

Both elements of the saw-wort's scientific name have medieval overtones: *Serratula* is Latin for 'little saw', after the saw-like edge of the leaves, which were once supposed to heal wounds; while *tinctoria*, meaning 'pertaining to dyers', shows that the plant was used in dyeing cloth. Saw-wort is often confused with knapweed, but it differs in having no feathery rim on the overlapping bracts round the flower-heads.

Woods and damp grassland, England and Wales; rare in Scotland and Ireland.

J F M A M J **J A S** O N D   J F M A M J **J A S** O N D   J F M A M J **J A S** O N D   J F M A M J J **A S** O N D

All thistles have prickly leaves and dense heads of purplish flowers, but they differ in their height and in the size and exact form of their flower-heads. Knapweeds have thistle-like heads but no prickles; burdocks have prickly heads but large, spineless leaves.

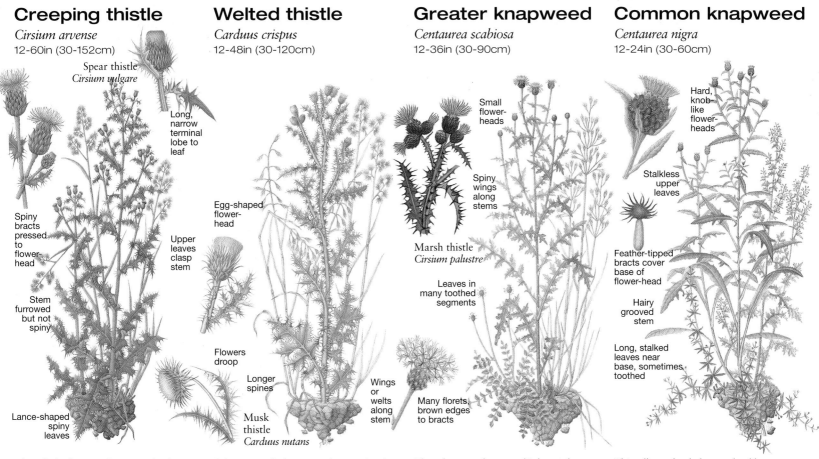

Creeping thistle

Cirsium arvense
12-60in (30-152cm)

Spear thistle
Cirsium vulgare

Long, narrow terminal lobe to leaf

Spiny bracts pressed to flower-head

Stem furrowed but not spiny

Upper leaves clasp stem

Lance-shaped spiny leaves

Welted thistle

Carduus crispus
12-48in (30-120cm)

Egg-shaped flower-head

Flowers droop

Longer spines

Musk thistle
Carduus nutans

Wings or welts along stem

Greater knapweed

Centaurea scabiosa
12-36in (30-90cm)

Small flower-heads

Spiny wings along stems

Marsh thistle
Cirsium palustre

Leaves in many toothed segments

Many florets, brown edges to bracts

Common knapweed

Centaurea nigra
12-24in (30-60cm)

Hard, knob-like flower-heads

Stalkless upper leaves

Feather-tipped bracts cover base of flower-head

Hairy grooved stem

Long, stalked leaves near base, sometimes toothed

Though the flowers of creeping thistle have a sweet musky odour that attracts butterflies, its root system makes it unpopular with farmers and gardeners. The plants are normally either male or female, so both sexes have to be growing close together for successful pollination, and the thistles usually grow in large clumps.
Spear thistle, the emblem of Scotland, has larger flower-heads and spear-like leaf tips.

Fields, roadsides and waste places.

Spine-covered wings, or welts, running the length of the stem make this species easier to identify than some thistles. The welts start from the spiny base of each leaf. The blades into which the leaves are separated are wide and soft. Butterflies, hoverflies and bees pollinate this tall prickly plant. The mush thistle is distinguished by its drooping, cup-shaped flowers, which have a musky odour. It normally grows on drier ground than the welted thistle.

Damp grassy areas, scarcer in north and west.

Though greater knapweed is larger than hardheads, or common knapweed, the easiest way to distinguish them is to compare the bracts enclosing the flower-heads. On hardheads the dark brown appendages at the top of the bracts are triangular or circular, whereas on greater knapweed they are horseshoe-shaped. Marsh thistle grow to 72in (180cm) high, but its flower-heads are less than 3/4in (20mm)across.

Roadsides and dry grassland, rarer in north.

This tall grassland plant, rather like a thistle with no prickles, gets its alternative name of hardheads from its knob-like flower-heads; it is also known as black knapweed. The tough stems, though lacking prickles, are nevertheless avoided by cattle, and this knapweed has become a grassland menace in some regions. The heads owe their hardness to the rows of overlapping blackish scales, which surround the reddish-purple florets.

Widespread on grassland, roadsides and cliffs.

| J | F | M | A | M | J | J | A | S | O | N | D |

| J | F | M | A | M | J | J | A | S | O | N | D |

| J | F | M | A | M | J | J | A | S | O | N | D |

| J | F | M | A | M | J | J | A | S | O | N | D |

Green, brown or colourless flowers

Up to four petals
Regular flowers

Opposite-leaved golden saxifrage

Chrysosplenium oppositifolium
2-6in (5-15cm)

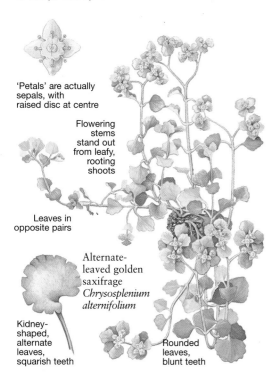

'Petals' are actually sepals, with raised disc at centre

Flowering stems stand out from leafy, rooting shoots

Leaves in opposite pairs

Alternate-leaved golden saxifrage
Chrysosplenium alternifolium

Kidney-shaped, alternate leaves, squarish teeth

Rounded leaves, blunt teeth

A pale green carpet speckled with gold shows that opposite-leaved golden saxifrage has taken up residence along the banks of a bubbling stream or rushing mountain rill. In the dampest surroundings, the plant's leafy shoots take root at intervals to create large patches; in drier areas, the plant is restricted to small but dense tufts. The leaves are rounded, blunt-toothed and borne in opposite pairs.
Alternate-leaved golden saxifrage differs from its relative in the spacing of its kidney-shaped leaves.

Wet and shady places throughout British Isles.

Perennial glasswort

Sarcocornia perennis
To 12in (30cm)

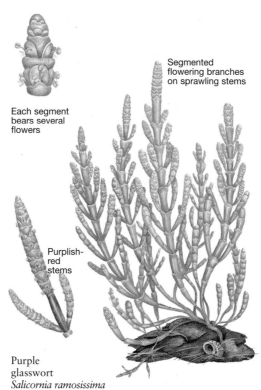

Segmented flowering branches on sprawling stems

Each segment bears several flowers

Purplish-red stems

Purple glasswort
Salicornia ramosissima

This glasswort clings tenaciously to its tussocks on pebble shores and salt-marshes. The sprawling stems bear segmented flowering branches. Each segment is formed by a pair of fleshy leaves fused around a woody stem; it is dark green at first but later turns yellow. The segments carry several flowers, each with two stamens and two feathery styles.
Purple glasswort often has purplish-red stems. Each flower has only one stamen but three styles.

Seashores and salt-marshes in southern England and south-east Ireland.

Procumbent pearlwort

Sagina procumbens
To 8in (20cm)

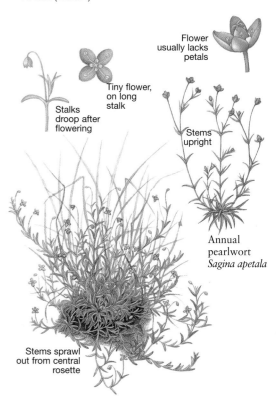

Flower usually lacks petals

Tiny flower, on long stalk

Stalks droop after flowering

Stems upright

Annual pearlwort
Sagina apetala

Stems sprawl out from central rosette

Look at a grassy bank, verge or footpath anywhere in Britain and procumbent pearlwort will probably be found, its many stems sprawling outwards from a dense rosette of leaves. They bear tiny flowers on long stalks. The flower has sepals and petals arranged in whorls, the petals being tiny and sometimes absent altogether. The slender flower-stalks droop after flowering, becoming upright again as the fruit ripens.
Annual pearlwort forms upright stalks and has no sprawling stems. It is often found rooted in gravel paths.

Grassy places throughout British Isles.

J F M A M J J A S O N D

J F M A M J J A S O N D

J F M A M J J A S O N D

Lacking the bright colours to attract insect pollinators, plants such as nettles and plantains rely mainly on the wind to carry pollen from one flower to another. The unusual mistletoe and butcher's-broom atone for their lack of bright petals by their showy autumn berries.

Mistletoe

Viscum album
To 36in (90cm)

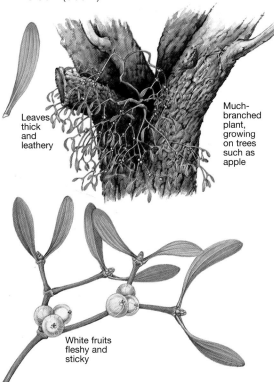

Leaves thick and leathery

Much-branched plant, growing on trees such as apple

White fruits fleshy and sticky

Mistletoe has perhaps more fanciful attributes than any other plant – from its association with druid rituals to its use as decoration at Christmas. It is a semi-parasitic plant, living partly off its host tree and partly off food obtained by its own chlorophyll. Each of its stems ends in a flower-head and two side buds which then repeatedly fork into two. The semi-translucent white fruits contain sticky seeds which are spread by birds. Mistletoe is particularly common on apple, hawthorn, poplar, lime and maple.

Common only in southern England and parts of Midlands.

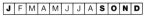

Common water starwort

Callitriche stagnalis
To 24in (60cm) in water, 6in (15cm) in mud

Water starwort's upper leaves, almost round, form a floating rosette in ponds and ditches or on wet mud at their edges. But below the surface the leaves are much longer and narrower. Plants growing in deeper water are anchored to the bottom by roots from the lower part of the stem; the higher roots trail in the water. Of some half-dozen species of water starwort, hard to tell apart, common water starwort is distinguished by its winged fruits and its more rounded leaves. The tiny green flowers are borne singly at the base of the leaves in the floating rosettes. They have no petals, and separate male and female flowers grow on the same plant; pollen is blown by the wind or carried along the surface of the water.

Common in ponds, ditches, slow streams and in mud throughout Britain and Ireland.

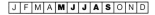

Upper leaves, almost round, form floating rosette

Tiny green flowers have no petals

Fruit grooved to form four oval segments

Submerged leaves narrower and less rounded

Higher roots trail in water

Lower roots anchor plant to bottom

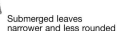

Green, brown or colourless flowers

Up to four petals
Regular flowers

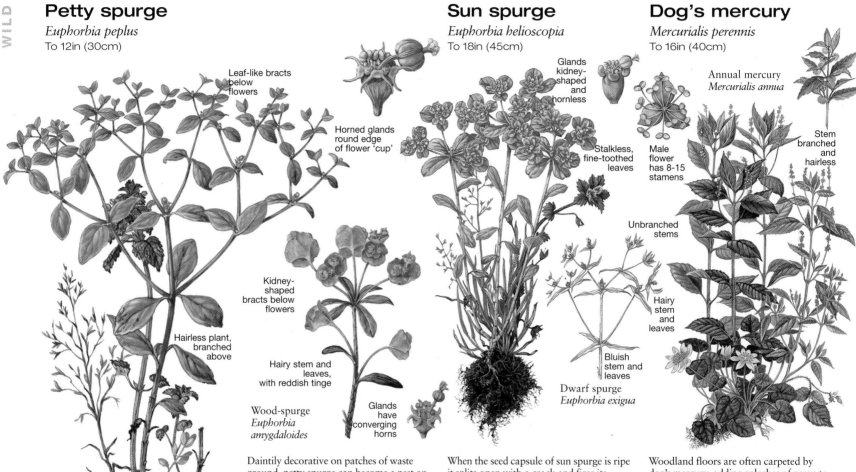

Petty spurge
Euphorbia peplus
To 12in (30cm)

Leaf-like bracts below flowers

Horned glands round edge of flower 'cup'

Kidney-shaped bracts below flowers

Hairless plant, branched above

Hairy stem and leaves, with reddish tinge

Wood-spurge
Euphorbia amygdaloides

Glands have converging horns

Sun spurge
Euphorbia helioscopia
To 18in (45cm)

Glands kidney-shaped and hornless

Stalkless, fine-toothed leaves

Bluish stem and leaves

Dwarf spurge
Euphorbia exigua

Dog's mercury
Mercurialis perennis
To 16in (40cm)

Annual mercury
Mercurialis annua

Stem branched and hairless

Male flower has 8-15 stamens

Unbranched stems

Hairy stem and leaves

Daintily decorative on patches of waste ground, petty spurge can become a pest on cultivated land. As in other spurges, the flowers have neither sepals nor petals. Instead the male and female flowers are contained in cup-like structures, surrounded by four or five glands, which are often horned.
Wood spurge is hairy and has kidney-shaped bracts joined around the stem. The flower glands have converging horns.

Widespread on waste ground and farmland.

When the seed capsule of sun spurge is ripe it splits open with a crack and fires its contents like a bursting hand grenade. The plant has a single stem, or is branched near the base. The bracts below the flowers resemble the oval, stalkless leaves, which are finely toothed. As in petty spurge, the stamens are flanked by a stalked ovary. Dwarf spurge is bluish, with narrow leaves and bracts. Its flower glands have long slender horns.

Widespread on cultivated ground.

Woodland floors are often carpeted by dog's mercury, adding splashes of green to the shadows. The plant is highly poisonous and gives off a strongly fetid smell, which attracts the midges which pollinate the female flowers. The flowers are carried on long spikes.
Annual mercury is a hairless plant with branched stems. It grows on waste ground, largely in the south.

Woods and shady mountain places; absent from Orkney and Shetland, rare in Ireland.

J F M A M J J A S O N D   J F M A M J J A S O N D   J F M A M J J A S O N D

Stinging nettle

Urtica dioica
To 60in (152cm)

Plant bristles with stinging hairs

On male plant, flower clusters stick out

On female, flower clusters hang

Male and female flowers in cluster

Small nettle
Urtica urens

Bog myrtle

Myrica gale
24-54in (60-137cm)

Leaves greyish-green and stalkless, toothed towards tip

Female catkins

Male catkins

Upright, reddish twigs

Shrubby plant with sucker shoots

Herb paris

Paris quadrifolia
6-16in (15-40cm)

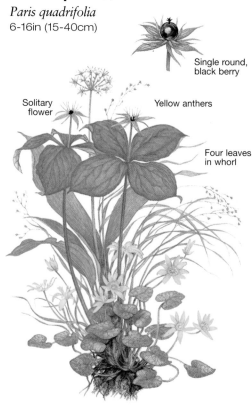

Single round, black berry

Solitary flower

Yellow anthers

Four leaves in whorl

Perhaps the plant which more than any other symbolises wasteland, the stinging nettle is one of Britain's most prolific growers. It is the curse of farmers and gardeners and the bane of bare knees. The nettle's creeping, rooting stems produce upright, leafy stems, and the whole plant bristles with stinging hairs. The stinging mechanism is simple: on being touched, the hair-tip breaks off and releases an acid which causes a painful rash.
The small nettle is less common, and seldom reaches more than about 12in (30cm). It has more deeply toothed leaves than its larger relative.

Waste ground, hedgerows and woods in all areas.

Large clumps of bog myrtle add fragrant perfume to wet heathland. The eucalyptus-like aroma is produced by the resinous substance exuded by hundreds of tiny yellow glands all over the plant. Before the introduction of hops into Britain, bog myrtle was added to beer during brewing to give it a tang. Bog myrtle's male and female catkins usually grow on separate plants. The twigs are reddish, and the stalkless, toothed leaves are greyish green. The older leaves are hairy underneath. Beware trying to approach a clump of bog myrtle too closely, for the ground in which it thrives is usually very damp.

Bogs, fens and wet heaths; commonest in north and west.

The clearing of large areas of Britain's woodlands has driven many woodland plants close to extinction. Herb paris is one that survives, but it is becoming increasingly rare. It is found only in isolated patches. It is a distinctive plant, with a single whorl of four leaves – the *quadrifolia* of its scientific name – and a single large pale green flower. The resemblance of its leaves to a love-knot explains the plant's common name, from the Latin *herba paris*, 'herb of a pair', and it is also sometimes called herb true-love. The flower has an unpleasant smell, and the dark purple berry into which it develops is poisonous.

Rare. Damp woods over most of Britain, but mainly in east.

J F M A M J J A S O N D

J F M A M J J A S O N D

J F M A M J J A S O N D

121

Green, brown or colourless flowers

Up to four petals
Regular flowers

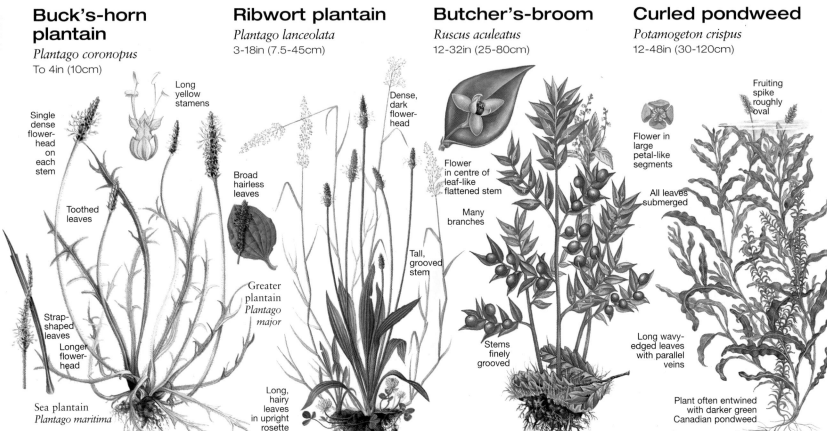

Buck's-horn plantain
Plantago coronopus
To 4in (10cm)

Single dense flower-head on each stem

Long yellow stamens

Toothed leaves

Strap-shaped leaves

Longer flower-head

Sea plantain
Plantago maritima

Ribwort plantain
Plantago lanceolata
3-18in (7.5-45cm)

Dense, dark flower-head

Broad hairless leaves

Greater plantain
Plantago major

Long, hairy leaves in upright rosette

Butcher's-broom
Ruscus aculeatus
12-32in (25-80cm)

Flower in centre of leaf-like flattened stem

Many branches

Tall, grooved stem

Stems finely grooved

Curled pondweed
Potamogeton crispus
12-48in (30-120cm)

Fruiting spike roughly oval

Flower in large petal-like segments

All leaves submerged

Long wavy-edged leaves with parallel veins

Plant often entwined with darker green Canadian pondweed

The numerous, many-toothed leaves of buck's-horn plantain at first grow flat against the ground in a star-like rosette, then turn up at the ends to form a bowl shape. The cylindrical hairy stems also curve upwards. Sea plantain grows nearer the salt water. Its leaves are fleshy and strap-shaped, with only one or two teeth, and have up to seven faint veins as compared with the one to three veins of buck's-horn plantain.

Sand, gravel and rocks, especially near sea.

The long, dark flower-head at the tip of a furrowed stem makes ribwort distinctive. The flowers have four white sepals, each with a brown central keel and four brownish petals. The long stamens are either white or pale yellow – unlike the purple stamens of hoary plantain (p. 89) which give the plant a reddish appearance. Greater plantain, common in garden paths, has broad, almost hairless leaves. Its flower-heads are long and pale green.

Pastures, lawns and roadsides.

This curious plant has no true leaves, but only leaf-like growths which are in reality flattened stems. These are dark green, thick and rigid and carry, in the angles of small papery bracts, the tiny greenish male and female flowers. On the female flowers, the petals are shorter than the sepals and there is a cup in place of the anthers of the male flower. If the female flowers are pollinated, large bright red berries develop.

Rocks and dry woods in southern England and Wales.

An adornment to still or slow-moving water, curled pondweed is easily recognised by its twin rows of delicate, shiny and translucent leaves. The much branched stems bear only submerged leaves, which are a favourite food and hiding place for fish and crustaceans. The short and rather loose flower spikes emerge above the water. The petal-like segments of each flower are larger than in broad-leaved pondweed.

Common in ponds, streams and canals.

J F M **A M J J A** S O N D

J F M **A M J J A** S O N D

J F M **A** M J J **A** S O N D

J F M **A** M J J **A** S O N D

Broad-leaved pondweed

Potamogeton natans
Usually 40in (100cm)

Flowers in small petal-like segments

Fruiting spike cylindrical

Broad, leathery leaves, submerged and floating

Leaf margins continue down stalk

A common plant of pond and river, broad-leaved pondweed seldom grows in water more than 3ft (1m) deep. Its sparsely branched stems bear submerged and floating leaves which are broad, coarse and leathery with many parallel veins. The flowers do not have separate petals and sepals but have four rounded segments around the stamens and styles. The pollen is carried by the wind – curiously, for a water plant, the pollen is quickly made sterile by contact with water. The fruiting spike rises above the surface and carries many green, flattened fruits.

Rivers and ponds throughout Britain and Ireland.

JFMAMJJASOND

Traveller's-joy

Clematis vitalba
To 100ft (30m)

Leaf has three to five leaflets and twists around other plants

Flower hairy on outside, with long stamens

Many plumed seeds in each head

Woody plant, climbing on dogwood

Long, feathery plume on seed

The thick, woody climbing stems of traveller's-joy adorn hedgerows and woodlands, hanging from the treetops like jungle lianas and made instantly recognisable by their white or greenish flowers which smell strongly of vanilla. The small, hairy flowers do not have true petals. The ripe seeds hang in clusters, each resplendent with a long hairy plume which gives the plant its other popular name, old man's beard. The leaves have between three and five leaflets and the leaf-stalks twist around the branches of other plants.

Wood margins and scrub in southern England and Wales.

JFMAMJJASOND

Lady's mantle

Alchemilla vulgaris
2-18in (5-45cm)

Sepals in two rings of four

Leaf has 7-11 toothed lobes

Leaves more deeply divided

Alpine lady's mantle
Alchemilla alpina

The greater the altitude, the more luxuriantly grows lady's mantle. It is often found alongside hilly roads, but its low-growing sprawling stems make it difficult to see among the surrounding vegetation. Each of the leaves which rise from the roots has between seven and eleven toothed lobes. The sepals of the flower are grouped in two rings of four; there are no petals.
Alpine lady's mantle is smaller, with greyer, star-shaped leaves, deeply divided to the base. The leaves are densely covered with silky silver hairs below. It grows in rocky areas mainly in the north.

Grassy roadsides, but rare in south-east.

JFMAMJJASOND

123

Green, brown or colourless flowers

Up to four petals
Regular flowers

Parsley piert

Aphanes arvensis
$3/4$-8in (2-20cm)

Scarlet pimpernel

Stipules form leaf-like cup

A sprawling, downy plant

Green sepals form flower

Leaves have three segments, each lobed at tip

Salad burnet

Sanguisorba minor ssp. *minor*
To 12in (30cm)

Female flowers above – purple-red stigmas

Hermaphrodite flowers

Male flowers below – yellow stamens, purplish filaments

Flowers in balls

Small, toothed leaflets

Spiked water-milfoil

Myriophyllum spicatum
Stems 20-100in (50-250cm) long

Upper flowers male – dull red petals

Leaves divided into fine segments, set in whorls

Lower flowers female – no petals

Only flowers above water

Whorls of leaves above and below water

Mare's-tail
Hippurus vulgaris

A lover of shallow, stony soil, parsley piert is a familiar sight in arable fields and bare wastelands. It has tiny green flowers with four sepals and no petals, and the stipules form a cup which encloses the flower. The short-stalked leaves have three segments, each lobed at the tip. The distinctive fruit is oval. Because it often grows between stones, parsley piert was once thought to be able to break rocks, and was used to treat gall-bladder and kidney stones.

Widespread on farmland and waste ground.

Though smaller than great burnet (p. 92) and having much less conspicuous flowers, salad burnet does have one claim to distinction. This is the cucumber-like smell that comes from its crushed leaves, which often pervades chalk grasslands on a summer's day. The leaves can in fact be eaten in salads, and in ancient days they were used to flavour wine. The flowers are in green balls with separate male, female and bisexual flowers.

Lime-rich soils; rare in Scotland and Ireland.

Only the spikes of tiny flowers of spiked water-milfoil can be seen above the surface of the still or slow-moving water where it makes its home. Under water, its many feathery leaves are divided into 15 to 35 fine segments; the plant's common and botanical names come from Latin and French words for 'thousand leaves'. The related mare's-tail has dark green leaves above the water as well as paler green, drooping leaves beneath.

Lakes, ditches and streams in lowland areas.

J F M **A M J J A S** O N D

J F M A M **J J** A S O N D

J F M A M **J J** A S O N D

Green, brown or colourless flowers

Up to four petals
Irregular flowers

Wood sage

Teucrium scorodonia
6-12in (15-30cm)

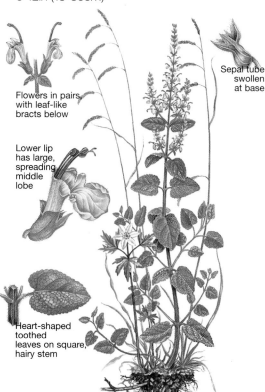

Flowers in pairs, with leaf-like bracts below

Sepal tube swollen at base

Lower lip has large, spreading middle lobe

Heart-shaped toothed leaves on square, hairy stem

Spikes of yellow-green flowers and wrinkled, toothed leaves make the wood sage easily recognisable. The flowers are borne towards the tip of the upright stem, and below them the leaves grow in opposite pairs. Each flower has a five-lobed lower lip with a large, spreading middle lobe; there is no upper lip, and the brown stamens are exposed. The fruit is formed by four smooth nutlets. This plant is not a true sage although, like the sages, it belongs to the mint family. The 19th-century gardener Gertrude Jekyll thought the wood sage worthy of a place in the garden. In taste and smell it resembles hops.

Dry soils throughout British Isles.

J F M A M J **J A S** O N D

Wild liquorice

Astragalus glycyphyllos
24-40in (60-100cm)

No tendril at tip of leaf

Zigzag stems

Many flowers on stalk

Straggling stems

Smooth, curving pods

Growing in tall grass, often on a woodland floor, this plant – also called milk-vetch – can easily be overlooked because its greenish-cream flowers blend with the colour of its foliage. It can also be mistaken for other sorts of vetch, but it is distinguished by having no tendrils at its leaf tips. Its straggling stems bend when they give rise to a leaf, and the result is a zigzag appearance. Many flowers grow on one stalk. The name milk-vetch comes from an old claim that goats eating it would yield more milk. It is not the plant used medicinally or for flavouring sweets.

Rough, grassy places and woodlands on chalk and limestone; not in Ireland.

J F M A M J **J A S** O N D

Moschatel

Adoxa moschatellina
2-4in (5-10cm)

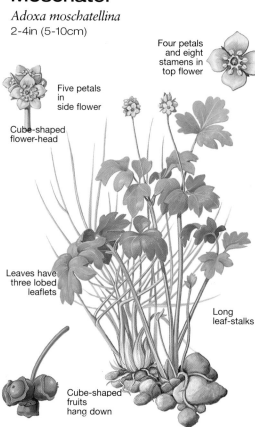

Four petals and eight stamens in top flower

Five petals in side flower

Cube-shaped flower-head

Leaves have three lobed leaflets

Long leaf-stalks

Cube-shaped fruits hang down

Some botanists have treated this little plant with disdain – the generic name *Adoxa* means 'without glory'. Yet there is much about moschatel that is fascinating – not least its other name, 'town-hall clock'. This refers to the neat arrangement of five flowers together – four facing outwards and one upwards; the fruits are arranged in a similar way. The plant owes the name moschatel to its musk-like scent, which is stronger at dusk and in dampness. This attracts insects which suck nectar from a ring at the base of the stamens. The plants usually spread by sending up shoots from their rootstock.

Widespread throughout Britain; not in Ireland.

J F M A **M** J J A S O N D

125

Green, brown or colourless flowers

Five to seven petals
Regular flowers

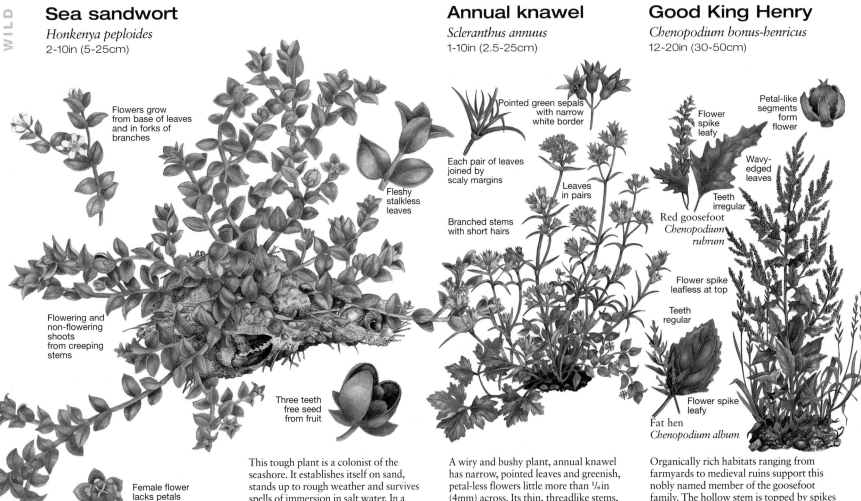

Sea sandwort

Honkenya peploides
2-10in (5-25cm)

Flowers grow from base of leaves and in forks of branches

Fleshy stalkless leaves

Flowering and non-flowering shoots from creeping stems

Three teeth free seed from fruit

Female flower lacks petals

Male flower has petals and sepals

Annual knawel

Scleranthus annuus
1-10in (2.5-25cm)

Pointed green sepals with narrow white border

Each pair of leaves joined by scaly margins

Leaves in pairs

Branched stems with short hairs

Good King Henry

Chenopodium bonus-henricus
12-20in (30-50cm)

Petal-like segments form flower

Flower spike leafy

Wavy-edged leaves

Teeth irregular

Red goosefoot
Chenopodium rubrum

Flower spike leafless at top

Teeth regular

Flower spike leafy

Fat hen
Chenopodium album

This tough plant is a colonist of the seashore. It establishes itself on sand, stands up to rough weather and survives spells of immersion in salt water. In a setting where plants tend to be dull and greyish, the sea sandwort spreads carpets of glossy green among the dunes. Once the plant has taken hold, sand builds up on its windward side and starts forming a dune. As the plant spreads it stabilises the sand under its dense carpet of stems and leaves.

All around coast of British Isles.

A wiry and bushy plant, annual knawel has narrow, pointed leaves and greenish, petal-less flowers little more than $\frac{1}{6}$in (4mm) across. Its thin, threadlike stems, bearing short hairs, give it the name knawel, which comes from the German and means 'a tangle of threads'. The flowers come in clusters or singly. The fruit is a dry nut, hidden by pointed sepals with a narrow white border.

Sandy or gravelly soils throughout British Isles except extreme north.

Organically rich habitats ranging from farmyards to medieval ruins support this nobly named member of the goosefoot family. The hollow stem is topped by spikes of greenish-yellow flowers. The broad, arrowhead-shaped leaves used to be boiled and eaten as a kind of spinach.
Fat hen, another goosefoot, is tastier but not as handsome; it has a leafy flower spike and broad, toothed lower leaves. Red goosefoot has irregularly toothed leaves.

Rich soil throughout British Isles.

J F M A **M** **J** **J** **A** S O N D

J F M A M **J** **J** **A** S O N D

J F M A **M** **J** **J** **A** S O N D

The tiny flowers of seashore plants such as sea sandwort and sea beet are hidden among fleshy leaves. Inland, ivy and hops are easily recognised despite their lack of showy flowers. Docks and sorrels differ in leaf shape, and rushes can be differentiated by the placing of their flower tufts.

Common orache

Atriplex patula
To 40in (100cm)

Female flower has green leaf-like bracts in place of petals

Stem branched, often reddish

Toothed, tapering leaves

Male flower has petals and sepals – both alike

Triangular leaf

Toothed bracts around fruit

Spear-leaved orache
Atriplex prostrata

The oraches and their close relatives the goosefoots are distinguished from one another only by their flower structure. Goosefoot flowers are usually bisexual, while those of the orache are male or female. Each goosefoot flower is surrounded, but not enclosed, by about five small scales; the female flower of the orache is enclosed by two leaf-like bracts. Some varieties in both groups have in the past been cooked and eaten.
Spear-leaved orache often grows alongside the common species. Leaf-like, toothed bracts surround its fruit.

Widespread around British coasts.

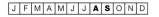

Sea beet

Beta vulgaris ssp. maritima
12-50in (30-127cm)

Flowers in clusters of three form long spike

Greenish segments contain stamens and styles

Leathery leaves

Reddish stems sprawl

This sprawling plant, with small flowers and large leathery leaves, may not have a very striking appearance but it is a tough seaside dweller – one of the few shore plants that survive in salt spray. Its leaves and flowering stems are usually tinged with red; the flowers – each with five greenish segments – grow in clusters of three. Sea beet belongs to the goosefoot family – which includes beetroot, sugar beet and spinach – and was cultivated as food in the Middle East as long as 2000 years ago. Its leaves have been eaten in Britain as 'sea spinach'.

Most shores around British Isles.

Sea-purslane

Atriplex portulacoides
To 32in (80cm)

Female flower has styles projecting from bracts

Male flower a whorl of sepal-like segments

Leaves elliptical

Stems splayed out at base

The hardy, attractive sea-purslane often draws attention to itself by a silvery sheen on its leaves which can be seen from some distance. This is caused by tiny, papery scales which protect the young leaf from the drying effect of salt water. The plant, which survives in the tough conditions of salt-marshes, has sprawling stems and elliptical, untoothed leaves. The male flower is easily distinguished from the female, which has styles projecting from enclosing bracts – creating the effect of a tiny, horned animal's head.

Salt-marshes and coastal areas of England, Wales and eastern Ireland.

Green, brown or colourless flowers

Five to seven petals
Regular flowers

Annual seablite

Suaeda maritima
3-12in (7.5-30cm)

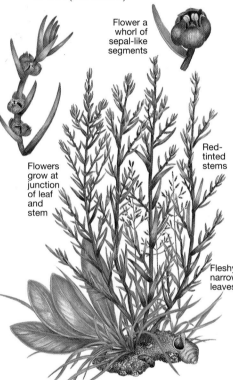

Flower a whorl of sepal-like segments

Flowers grow at junction of leaf and stem

Red-tinted stems

Fleshy narrow leaves

The red tinting of annual seablite is more suggestive of blood than of salt – which is what it thrives on. It is a true halophyte, or 'salt lover', which, when growing on the seashore, may be covered by the tide twice daily. Its leaves are stalkless and fleshy, flat on top and rounded at the back. The plant is a member of the goosefoot family, which includes several plants with edible leaves; 'blite' is an Old English word from the Latin for spinach.

Seashores and salt-marshes on most coasts.

Navelwort

Umbilicus rupestris
4-16in (10-40cm)

Flowers bell-shaped and drooping

Long straight stalks

Dimpled round leaves

This plant gets its name from the navel-like dimple in the centre of the leaf; its alternative name of pennywort refers to the leaf's coin-like shape. Above the leaves, drooping bell-like flowers grow abundantly on tall stalks. Navelwort is a familiar sight in Devon and Cornwall, growing between stones on high roadside banks. It is also found in rock crevices and on cliffs, and grows at altitudes of up to 1800ft (550m). Its size depends on the location.

Mostly in western Britain and Ireland.

Water purslane

Lythrum portula
1½-10in (4-25cm)

Capsule globe-shaped

Pink petals, brown sepals

Annual meadow-grass

Paired leaves, sometimes reddish

Reddish stems

One of nature's thrusting creepers, water purslane can easily be overlooked as it pursues its lowly journeys. Its stems, often reddish, send out roots to form whole strings of connected plants. Dense creeping mats are created, and these often spread into water. The tiny, stalkless flowers normally have six pink petals surrounded by brown sepals. Leaves are paired, and the flowers grow singly at the junction of leaf and stem.

Scattered, on non-chalky soil.

Ivy

Hedera helix
To 100ft (30m)

Wasp pollinates globe-shaped flower cluster

Leaves on non-flowering stems have three to five lobes

Stems with adhesive-covered roots

The clinging ivy is one of Britain's few climbers to reach any great size – to 100ft (30m) with support. Its stems have many fibrous, adhesive-covered roots which enable it to clamber vigorously over walls, rocks and trees. The familiar lobed leaves borne on the non-flowering stems are dark green and shiny; they may be purplish in winter. Leaves on the flowering stems are usually without lobes. In autumn, the flowers have plentiful nectar.

Common throughout British Isles.

J F M A M J **J A S** O N D   J F M A M J **J A S** O N D   J F M A M J **J A S** O N D   J F M A M J **J A** S **O N** D

Marsh pennywort

Hydrocotyle vulgaris
½-10in (1.3-25cm)

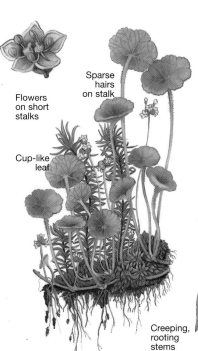

Flowers on short stalks

Sparse hairs on stalk

Cup-like leaf

Creeping, rooting stems

The leaf of the marsh pennywort is cup-like – *hydrocotyle* in its Latin name means 'water cup'. Its pinkish-green flowers are tiny and often hidden under the leaves. Marsh pennywort is a prostrate, creeping plant carpeting marshy and boggy areas. It sometimes grows on sphagnum moss. Once it was blamed for killing sheep – perhaps because it thrives in the same marshy places as the sheep disease liver fluke.

In damp places throughout British Isles.

J F M A M **J J A S** O N D

Water-pepper

Persicaria hydropiper
9-30in (23-76cm)

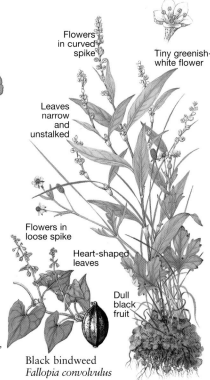

Flowers in curved spike

Tiny greenish-white flower

Leaves narrow and unstalked

Flowers in loose spike

Heart-shaped leaves

Dull black fruit

Black bindweed
Fallopia convolvulus

The narrow, unstalked leaves of water-pepper have a strong, acrid taste which gives the plant its name. This keeps the plant safe from grazing animals, and at one time it was put into beds to repel fleas. The tiny flowers are white, tinged green or pink, and grow in a slender, curved spike. Black bindweed is distinguished from common bindweed (p. 106) by its black fruits and its less conspicuous pinkish-green flowers.

Wet ground throughout British Isles.

J F M A M J **J A S** O N D

Sheep's sorrel

Rumex acetosella
To 12in (30cm)

Branched, almost leafless flower spikes

Male flower has yellow anthers

Female flower has feathery style

Spearhead-shaped leaf

The tiny flowers of sheep's sorrel are not much to look at and have no noticeable scent. They do not have to be enticing to insects: they are not pollinated by them, but by the wind. Male and female flowers are on separate plants; the males produce abundant pollen which is blown to the females. The leaf, shaped like a spearhead, has lobes at its base. The juice of the leaves was once used in the form of an extract to treat kidney and bladder ailments.

Heaths and grasslands in all areas.

J F M A M **J J A S** O N D

Sorrel

Rumex acetosa
To 40in (100cm)

Dense, branched flower spikes

Male flower has yellow anthers

Female flower has bent-back sepals

Veined petals enclose ovary

Lance-shaped leaf with two lobes at base

Once sorrel graced royal tables as a food; now it is more appreciated for bringing a splash of colour to the grassy places where it grows. In late summer its leaves and stems turn a lovely crimson, and other parts of the plant become tinged with red. Sorrel is sturdy and upright, with branched spikes of tiny flowers. The lance-shaped leaf has two backward-pointing lobes at the base. In Tudor times sorrel was England's most prized vegetable.

Common throughout British Isles.

J F M A M **J J** A S O N D

Green, brown or colourless flowers

Five to seven petals
Regular flowers

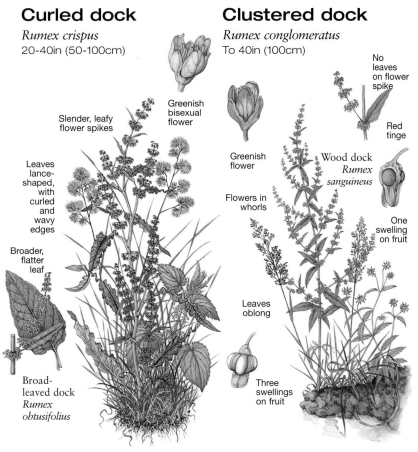

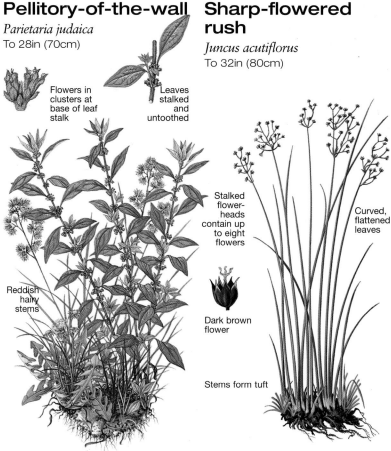

Curled dock

Rumex crispus
20-40in (50-100cm)

Greenish bisexual flower

Slender, leafy flower spikes

Leaves lance-shaped, with curled and wavy edges

Broader, flatter leaf

Broad-leaved dock
Rumex obtusifolius

Clustered dock

Rumex conglomeratus
To 40in (100cm)

No leaves on flower spike

Greenish flower

Wood dock
Rumex sanguineus

Red tinge

Flowers in whorls

One swelling on fruit

Leaves oblong

Three swellings on fruit

Pellitory-of-the-wall

Parietaria judaica
To 28in (70cm)

Flowers in clusters at base of leaf stalk

Leaves stalked and untoothed

Reddish hairy stems

Sharp-flowered rush

Juncus acutiflorus
To 32in (80cm)

Stalked flower-heads contain up to eight flowers

Curved, flattened leaves

Dark brown flower

Stems form tuft

In places where nettles thrive, nature often provides docks as well – for 'medicine'. Country walkers still use dock leaves to neutralise nettle stings. It is an upright plant with slender flower spikes; the leaves have curled, wavy edges and grow to a length of 12in (30cm).
The name of broad-leaved dock indicates its main feature. The leaf is also long-stalked, with wavy but uncurled edges. The plant grows to 36in (90cm).

Common throughout British Isles.

Also known as sharp dock, this species favours damp grassy places and woods. It has spreading branches and a stem that is inclined to zig-zag. Its leaves are narrow and roughly oblong; the flowers grow in whorls on a leafy spike. Clustered dock has fruit segments with three prominent swellings.
Wood dock, also called red-veined dock, is similar to clustered dock except in the red or purple tinge of its leaves and stems.

Throughout British Isles; rarer in Scotland.

This non-stinging relative of the nettle spreads a cheerful pale pink coating over walls and rocks, where it flourishes in cracks and crevices. Its leaves, stalked and untoothed, are similar in shape and arrangement to those of the nettle, but they lack the stinging hairs. Flowers grow in clusters of three or more at the base of each leaf-stalk, with the female flowers towards the centre.

Locally common in England, Wales and Ireland; rare in Scotland.

Rushes look like grasses or sedges but are related to the lily family. Like lilies, they have three petals, three sepals and six stamens to each flower; but the petals and sepals are like small papery scales and much less spectacular than the lily's. The sharp-flowered rush has chestnut-brown flowers; the fruit tapers to a sharp point. The rhizomes, or underground stems, are long. Like most rushes, the plant grows best in wet acid soil.

Wet moors and woods.

J F M A M J J A S O N D

J F M A M J J A S O N D

J F M A M J J A S O N D

J F M A M J J A S O N D

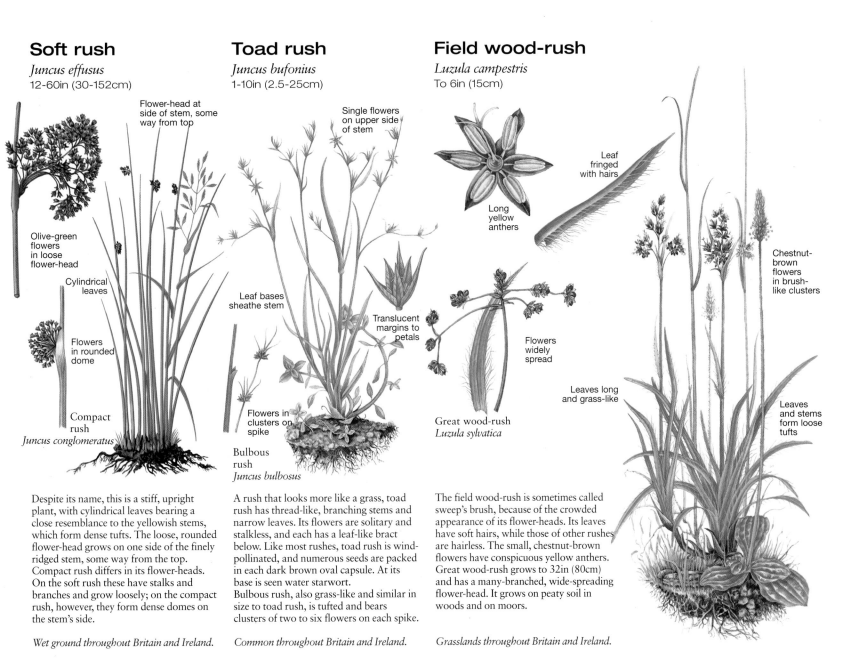

Soft rush

Juncus effusus
12-60in (30-152cm)

Flower-head at
side of stem, some
way from top

Olive-green
flowers
in loose
flower-head

Cylindrical
leaves

Flowers
in rounded
dome

Compact
rush
Juncus conglomeratus

Despite its name, this is a stiff, upright
plant, with cylindrical leaves bearing a
close resemblance to the yellowish stems,
which form dense tufts. The loose, rounded
flower-head grows on one side of the finely
ridged stem, some way from the top.
Compact rush differs in its flower-heads.
On the soft rush these have stalks and
branches and grow loosely; on the compact
rush, however, they form dense domes on
the stem's side.

Wet ground throughout Britain and Ireland.

Toad rush

Juncus bufonius
1-10in (2.5-25cm)

Single flowers
on upper side
of stem

Leaf bases
sheathe stem

Translucent
margins to
petals

Flowers in
clusters on
spike

Bulbous
rush
Juncus bulbosus

A rush that looks more like a grass, toad
rush has thread-like, branching stems and
narrow leaves. Its flowers are solitary and
stalkless, and each has a leaf-like bract
below. Like most rushes, toad rush is wind-
pollinated, and numerous seeds are packed
in each dark brown oval capsule. At its
base is seen water starwort.
Bulbous rush, also grass-like and similar in
size to toad rush, is tufted and bears
clusters of two to six flowers on each spike.

Common throughout Britain and Ireland.

Field wood-rush

Luzula campestris
To 6in (15cm)

Leaf
fringed
with hairs

Long
yellow
anthers

Chestnut-
brown
flowers
in brush-
like clusters

Flowers
widely
spread

Great wood-rush
Luzula sylvatica

Leaves long
and grass-like

Leaves
and stems
form loose
tufts

The field wood-rush is sometimes called
sweep's brush, because of the crowded
appearance of its flower-heads. Its leaves
have soft hairs, while those of other rushes
are hairless. The small, chestnut-brown
flowers have conspicuous yellow anthers.
Great wood-rush grows to 32in (80cm)
and has a many-branched, wide-spreading
flower-head. It grows on peaty soil in
woods and on moors.

Grasslands throughout Britain and Ireland.

JFMAM**JJAS**OND   JFMA**MJJAS**OND   JFM**AMJ**JASOND   131

Green, brown or colourless flowers

Five to seven petals
Regular flowers

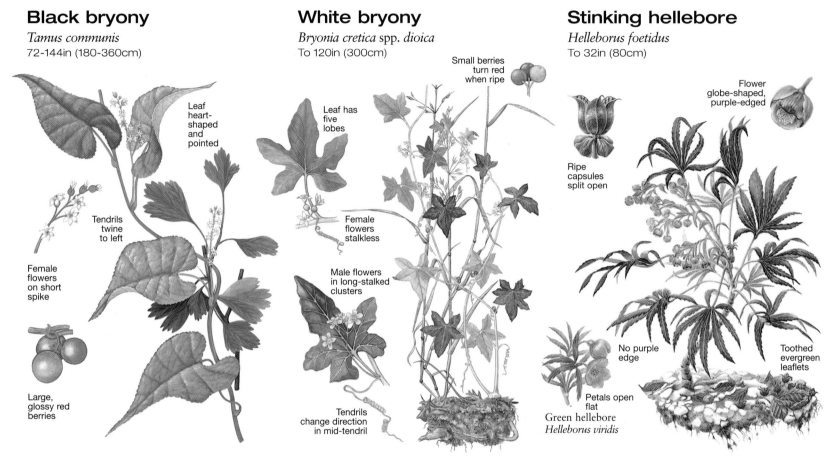

Black bryony

Tamus communis
72-144in (180-360cm)

Leaf heart-shaped and pointed

Tendrils twine to left

Female flowers on short spike

Large, glossy red berries

White bryony

Bryonia cretica spp. *dioica*
To 120in (300cm)

Small berries turn red when ripe

Leaf has five lobes

Female flowers stalkless

Male flowers in long-stalked clusters

Tendrils change direction in mid-tendril

Stinking hellebore

Helleborus foetidus
To 32in (80cm)

Flower globe-shaped, purple-edged

Ripe capsules split open

No purple edge

Petals open flat
Green hellebore
Helleborus viridis

Toothed evergreen leaflets

This climbing perennial is the only British member of the yam family – a group of tropical plants with edible tubers. Black bryony's tubers, however, are poisonous unless they are boiled. The plant has tiny greenish flowers and red poisonous berries; it gets its name from the colour of its fleshy underground tubers. The stem is slender, unbranched and hairless; the glossy, heart-shaped leaves are on long stalks. Female and male flowers are on different plants, the female flowers on a short spike and the male flowers forming long, upright spikes. Pollination is by insects.

Common in England and Wales, north to Durham.

This climber is unrelated to black bryony, but shares with it the distinction of being an 'odd one out', being the only British representative of the tropical gourd family. Tendrils help it to climb in hedges and on other plants. The tendrils and the flower clusters shoot from the junction of the leaf and the main stem. Male and female flowers are borne on separate plants, the male flowers in long-stalked clusters and the female clusters almost stalkless. The white roots at one time were sold by fairground charlatans as the reputedly magical mandrake. All parts of white bryony are poisonous.

Hedgerows in England; very rare elsewhere.

The smell of this rare woodland plant is, as its name implies, unpleasantly fetid, becoming oppressive when the stem or leaves are crushed. However, it serves its purpose by attracting bees and other insects to the nectar. The plant has numerous bell-shaped flowers, tipped with purple, which grow in clusters. The dark, evergreen leaves are hand-shaped, with many toothed leaflets.
The even rarer green hellebore is a smaller plant, growing to 24in (60cm), with flatter flowers. All parts of both species are poisonous.

Rare. Scattered in woods in England and Wales.

J F M A M J J A S O N D

J F M A M J J A S O N D

Alexanders

Smyrnium olusatrum
18-60in (45-152cm)

Many-branched flower-head

Curved petals

Toothed, diamond-shaped leaflets

Grooved stem

Leaf-stalks sheathe stem

A stout, bushy plant with solid stems, alexanders often colonises waste ground near the sea, decorating the cliffs with its greenish-yellow flowers from spring onwards. Its unusual name is a reference to its origins in Macedonia, the country of Alexander the Great. All the plant is edible. The dark green leaves can be used as a herb, or to flavour a white sauce; the young stems can be cooked and eaten like asparagus; the flower-buds can be used in salads and even the roots can be cooked and eaten as a substitute for parsnip.

Widespread in hedgebanks and on waste ground near sea.

Rock samphire

Crithmum maritimum
6-12in (15-30cm)

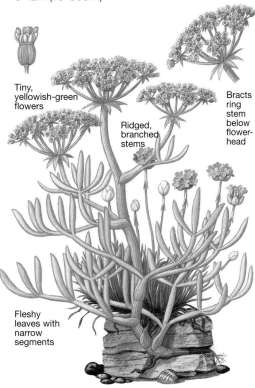

Tiny, yellowish-green flowers

Ridged, branched stems

Bracts ring stem below flower-head

Fleshy leaves with narrow segments

The squat, grey rock samphire needs its thick, ridged skin to protect it from the drying salt winds which blow round its sea-cliff home. In Shakespeare's time it was a popular vegetable, cooked in the same way as asparagus; the leaves were also pickled in vinegar. In many ways it resembles a desert plant, the main purpose of its design being the retention of moisture. It has thick, fleshy leaves with narrow, untoothed segments and a solid stem, enfolded by a membrane-like sheath. The umbrella-heads of flowers are pale greenish-yellow, with many bracts.

Locally common on sea-cliffs and rocks along coasts of Wales, Ireland, south and west England and south-west Scotland.

Hop

Humulus lupulus
To 20ft (6m)

Twisting stems

Scales paper when fruit ripe

Ripe fruiting heads used in brewing

Toothed leaves

Male flower

Female flower-heads in many-branched clusters

Since the Middle Ages when, in the absence of any other suitable beverage, beer was drunk even for breakfast, hops have been used to clarify, preserve and flavour it. The hop is a native of woods in southern England, but its use in brewing originated on the Continent, and it only began to be cultivated on a large scale in Britain from the end of the 16th century. Many of the hops now growing wild have escaped from cultivation. The woody stems of the hop wind themselves in a clockwise direction up any convenient tree, shrub or telegraph pole. The female flowers grow in a green cone that ripens to hops.

Widespread in hedges and thickets in England and Wales.

J F M **A M J J A** S O N D   J F M A M **J J A** S O N D   J F M A M **J J A** S O N D   133

WILD FLOWERS

Green, brown or colourless flowers

Five to seven petals
Irregular flowers

Common twayblade

Listera ovata
8-12in (20-30cm)

Bird's-nest orchid

Neottia nidus-avis
8-18in (20-45cm)

Lords and ladies

Arum maculatum
12-18in (30-45cm)

Branched bur-reed

Sparganium erectum
12-60in (30-152cm)

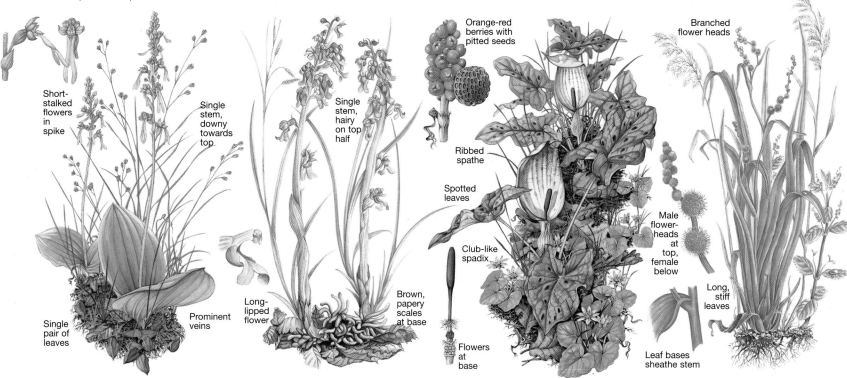

Short-stalked flowers in spike

Single stem, downy towards top

Single pair of leaves

Prominent veins

Single stem, hairy on top half

Long-lipped flower

Brown, papery scales at base

Orange-red berries with pitted seeds

Ribbed spathe

Spotted leaves

Club-like spadix

Flowers at base

Branched flower heads

Male flower-heads at top, female below

Long, stiff leaves

Leaf bases sheathe stem

Britain's most widespread orchid may only have two leaves, or 'tway blades', but they cannot pass unnoticed. They open like the jaws of a capacious mouth to release a single, slender, erect stem bearing a multitude of flowers – as many as 100 on a single plant. The flowers have a long, hanging lip, with a groove in the middle covered with nectar. Insects follow this nectar trail to the top, where they touch the modified stamens.

Common in damp woods and meadows.

The appearance of its densely matted roots inspired this plant's name. It lives in darkest woodland and depends for its survival on a fungus which surrounds its roots, absorbing nutrients from rotting leaves and passing them on to the plant. In return, the plant gives the fungus some of its nutrients. Bird's-nest orchid is honey-coloured, lacking any chlorophyll. Its flowers have a sickly fragrance.

Widespread but local in shaded woodland throughout Britain and Ireland.

The flowers of lords and ladies – also known as cuckoo pint – are hidden within a sheath-like hood called a spathe, which is ribbed like a seashell. A long purple finger called a spadix beckons insects towards the tiny flowers, which cluster at its base. The spadix is warm and gives off a slight smell attractive to insects. It is covered in backward-pointing hairs which trap the insect and ensure that it is well covered in pollen. The berries are highly poisonous.

Woods and hedgerows, except far north.

Each of the globe-like flower-heads of branched bur-reed contains numerous flowers of one sex only. Male and female flower-heads grow on each stem, the smaller male flowers clustered at the top and the larger female flower-heads below; and the wind does the pollinating. Once pollinated, the female flowers swell into bur-like heads of long-beaked fruit. The leaves are erect and spear-like.

Widespread in mud or shallow water in ponds, ditches, slow rivers and marshes.

134   J F M A M J J A S O N D   J F M A M J J A S O N D   J F M A M J J A S O N D   J F M A M J J A S O N D

With their slender leaves, hollow stems and much-simplified flowers, grasses appear superficially similar, but they differ in the shape and arrangement of the spikelets that hold the flowers. Sedges have solid, often three-sided stems, and no flap of tissue at the junction of leaf and sheath.

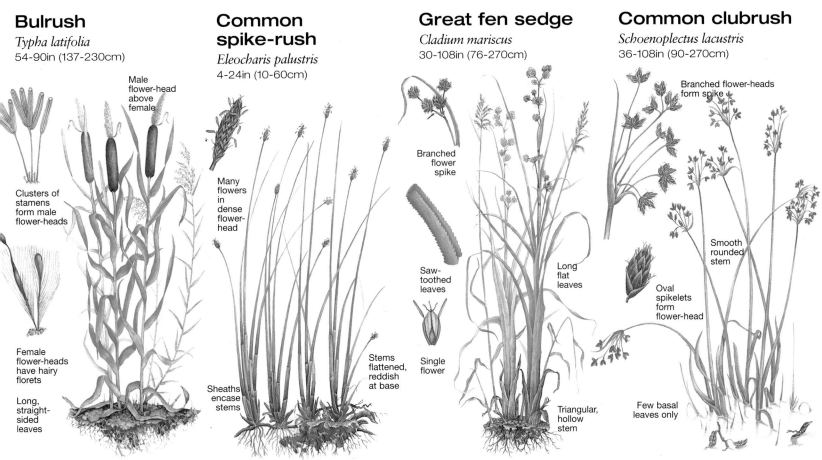

Bulrush

Typha latifolia
54-90in (137-230cm)

Male flower-head above female

Clusters of stamens form male flower-heads

Female flower-heads have hairy florets

Long, straight-sided leaves

Common spike-rush

Eleocharis palustris
4-24in (10-60cm)

Many flowers in dense flower-head

Sheaths encase stems

Stems flattened, reddish at base

Great fen sedge

Cladium mariscus
30-108in (76-270cm)

Branched flower spike

Saw-toothed leaves

Single flower

Long flat leaves

Triangular, hollow stem

Common clubrush

Schoenoplectus lacustris
36-108in (90-270cm)

Branched flower-heads form spike

Smooth rounded stem

Oval spikelets form flower-head

Few basal leaves only

Long known as reedmace, this plant has been officially re-named the bulrush – a name originally belonging to clubrush – a century after the Victorian artist Sir Lawrence Alma-Tadema depicted it in his painting 'Moses in the Bulrushes'. The male flower-head sits like a golden plume at the top of the stiff erect stem. Immediately below is the even more striking female flower-head, a long, brown sausage packed with tiny florets.

Widespread in swampy ground in lowlands.

The common spike-rush is an example of the way in which plants adapt general characteristics to specific needs. In this case, the plant is devoid of conventional leaves: they have been reduced to sheaths protecting the lower part of the stems from the damp ground in which they grow. They are yellowish-brown, compared with the bright green of the stems, and have squared-off ends. The flowers grow in small terminal spikelets.

Widespread in ditches and marshes.

The leaves of the great fen sedge need to be treated with caution because they can cut, having sharp saw-edges on both leaf margins and also along the leaf's keel. The leaves themselves grow up to 9ft (2.75m) long and can live for two or three years. Great fen sedge has hollow, round or triangular stems at the top of which are sprays of flowers in many-branched clusters. It has been used as a litter for farm animals and as a ridging on reed-thatched roofs.

Widespread but local in swamps and fens.

A distinctive feature of common clubrush – formerly known as bulrush – is the long, pointed, leaf-like bract which stands erect behind the dense mass of rust-coloured flower-heads on each stem. Unlike the bulrush, common clubrush stems usually have no leaves; the creeping stem or rhizome may however produce tufts of floating or submerged leaves. The flowers grow in oval spikelets, and instead of petals the flowers have short, barbed bristles.

Widespread but local in ponds and rivers.

J F M A M J J A S O N D   J F M A M J J A S O N D   J F M A M J J A S O N D   J F M A M J J A S O N D   135

Green, brown or colourless flowers

Five to seven petals
Irregular flowers

Common sedge

Carex nigra
3-30in (7.5-76cm)

Glaucous sedge

Carex flacca
4-18in (10-45cm)

Long-stalked yellow sedge

Carex viridula ssp. *brachyrrynchus*
8-24in (20-60cm)

Pendulous sedge

Carex pendula
36-54in (90-137cm)

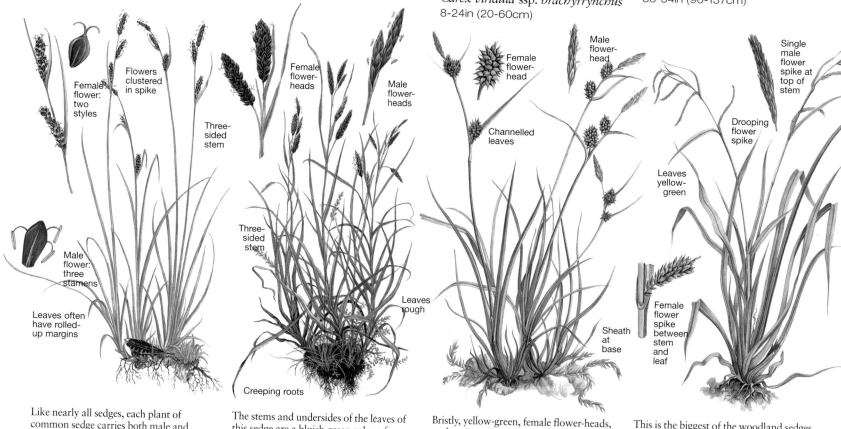

Female flower: two styles

Flowers clustered in spike

Three-sided stem

Male flower: three stamens

Leaves often have rolled-up margins

Female flower-heads

Male flower-heads

Three-sided stem

Leaves rough

Creeping roots

Female flower-head

Male flower-head

Channelled leaves

Sheath at base

Single male flower spike at top of stem

Drooping flower spike

Leaves yellow-green

Female flower spike between stem and leaf

Like nearly all sedges, each plant of common sedge carries both male and female flowers, normally bunched together on separate spikes. The flowers are reduced to a single scale called a glume which in the case of common sedge is blackish in colour, unlike most other sedges. Common sedge is a creeping plant. The leaf-like bracts are usually at least as long as the flower spike.

Common in wet, grassy areas, or beside water.

The stems and undersides of the leaves of this sedge are a bluish-green colour, for which the botanical name is glaucous. The male flower-heads grow on two or three spikes at the top of the flower stem, with two and occasionally three female spikes below. The leaves are rough and shorter than the stem, and have a slight keel. Towards the top they become flat, pale green above and grey-green underneath.

Lime-rich grassland and marshes.

Bristly, yellow-green, female flower-heads, as they ripen into fruit, are the most distinctive feature of this plant. They grow on the lower, usually stalkless spikes, and are held at the base between the flower stem and a long, thin, leaf-like bract. The beaked nutlets give the flower-heads their bristly appearance. The male flower-head grows on a long stalk at the very top of the flower stem.

Lime-rich wet places, especially fens.

This is the biggest of the woodland sedges and a handsome plant, particularly when seen drooping gracefully over the edge of a pond or slow-moving stream. The hanging flower spikes, when fully developed, look like reddish-brown catkins. They are supported on stout, triangular, sharp-edged stems. Long, rough-edged leaves hang like buckled swords.

River banks and damp woodlands, commonest in southern England.

J F M A M J J A S O N D   J F M A M J J A S O N D   J F M A M J J A S O N D   J F M A M J J A S O N D

Perennial rye grass

Lolium perenne
4-36in (10-90cm)

Wall barley

Hordeum murinum
2-24in (5-60cm)

Sheep's-fescue

Festuca ovina
2-24in (5-60cm)

Common reed

Phragmites australis
60-120in (152-300cm)

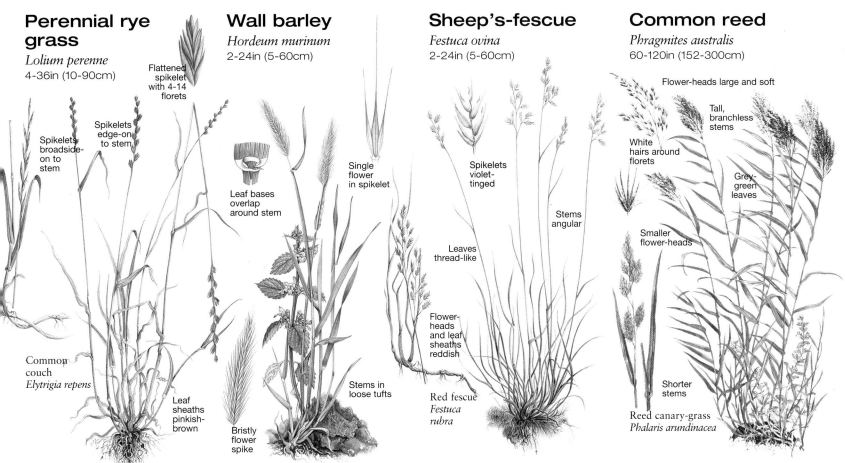

Perennial rye grass: Flattened spikelet with 4-14 florets · Spikelets edge-on to stem · Spikelets broadside-on to stem · Common couch *Elytrigia repens* · Leaf sheaths pinkish-brown

Wall barley: Leaf bases overlap around stem · Single flower in spikelet · Bristly flower spike

Sheep's-fescue: Spikelets violet-tinged · Leaves thread-like · Stems angular · Flower-heads and leaf sheaths reddish · Red fescue *Festuca rubra* · Stems in loose tufts

Common reed: Flower-heads large and soft · Tall, branchless stems · White hairs around florets · Grey-green leaves · Smaller flower-heads · Shorter stems · Reed canary-grass *Phalaris arundinacea*

This is a popular grass in the farming world because it produces an abundance of foliage for grazing and hay. The top third of the plant is taken up with the flower spike. The spikelets – groups of narrow, green scales enclosing the tiny flowers – are stalkless and grow alternately up the stem. The spikelets of common couch are flatter and set broadside-on. It is a creeping plant flowering from June to September.

Widespread on damp rich soils.

At first glance it looks as if wall barley, seen on a patch of waste ground, might have escaped from the cereal crop in a nearby cultivated field. It is a pale green plant, growing in loose tufts, with dense, prickly spikes which children put up their sleeves to feel them creep upwards. The lower part of the spike is often covered by a smooth leaf-sheath. The single-flowered spikelets are grouped in alternating clusters of three.

Waste ground and margins of cultivated land.

A grass which can grow abundantly on poor soils all the way from sea-level to mountain-side is popular with sheep farmers. Sheep's-fescue is a close-tufted, non-creeping grass, with short leaves which are often almost hair-like. Its flower-heads are branched, each stem being tipped with a violet-tinged spikelet.
Red fescue is slightly hairy, with flat upper leaves and creeping runners.

Widespread, especially on chalk and limestone.

The huge feathery heads of Britain's tallest grass look wild and dramatic against a stormy skyline. The large, dark purple flower-heads are many-flowered, the spikelets being covered in long, silky hairs. The long, rough-edged leaves sheathe the tough stems, which are used for thatching. Reed canary-grass is shorter and has smaller, hairless flower-heads.

Widespread at edge of rivers, lakes and brackish waters throughout British Isles.

J F M A **M** J **J** A S O N D   J F M A M **J** **J** A S O N D   J F M A **M** J **J** A S O N D   J F M A M J J **A** S **O** N D   **137**

Green, brown or colourless flowers

Five to seven petals
Irregular flowers

Annual meadow-grass

Poa annua
1½-12in (4-30cm)

Cock's-foot

Dactylis glomerata
6-54in (15-137cm)

Crested dog's-tail

Cynosurus cristatus
2-30in (5-76cm)

Quaking grass

Briza media
8-20in (20-50cm)

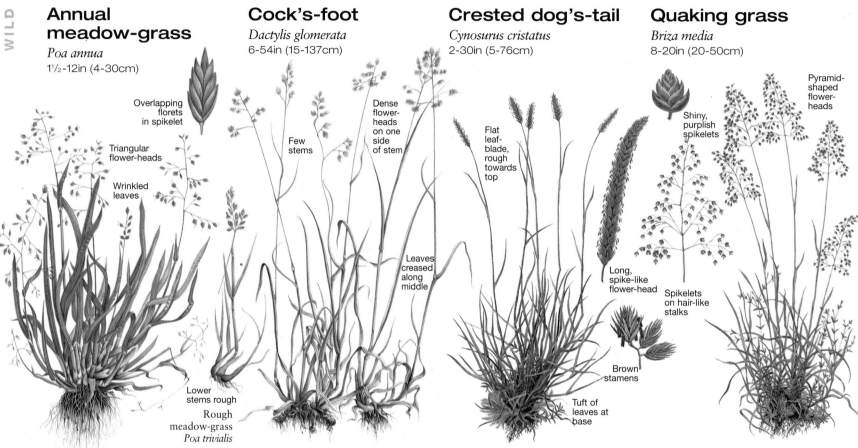

Overlapping florets in spikelet

Triangular flower-heads

Wrinkled leaves

Few stems

Dense flower-heads on one side of stem

Leaves creased along middle

Lower stems rough

Rough meadow-grass
Poa trivialis

Flat leaf-blade, rough towards top

Long, spike-like flower-head

Brown stamens

Tuft of leaves at base

Shiny, purplish spikelets

Pyramid-shaped flower-heads

Spikelets on hair-like stalks

Because it is an annual, dying every year, this is not on its own an ideal species of grass for lawns. However, it is often sown there in mixtures with other seeds, because its production of seed throughout the year enables it quickly to cover bare patches. This facility has made it one of the commonest grasses in the world.
Rough meadow-grass can be distinguished by the roughness of its stems to the touch. It flowers only in June and July.

Widespread on cultivated and waste ground.

The branching, one-sided flower-heads of this grass, which might fancifully be thought to resemble a cockerel's foot, make it one of the most distinctive of Britain's common grasses. Although tough and coarse, it is cultivated as a pasture and hay grass. The green or purplish spikelets of two to five flowers are in oval clusters, the lowest on a long stalk almost at right angles to the stem. The leaves have a keel along the middle.

Widespread on meadows and roadsides.

The narrow, dense flower-head of crested dog's-tail is made up of two different sorts of spikelet. One sort is sterile and gives the flower-head its spiky appearance. The other is fertile, bearing brown stamens which give the flower-head a brownish fringe. It is a short grass with upright, wiry stems and is leafy at the base, which makes it suitable only for grazing by sheep.

Widespread on meadowland and disused pasture.

The grass does not so much quake as tremble in the wind, the little spikelets emitting a distinctive rattling sound as their papery scales jostle together. The shiny, purplish spikelets are flat and oval and delicately hung on hair-like stalks. The flowers are often dried and used for floral decorations. Although often found in pasture, the grass is of little value as fodder because it has so little foliage.

Widespread on pasture and dry, grassy banks.

J F M A M J J A S O N D

J F M A M J J A S O N D

J F M A M J J A S O N D

J F M A M J J A S O N D

Barren brome

Anisantha sterilis
6-31in (15-80cm)

False oat-grass

Arrhenatherum elatius
20-60in (50-152cm)

Yorkshire fog

Holcus lanatus
8-24in (20-60cm)

Sweet vernal grass

Anthoxanthum odoratum
8-40in (20-100cm)

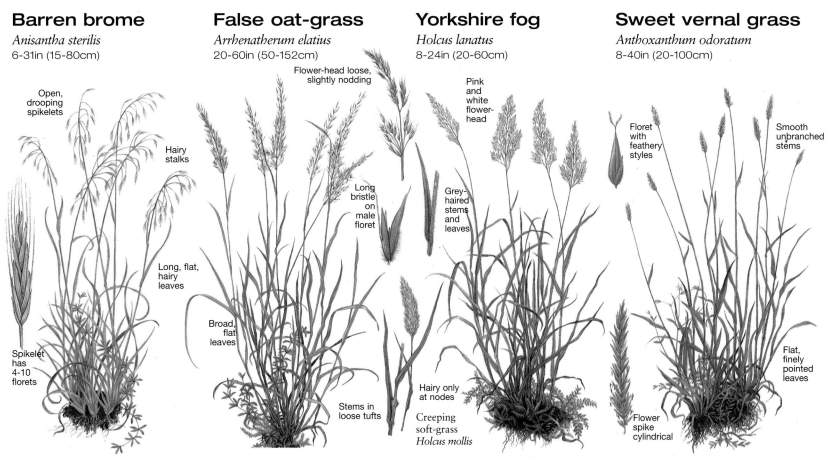

Open, drooping spikelets

Hairy stalks

Long, flat, hairy leaves

Spikelet has 4-10 florets

Flower-head loose, slightly nodding

Long bristle on male floret

Broad, flat leaves

Stems in loose tufts

Pink and white flower-head

Grey-haired stems and leaves

Hairy only at nodes

Creeping soft-grass
Holcus mollis

Floret with feathery styles

Smooth unbranched stems

Flat, finely pointed leaves

Flower spike cylindrical

Despite its name, barren brome produces fertile seed. But it has a thin, drooping, undernourished appearance. Its leaves are long, flat, hairy and often purplish, with sheaths that are hairy below. The flower-heads are very loose, open and drooping, with long branches bearing a single spikelet which contains between four and ten flowers.

Waste places and by roadsides throughout lowland areas of British Isles; rare in north and west.

The flower-bearing spikelets of this grass, which is one of the most abundant of all British wild grasses, are similar in structure to those of the true oat. But they stand erect on their stems rather than hanging down as those of oats do. The spikelets make up a lance-shaped green or purplish flower-head, with densely clustered branches. The leaves are wide, flat and rough, and the stems are smooth.

Roadsides, waste ground and rough grassland throughout British Isles.

A field filled with this grass looks more like a pink mist than a Yorkshire fog when the flowers are first out. As they mature they turn purple, finally fading to greyish-white in autumn.
Creeping soft-grass is similar to Yorkshire fog, but smooth-stemmed except for long hairs at the nodes. It is a vigorous underground creeper, most often found in woods. It flowers from May to August.

Meadows, pastures and waste ground throughout British Isles.

This grass produces the fine country smell of new-mown hay. Ironically, however, it is no longer sown for fodder, for the chemical called coumarin that produces the smell has a bitter taste which makes the grass unpalatable. The flower-head is in the form of a spike with many short branches, which turn from green or purplish to yellow. The leaf sheaths are bearded at the top.

Heaths, moors, pastures and woods throughout British Isles.

| J | F | M | A | **M** | **J** | **J** | A | S | O | N | D |

| J | F | M | A | M | **J** | **J** | **A** | S | O | N | D |

| J | F | M | A | **M** | **J** | **J** | **A** | S | O | N | D |

| J | F | M | A | **M** | **J** | **J** | **A** | S | O | N | D |

139

Green, brown or colourless flowers

Five to seven petals
Irregular flowers

Marram
Ammophila arenaria
24-48in (60-120cm)

Spikelet has single floret

Spikelets in long, dense spike

Rolled leaf-blades

Overlapping sheaths

Creeping stems

Big hedgehog-like humps of marram are a common sight around the coast. The grass is coarse, and can prick and scratch. But marram also has very long roots which help to bind the dunes and so prevent wind erosion. The leaves are long and narrow and, in dry weather, curl into a thin tube, to reduce water loss. The upper surface of the leaves – which is on the inside when they curl – is ribbed and grooved, with hairs along the ribs, which also helps to conserve water. The name marram comes from two old Norse words: *marr* (sea) and *halmr* (reed).

Sand-dunes all round coasts of Britain and Ireland.

J F M A M **J J A** S O N D

Creeping bent
Agrostis stolonifera
3-16in (7.5-40cm)

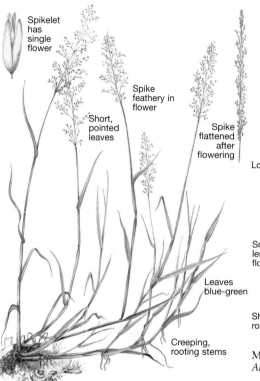

Spikelet has single flower

Short, pointed leaves

Spike feathery in flower

Spike flattened after flowering

Leaves blue-green

Creeping, rooting stems

Creeping bent spreads by leafy rooting stems, forming a close turf which is often seen in circular patches on lawns. The leaves are greyish or bluish-green, with smooth sheaths, their blades flat and finely pointed. An unusual feature of the flower-heads is the way they close up after flowering. The name 'bent' comes from an Old English word *beonet* which described various kinds of stiff, coarse grass.

Widespread, particularly in wet lowland areas.

J F M A M **J J A** S O N D

Timothy
Phleum pratense
20-40in (50-100cm)

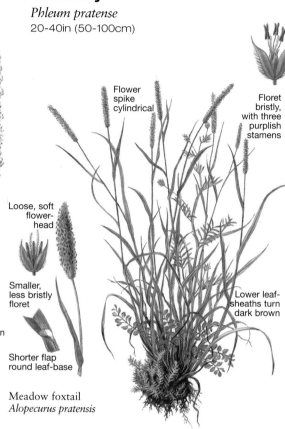

Flower spike cylindrical

Floret bristly, with three purplish stamens

Loose, soft flower-head

Smaller, less bristly floret

Shorter flap round leaf-base

Lower leaf-sheaths turn dark brown

Meadow foxtail
Alopecurus pratensis

An alternative name for timothy is meadow cat's-tail, after its long, dense, cylindrical flower spikes. It is a very hardy grass, widely grown for grazing and hay. The leaves are long, flat and hairless, and the flower-heads packed with single-flowered spikelets.
Meadow foxtail is a nutritious grass and among the first grasses to flower, from April to June. It provides valuable grazing and hay, and differs from timothy in that the outer scales of the spikelets are not tipped with a bristle. Meadow foxtail also has a looser, softer-looking flower-head.

Wild in grassland, on roadsides and waste ground.

J F M A M **J J A** S O N D

Green, brown or colourless flowers *Eight or more petals*

Mugwort
Artemisia vulgaris
24-48in (60-120cm)

Flower-heads bunched near ends of branches

Stem reddish and grooved

Hairy leaves clasp stem

The poet Edward Thomas compared the scent of mugwort to honeycomb. The mild, aromatic smell is one asset, but mugwort has other subtle but appealing features too. Chief amongst them is the way the reddish-to-purple stem, which is angled and grooved, matches the colour of the little flowers, shaped like bird's eggs, that crowd along the branches. The leaves are attractively divided and covered with short, white, velvety hairs.

Roadsides, waste places and hedgerows.

Pineapple weed
Matricaria discoidea
2-12in (5-30cm)

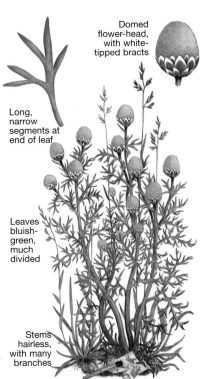

Domed flower-head, with white-tipped bracts

Long, narrow segments at end of leaf

Leaves bluish-green, much divided

Stems hairless, with many branches

Although very common, and undeniably a weed, this plant is always a pleasure to come across, with its clusters of greenish-yellow flower-heads looking like tiny pineapples. It is, in fact, as exotic as its appearance suggests, since it is probably a native of north-east Asia. To add to their appeal, the flower-heads have a strong pineapple or apple smell. They are surrounded by a ring of green bracts which have white, papery tips.

Roadsides and waste places.

Marsh cudweed
Gnaphalium uliginosum
1½-8in (4-20cm)

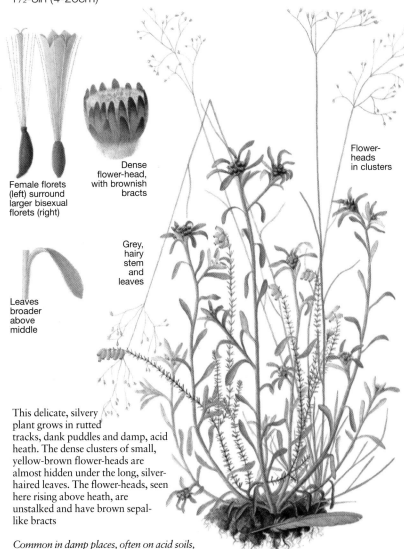

Female florets (left) surround larger bisexual florets (right)

Dense flower-head, with brownish bracts

Leaves broader above middle

Grey, hairy stem and leaves

Flower-heads in clusters

This delicate, silvery plant grows in rutted tracks, dank puddles and damp, acid heath. The dense clusters of small, yellow-brown flower-heads are almost hidden under the long, silver-haired leaves. The flower-heads, seen here rising above heath, are unstalked and have brown sepal-like bracts

Common in damp places, often on acid soils, throughout Britain and Ireland.

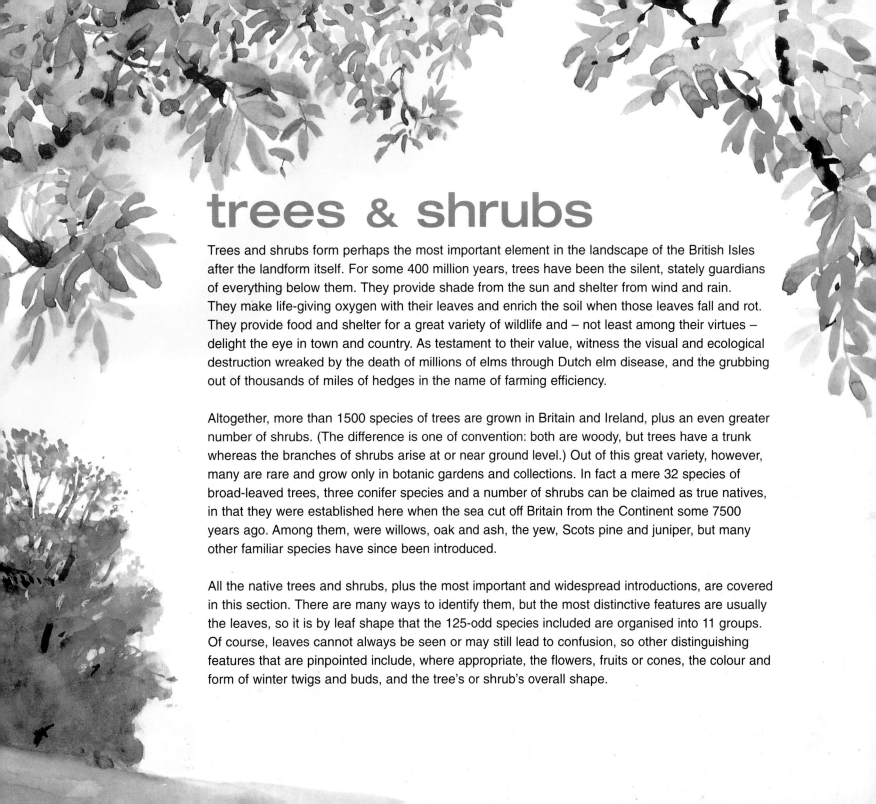

trees & shrubs

Trees and shrubs form perhaps the most important element in the landscape of the British Isles after the landform itself. For some 400 million years, trees have been the silent, stately guardians of everything below them. They provide shade from the sun and shelter from wind and rain. They make life-giving oxygen with their leaves and enrich the soil when those leaves fall and rot. They provide food and shelter for a great variety of wildlife and – not least among their virtues – delight the eye in town and country. As testament to their value, witness the visual and ecological destruction wreaked by the death of millions of elms through Dutch elm disease, and the grubbing out of thousands of miles of hedges in the name of farming efficiency.

Altogether, more than 1500 species of trees are grown in Britain and Ireland, plus an even greater number of shrubs. (The difference is one of convention: both are woody, but trees have a trunk whereas the branches of shrubs arise at or near ground level.) Out of this great variety, however, many are rare and grow only in botanic gardens and collections. In fact a mere 32 species of broad-leaved trees, three conifer species and a number of shrubs can be claimed as true natives, in that they were established here when the sea cut off Britain from the Continent some 7500 years ago. Among them, were willows, oak and ash, the yew, Scots pine and juniper, but many other familiar species have since been introduced.

All the native trees and shrubs, plus the most important and widespread introductions, are covered in this section. There are many ways to identify them, but the most distinctive features are usually the leaves, so it is by leaf shape that the 125-odd species included are organised into 11 groups. Of course, leaves cannot always be seen or may still lead to confusion, so other distinguishing features that are pinpointed include, where appropriate, the flowers, fruits or cones, the colour and form of winter twigs and buds, and the tree's or shrub's overall shape.

Triangular or round leaves

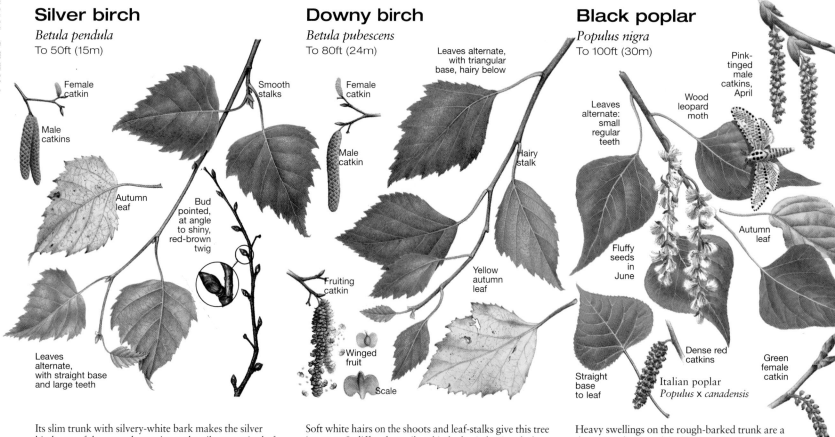

Silver birch
Betula pendula
To 50ft (15m)

Female catkin

Male catkins

Smooth stalks

Autumn leaf

Bud pointed, at angle to shiny, red-brown twig

Leaves alternate, with straight base and large teeth

Downy birch
Betula pubescens
To 80ft (24m)

Leaves alternate, with triangular base, hairy below

Female catkin

Male catkin

Hairy stalk

Yellow autumn leaf

Fruiting catkin

Winged fruit

Scale

Black poplar
Populus nigra
To 100ft (30m)

Pink-tinged male catkins, April

Wood leopard moth

Leaves alternate: small regular teeth

Fluffy seeds in June

Autumn leaf

Straight base to leaf

Italian poplar
Populus x canadensis

Dense red catkins

Green female catkin

Its slim trunk with silvery-white bark makes the silver birch one of the most decorative and easily recognised of Britain's native trees. It is one of the first trees to establish itself on newly cleared land or burned heaths and, with the rowan, it grows higher up mountains than any other deciduous tree. The long, whip-like twigs are shiny and red-brown, and the branches usually droop, hence the name *pendula*, 'hanging'. On mature trees the bark is covered with black, diamond-shaped patches. It is not a long-lived tree.

Light dry soils in all areas.

Branches droop in delicate tracery

Soft white hairs on the shoots and leaf-stalks give this tree its name. It differs from silver birch also in having darker brown twigs and reddish-brown bark, without black patches. It prefers wetter areas, and is a quick coloniser of poor soils. In winter, the male catkins, like those of silver birch, hang from leafless trees like grey lambs' tails. In April, as the leaves expand, the catkins shed golden pollen which the wind carries to female catkins – short, scaly green spikes on the same tree. After fertilisation they droop.

Damp uplands, especially in north and west.

Twisting branches, seldom drooping

Heavy swellings on the rough-barked trunk are a distinctive feature of the black poplar. The young twigs are yellowish-brown and ridged, and the buds chestnut-brown. Like most poplars, it has male and female flowers on separate trees.
The Italian poplar is one of the commonest of several fast-growing hybrids between black poplar and two American species, which are often planted as a screen or windbreak.

River valleys in south.

Branches arch down; bosses on trunk

The woodland birch and the suburban lilac have one feature in common – a triangular leaf. Trees having round leaves include the grey poplar and the aspen. Leaves of different species may have plain or toothed margins, and be set alternately up the stem or in opposite pairs.

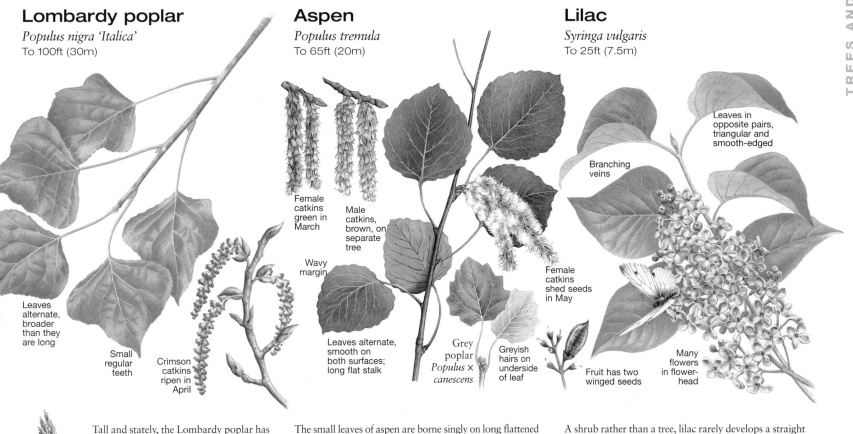

Lombardy poplar

Populus nigra 'Italica'
To 100ft (30m)

Leaves alternate, broader than they are long

Small regular teeth

Crimson catkins ripen in April

Aspen

Populus tremula
To 65ft (20m)

Female catkins green in March

Male catkins, brown, on separate tree

Wavy margin

Leaves alternate, smooth on both surfaces; long flat stalk

Grey poplar *Populus × canescens*

Greyish hairs on underside of leaf

Lilac

Syringa vulgaris
To 25ft (7.5m)

Leaves in opposite pairs, triangular and smooth-edged

Branching veins

Female catkins shed seeds in May

Fruit has two winged seeds

Many flowers in flower-head

Tall and stately, the Lombardy poplar has become a familiar sight since it was introduced to Britain from northern Italy in the mid-18th century. The branches all follow the trend of the main trunk, instead of spreading outwards, and give the tree a tall, column-like profile. Most trees are males, so there are few of the white woolly seeds to be swept up from lawns and pavements; the trees are propagated by cuttings. Like all poplars, it grows quickly and is often planted as a screen or windbreak. Its grey-brown bark has low ridges.

Planted as windbreak all over Britain.

Upright branches; trunk base often fluted

The small leaves of aspen are borne singly on long flattened stalks and flutter in the slightest breeze. They are smooth on both sides, the underside being paler, and turn amber-yellow in autumn. In the wild, aspens quickly colonise new ground and rarely grow singly; they send up suckers from their roots which often form dense thickets. The silvery-green bark becomes furrowed with age. The grey poplar is more tolerant of shade than other poplars and grows in damper places, reaching 75ft (23m), with massive, spreading branches. The leaves have wide teeth and the bark is diamond patterned.

Widespread in all areas.

Leaves tremble in wind

A shrub rather than a tree, lilac rarely develops a straight central stem. Its growing shoots end in two hard green buds which in spring sprout with equal vigour, so that the shoots fork repeatedly. The light brown shaggy bark peels off in long strips. The lilac is a native of eastern Europe and was brought to England in 1621; since then it has become established in the wild in many parts of Britain. Its flowers open in May in sweet-scented, showy clusters and attract butterflies. The flowers are naturally mauve, but white, red and purple 'sports' are reproduced by grafting. By October the flowers have each produced a leathery seed-pod.

Thinly scattered in wild.

Many-stemmed shrub, or small suckering tree

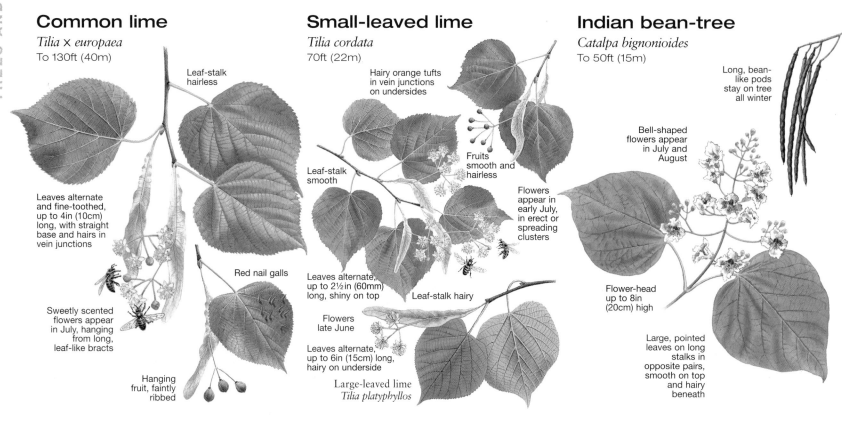

Heart-shaped leaves

Commonest of the trees with heart-shaped leaves are the tall stately limes, which vary in size of leaf and in the time at which they flower. Like the Indian bean-tree, they are planted as ornamentals.

Common lime

Tilia × europaea
To 130ft (40m)

Leaf-stalk hairless

Leaves alternate and fine-toothed, up to 4in (10cm) long, with straight base and hairs in vein junctions

Sweetly scented flowers appear in July, hanging from long, leaf-like bracts

Red nail galls

Hanging fruit, faintly ribbed

Small-leaved lime

Tilia cordata
70ft (22m)

Hairy orange tufts in vein junctions on undersides

Leaf-stalk smooth

Fruits smooth and hairless

Flowers appear in early July, in erect or spreading clusters

Leaves alternate, up to 2½in (60mm) long, shiny on top

Leaf-stalk hairy

Flowers late June

Leaves alternate, up to 6in (15cm) long, hairy on underside

Large-leaved lime
Tilia platyphyllos

Indian bean-tree

Catalpa bignonioides
To 50ft (15m)

Long, bean-like pods stay on tree all winter

Bell-shaped flowers appear in July and August

Flower-head up to 8in (20cm) high

Large, pointed leaves on long stalks in opposite pairs, smooth on top and hairy beneath

---

Full-bodied and stately as a galleon, the common lime can live for as long as 500 years. It is probably the tallest broad-leaved tree in Britain, familiar in streets and distinguished from other limes by the bushy side-shoots that start from near the ground. The bark is smooth and dull grey when young, becoming fissured later. The dull green leaves grow up to 4in (10cm) long and often bear the tiny red pimples of nail gall blight. In summer, greenfly on the leaves drop sticky honeydew.

Streets and parks everywhere.

Long, slender branches start near base

Apart from the smallness of its leaves, which only reach 2½in (60mm) long, this species of lime is also readily recognised from the way its little greenish-yellow flowers stand erect or at various angles on their bracts, rather than hanging down. Its smooth grey bark cracks into shallow plates. The fruits are neither ribbed nor hairy as with other limes. The leaves of large-leaved lime are up to 6in (15cm) long. It is the first lime to flower, attracting many bees.

Native of limestone soils in England and Wales.

Dense crown; branches arch downwards

This tree from the southern United States has many unmistakable features. Its seeds come in long, bean-like pods which turn from green to dark brown as they hang on the tree all winter; they are not edible. The leaves appear late in the season but grow up to 10in (25cm) long. The white flowers, flecked with yellow or purple, are carried on an upright head. The bark is rough and light brown.

Parks and gardens in warmer areas.

Wide, spreading crown; old trees often lean

Oval leaves

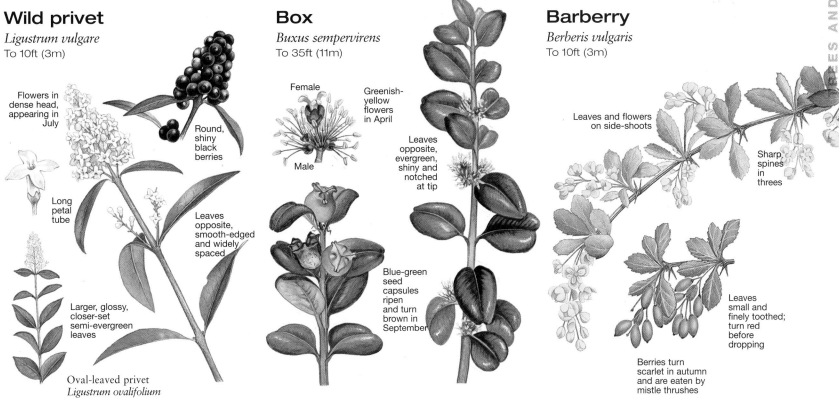

Wild privet
Ligustrum vulgare
To 10ft (3m)

Flowers in dense head, appearing in July

Round, shiny black berries

Long petal tube

Leaves opposite, smooth-edged and widely spaced

Larger, glossy, closer-set semi-evergreen leaves

Oval-leaved privet
Ligustrum ovalifolium

Box
Buxus sempervirens
To 35ft (11m)

Female

Greenish-yellow flowers in April

Male

Leaves opposite, evergreen, shiny and notched at tip

Blue-green seed capsules ripen and turn brown in September

Barberry
Berberis vulgaris
To 10ft (3m)

Leaves and flowers on side-shoots

Sharp spines in threes

Leaves small and finely toothed; turn red before dropping

Berries turn scarlet in autumn and are eaten by mistle thrushes

---

This bush if often found in hedgerows or on waste ground. Unlike its near relative oval-leaved or garden privet, it is not fully evergreen. Its leaves are narrower and more widely spaced; they are shiny on top and lighter beneath, sometimes tinged with red. The white flowers are set in little spikes near branch ends and have a heavy, sickly scent, attractive to insects.

Common privet has been largely supplanted as a hedging plant by oval-leaved privet, which came from Japan in the 1840s. This species retains at least some of its broader leaves in winter.

Hedgerows, woodlands and wasteland everywhere, but rarer in north.

A many-stemmed hedgerow shrub

Places where this elegant tree can be seen in the wild are now few, and often proclaimed in the name – Box Hill in Surrey, Boxwell in Gloucestershire and Boxley in Kent, for example. Left to its own devices it can reach a considerable height, but it is most commonly seen in a severely trimmed form, often as a dense hedge. The leaves are shiny and leathery, with a prominent mid-rib and notched tip. The flowers cluster at the base of the leaves.

The bark is light brown, turning grey with age, covering a dense wood which is a prized material for carving.

Very local in wild on chalk and limestone in south.

Small, dense, rounded tree

A popular and colourful garden shrub, barberry can also be found in the wild in woods and hedges. It produces many branches, which are grooved and covered with a yellow bark. The branches are marked at regular intervals by clusters of long thorns, growing in threes.

The brilliant yellow flowers appear in early summer, hanging in clusters from short side-shoots. Later come the scarlet oval berries which are rich in vitamin C but tart, and are avoided by many birds.

Wild in woods and hedges; also grown in gardens.

Many-stemmed shrub can form a hedge

147

Oval leaves

Blackthorn
Prunus spinosa
To 13ft (4m)

Thorny twigs

Black hairstreak butterfly lays its eggs on blackthorn

White flowers appear before leaves

Five-petalled flowers with red-tipped anthers

Fruit (sloe) round and blue-black with 'bloom' on it

Leaves alternate, hairy beneath

Holly
Ilex aquifolium
To 65ft (20m)

Cultivated varieties have different leaf colours and shapes

'Ferox'

'Lawsonia'

Red berries on female trees

'Auremarginata'

Caterpillar of holly blue butterfly

Flowers appear in May

Male

Female

Leaves alternate, dark and glossy on top, paler beneath

Strawberry tree
Arbutus unedo
To 30ft (9m)

Alternate leaves, evergreen and toothed

Main leaf vein white

Red warty fruits in clusters

Bell-shaped flowers appear in autumn

In spring the brilliant white flowers of blackthorn hide for a time the long, cruel thorns on which they grow. It is a dense and, because of its thorns, almost impenetrable, much-branched shrub which can grow to the height of a small tree and is often seen in hedgerows.

The bark is a rough, blackish-brown and the leaves small, oval and tooth-edged, set alternately on short, red-tinged stalks. The fruit is the blue-black, bitter sloe, often used to give a piquant flavour to gin. The tough, yellow wood is used to make walking-sticks and, in Ireland, the club-like shillelagh.

Common in hedgerows and thickets.

Thorny shrub that spreads to form thickets

Only the female holly tree bears the bright red berries which are so much a part of traditional Christmas decorations. As one of Britain's few native evergreen trees, the holly's leaves and berries were almost certainly used as a winter decoration in pre-Christian times. Being thick, with a waxy surface, the leaves resist water loss and last up to four years on the tree. Sharp spines on the lower leaves protect them from browsing animals, while the upper leaves are spineless. The bark is green when young, smooth and grey later.

Common everywhere, except on wet soils, throughout British Isles.

A narrow-crowned, conical tree

Its fruit is acid-tasting and no match for a real strawberry. However, the strawberry tree's cinnamon-red bark and rich red fruit make it a distinctive and attractive bush. The shiny leaves are set on pink, hairy stalks. The little bell-shaped flowers hang in clusters and are often tinged pink or green. They appear in the autumn, when the previous year's fruits are still ripening. The tree's patchy distribution – it grows wild in western Ireland, western France and the Mediterranean coast – reflects its pattern of spread after the Ice Age.

Wild in western Ireland; planted in parks and gardens.

Twisting branches; short trunk

Species with oval leaves range from such shrubs as barberry, blackthorn and dogwood to some of our most majestic trees, including beech and elm. Points of differentiation include the size and shape of the teeth on the leaves, the shape of the leaf tips and the arrangement of the veins.

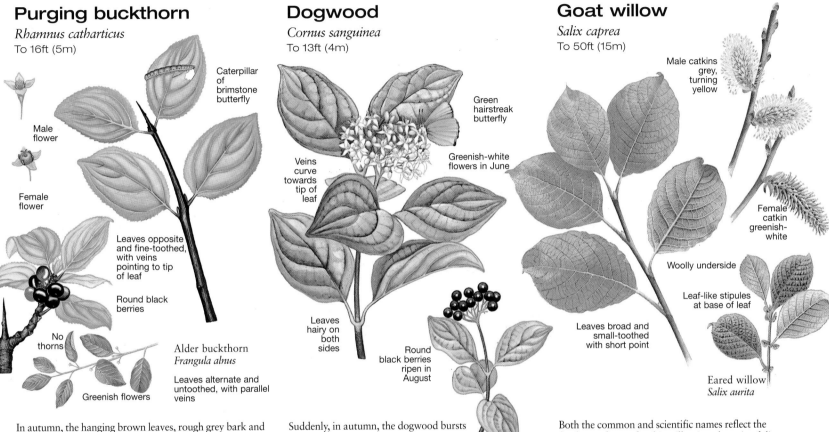

Purging buckthorn

Rhamnus catharticus
To 16ft (5m)

Male flower

Female flower

Caterpillar of brimstone butterfly

Leaves opposite and fine-toothed, with veins pointing to tip of leaf

Round black berries

No thorns

Greenish flowers

Alder buckthorn
Frangula alnus

Leaves alternate and untoothed, with parallel veins

Dogwood

Cornus sanguinea
To 13ft (4m)

Green hairstreak butterfly

Greenish-white flowers in June

Veins curve towards tip of leaf

Leaves hairy on both sides

Round black berries ripen in August

Goat willow

Salix caprea
To 50ft (15m)

Male catkins grey, turning yellow

Female catkin greenish-white

Woolly underside

Leaf-like stipules at base of leaf

Leaves broad and small-toothed with short point

Eared willow
Salix aurita

In autumn, the hanging brown leaves, rough grey bark and shiny black berries of purging buckthorn add striking colours to the woodland scene. The thorny shoots are of two kinds: long shoots, which extend growth, and short ones which bear bunches of leaves, flowers and fruit. The short ones, with the inconspicuous green flowers, look like a roebuck's antlers. The leaves are oval, finely toothed and set in pairs, as are the spurs and branches. Unlike purging buckthorn, alder buckthorn does not have thorns and its bark is smooth and dark; the leaves are alternate and untoothed. The slender branches form an erect, bushy shrub.

Common in woodland on lime-rich soils and in hedgerows, except in north.

Suddenly, in autumn, the dogwood bursts into colour with blood-red shoots and crimson leaves. After a year spent in dull obscurity, it is a spectacular show. The leaves are deep green in summer, pointed and hairy on both sides, with strongly impressed veins. The tiny greenish or creamy-white flowers appear in June and July in large, umbrella-shaped heads. They are unpleasantly scented. The fruit, which grows in clusters, is round, black and bitter. The bark is ridged and gives off a fetid smell when bruised. But in autumn all this is forgotten in the blaze of colour.

Locally common on chalk and limestone in southern Britain and Ireland.

Both the common and scientific names reflect the fondness of goats for this willow's early spring foliage. But most people know it as 'pussy willow' after its soft, silky catkins. Young twigs are covered in light grey hairs, but by winter they are shiny and hairless. The oval, fluffy catkins open in March before the leaves break. Eared willow is one of the smallest of willows, growing to little more than 72in (180cm) high, on damp ground. Its 'ears' are pairs of tiny stipules which grow at the base of the leaves.

Widespread in damp woods and coppices.

Many-branched tree; foliage to ground

Tall suckering shrub may form dense thickets

Small, many-stemmed shrubby tree

149

Oval leaves

Bay willow
Salix pentandra
To 40ft (12m)

Female catkins green and hairless

Shiny, pointed leaves, finely toothed and hairless

Downy covering on fruits

Male catkins, yellow and hairy, open in May

The bay willow's leaves bear a remarkable resemblance to those of the sweet bay tree, and both leaves and flowers are aromatic. The smooth, reddish-brown branches shine as if varnished. When the seed is being dispersed the female bay willow is covered in what looks, at a distance, like cotton wool; this is in fact the down which covers the fruits and helps the wind to disperse them.

Common especially in fens and beside streams in northern areas.

Small tree, with shiny brown branches

Creeping willow
Salix repens
To 36in (90cm)

Male catkins appear in April

Small leaves, with silvery hairs beneath

Female catkins

This is one of the most locally abundant willows in Britain, but hardly one of the most noticeable, since it normally grows to no more than 12in (30cm). It is all too easily overlooked, particularly when mixed with a mass of other vegetation. It is a creeping dwarf shrub which is sometimes erect, or almost erect, and sometimes prostrate. The small leaves are covered with fine silvery hairs when young, but as they mature the upper surfaces shed their hairs, becoming dull green. Flowering starts in April, the catkins appearing before the leaves.

Locally common in wet heaths and sand-dunes throughout British Isles.

A dwarf shrub

Common alder
Alnus glutinosa
To 70ft (22m)

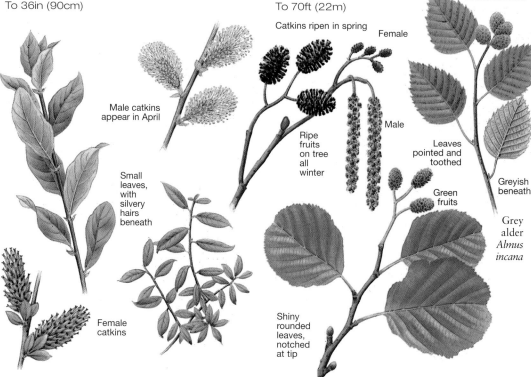

Catkins ripen in spring

Female

Fruits stalkless

Male

Ripe fruits on tree all winter

Leaves pointed and toothed

Greyish beneath

Green fruits

Grey alder *Alnus incana*

Shiny rounded leaves, notched at tip

The common alder is one of the few trees whose leaves stay green well into the autumn. But it is the catkins that give the tree its distinctive appearance. After the fruits of the female catkins have ripened in October they stay on the tree all winter as brown, woody cones, until the seeds are dispersed in spring.

The grey alder has pointed leaves, lighter in colour, and smooth, dark grey bark. Its suckers make it useful for binding river banks and new soil. It grows to 80ft (24m).

Widespread on wet ground.

Arched crown of regular, crooked branches

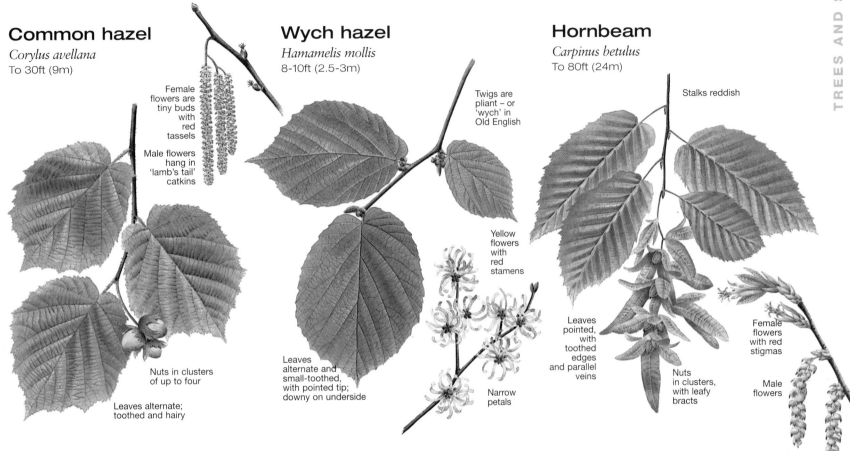

Common hazel

Corylus avellana
To 30ft (9m)

Female flowers are tiny buds with red tassels

Male flowers hang in 'lamb's tail' catkins

Nuts in clusters of up to four

Leaves alternate; toothed and hairy

Wych hazel

Hamamelis mollis
8-10ft (2.5-3m)

Twigs are pliant – or 'wych' in Old English

Yellow flowers with red stamens

Leaves alternate and small-toothed, with pointed tip; downy on underside

Narrow petals

Hornbeam

Carpinus betulus
To 80ft (24m)

Stalks reddish

Leaves pointed, with toothed edges and parallel veins

Nuts in clusters, with leafy bracts

Female flowers with red stigmas

Male flowers

Clusters of brown nuts, each surrounded by deeply toothed green bracts, identify the common hazel in autumn. The most distinctive features earlier in the year are the long, yellow, male catkins, which shed their pollen in February, before the leaves come out. The small female flowers are contained within the leaf-buds, and are tipped with red stigmas. Slender hazel rods obtained by coppicing – regular cutting back to ground level – are very pliant and have traditionally been used for making many things, from baskets to boats and houses.

Common in woods and hedgerows.

January parks and gardens, normally bereft of colour, spring to life with wych (or witch) hazel, which starts to bear clusters of bright yellow, sweet-scented flowers soon after Christmas. For this reason it has become a very popular garden shrub, and numerous colour varieties have been developed. It is a fairly small shrub whose distinctive flowers have narrow, strip-like petals. Wych hazel was introduced from China in 1879.

Common in gardens and shrubberies.

The smooth, light grey bark of hornbeam is similar to that of beech, but the trunk is fluted. Hornbeam also has a unique feature – hanging clusters of triangular, ribbed nutlets surrounded by long, three-lobed bracts. The leaf-buds are shorter and broader than those of the beech, and the leaves are double-toothed. Male flowers form in drooping catkins, female in leafy buds.

Locally common in the wild only in southern England.

Coppiced tree produces many stems

Small shrub with upward-pointed branches

Steeply rising branches form rounded crown

151

Oval leaves

Common beech
Fagus sylvatica
To 120ft (36m)

English elm
Ulmus procera
120ft (36m)

Wych elm
Ulmus glabra
To 100ft (30m)

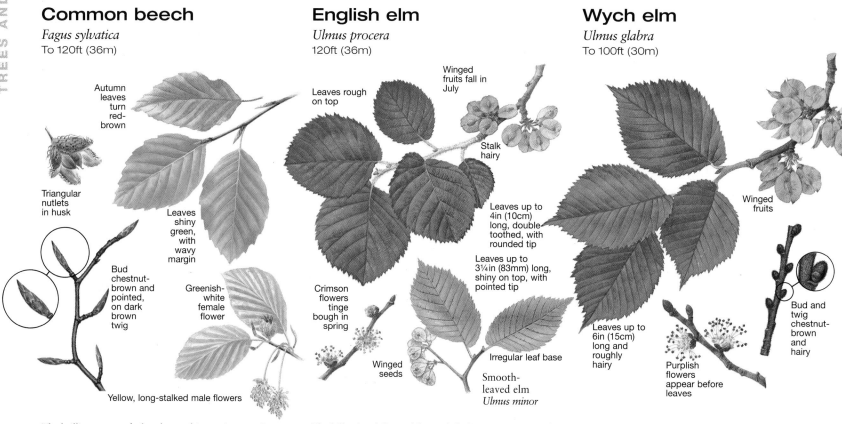

Autumn leaves turn red-brown

Triangular nutlets in husk

Bud chestnut-brown and pointed, on dark brown twig

Leaves shiny green, with wavy margin

Greenish-white female flower

Yellow, long-stalked male flowers

Leaves rough on top

Winged fruits fall in July

Stalk hairy

Leaves up to 4in (10cm) long, double-toothed, with rounded tip

Leaves up to 3¼in (83mm) long, shiny on top, with pointed tip

Crimson flowers tinge bough in spring

Winged seeds

Irregular leaf base

Smooth-leaved elm
Ulmus minor

Winged fruits

Bud and twig chestnut-brown and hairy

Leaves up to 6in (15cm) long and roughly hairy

Purplish flowers appear before leaves

The brilliant green of a beech wood just as it comes into full leaf is a memorable sight. Inside the wood all is shady even on the sunniest of days, because of the dense leaf canopy, and little else in the way of plant life manages to grow. But this is a small price to pay for the glory of the trees themselves, which present another impressive spectacle in autumn when the leaves turn orange then rich red-brown. The fruit of 'mast' contains pairs of nuts eaten by squirrels.

Widespread throughout British Isles.

The billowing foliage of the English elm was once one of the characteristic features of the English landscape. Since Dutch elm disease re-entered the country in 1967, more than 12 million English elms have been destroyed. The tragedy is all the greater in that the elm takes at least 150 years to reach its full glory. The English elm has a massive trunk and branches, and bark with long fissures.
The smooth-leaved elm, seen only in southern and eastern England, is narrower and grows only to 90ft (27m).

Formerly widespread in English lowlands, now rarer.

This is the most common elm found in Scotland, Wales and northern England, and has proved more resistant than most elms to Dutch elm disease. Unlike most elms wych elm reproduces itself by seed, rather than by suckers from the roots of the parent tree. Its canopy spreads more broadly than the English elm, although the tree is not so tall. Small flowers with long, purple-tipped stamens appear in conspicuous clusters in February and March.

Widespread, mainly in lowland areas; most abundant in west and north of British Isles.

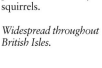

Older trees have massive, many-branched dome

Tall, narrow-crowned; with billowing foliage

Broad, fan-shaped crown

Crab apple

Malus sylvestris
To 30ft (9m)

Wild pear

Pyrus pyraster
To 50ft (15m)

Cherry plum

Prunus cerasifera
To 25ft (7.5m)

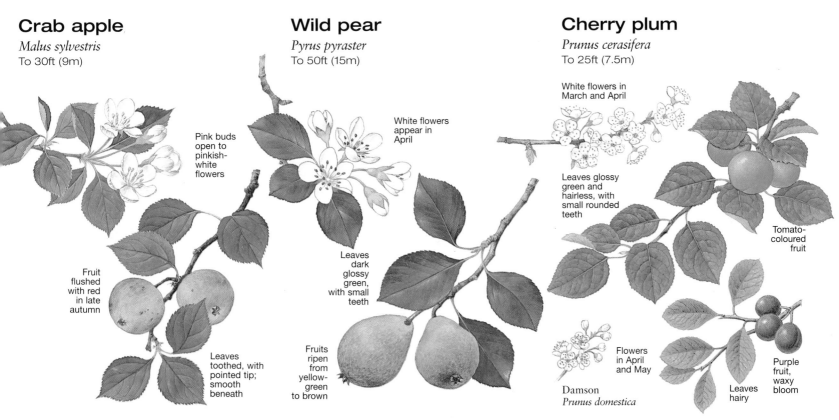

Pink buds open to pinkish-white flowers

White flowers appear in April

White flowers in March and April

Leaves glossy green and hairless, with small rounded teeth

Tomato-coloured fruit

Fruit flushed with red in late autumn

Leaves dark glossy green, with small teeth

Leaves toothed, with pointed tip; smooth beneath

Fruits ripen from yellow-green to brown

Flowers in April and May

Damson
Prunus domestica

Leaves hairy

Purple fruit, waxy bloom

It is hard to believe that the bitter little crab apple is the ancestor of the Cox's Orange Pippin, Bramley's Seedling and many other delicious apples, all the result of centuries of selective breeding and improvement. The crab apple still plays a vital part in raising cultivated apples, all of which are grafted onto its hardy rootstock. The fruit makes a pretty display on the tree, as do the flowers, which are paler pink than those of cultivated apples. The leaves are smooth rather than hairy beneath.

Hedges and thickets, except in northern Scotland.

Pears have been eaten by man for thousands of years. Unlike the crab apple, the fruit of the pear becomes sweet when ripe, but then quickly rots. The tree differs, too, in having a narrow outline, with sparse branches and usually one continuous, often leaning, main stem. It suckers freely, often forming a small thicket. The flowers have conspicuous red to purple anthers and grow in domed clusters on woolly stalks. The leaves have smaller teeth than those of the crab apple and a longer stalk.

In woods and hedges and along roadsides in England and Wales.

The spreading, open-crowned cherry plum looks at a distance like the blackthorn, but it blooms earlier and its foliage is not so thick. Its branches are normally smooth, although occasionally they have thorns. In March and April the cherry plum is smothered in small white flowers as the leaves unfold. The flower buds are on long shoots on either side of a leaf-bud. The fruit is small and round, with a good flavour. The damson, developed from crosses between the blackthorn and cherry plum, spreads over much of Britain.

Local in hedges and thickets in south and east.

A shrub-like tree with tangled, spiny branches

A narrow tree with sparse branches, often leaning

Spreading, open-crowned tree

153

Oval leaves

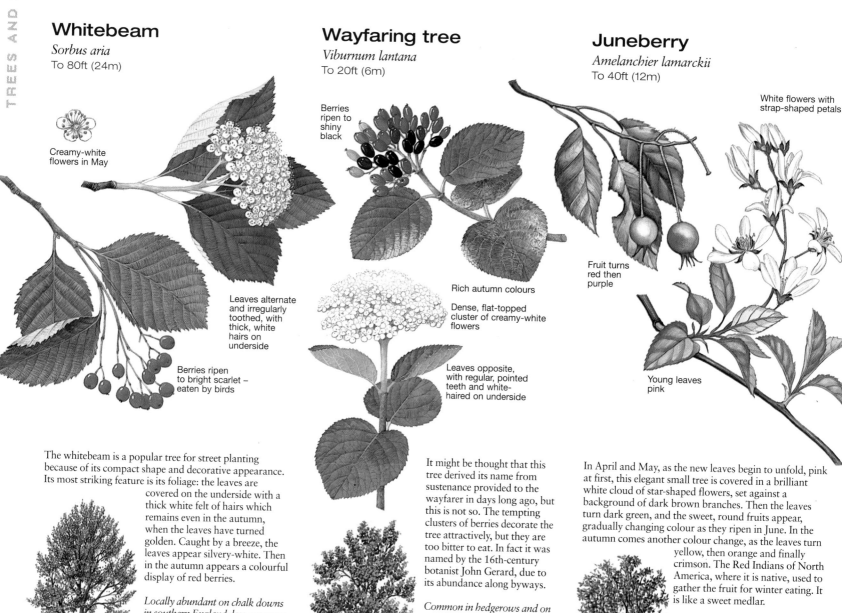

Whitebeam

Sorbus aria
To 80ft (24m)

Creamy-white
flowers in May

Leaves alternate
and irregularly
toothed, with
thick, white
hairs on
underside

Berries ripen
to bright scarlet –
eaten by birds

The whitebeam is a popular tree for street planting because of its compact shape and decorative appearance. Its most striking feature is its foliage: the leaves are covered on the underside with a thick white felt of hairs which remains even in the autumn, when the leaves have turned golden. Caught by a breeze, the leaves appear silvery-white. Then in the autumn appears a colourful display of red berries.

Locally abundant on chalk downs in southern England; less common elsewhere, except where planted.

A compact, domed tree with upswept branches

Wayfaring tree

Viburnum lantana
To 20ft (6m)

Berries
ripen to
shiny
black

Rich autumn colours

Dense, flat-topped
cluster of creamy-white
flowers

Leaves opposite,
with regular, pointed
teeth and white-
haired on underside

It might be thought that this tree derived its name from sustenance provided to the wayfarer in days long ago, but this is not so. The tempting clusters of berries decorate the tree attractively, but they are too bitter to eat. In fact it was named by the 16th-century botanist John Gerard, due to its abundance along byways.

Common in hedgerows and on wood edges in southern England; less frequent in north.

A spreading shrub or small tree

Juneberry

Amelanchier lamarckii
To 40ft (12m)

White flowers with
strap-shaped petals

Fruit turns
red then
purple

Young leaves
pink

In April and May, as the new leaves begin to unfold, pink at first, this elegant small tree is covered in a brilliant white cloud of star-shaped flowers, set against a background of dark brown branches. Then the leaves turn dark green, and the sweet, round fruits appear, gradually changing colour as they ripen in June. In the autumn comes another colour change, as the leaves turn yellow, then orange and finally crimson. The Red Indians of North America, where it is native, used to gather the fruit for winter eating. It is like a sweet medlar.

Locally common in open woodland in south-east England.

A shrub with several stems, or small tree

Long leaves

Snowberry

Symphoricarpos rivularis
To 72in (180cm)

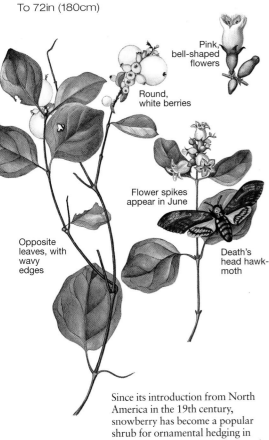

Pink, bell-shaped flowers

Round, white berries

Flower spikes appear in June

Death's head hawk-moth

Opposite leaves, with wavy edges

Since its introduction from North America in the 19th century, snowberry has become a popular shrub for ornamental hedging in gardens, from where it has escaped to the woods and formed dense thickets. Large, spongy white berries decorate the dense tangle of branches for most of the winter. Although pretty to look at they are not edible, and even birds ignore them. The pink bell-shaped flowers last from June to September.

By roadsides, and in woods and parks.

A shrub that forms dense tangled thickets

Wild cherry

Prunus avium
To 100ft (30m)

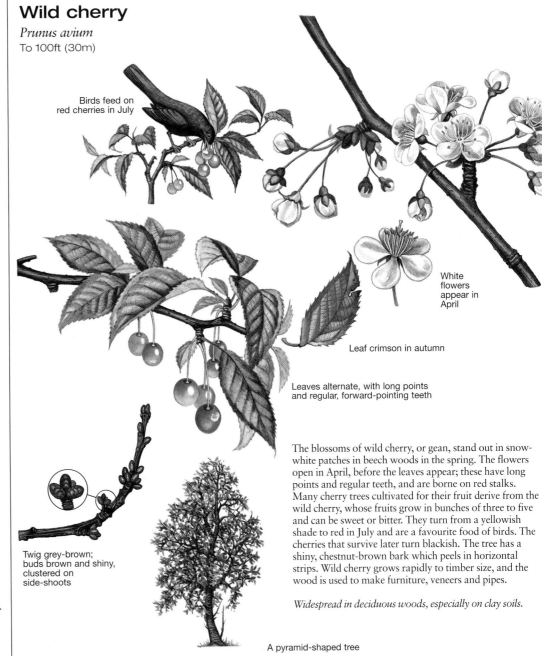

Birds feed on red cherries in July

White flowers appear in April

Leaf crimson in autumn

Leaves alternate, with long points and regular, forward-pointing teeth

Twig grey-brown; buds brown and shiny, clustered on side-shoots

The blossoms of wild cherry, or gean, stand out in snow-white patches in beech woods in the spring. The flowers open in April, before the leaves appear; these have long points and regular teeth, and are borne on red stalks. Many cherry trees cultivated for their fruit derive from the wild cherry, whose fruits grow in bunches of three to five and can be sweet or bitter. They turn from a yellowish shade to red in July and are a favourite food of birds. The cherries that survive later turn blackish. The tree has a shiny, chestnut-brown bark which peels in horizontal strips. Wild cherry grows rapidly to timber size, and the wood is used to make furniture, veneers and pipes.

Widespread in deciduous woods, especially on clay soils.

A pyramid-shaped tree

155

Long leaves

Cherry laurel

Prunus laurocerasus
To 20ft (6m)

Flowers in upright spikes

Small white flowers with long stamens

Leaves alternate, leathery and small-toothed

Berries ripen from red to black

This hardy shrub was introduced to Britain from south-east Europe in the late 16th century. When growing wild it forms dense cover for game and many forms of wildlife, its large evergreen leaves keeping the ground warm for birds in winter. These leaves are leathery, with small, wavy teeth, and grow on thick green stalks. Upright spikes of white flowers which smell of marzipan appear in June, and the cherry-like fruits, each containing one hard seed, ripen in September and are eaten by birds. The leaves contain a poison allied to prussic acid.

Naturalised on chalk-free soils.

Evergreen bush spreads to form hedge

Bird cherry

Prunus padus
20-30ft (6-9m)

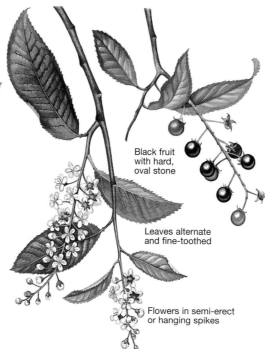

Black fruit with hard, oval stone

Leaves alternate and fine-toothed

Flowers in semi-erect or hanging spikes

Few trees can match the bird cherry's display in late May, when its white blossom hangs in long bunches and spreads strong almond scent. The fruit, which ripens in July and August, is bitter and edible only to birds. The bark peels and has scattered pores; its smell is unpleasant, but in the Middle Ages it was the source of a medicine for upset stomachs. Bird cherry is often planted as an ornamental tree, and flourishes farther north than wild cherry (p. 155).

Common in woods and by streams in limestone areas, especially in north.

A small, compact tree

Almond

Prunus dulcis
To 30ft (9m)

Green fruits rarely ripen

Leaves alternate, creased in V-shape

Oval, pitted nut and kernel

Flowers appear in March or April, before leaves

Brown buds in clusters on green side-shoots

Almond blossom is an early sign of spring in Britain. The pink flowers, which precede the leaves, can appear even before March in some areas; they measure up to 2in (50mm) across and grow singly or in pairs. The long, pointed leaves are unusual in folding about the midrib to form a 'valley'. In Britain the weather is rarely warm enough for the yellowish-green fruits to ripen and produce nuts.

Rarely naturalised.

An open-branched tree, with hanging leaves

Familiar species that have leaves longer than they are broad include most shrubs, many willows, some oaks and the sweet chestnut. Identifying features include the length of the leaves and whether they are thick and evergreen, toothed or untoothed, and smooth or hairy beneath.

Holm oak

Quercus ilex
To 90ft (27m)

Leaves on young trees broad and spiny

On older trees, leaves are narrower and toothless, felted white on underside

Male flowers in long catkins in June

Acorns two-thirds enclosed in scaly cups

A tough tree, brought to Britain from the Mediterranean over 400 years ago, holm or evergreen oak is unharmed by salty sea winds and thrives on exposed coastal sites. The spiky appearance of the leaves on young trees gives the tree its English and Latin names of holm and *ilex* – both refer to holly.

Coasts and estuaries, and in parks and gardens.

Evergreen, round-headed tree

Rhododendron

Rhododendron ponticum
To 20ft (6m)

Flowers in clusters of 10-15

Flowers appear in May and June

Leaves alternate and untoothed, up to 2in (50mm) across, with sharp point

Long capsule has many small, flattened seeds

This evergreen shrub, brought to Britain from Asia Minor more than 200 years ago, makes a magnificent display in May and June when its funnel-shaped purple flowers are fully open. When the rhododendron was introduced, it was planted extensively in woods as game cover. Now some 200 varieties grow in gardens and parks, but *R. ponticum* is the only one growing wild. The rhododendron can survive under heavy tree shade; in some areas it forms a dense, almost impenetrable shrub layer beneath the trees.

On sandy and peaty soil in many parts of the British Isles.

A spreading, smothering shrub

Entire-leaved cotoneaster

Cotoneaster integrifolius
To 12in (30cm)

Wide-spreading branches

Berries scarlet

Pink flowers in clusters of two to four

Berries orange-red

White flowers in May and June

Leaves rounded at tip, grey and hairy beneath

Himalayan cotoneaster
Cotoneaster simonsii

This small-leaved evergreen species of cotoneaster is a robust dwarf shrub from the Himalayas with wide-spreading, slender, rigid branches, ideal for covering bare banks, walls or unsightly waste ground. It has escaped from gardens and grows wild on limestone rocks. The scarlet, globe-shaped berries which cluster along the branches are popular with birds. The berries of Himalayan cotoneaster are egg-shaped and orange-red. This semi-evergreen hedging shrub sometimes grows wild, up to 13ft (4m).

Wild on rough ground, especially limestone.

A dwarf, spreading shrub

157

Long leaves

Sweet chestnut

Castanea sativa
To 100ft (30m)

Female flowers at base of catkin

Spiky husk has up to three nuts

Leaves alternate, with saw-like teeth and parallel veins

Large, pointed, red-brown buds

Tassel-like male flowers at tip of catkin

Although the sweet chestnut has grown in Britain since Roman times, British summers are too cool to allow its nuts to ripen to full size, and most chestnuts eaten in this country come from southern Europe. However, chestnut trees are widely grown for coppicing, or cutting to the ground every 12-14 years to provide fence paling. Large trees grow in parks or former parkland. Some catkins bear only male flowers, while others have both male and female.

Widespread in coppices and parklands; rare in north and west.

A narrow tree with many low branches

Spindle

Euonymus europaeus
To 20ft (6m)

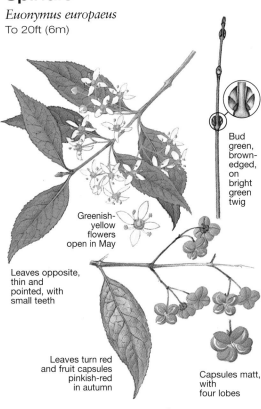

Bud green, brown-edged, on bright green twig

Greenish-yellow flowers open in May

Leaves opposite, thin and pointed, with small teeth

Leaves turn red and fruit capsules pinkish-red in autumn

Capsules matt, with four lobes

Inconspicuous for most of the year, the spindle asserts itself in autumn with a fine show of dark red leaves and pinkish-red fruits. On the strength of this display it is cultivated in parks and gardens. The colour of the fruit is attractive to birds, which spread the seeds. The wood was used for centuries to make spindles for wool spinning.

Common in southern half of England, rarer elsewhere and absent from northern Scotland.

Small tree in woods and hedgerows

Buddleia

Buddleia davidii
To 12ft (3.65m)

Scented flower spike attracts butterflies

Flowers purple or lavender-blue

Leaf-buds in bunches, unprotected by scales, on fawn twig

Peacock

Red admiral

Fruits stay on tree during winter

Leaves opposite and pointed, dark green above and felted on underside

This buddleia is also known as the butterfly bush, because its scented flowers are a powerful attraction to butterflies in midsummer. The deciduous shrub was introduced to Britain from China late in the 19th century and has proved itself a hardy resident, both cultivated and wild. The type most often seen in the wild has lavender-blue or purple flowers, growing in long spikes. Its winged seeds, borne by the wind, will quickly sprout on any sunny patch of waste ground where the soil is not too heavy.

Widespread throughout British Isles, wild and cultivated.

A fast-growing, wide-spreading shrub

White willow

Salix alba
To 60ft (18m)

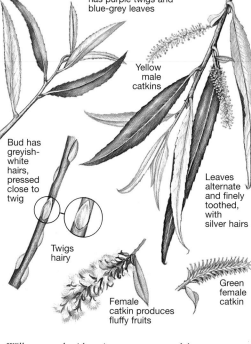

'Coerulea', cricket-bat willow, has purple twigs and blue-grey leaves

Yellow male catkins

Bud has greyish-white hairs, pressed close to twig

Twigs hairy

Leaves alternate and finely toothed, with silver hairs

Female catkin produces fluffy fruits

Green female catkin

Willow trees beside a river present one of the most characteristic pictures of the British countryside. The white willow is named after its leaves, which have silver hairs – denser on the underside. Male and female catkins form on separate trees in spring, and females ripen to produce fluffy fruits which are distributed by the wind. The dark grey bark forms a network of thick ridges. A variety of white willow called 'Coerulea' is cultivated in East Anglia as a source of wood for cricket bats.

Widespread by water in low-lying areas.

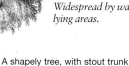

A shapely tree, with stout trunk

Golden weeping willow

Salix × sepulcralis 'Chrysocoma'
To 65ft (20m)

Leaves alternate, long and narrow, with silky hairs

Bud yellowish-green, close to twig

Bright yellow branches and twigs

This graceful tree is the commonest of a number of cultivated weeping willows, and is so widely planted in parks and gardens that it fits the landscape like a native tree. It thrives on damp or dry soil and grows fast. From its early days it produces a distinctive round cascade of yellow twigs and green leaves, reaching almost to the ground. The catkins, which open in April, are predominantly male; sometimes flowers of both sexes grow on the same catkin. The greyish-brown bark is criss-crossed by shallow ridges.

Widespread, especially by water, throughout Britain.

Cascade of foliage almost to ground

Crack willow

Salix fragilis
To 80ft (24m)

Green female catkins appear in May

Leaves alternate, often twisted

White woolly fruits released in summer

Catkins upright on stalk

Rounded leaf-base

Yellow male catkins

Almond willow
Salix triandra

Frequently seen alongside rivers and streams are crack willows that have been pollarded to induce dense new growth to provide poles for hurdles. This native willow's names refer to the brittleness of its twigs, which snap off easily. The bark is grey, deeply ridged and cracked. The tree's fragility helps propagation; broken twigs, carried away by water, often take root.

Almond willow has shorter leaves, more rounded at the base, on olive-brown twigs. Its greyish bark peels off in irregular patches to expose a reddish-brown underlayer.

Common, usually near water.

A tall, rounded tree if not pollarded

Long leaves

Grey willow
Salix cinerea
To 33ft (10m)

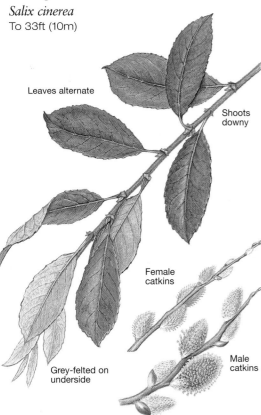

Leaves alternate

Shoots
downy

Female
catkins

Grey-felted on
underside

Male
catkins

Also known as grey sallow, this native shrub or tree is distinguished by its slender shoots, densely covered with brown down. Its leaves have a dark green upper surface, while the underside is closely grey-felted and sometimes bears short rusty-red hairs. Bushy catkins open in March or April, and fruit capsules release seed in May. Grey willow hybridises freely with goat willow (p. 149), and a number of intermediate forms have evolved. The handsome male catkins are much in demand for table decorations.

Common in marshes, heaths and wet woods.

A shrub or tree, with fissured grey bark

Purple willow
Salix purpurea
To 16ft (5m)

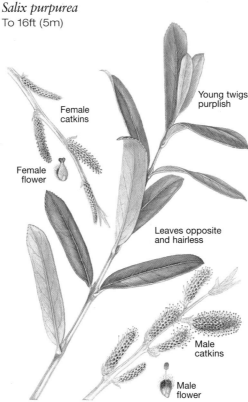

Female
catkins

Young twigs
purplish

Female
flower

Leaves opposite
and hairless

Male
catkins

Male
flower

Twigs that are purplish when young, turning later to olive or yellowish-grey, give this species its name. It may be a spreading shrub no more than 60in (152cm) high, or a much taller bush or small tree. Its leaves are waxy, bluish-green and hairless, varying considerably in width. Catkins normally appear before the leaves in March or April – later in some northern areas. The male catkins are especially attractive, with black bracts, silvery hairs and red anthers. Purple willow is sometimes called purple osier, and hybridises with the osier.

Locally common in wet places.

A spreading shrub, or slender tree

Osier
Salix viminalis
To 33ft (10m)

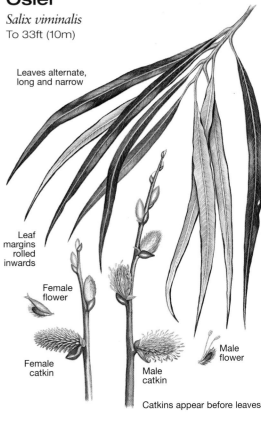

Leaves alternate,
long and narrow

Leaf
margins
rolled
inwards

Female
flower

Female
catkin

Male
catkin

Male
flower

Catkins appear before leaves

This willow shrub still serves a commercial purpose by supplying long, pliant stems for basket-work and other purposes. When cultivated with this in view, the osier is coppiced – that is, cut down to ground level once a year so that it grows a mass of shoots, called withies. Apart from making baskets, withies have been used over the years for lobster-pots, chairs and fencing. The leaves of the osier are green on top, with dense silvery hairs beneath. They are toothless, but come to a narrow point. Male and female catkins grow on separate twigs.

Watersides in lowland areas.

Many-branched shrub, often coppiced

Sea buckthorn

Hippophae rhamnoides
5-8ft (1.5-2.5m)

Berries stay on tree all winter

Willow-like alternate leaves, silver scale-like hairs on underside

Thorny twig, with flowers in clusters

Small green flowers appear in March and April

Female flower

Male flower

Found growing naturally by the sea, this ancient native shrub has leaves adapted to retain moisture and reflect the sun's heat. These leaves are willow-like, slender and untoothed, green on top and with silver scale-like hairs beneath. The flowers grow in clusters on separate male and female bushes. The berries, which ripen in October and remain throughout the winter, have an acid taste, but their vitamin C content is high and they can be used in making marmalade. On dunes, the mesh of tough roots makes its thickets all but indestructible, and helps to stabilise the sand.

Sand-dunes around most of coastline.

A spiny, sprawling shrub

Spurge laurel

Daphne laureola
To 40in (100cm)

Berries black and poisonous

Leaves evergreen, long and thick, clustered near top of plant

Long-tubed flowers appear between February and April

Dwarfed by trees in the woodlands, this delicate-looking shrub has to make the most of what little sunshine it gets. Evergreen leaves, thick and tough enough to withstand the repeated dripping of rainwater, help it to absorb much of the light it needs for growth during the time when deciduous trees are bare. It starts to blossom in winter, before leaves grow on the big trees and plunge it into summer shadow. The greenish, sweet-scented flowers are bisexual; after pollination by insects, they turn into black fruits which are poisonous to humans but eaten by birds.

Woodlands, mainly in England and Wales.

Branches upright, bare except near top

Cider gum

Eucalyptus gunnii
To 100ft (30m)

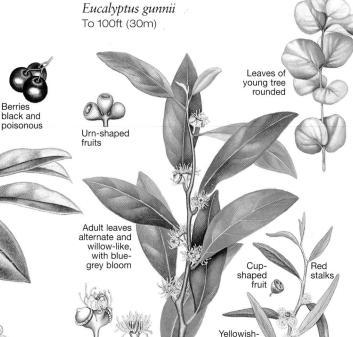

Leaves of young tree rounded

Urn-shaped fruits

Adult leaves alternate and willow-like, with blue-grey bloom

Cup-shaped fruit

Red stalks

Yellow flowers bisexual, appear in clusters of three in July and August

Yellowish-white flowers in clusters of ten

Long grey-green leaves

Snow gum
Eucalyptus niphophila

Of all the gum trees brought from Australasia for their ornamental qualities, the cider gum is the one most frequently seen in Britain. For the first four years of the tree's life its leaves are rounded and stalkless, clasping the stem in pairs. The adult leaves are long, willow-like and evergreen, but do not give off a eucalyptus smell when crushed. The smooth, reddish bark peels off in patches.
The snow gum has green juvenile leaves, replaced by long, sickle-shaped grey-green adult leaves.

Parklands, except in far north.

A graceful, open evergreen tree

161

Maple-like leaves

London plane

Platanus × hispanica
To 100ft (30m)

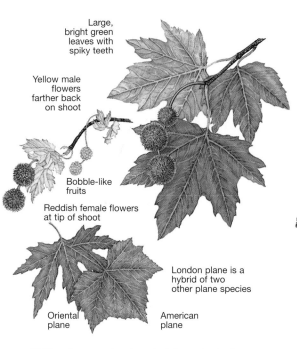

Large, bright green leaves with spiky teeth

Yellow male flowers farther back on shoot

Bobble-like fruits

Reddish female flowers at tip of shoot

London plane is a hybrid of two other plane species

Oriental plane

American plane

Unlike other trees the plane sheds its ageing bark regularly in large flakes, leaving creamy patches. This prevents the tree from becoming stifled under sulphurous grime, and so enables it to thrive even in the polluted atmosphere of big towns. Though the leaves of the plane are shaped like those of maples, they are up to 6in (15cm) or more across, and set alternately on the stem and not in opposite pairs.

Widespread in streets, parks and gardens.

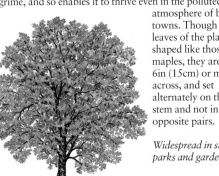

Thick twisting branches and domed crown

White poplar

Populus alba
To 65ft (20m)

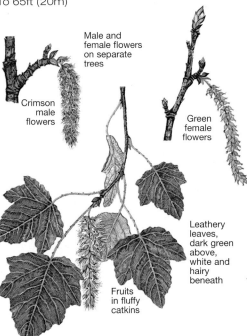

Male and female flowers on separate trees

Crimson male flowers

Green female flowers

Leathery leaves, dark green above, white and hairy beneath

Fruits in fluffy catkins

Dense pale hairs on the underside of the white poplar's leaves flash white in the slightest breeze. The bark is light grey, dotted with diamond-shaped black pores. Though smaller than other poplars, the white poplar is hardy, and often grown to give shelter from winds. The flowers, like those of other poplars, appear in March or April, before the leaves, and hang in catkins. The fruits release white cottony fruits in June. But the seeds seldom germinate and the tree usually reproduces itself from suckers, or is planted.

Widespread in streets and parks.

Twisting branches spread widest near top

Sycamore

Acer pseudoplatanus
To 115ft (35m)

Fruit wings closer than in Norway maple

Large leaves, dark green and leathery, with regular teeth

Greenish-yellow flowers appear with leaves

Leaf stalks tinged red

The massive domed outline of the sycamore, tallest and commonest member of the maple family, has become a familiar sight in Britain's gardens and woods since it was introduced from France in the Middle Ages. The sycamore proved hardy even in the bleakest conditions, and spread vigorously on waste ground by winged fruits which in October twirl through the air like helicopter blades. The leaves, often spotted by disease, are set in opposite pairs and the branches spread out evenly. Flowers appear in May and June.

Widespread in all areas.

Massive domed outline; dense foliage

Many trees outside the maple family have leaves closely similar in shape. They include the familiar hawthorn and London plane, as well as the less common guelder rose, wild service tree and tulip tree. The shape and number of the lobes of the leaf are clues to identification.

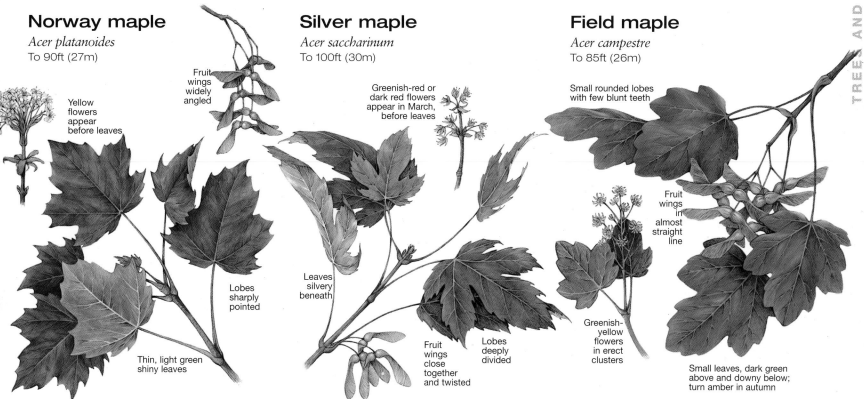

Norway maple
Acer platanoides
To 90ft (27m)

Yellow flowers appear before leaves

Fruit wings widely angled

Lobes sharply pointed

Thin, light green shiny leaves

Silver maple
Acer saccharinum
To 100ft (30m)

Greenish-red or dark red flowers appear in March, before leaves

Leaves silvery beneath

Fruit wings close together and twisted

Lobes deeply divided

Field maple
Acer campestre
To 85ft (26m)

Small rounded lobes with few blunt teeth

Fruit wings in almost straight line

Greenish-yellow flowers in erect clusters

Small leaves, dark green above and downy below; turn amber in autumn

Clusters of bright yellow flowers appearing in March and leaves that turn golden-orange in autumn make the Norway maple a popular ornamental tree in streets and parks. It is very hardy, as might be expected of a species whose native home is Scandinavia, from where it was introduced to Britain in the 17th century. The Norway maple is a shorter and more slender tree than the sycamore, with a more open crown and less dense foliage.

Common in parks and gardens in most areas.

Slender tree with fissured grey bark

The *saccharinum* in the silver maple's scientific name refers to the sweet sap which, in its North American homeland, is tapped for maple syrup and sugar. In Britain the tree yields little sugar, but it flourishes as an ornamental. The leaves, which show their silvery undersides in summer, turn a variety of attractive shades from yellow to brilliant red in autumn. The lobes are more deeply divided than in other maples and bear the closest resemblance to the leaf on Canada's flag.

Common in parks and at roadsides.

Tall open shape, with grey-brown bark

Britain's only native maple is a small round-headed tree often found growing in hedgerows and trimmed back to form part of the hedge. In woodlands on chalky soils in the south-east it can grow taller but seldom as high as other maples. The tree is easily recognised by the rounded lobes of its leaves, its corky twigs and its ribbed grey or light brown bark, with fine shallow fissures. Small flowers appear in upright spikes in May or June. Fruits form in pairs, with wings set straighter than in other maples.

Widespread; rare in Scotland and Ireland.

Round head, with sinuous trunk

163

Maple-like leaves

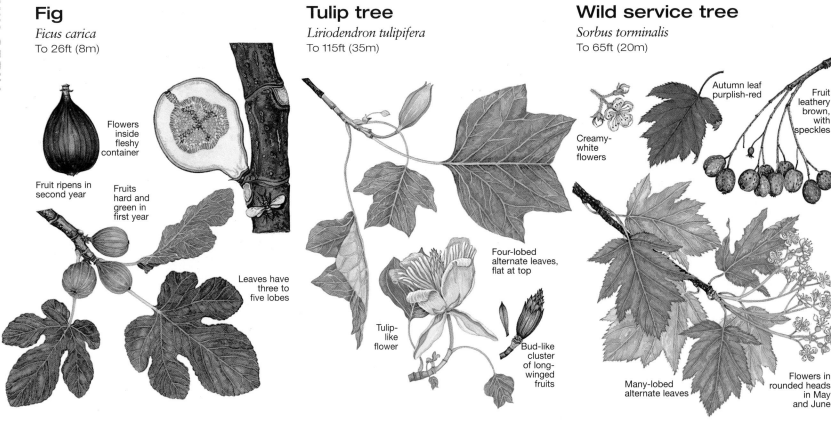

Fig
Ficus carica
To 26ft (8m)

Flowers inside fleshy container

Fruit ripens in second year

Fruits hard and green in first year

Leaves have three to five lobes

Tulip tree
Liriodendron tulipifera
To 115ft (35m)

Four-lobed alternate leaves, flat at top

Tulip-like flower

Bud-like cluster of long-winged fruits

Wild service tree
Sorbus torminalis
To 65ft (20m)

Autumn leaf purplish-red

Fruit leathery brown, with speckles

Creamy-white flowers

Many-lobed alternate leaves

Flowers in rounded heads in May and June

This native of western Asia is often seen in British parks, although the fruit does not often ripen. The tree's leaves, reputed to have covered Adam's nakedness, have featured in innumerable sculptures of gods and heroes. They are distinctive in appearance – leathery, hairy and up to 12in (30cm) long. The flowers are borne inside a fleshy receptacle which develops into the fruit. Pollination is usually by a fig wasp which enters the receptacle to lay its eggs, although cultivated figs may fruit without pollination.

Scattered; commonest in south.

Widely spreading tree, with pale grey bark

When this graceful tree was introduced in the 17th century it was called a poplar, because of its general appearance and its leaves that flutter on long, slender stems. These leaves, distinctive in shape, turn rich yellow or russet in autumn and remain until late November. In a good summer, the tulip-like flowers make an impressive show.

Parks and gardens, rarer in north and in Ireland.

Tall and narrow, with trembling leaves

One of the rarest British native trees, the wild service can easily be mistaken for a maple because of its leaf shape and autumn colouring. Once it was widespread, but now it is found only in patches of old woodland in southern England, having been widely used by charcoal burners. The dark grey bark peels off in rectangular strips, producing a chequered effect which explains the tree's alternative name of chequer tree. The name 'service' is thought to be a corruption of the Latin generic name *Sorbus*.

Uncommon, surviving mainly in south and east.

Many ascending branches form dense head

Guelder rose

Viburnum opulus
To 13ft (4m)

Flattish
flower-head

Leaves opposite,
with three
to five lobes

Small fertile
flowers
surrounded by
ring of bigger,
infertile flowers

Berries stay on tree
after leaves have fallen

The posy-like flower clusters of guelder rose decorate woodlands on damp ground in the wild, and in cultivated forms the shrub brightens gardens. The leaves, smooth on top and hairy below, turn a dull red in autumn. The fragrant flowers appear in May and June, in broad clusters. The larger ones on the outside are showy but sterile, attracting pollinating insects to the smaller fertile flowers farther in; there is also some self-pollination. The leaves, bark and berries are all poisonous. One variety, called the snowball tree, has spherical flower-heads.

Widespread in most of British Isles, but rarer in Scotland.

Spreading shrub with few branches

Hawthorn

Crataegus monogyna
To 45ft (14m)

Scented
flowers

Stipules at
base of leaf

Leaves alternate,
with five to
seven lobes

Fruit contains
single seed

Haws ripen
to dark red in
September

For centuries the fast-growing, sturdy hawthorn has divided fields across mile upon mile of British countryside. It will grow almost anywhere, and a well-made hedge can be practically impenetrable. The white flowers, with their sickly sweet scent, are borne in May – which gives the hawthorn its other familiar name. They grow in clusters on trees, but seldom appear on close-clipped hedges. The greyish-brown bark has many small scales. A variety called *biflora* flowers in winter in mild years.

Common except in northern Scotland.

Tree fans out from low level

Midland hawthorn

Crataegus laevigata
To 25-30ft (7.5-9m)

Leaves alternate,
with lobes shallower
and more rounded than
on common hawthorn

Fruit contains
two or
three seeds

Red-flowered
garden variety

Fruits
rounder
and brighter
red than on
common hawthorn

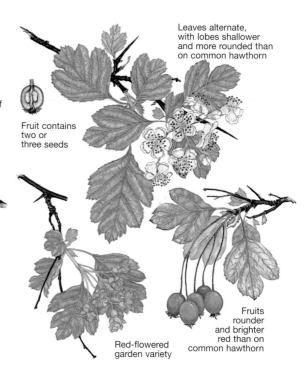

Unlike the common hawthorn, the Midland species – also called English hawthorn – is usually found as a woodland tree rather than in hedge form. White flowers open in May, and later there are fruits which are rounder and redder than those of the common hawthorn. Midland hawthorn is probably best known for its cultivated garden variants, many of which are hybrids with the common hawthorn. Their flowers may be double or single, white, pink or red.

Locally common in parts of southern England; rare introduction in Scotland.

Densely branched tree with fluted bole

165

Lobed leaves

Swedish whitebeam

Sorbus intermedia
To 50ft (15m)

Tree in winter

Flowers in dense clusters, with pink anthers

Leaves alternate, with lobes larger at base

Bright red berries in autumn

Upper surface of leaf dark green and shiny

Underside grey-green and hairy

English oak

Quercus robur
To 115ft (35m)

Tree in winter

Female flowers on long upright stalks

Male flowers are long hanging catkins

Acorns on long stalks, often in pairs

Leaves alternate and stalkless; ear-like lobes at base, four or five lobes on each side

Sessile oak

Quercus petraea
To 130ft (40m)

Tree in winter

Female flowers

Male flowers like those of English oak

Acorns, stubby, with no stalks

Leaves alternate, with stalks; no lobes at base

A native of Scandinavia, the Swedish whitebeam has become a confirmed town-dweller in Britain. Being small and compact, and tolerant of pollution, it is well-suited to street planting; and the tomato-red berries are so quickly taken by birds that few are left to litter the pavements. In autumn the leaves turn to beautiful shades of yellow and brown. The bark is grey.

Widespread in towns and parks in all areas.

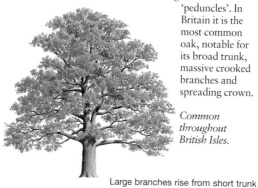

Compact, with broad crown on short trunk

The alternative name of English oak is pedunculate oak, because it bears its fruit or acorns on long stalks or 'peduncles'. In Britain it is the most common oak, notable for its broad trunk, massive crooked branches and spreading crown.

Common throughout British Isles.

Large branches rise from short trunk

Britain's second native oak is the one more commonly found in upland, less-fertile areas. It tends to be taller than its lowland relative, the English oak, with a longer, straighter trunk. The acorns are sessile, or stalkless, and are shorter and rounder.

Dominant native oak in northern and western uplands.

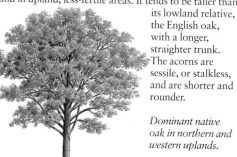

Long straight trunk and fan-shaped crown

The indenting of the margin of each leaf to create a series of lobes is a characteristic feature of many oaks, shared also by the Swedish whitebeam and the unusual maidenhair tree. The lobes can be rounded or sharp-pointed. The leaves themselves are usually arranged alternately.

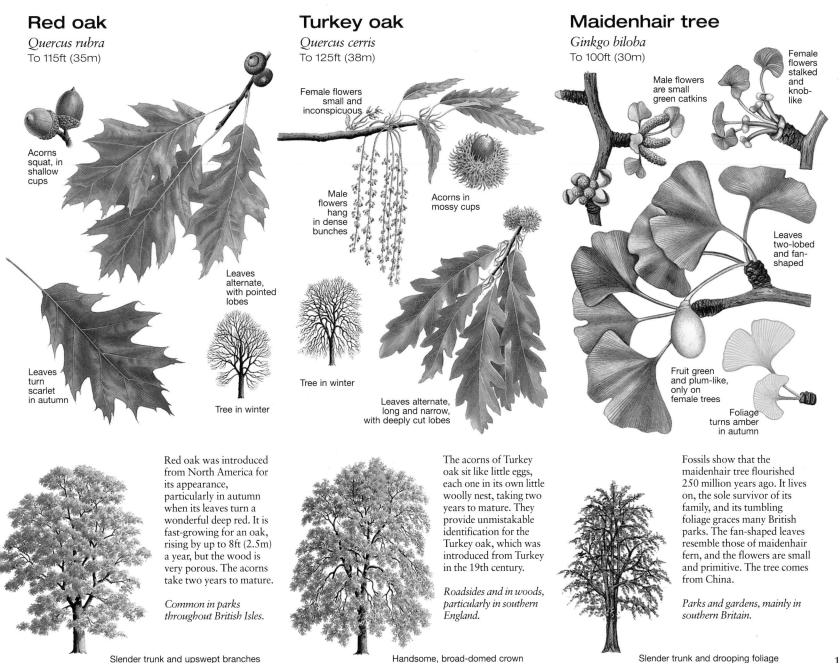

Red oak
Quercus rubra
To 115ft (35m)

Acorns squat, in shallow cups

Leaves turn scarlet in autumn

Leaves alternate, with pointed lobes

Tree in winter

Turkey oak
Quercus cerris
To 125ft (38m)

Female flowers small and inconspicuous

Male flowers hang in dense bunches

Acorns in mossy cups

Tree in winter

Leaves alternate, long and narrow, with deeply cut lobes

Maidenhair tree
Ginkgo biloba
To 100ft (30m)

Male flowers are small green catkins

Female flowers stalked and knob-like

Leaves two-lobed and fan-shaped

Fruit green and plum-like, only on female trees

Foliage turns amber in autumn

Red oak was introduced from North America for its appearance, particularly in autumn when its leaves turn a wonderful deep red. It is fast-growing for an oak, rising by up to 8ft (2.5m) a year, but the wood is very porous. The acorns take two years to mature.

Common in parks throughout British Isles.

Slender trunk and upswept branches

The acorns of Turkey oak sit like little eggs, each one in its own little woolly nest, taking two years to mature. They provide unmistakable identification for the Turkey oak, which was introduced from Turkey in the 19th century.

Roadsides and in woods, particularly in southern England.

Handsome, broad-domed crown

Fossils show that the maidenhair tree flourished 250 million years ago. It lives on, the sole survivor of its family, and its tumbling foliage graces many British parks. The fan-shaped leaves resemble those of maidenhair fern, and the flowers are small and primitive. The tree comes from China.

Parks and gardens, mainly in southern Britain.

Slender trunk and drooping foliage

167

Feathery or spiny leaves

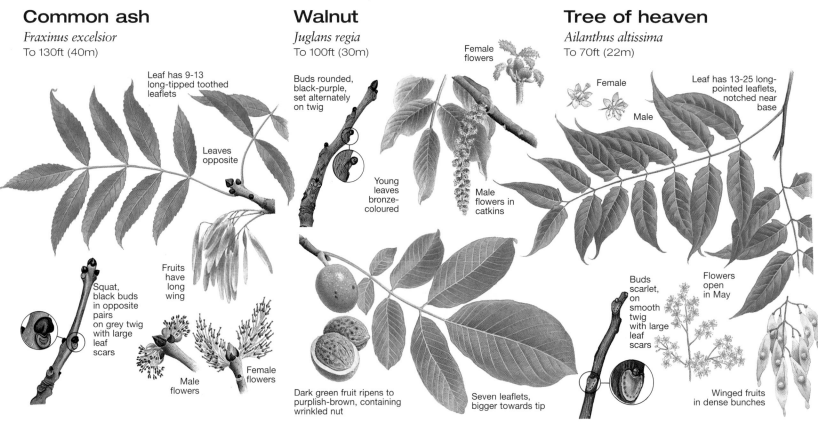

Common ash
Fraxinus excelsior
To 130ft (40m)

Leaf has 9-13 long-tipped toothed leaflets

Leaves opposite

Squat, black buds in opposite pairs on grey twig with large leaf scars

Fruits have long wing

Male flowers

Female flowers

Walnut
Juglans regia
To 100ft (30m)

Buds rounded, black-purple, set alternately on twig

Young leaves bronze-coloured

Female flowers

Male flowers in catkins

Dark green fruit ripens to purplish-brown, containing wrinkled nut

Seven leaflets, bigger towards tip

Tree of heaven
Ailanthus altissima
To 70ft (22m)

Female

Male

Leaf has 13-25 long-pointed leaflets, notched near base

Buds scarlet, on smooth twig with large leaf scars

Flowers open in May

Winged fruits in dense bunches

This tall, handsome tree provides one of the most valuable of timbers, the raw material of a variety of products from oars and axe handles to farm carts and furniture. The leaves of ash are particularly distinctive, made up of 9-13 pale green leaflets. The branches are smooth, grey or ash-coloured, with conspicuous hard black velvety buds.

Widespread, particularly on lime-rich soil.

Domed tree, with widely spaced branches

The Romans planted the walnut tree wherever they conquered, prizing it as a source of both food and cooking oil. It is a tall, broad and stately tree, with a rounded crown and bright green foliage. The leaves are fragrant when crushed. The smooth, rounded fruit contains the familiar nut. The timber is also in great demand.

Common in southern England; infrequent elsewhere.

Broad, rounded crown and thick trunk

The steeply rising branches may not ascend all the way to paradise, as the Chinese name for the tree implies, but laden with deep-green leaflets they make a good job of brightening many city squares and streets. The fruit hangs in large clusters, red at first and brown later, each seed embedded in twisted, propeller-like wing.

Common in city streets in southern England; rare elsewhere.

Straight trunk with ascending branches

The feather-like appearance of many leaves arises from the division of each leaf into numerous leaflets on the same stalk. The number of these leaflets varies. They may grow in opposite pairs or alternately up the stalk, and may have straight or toothed edges. Gorse 'leaves' are rigid spines.

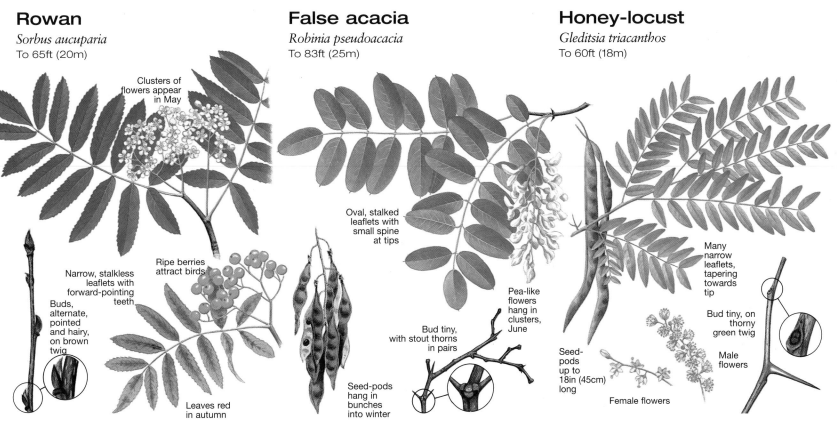

Rowan
Sorbus aucuparia
To 65ft (20m)

Clusters of flowers appear in May

Narrow, stalkless leaflets with forward-pointing teeth

Ripe berries attract birds

Buds, alternate, pointed and hairy, on brown twig

Leaves red in autumn

False acacia
Robinia pseudoacacia
To 83ft (25m)

Oval, stalked leaflets with small spine at tips

Pea-like flowers hang in clusters, June

Bud tiny, with stout thorns in pairs

Seed-pods hang in bunches into winter

Honey-locust
Gleditsia triacanthos
To 60ft (18m)

Many narrow leaflets, tapering towards tip

Bud tiny, on thorny green twig

Male flowers

Seed-pods up to 18in (45cm) long

Female flowers

Although the rowan is also called mountain ash and has leaves like those of the ash family, it is in fact related to the whitebeam and wild service tree. It is very hardy and can be found, with the birch, higher up on mountainsides than any other tree in Britain; but it is also common in gardens. Clusters of white blossoms spread evenly over the crown in May. By September they are transformed into bunches of red berries which birds find irresistible and which make a jelly to accompany game.

Widespread in high and lowlands.

This is also called the locust tree, but it is not the one whose locust beans fed John the Baptist. Its pods are similar but the seeds inedible. The leaves look like those of some true acacias. It is a graceful North American native with a deeply furrowed and twisted trunk. The flowers are rich in nectar.

Common in streets and parks; wild mainly in south.

Clusters of long, vicious branching thorns girdle the trunk and branches of the honey-locust, guarding them from browsing animals. The seed-pods of this North American tree are long and often twisted. The leaves are light green and glossy, turning a rich gold in autumn. Female and male flowers grow separately on one tree.

Parks and gardens, mainly in East Anglia and south.

A graceful, open tree; smooth grey-brown bark

Twisting branches and open crown

Flat top; long spines on trunk and branches

Feathery or spiny leaves

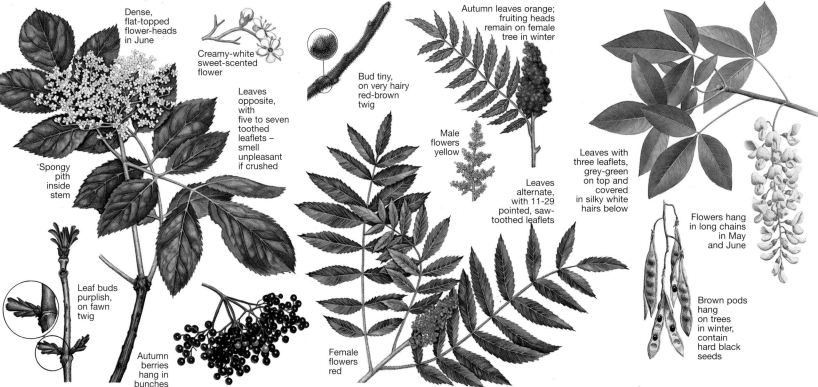

Elder
Sambucus nigra
To 30ft (9m)

Dense, flat-topped flower-heads in June

Creamy-white sweet-scented flower

Leaves opposite, with five to seven toothed leaflets – smell unpleasant if crushed

Spongy pith inside stem

Leaf buds purplish, on fawn twig

Autumn berries hang in bunches

Stag's horn sumac
Rhus typhina
To 26ft (8m)

Autumn leaves orange; fruiting heads remain on female tree in winter

Bud tiny, on very hairy red-brown twig

Male flowers yellow

Leaves alternate, with 11-29 pointed, saw-toothed leaflets

Female flowers red

Laburnum
Laburnum anagyroides
To 23ft (7m)

Leaves with three leaflets, grey-green on top and covered in silky white hairs below

Flowers hang in long chains in May and June

Brown pods hang on trees in winter, contain hard black seeds

A famous and ancient source of wines and jams, the elder is one of the first trees to come into leaf – in mild years often starting in mid-winter. It is usually a bush made up of many stems, but given sufficient light and space it can grow into a tall tree, with deeply furrowed bark. The young shoots contain thick white pith. Clusters of blue-black berries ripen in August and September.

Woodland, hedges, wasteland.

Shrub or small tree

A slender, graceful tree at any time of year, the stag's horn sumac looks its best in the autumn, when its leaves turn orange and red and the big fruiting heads start to glow deeply crimson. Bare of foliage, the inspiration for this North American tree's name is immediately apparent: the branches are sparse and widely spread, and covered in brown, velvety down, like antlers.

Common in gardens.

Curved twigs and open crown

Cascades of scented yellow flowers in May and June justify laburnum's alternative name of golden rain. It is one of the most poisonous trees grown in Britain in all its parts – leaves, flowers and seeds.

Occasionally wild; common in gardens.

Branches arched and ascending; bark smooth

Tamarisk

Tamarix gallica
To 10ft (3m)

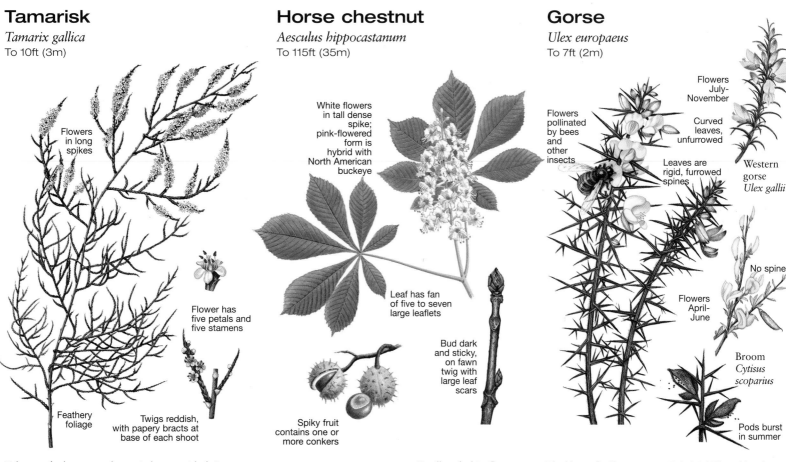

Flowers in long spikes

Flower has five petals and five stamens

Feathery foliage

Twigs reddish, with papery bracts at base of each shoot

Horse chestnut

Aesculus hippocastanum
To 115ft (35m)

White flowers in tall dense spike; pink-flowered form is hybrid with North American buckeye

Leaf has fan of five to seven large leaflets

Bud dark and sticky, on fawn twig with large leaf scars

Spiky fruit contains one or more conkers

Gorse

Ulex europaeus
To 7ft (2m)

Flowers pollinated by bees and other insects

Flowers July-November

Curved leaves, unfurrowed

Leaves are rigid, furrowed spines

Western gorse *Ulex gallii*

No spines

Flowers April-June

Broom *Cytisus scoparius*

Pods burst in summer

Tolerant of salt spray and sea winds, tamarisk thrives along Britain's coast. To minimise the water loss caused by salt's drying effect, its leaves are reduced to small scales, giving the shrub a feathery appearance. Spikes of pink or white flowers are borne from July to September – and in mild areas sometimes up to Christmas. Tamarisk was introduced from the Middle East in the 16th century. It grows well on dunes, where its roots help to bind the sand.

Naturalised on English and Welsh coasts.

Candles of white flowers in May are followed by spiky fruits in autumn containing the conkers treasured by children. Sticky terminal winter buds and horseshoe-like leaf scars are other distinctive features.

Common in parks; sometimes wild in woods.

The blaze of yellow gorse on Britain's hills and heaths begins in March, or even earlier after a mild winter. The 'leaves' of this evergreen shrub, also known as furze or whin, are rigid, furrowed spines, with smaller, true leaves on the new growth only. In high summer, the blackish seed-pods can be heard bursting. Common gorse ceases to flower in June, but at almost any time of the year one species or another is in bloom. Western gorse has curved, unfurrowed spiny leaves, but broom has no spines.

Common on heathland.

Shrub often bent by prevailing winds

Stately tree, with branches turned up at ends

Dense, prickly shrub

Needles in rosettes

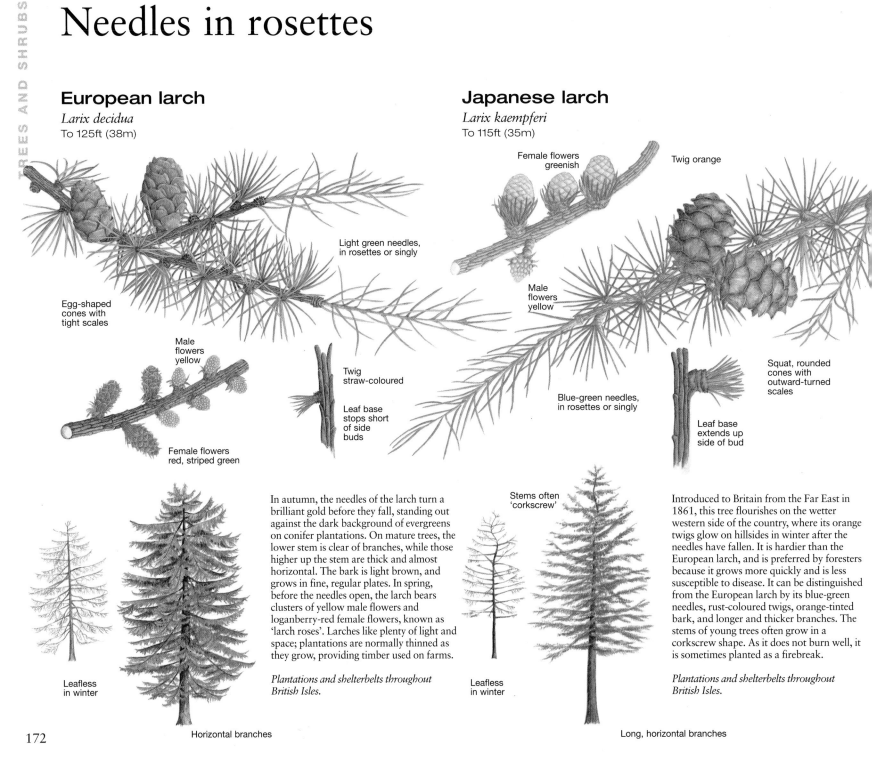

European larch
Larix decidua
To 125ft (38m)

Egg-shaped
cones with
tight scales

Light green needles,
in rosettes or singly

Male
flowers
yellow

Female flowers
red, striped green

Twig
straw-coloured

Leaf base
stops short
of side
buds

Leafless
in winter

Horizontal branches

Japanese larch
Larix kaempferi
To 115ft (35m)

Female flowers
greenish

Twig orange

Male
flowers
yellow

Blue-green needles,
in rosettes or singly

Squat, rounded
cones with
outward-turned
scales

Leaf base
extends up
side of bud

Stems often
'corkscrew'

Leafless
in winter

Long, horizontal branches

In autumn, the needles of the larch turn a brilliant gold before they fall, standing out against the dark background of evergreens on conifer plantations. On mature trees, the lower stem is clear of branches, while those higher up the stem are thick and almost horizontal. The bark is light brown, and grows in fine, regular plates. In spring, before the needles open, the larch bears clusters of yellow male flowers and loganberry-red female flowers, known as 'larch roses'. Larches like plenty of light and space; plantations are normally thinned as they grow, providing timber used on farms.

Plantations and shelterbelts throughout British Isles.

Introduced to Britain from the Far East in 1861, this tree flourishes on the wetter western side of the country, where its orange twigs glow on hillsides in winter after the needles have fallen. It is hardier than the European larch, and is preferred by foresters because it grows more quickly and is less susceptible to disease. It can be distinguished from the European larch by its blue-green needles, rust-coloured twigs, orange-tinted bark, and longer and thicker branches. The stems of young trees often grow in a corkscrew shape. As it does not burn well, it is sometimes planted as a firebreak.

Plantations and shelterbelts throughout British Isles.

Larches are unusual among conifers in having leaves, or needles, that fall in winter. They are in rosettes on the older twigs, as are those of cedars. The seed-bearing cones of cedar are erect and barrel-shaped, while those of larch are round or egg-shaped, and scattered on the twig.

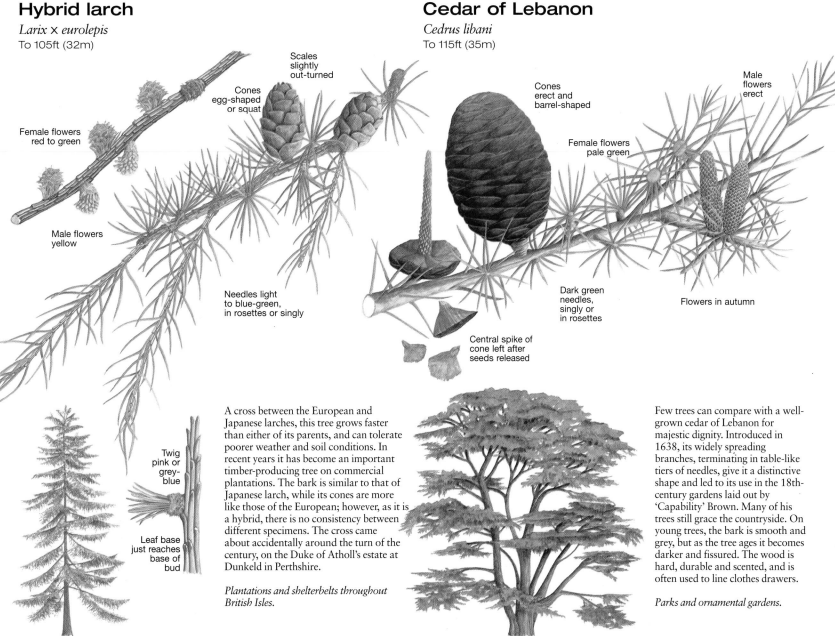

Hybrid larch

Larix × *eurolepis*
To 105ft (32m)

Female flowers red to green

Male flowers yellow

Cones egg-shaped or squat

Scales slightly out-turned

Needles light to blue-green, in rosettes or singly

Cedar of Lebanon

Cedrus libani
To 115ft (35m)

Cones erect and barrel-shaped

Male flowers erect

Female flowers pale green

Central spike of cone left after seeds released

Dark green needles, singly or in rosettes

Flowers in autumn

Twig pink or grey-blue

Leaf base just reaches base of bud

A cross between the European and Japanese larches, this tree grows faster than either of its parents, and can tolerate poorer weather and soil conditions. In recent years it has become an important timber-producing tree on commercial plantations. The bark is similar to that of Japanese larch, while its cones are more like those of the European; however, as it is a hybrid, there is no consistency between different specimens. The cross came about accidentally around the turn of the century, on the Duke of Atholl's estate at Dunkeld in Perthshire.

Plantations and shelterbelts throughout British Isles.

Few trees can compare with a well-grown cedar of Lebanon for majestic dignity. Introduced in 1638, its widely spreading branches, terminating in table-like tiers of needles, give it a distinctive shape and led to its use in the 18th-century gardens laid out by 'Capability' Brown. Many of his trees still grace the countryside. On young trees, the bark is smooth and grey, but as the tree ages it becomes darker and fissured. The wood is hard, durable and scented, and is often used to line clothes drawers.

Parks and ornamental gardens.

Horizontal branches

Level branches, with foliage in tables

173

Needles in rosettes

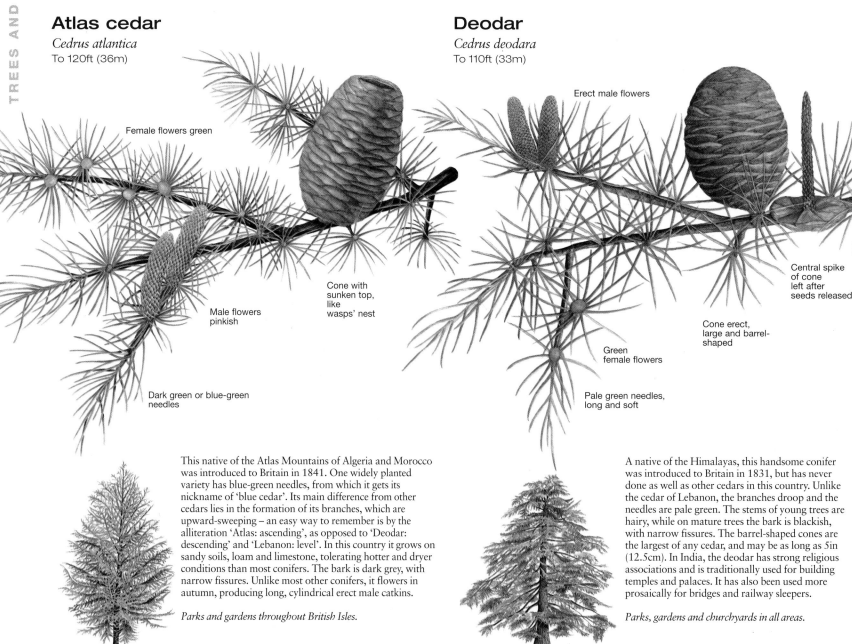

Atlas cedar
Cedrus atlantica
To 120ft (36m)

Female flowers green

Male flowers
pinkish

Cone with
sunken top,
like
wasps' nest

Dark green or blue-green
needles

Deodar
Cedrus deodara
To 110ft (33m)

Erect male flowers

Central spike
of cone
left after
seeds released

Cone erect,
large and barrel-
shaped

Green
female flowers

Pale green needles,
long and soft

This native of the Atlas Mountains of Algeria and Morocco was introduced to Britain in 1841. One widely planted variety has blue-green needles, from which it gets its nickname of 'blue cedar'. Its main difference from other cedars lies in the formation of its branches, which are upward-sweeping – an easy way to remember is by the alliteration 'Atlas: ascending', as opposed to 'Deodar: descending' and 'Lebanon: level'. In this country it grows on sandy soils, loam and limestone, tolerating hotter and dryer conditions than most conifers. The bark is dark grey, with narrow fissures. Unlike most other conifers, it flowers in autumn, producing long, cylindrical erect male catkins.

Parks and gardens throughout British Isles.

Ascending branches

A native of the Himalayas, this handsome conifer was introduced to Britain in 1831, but has never done as well as other cedars in this country. Unlike the cedar of Lebanon, the branches droop and the needles are pale green. The stems of young trees are hairy, while on mature trees the bark is blackish, with narrow fissures. The barrel-shaped cones are the largest of any cedar, and may be as long as 5in (12.5cm). In India, the deodar has strong religious associations and is traditionally used for building temples and palaces. It has also been used more prosaically for bridges and railway sleepers.

Parks, gardens and churchyards in all areas.

Long, drooping branches

Flat needles

Common silver fir
Abies alba
To 150ft (46m)

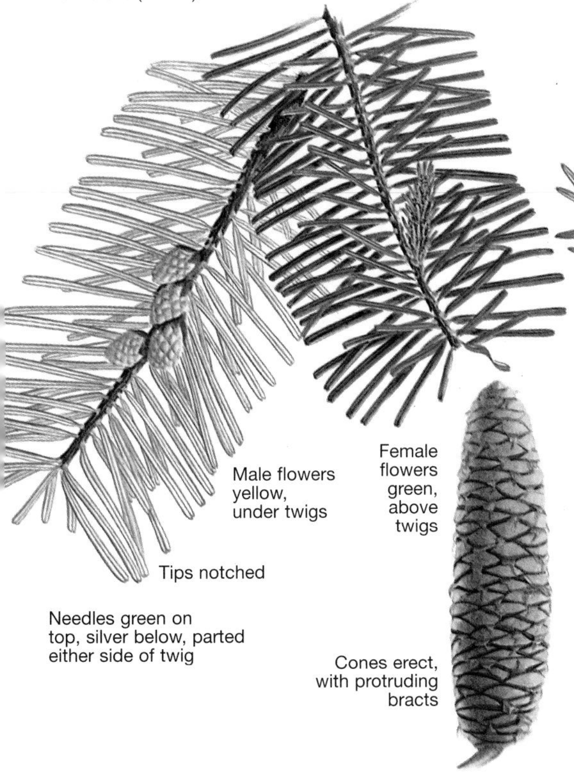

Male flowers yellow, under twigs

Female flowers green, above twigs

Tips notched

Needles green on top, silver below, parted either side of twig

Cones erect, with protruding bracts

A native of the mountains of southern and central Europe, this lofty conifer was widely planted in Britain in the 19th century. Though it is a good timber tree, it is liable to attack by sap-sucking aphids, and has largely been replaced in this country by the fast-growing grand fir. It still grows widely on the Continent and was the original Christmas tree.

Old conifer plantations throughout British Isles.

Narrow and conical in shape

Grand fir
Abies grandis
To 200ft (61m)

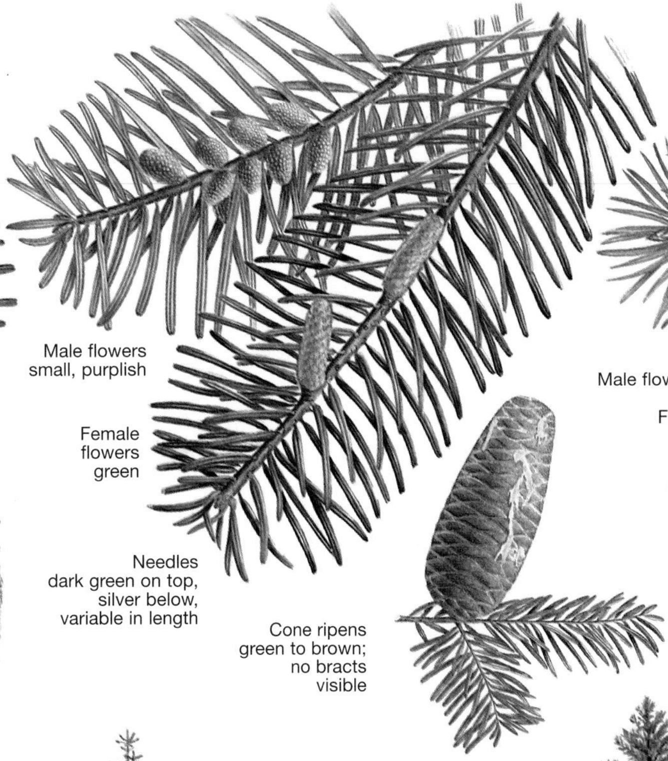

Male flowers small, purplish

Female flowers green

Needles dark green on top, silver below, variable in length

Cone ripens green to brown; no bracts visible

This mighty conifer, also known as giant fir, was brought from western Canada in 1852. At over 200ft (62m) it is Britain's tallest tree. It can reach 55ft (16m) in 20 years, and is widely planted as a timber tree. It has taken the place of the silver fir, as it resists sap-sucking aphids.

Plantations with high rainfall and good soil.

Symmetrical and conical in shape

Douglas fir
Pseudotsuga menziesii
To 200ft (61m)

Male flowers yellow

Female flowers red, tassel-shaped

Needles all around shoot, parted on upper and underside; two white bands below

Cones hang down, with three-pronged bracts

Two of the tallest trees in Britain, both over 180ft (55m), are Douglas firs; but these are pygmies compared with specimens found in the tree's native habitat on the west coast of North America, where Douglas firs grow to over 400ft (123m). Unlike the silver fir, its cones are egg-shaped and hang downwards. As the tree ages, the bark becomes dark-grey or purple, with a corky texture.

Plantations throughout British Isles.

Tall, narrow and conical

175

Flat needles

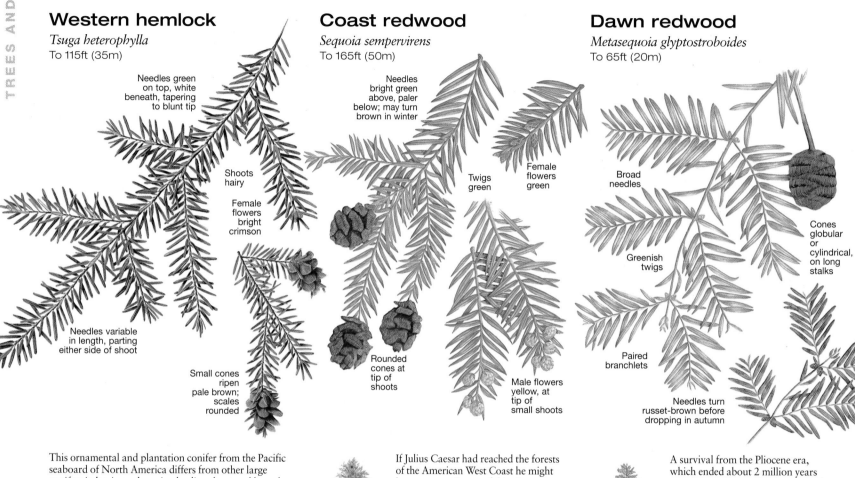

Western hemlock

Tsuga heterophylla
To 115ft (35m)

Needles green on top, white beneath, tapering to blunt tip

Shoots hairy

Female flowers bright crimson

Needles variable in length, parting either side of shoot

Small cones ripen pale brown; scales rounded

Coast redwood

Sequoia sempervirens
To 165ft (50m)

Needles bright green above, paler below; may turn brown in winter

Twigs green

Female flowers green

Rounded cones at tip of shoots

Male flowers yellow, at tip of small shoots

Dawn redwood

Metasequoia glyptostroboides
To 65ft (20m)

Broad needles

Greenish twigs

Paired branchlets

Cones globular or cylindrical, on long stalks

Needles turn russet-brown before dropping in autumn

This ornamental and plantation conifer from the Pacific seaboard of North America differs from other large conifers in having a drooping leading shoot and branch tips. The purplish-brown bark is fissured and may be flaky. In spite of its name, the tree bears no relation to the hemlock from which the Ancient Greeks made their poison. Early settlers in America gave it the name from the smell of its crushed foliage, which resembles that of the waterside hemlock found in Britain.

Plantations and gardens throughout British Isles.

Leading shoot and branches downturned

If Julius Caesar had reached the forests of the American West Coast he might have seen specimens of this giant conifer that are still alive today, as they can live for 2500 years or more. The redwood gets its name from its soft, fibrous, reddish bark, which is also fire-resistant. Columnar in shape, it has rounded grey-brown cones that grow at the tips of the larger shoots, and fronds of flattened needles.

Parks and large gardens throughout British Isles.

Tall, narrow and column-shaped

A survival from the Pliocene era, which ended about 2 million years ago, this handsome deciduous conifer was thought to be extinct until the 1940s, when living specimens were discovered in China. The needles are pale green in spring, turning bright green with a bluish tinge in summer, and brown or red in autumn. The bole of the tree is fluted, with shaggy, reddish bark.

Parks and gardens; infrequent in Scotland and Ireland.

Narrow, conical tree; branches sweep upwards

Flat needles are characteristic of the largest group of conifers, including all the firs. They can be sub-divided by their length, the way they are arranged on the twig, and contrast in colour between upper and lower surfaces. The shape of the cones is another aid to identification.

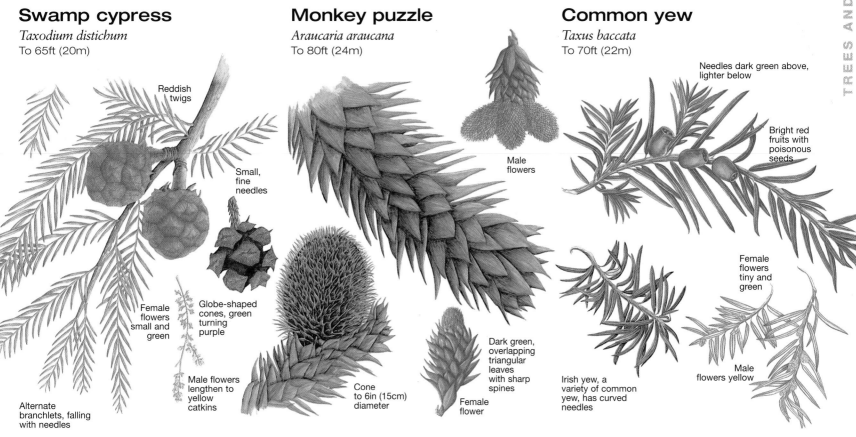

Swamp cypress

Taxodium distichum
To 65ft (20m)

Reddish twigs

Small, fine needles

Female flowers small and green

Globe-shaped cones, green turning purple

Male flowers lengthen to yellow catkins

Alternate branchlets, falling with needles

Monkey puzzle

Araucaria araucana
To 80ft (24m)

Male flowers

Cone to 6in (15cm) diameter

Dark green, overlapping triangular leaves with sharp spines

Female flower

Common yew

Taxus baccata
To 70ft (22m)

Needles dark green above, lighter below

Bright red fruits with poisonous seeds

Female flowers tiny and green

Irish yew, a variety of common yew, has curved needles

Male flowers yellow

Green, fern-like foliage that turns golden-orange in autumn makes the swamp cypress a favourite tree for ornamental planting. When growing in marshy conditions it increases its oxygen supply by sending up 'air-roots'. The tree is also known as the bald cypress because it sheds its leaves in winter. The green, globe-shaped cones turn purple in autumn.

In parks and by lakes in southern Britain.

Triangular outline, with 'air-roots'

Its distinctive shape and sharp-spined, close-set leaves made the monkey puzzle a favourite with Victorian gardeners, one of whom, it is said, gave it its name when he claimed that it would 'puzzle any monkey' to climb it. A native of Chile, the tree was introduced to Europe at the end of the 18th century. It tends to shed its lower branches as it matures, leaving horizontal rings on the thick grey bark.

Parks and gardens throughout British Isles.

Domed crown and clean lower stem

Slow-growing, hardy and durable, the yew is one of our most versatile trees, found as twisted, venerable specimens – some 1000 years old – in churchyards, wall-like hedges in formal gardens, and topiary birds and animals on close-shaven lawns. Many of the longbows that made English archers so formidable in the Middle Ages were cut from yew branches. All parts are poisonous except the fleshy red aril around the seeds.

Mainly in churchyards, parks and gardens.

Round-headed tree; often many trunks

Needles on pegs or in groups

Norway spruce

Picea abies
To 130ft (40m)

Needles light green, short and prickly; protrude all around shoot

Cones long, cigar-shaped and downward-pointing; scales rounded at end

Female flowers pink, erect

Male flowers yellow

The familiar Christmas tree, popularised in Britain by Prince Albert, grew here before the Ice Age, but was exterminated by the cold and only reintroduced about 1500. Its conical shape, with level or drooping lower branches and ascending higher ones, makes it ideal for displaying lights, tinsel and small presents. The tree is also valuable commercially; its timber is known as white wood or deal.

Plantations throughout British Isles.

Higher branches ascend, lower ones droop

Sitka spruce

Picea sitchensis
To 150ft (46m)

Long, thin, sharp needles, dark green above and blue-green below

Female flowers greenish-red

Male flowers yellow

Light brown cones with papery, blunt-tipped scales

Larger and faster-growing than the Norway spruce, the Sitka thrives in wet upland areas and is widely planted as a timber tree. It differs from the Norway spruce in having harder and more sharply pointed needles. The smooth, grey bark peels off in thin plates. In dry summers it may lose its needles from attacks by the spruce aphid.

Upland forestry plantations throughout Britain.

Conical, with long, heavy, lower branches

Scots pine

Pinus sylvestris
To 120ft (36m)

New buds in winter; last year's cones beneath

Blue-green needles in pairs, usually twisted

Two-year-old cones

Female flowers crimson, at end of shoot

Three-year-old cones

Male flowers yellow, at base of shoot

Growing singly or in small clumps on hills and high moorland, this fine conifer can provide a dramatic touch in a featureless landscape. As the tree matures, it loses its lower branches and forms a flat, spreading crown. Its bark is warm red on the upper part of the tree and deeply fissured lower down. The needles are shorter than those of other pines.

Native only in the Highlands; widely planted and naturalised elsewhere.

Older tree develops flat crown

The needles of spruces are sharp, and arise singly on pegs, while those of pines are in groups of two or three. The needles differ in size, and the buds vary in shape. Spruce cones are long, and hang all winter before shedding their seeds. Pine cones are round and hanging, and take two years to ripen.

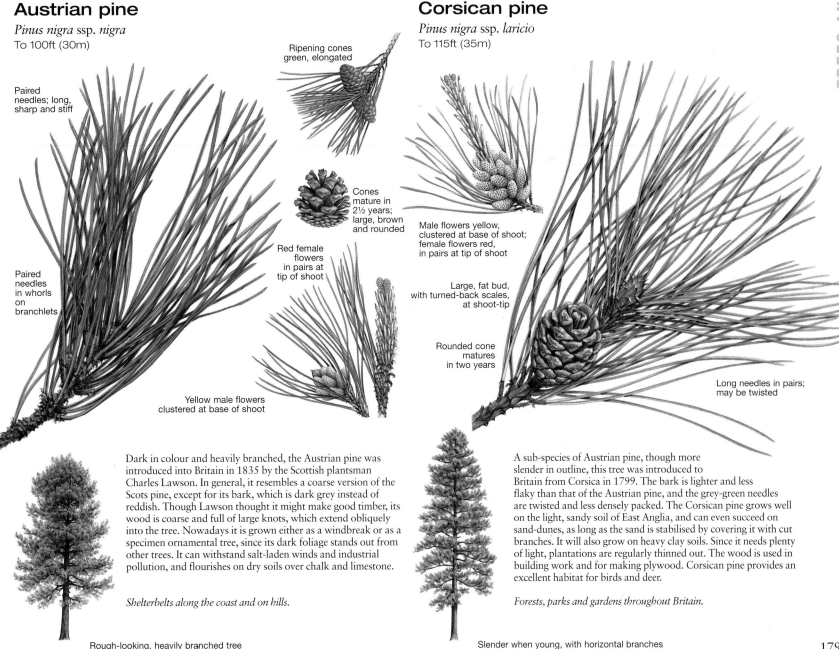

Austrian pine
Pinus nigra ssp. *nigra*
To 100ft (30m)

Paired needles; long, sharp and stiff

Paired needles in whorls on branchlets

Ripening cones green, elongated

Cones mature in 2½ years; large, brown and rounded

Red female flowers in pairs at tip of shoot

Yellow male flowers clustered at base of shoot

Corsican pine
Pinus nigra ssp. *laricio*
To 115ft (35m)

Male flowers yellow, clustered at base of shoot; female flowers red, in pairs at tip of shoot

Large, fat bud, with turned-back scales, at shoot-tip

Rounded cone matures in two years

Long needles in pairs; may be twisted

Dark in colour and heavily branched, the Austrian pine was introduced into Britain in 1835 by the Scottish plantsman Charles Lawson. In general, it resembles a coarse version of the Scots pine, except for its bark, which is dark grey instead of reddish. Though Lawson thought it might make good timber, its wood is coarse and full of large knots, which extend obliquely into the tree. Nowadays it is grown either as a windbreak or as a specimen ornamental tree, since its dark foliage stands out from other trees. It can withstand salt-laden winds and industrial pollution, and flourishes on dry soils over chalk and limestone.

Shelterbelts along the coast and on hills.

Rough-looking, heavily branched tree

A sub-species of Austrian pine, though more slender in outline, this tree was introduced to Britain from Corsica in 1799. The bark is lighter and less flaky than that of the Austrian pine, and the grey-green needles are twisted and less densely packed. The Corsican pine grows well on the light, sandy soil of East Anglia, and can even succeed on sand-dunes, as long as the sand is stabilised by covering it with cut branches. It will also grow on heavy clay soils. Since it needs plenty of light, plantations are regularly thinned out. The wood is used in building work and for making plywood. Corsican pine provides an excellent habitat for birds and deer.

Forests, parks and gardens throughout Britain.

Slender when young, with horizontal branches

Needles on pegs or in groups

Maritime pine

Pinus pinaster
To 110ft (33m)

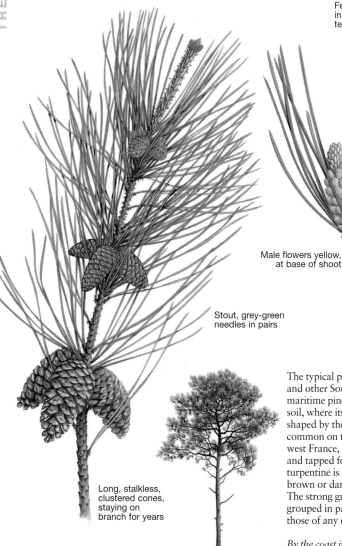

Female flowers red,
in clusters around
terminal bud

Male flowers yellow,
at base of shoot

Stout, grey-green
needles in pairs

Long, stalkless,
clustered cones,
staying on
branch for years

The typical pine tree of Bournemouth
and other South Coast resorts, the
maritime pine flourishes on sandy
soil, where its crown is flattened and
shaped by the prevailing wind. It is
common on the sand-dunes of south-
west France, where the bark is slit
and tapped for its resin, from which
turpentine is made. The bark is
brown or dark purple and fissured.
The strong grey-green needles,
grouped in pairs, are longer than
those of any other two-needled pine.

By the coast in southern England.

Older trees shed lower branches

Monterey pine

Pinus radiata
To 100ft (30m)

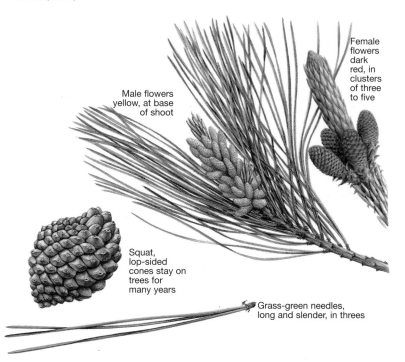

Female
flowers
dark
red, in
clusters
of three
to five

Male flowers
yellow, at base
of shoot

Squat,
lop-sided
cones stay on
trees for
many years

Grass-green needles,
long and slender, in threes

Unlike most other pines, the needles of the Monterey pine
come in threes, not in pairs. It was introduced into Britain
from Monterey in California in 1833. It grows quickly in
this country, especially in southern England, growing
throughout most of the year in the mild south-west. The
dark brown bark is thick, with deep fissures; the cones are
squat and lop-sided, and may remain on the tree for 20-30
years, which has led to the tree's alternative name of
'remarkable cone pine'. The close-set, grass-green foliage
forms an excellent windbreak and makes Monterey pine
an ideal tree for planting near the sea.

Milder parts of southern England.

Dense, high dome, with many branches

Scale-like leaves

Shore pine

Pinus contorta
To 80ft (24m)

Male flowers yellow, below tip of shoot

Female flowers red, at tip of shoot

Caterpillars of pine beauty moth strip plantations

Needles short, yellowish and twisted

Cones clustered, ripening to brown

Originating from the north-western seaboard of North America, this conifer flourishes near the sea. In this country it will grow in harsh climatic conditions and on peaty soil in Scotland and north-east England, where it suppresses the growth of heather. A variety of shore pine is the lodgepole pine (*P. contorta* var. *latifolia*), so called because North American Indians used it as the central pole of their lodges or tepees.

Coast and upland plantations throughout Britain.

A tall, erect tree, with dull green foliage

Common juniper

Juniperus communis
To 20ft (6m)

Spiky, blue-green needles, ½in (13mm) long, in groups of three

Berries green in first year, ripening to purple

Male flowers yellow

Female flowers green

Berries ripen purple in second year

The juniper may form a small conical tree or a sprawling shrub, according to its location. One of Britain's native conifers, it grows on chalk downs in the south, on limestone moorland farther north, and on the acid soils of Scottish pine forests. The blue-green needles grow in groups of three; male and female flowers are found on separate trees. The juniper's berries, ripening to purple in their second year, are used to flavour gin.

Widespread in British Isles; occasionally in gardens.

A small, conical tree or sprawling shrub

Wellingtonia

Sequoiadendron giganteum
To 150ft (46m)

Small, scale-like leaves curl away from twig; aniseed-scented

Cones remain many years

Female flowers green, at end of shoot

Cones large and corky, ripening after two years

Grooved scales

Male flowers yellow, at end of shoot

This stately giant of our parklands was found in the mountains of California about 1850 and introduced to Britain in 1853, where it was named after the Duke of Wellington. One specimen in California, 275ft (84m) tall, 83ft (26m) in circumference and about 4000 years old, is thought to be the world's largest tree. The reddish bark is thick and fibrous, with deep ridges, and is fire-resistant, as it does not contain resin.

Parks and avenues in lowland regions.

Narrow, with a pointed crown

181

Scale-like leaves

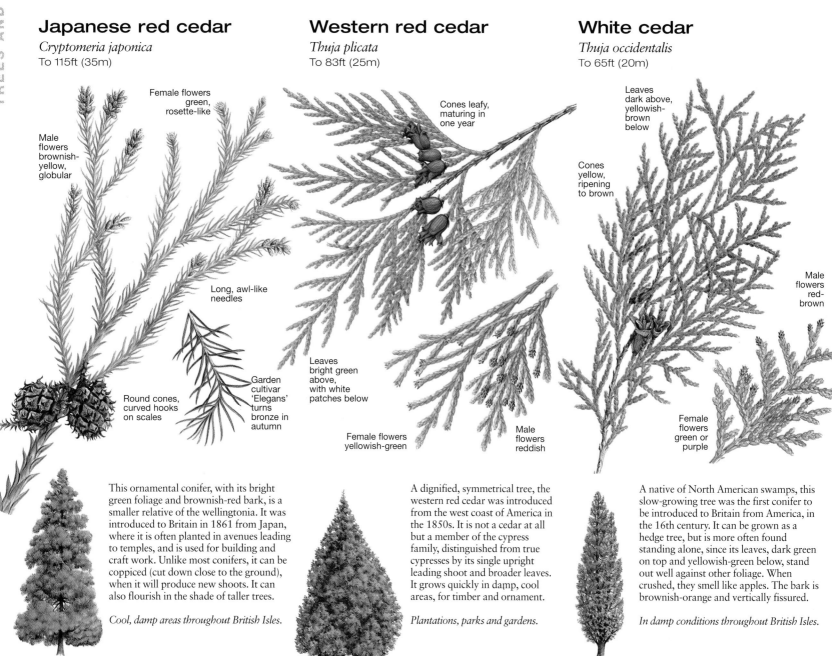

Japanese red cedar
Cryptomeria japonica
To 115ft (35m)

Male flowers brownish-yellow, globular

Female flowers green, rosette-like

Long, awl-like needles

Round cones, curved hooks on scales

Garden cultivar 'Elegans' turns bronze in autumn

Western red cedar
Thuja plicata
To 83ft (25m)

Cones leafy, maturing in one year

Leaves bright green above, with white patches below

Female flowers yellowish-green

Male flowers reddish

White cedar
Thuja occidentalis
To 65ft (20m)

Leaves dark above, yellowish-brown below

Cones yellow, ripening to brown

Male flowers red-brown

Female flowers green or purple

This ornamental conifer, with its bright green foliage and brownish-red bark, is a smaller relative of the wellingtonia. It was introduced to Britain in 1861 from Japan, where it is often planted in avenues leading to temples, and is used for building and craft work. Unlike most conifers, it can be coppiced (cut down close to the ground), when it will produce new shoots. It can also flourish in the shade of taller trees.

Cool, damp areas throughout British Isles.

Narrowly conical crown

A dignified, symmetrical tree, the western red cedar was introduced from the west coast of America in the 1850s. It is not a cedar at all but a member of the cypress family, distinguished from true cypresses by its single upright leading shoot and broader leaves. It grows quickly in damp, cool areas, for timber and ornament.

Plantations, parks and gardens.

Symmetrical, with single leading shoot

A native of North American swamps, this slow-growing tree was the first conifer to be introduced to Britain from America, in the 16th century. It can be grown as a hedge tree, but is more often found standing alone, since its leaves, dark green on top and yellowish-green below, stand out well against other foliage. When crushed, they smell like apples. The bark is brownish-orange and vertically fissured.

In damp conditions throughout British Isles.

Young tree narrowly conical, broadening with age

In some conifers, particularly cypresses, the adult leaves are reduced to scales closely pressed against the twig. This helps to reduce water loss. The juvenile leaves are sharp and awl-like. The cones of the cypresses and cedars in this group are small and roundish, scattered on the twig.

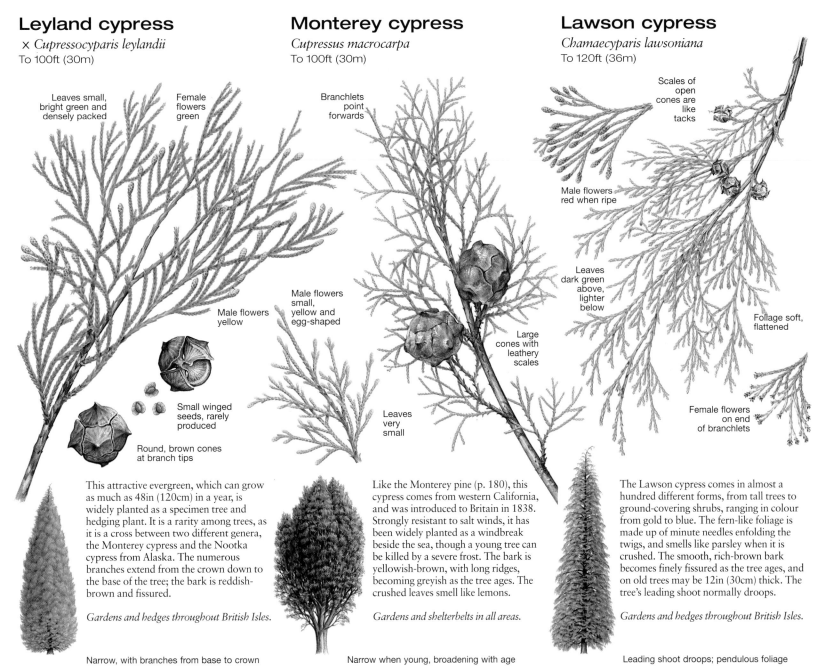

Leyland cypress

× *Cupressocyparis leylandii*
To 100ft (30m)

Leaves small, bright green and densely packed

Female flowers green

Male flowers yellow

Small winged seeds, rarely produced

Round, brown cones at branch tips

This attractive evergreen, which can grow as much as 48in (120cm) in a year, is widely planted as a specimen tree and hedging plant. It is a rarity among trees, as it is a cross between two different genera, the Monterey cypress and the Nootka cypress from Alaska. The numerous branches extend from the crown down to the base of the tree; the bark is reddish-brown and fissured.

Gardens and hedges throughout British Isles.

Narrow, with branches from base to crown

Monterey cypress

Cupressus macrocarpa
To 100ft (30m)

Branchlets point forwards

Male flowers small, yellow and egg-shaped

Large cones with leathery scales

Leaves very small

Like the Monterey pine (p. 180), this cypress comes from western California, and was introduced to Britain in 1838. Strongly resistant to salt winds, it has been widely planted as a windbreak beside the sea, though a young tree can be killed by a severe frost. The bark is yellowish-brown, with long ridges, becoming greyish as the tree ages. The crushed leaves smell like lemons.

Gardens and shelterbelts in all areas.

Narrow when young, broadening with age

Lawson cypress

Chamaecyparis lawsoniana
To 120ft (36m)

Scales of open cones are like tacks

Male flowers red when ripe

Leaves dark green above, lighter below

Foliage soft, flattened

Female flowers on end of branchlets

The Lawson cypress comes in almost a hundred different forms, from tall trees to ground-covering shrubs, ranging in colour from gold to blue. The fern-like foliage is made up of minute needles enfolding the twigs, and smells like parsley when it is crushed. The smooth, rich-brown bark becomes finely fissured as the tree ages, and on old trees may be 12in (30cm) thick. The tree's leading shoot normally droops.

Gardens and hedges throughout British Isles.

Leading shoot droops; pendulous foliage

183

non-flowering plants

The primitive plants that bear no flowers often play a vital role in the balance of nature. For example, fungi – which include moulds and yeasts as well as mushrooms and toadstools – lack the chlorophyll of green plants and so cannot make food substances from air, water and minerals as other plants do, so their food has to come from plant or animal sources, and in obtaining it the fungus breaks down living or dead organic matter and recycles it for eventual use by other creatures.

Ferns, mosses and the closely related liverworts do contain chlorophyll, but they are direct descendants of plants that thrived many millions of years before flowers first appeared on earth. They all need moisture, but their survival is an indication of their adaptability. Mosses in particular are important pioneers of rock surfaces where little else can grow; they hold moisture in their leaves and build up humus in which the seeds of more advanced flowering plants can germinate. Finally and most curious of all are the lichens, formed by a close association of two distinct types of plants – a fungus and a primitive alga (the group to which seaweeds belong). Lichens have no roots and are often no more than an encrustation on a rock or tree-trunk – again they are important pioneer species.

Most of these plants are so easy to overlook that it comes as a surprise to discover that some 50 species of ferns, 200 liverworts, 600 mosses, 1700 lichens and over 10,000 fungus species grow in the British Isles. A representative selection of the most notable types are covered in the following pages, concentrating on the more obvious and – to the lay observer – more interesting fungi and ferns.

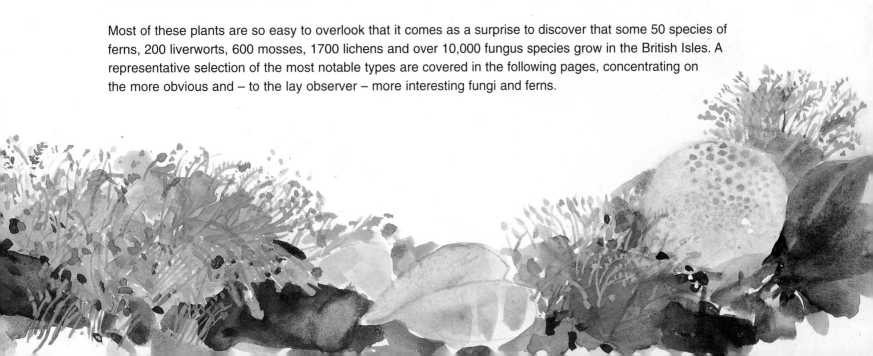

Mushrooms and toadstools

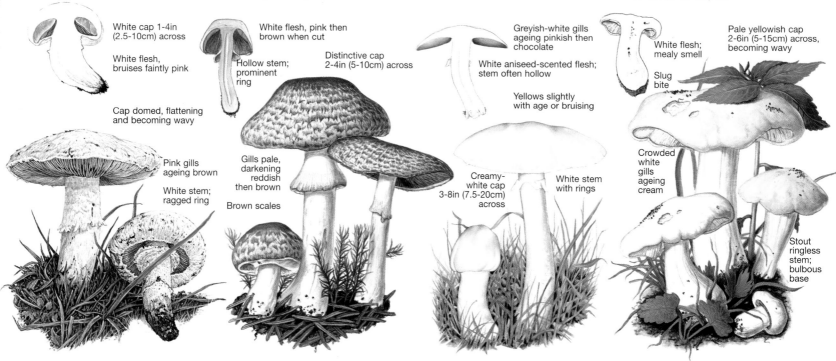

Field mushroom

Agaricus campestris
2-4in (5-10cm) tall

White cap 1-4in (2.5-10cm) across

White flesh, bruises faintly pink

Cap domed, flattening and becoming wavy

Pink gills ageing brown

White stem; ragged ring

Pine-wood mushroom

Agaricus silvaticus
2-4in (5-10cm) tall

White flesh, pink then brown when cut

Hollow stem; prominent ring

Distinctive cap 2-4in (5-10cm) across

Gills pale, darkening reddish then brown

Brown scales

Horse mushroom

Agaricus arvensis
3-6in (7.5-15cm) tall

Greyish-white gills ageing pinkish then chocolate

White aniseed-scented flesh; stem often hollow

Yellows slightly with age or bruising

Creamy-white cap 3-8in (7.5-20cm) across

White stem with rings

St George's mushroom

Calocybe gambosa
2-4in (5-10cm) tall

Pale yellowish cap 2-6in (5-15cm) across, becoming wavy

White flesh; mealy smell

Slug bite

Crowded white gills ageing cream

Stout ringless stem; bulbous base

This is the most frequently eaten wild mushroom, but it can be confused with the poisonous yellow stainer, which smells unpleasantly of ink and grows mainly in woods. The field mushroom's pink gills go dark brown with age and the cap has a silky texture and flattens as it matures. Like many other mushrooms it starts as a 'button', and leaves a characteristic ring on the stalk as the cap separates – but so do the poisonous amanitas which, however, have white gills.

Widespread on rich, natural pastures, especially where there are horses.

The cap of this mushroom, which grows in the needle litter of pine woods, is covered with brown scales. A key identification point is that the white flesh turns pink when cut; this may deter many people from eating it, but it is fairly good eating. It can be confused with the wood mushroom (*Agaricus silvicola*) which may be growing in the same habitat, but this has a creamy-white cap and yellow-staining flesh, smelling of aniseed; it too is good eating.

Common in pine woods.

As its name suggests, this edible mushroom grows best in rich pastures where farm animals graze. It may also be found in gardens, but never woods. Like other closely related mushrooms its cap and stem may turn yellow when bruised, but the base of its stalk never goes a deep chrome-yellow like the poisonous yellow stainer. The horse mushroom is a more heavily-built species than the field mushroom and has a characteristic 'cogwheel' beneath the ring.

Frequent, in meadows and pastures.

St George's Day (April 23) is the traditional date to see this species, but it may appear at any time in April and May. Its early appearance and strong mealy smell means that it is not likely to be confused unless some unseasonably early horse mushrooms are already up; the main mushroom season is in the autumn. The yellowish cap stands on a stout creamy-white stalk which turns buff-white with age; the gills are white, ageing to cream. It is good eating.

Forms rings in grassland and scrub on chalk and limestone soil.

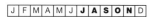
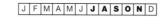
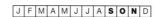
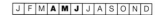

There is no simple demarcation between edible 'mushrooms' and inedible 'toadstools'. In making identifications, check all features labelled, including the size, shape and colour of the cap and stalk; the colour of the spores; gill characteristics; the presence of a ring on the stalk and a bag-like volva at its base; and the appearance and smell of the cut flesh.

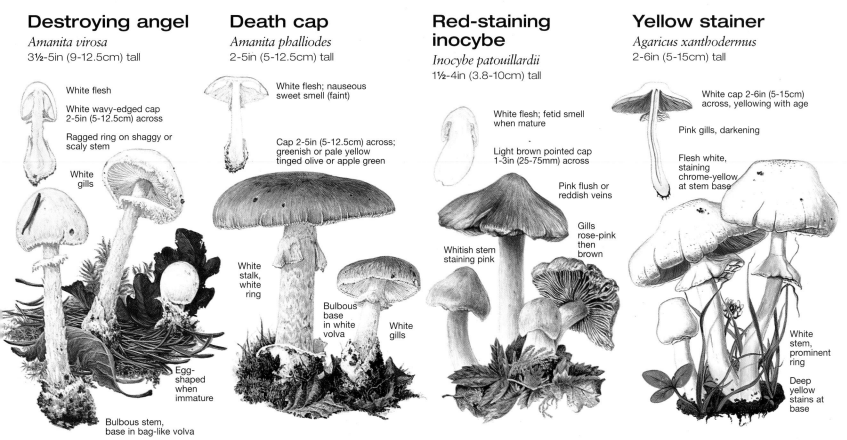

Destroying angel

Amanita virosa
3½-5in (9-12.5cm) tall

White flesh

White wavy-edged cap
2-5in (5-12.5cm) across

Ragged ring on shaggy or
scaly stem

White
gills

Egg-
shaped
when
immature

Bulbous stem,
base in bag-like volva

Death cap

Amanita phalliodes
2-5in (5-12.5cm) tall

White flesh; nauseous
sweet smell (faint)

Cap 2-5in (5-12.5cm) across;
greenish or pale yellow
tinged olive or apple green

White
stalk,
white
ring

Bulbous
base
in white
volva

White
gills

Red-staining
inocybe

Inocybe patouillardii
1½-4in (3.8-10cm) tall

White flesh; fetid smell
when mature

Light brown pointed cap
1-3in (25-75mm) across

Pink flush or
reddish veins

Gills
rose-pink
then
brown

Whitish stem
staining pink

Yellow stainer

Agaricus xanthodermus
2-6in (5-15cm) tall

White cap 2-6in (5-15cm)
across, yellowing with age

Pink gills, darkening

Flesh white,
staining
chrome-yellow
at stem base

White
stem,
prominent
ring

Deep
yellow
stains at
base

The pure white destroying angel is well named. It has caused many deaths on the Continent through confusion with field mushrooms, but the latter have pink to brownish gills and grow in the open rather than in woodland. Like many amanitas, the destroying angel has a ring attached to the stalk, whose base has a characteristic volva; this is the skin-like remains of the first part of the fungus to appear above ground. The cap is often assymetric.

Rare, mainly in Scotland in birch woods; very rare in the south.

The name is no exaggeration; this is the deadliest known fungus. Its poisons take 6-24 hours to have an effect, and there is no antidote. It has been confused with edible mushrooms, with fatal results. Its points of difference are the colour of the cap, the definite ring and ragged, bag-like volva on the stalk as well as the white gills. Unlike field mushrooms it grows in oak and beech woods, not in the open.

Not infrequent, in oak or mixed woods.

Some 90 species of *Inocybe* live in Britain. Most have a conical cap and are difficult to distinguish from one another. Some of them are poisonous, but this one is the deadliest. Its telltale signs are the pink flush on the cap (some specimens have distinctive red veins), and the way the flesh and particularly the gills turn red when bruised. It lives on the woodland floor, especially in beech woods on chalky soil.

Infrequent, mainly in beech woods.

Looking very much like a field or horse mushroom from a distance, the yellow-staining mushroom has a characteristic flat top and an unpleasant smell of ink. The best way to confirm recognition is to cut the swollen base of the stalk; this stains a deep chrome-yellow. The yellow stainer is not poisonous to everyone, but may cause vomiting or indigestion.

Occasional in woods, meadows and gardens.

J F M A M J **J A S O** N D   J F M A M J **J A S O N** D   J F M A **M J J A S O** N D   J F M A M J **J A S O** N D

Mushrooms and toadstools

Shaggy ink cap

Coprinus comatus
4-12in (10-30cm) tall

Grooved cap, no scales

Gills blacken and disintegrate

Drips 'ink'

Hollow white stem

Common ink cap
Coprinus atramentarius

Cap opens bell-shaped

Drips inky fluid when mature

Shaggy cylindrical cap 2-6in (5-15cm) tall

Often grows in clusters

also called the lawyer's wig, is feathered in white, hanging scale-like sections. It is very delicate, with a hollow stalk, and will break down if handled roughly. The white gills go pink, then black, with age. After a day the cap disintegrates into a black dripping mass, so gather unopened specimens and cook without delay. The common ink cap, lacking scales on the cap, is toxic if consumed with alcohol.

Very common everywhere, especially lawns and roadsides; also parks, rubbish tips.

J F M A **M J J A S O** N D

Parasol mushroom

Macrolepiota procera
6-12in (15-30cm) tall

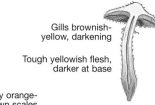

Hollow stem, white flesh

Cap 4-10in (10-25cm) across; conical centre

Dark scales

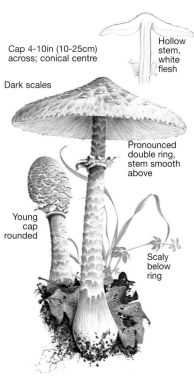

Pronounced double ring, stem smooth above

Young cap rounded

Scaly below ring

The large wide cap of this umbrella-shaped fungus, with its conical centre, flattens out with age, and scales on the cap create a pattern of light and dark areas. The parasol mushroom is much lighter and the cap is larger than its relative, the shaggy parasol (*Macrolepiota rhacodes*), whose cap is 2-6in (5-15cm) across; this has a smoother stalk and bruises red. The parasol mushroom is edible and good; avoid the shaggy parasol.

Fairly common, forming rings at woodland edge and in clearings and meadows.

J F M A M J **J A S O** N D

Shaggy pholiota

Pholiota squarrosa
2-5in (5-12.5cm) tall

Gills brownish-yellow, darkening

Tough yellowish flesh, darker at base

Many orange-brown scales on yellow cap and stem

Rounded cap 1-4in (2.5-10cm) across

Torn ring

Smooth above ring

Clustering habit on wood

A rounded cap and shaggy scales distinguish this inedible species that grows on dead stumps or at the base of a seemingly healthy tree. It grows on a wide variety of trees and is also called the scaly cluster fungus. As the cap grows older the rim breaks and inverts, showing off the yellow-brown gills which turn rust-brown. The stalk has a torn ring.

Common in deciduous woods.

J F M A M J **J A S O** N D

Wood blewit

Lepista nuda
3-5in (7.5-12.5cm) tall

Gills and stem violet

Flesh bluish-lilac; faint perfume

Cap up to 6in (15cm) across, lilac fading to pinkish-brown

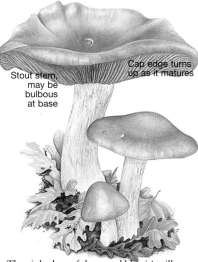

Cap edge turns up as it matures

Stout stem, may be bulbous at base

The violet hue of the wood blewit's gills and stem is quite unmistakable, but the colour fades after a day. Older specimens may be difficult to identify without their violet colour, and younger ones could be confused with the violet *Cortinarius* species; but the latter have a cortina – a web-like gill covering – when young. Apart from woodland, the wood blewit inhabits garden compost heaps. It is considered edible, but may cause an allergic reaction in some people. so eat only a little at first.

Common in all sorts of woods.

J F M A M J J A **S O N D**

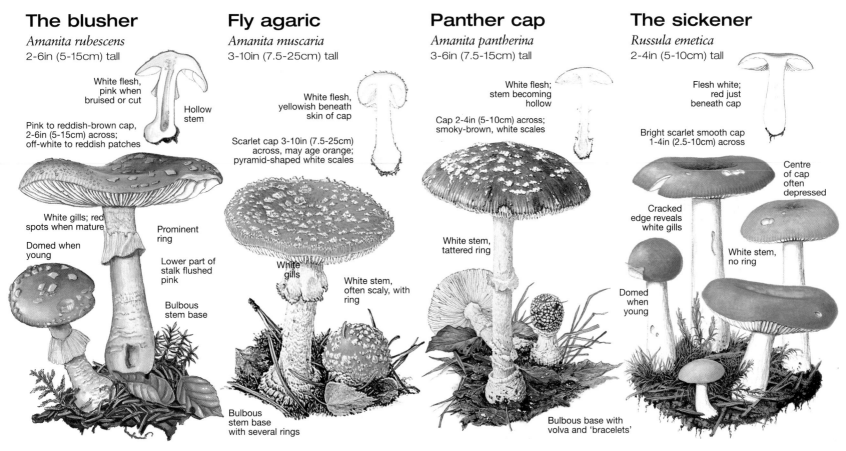

The blusher
Amanita rubescens
2-6in (5-15cm) tall

White flesh, pink when bruised or cut

Hollow stem

Pink to reddish-brown cap, 2-6in (5-15cm) across; off-white to reddish patches

White gills; red spots when mature

Domed when young

Prominent ring

Lower part of stalk flushed pink

Bulbous stem base

Fly agaric
Amanita muscaria
3-10in (7.5-25cm) tall

White flesh, yellowish beneath skin of cap

Scarlet cap 3-10in (7.5-25cm) across, may age orange; pyramid-shaped white scales

White gills

White stem, often scaly, with ring

Bulbous stem base with several rings

Panther cap
Amanita pantherina
3-6in (7.5-15cm) tall

White flesh; stem becoming hollow

Cap 2-4in (5-10cm) across; smoky-brown, white scales

White stem, tattered ring

Bulbous base with volva and 'bracelets'

The sickener
Russula emetica
2-4in (5-10cm) tall

Flesh white; red just beneath cap

Bright scarlet smooth cap 1-4in (2.5-10cm) across

Centre of cap often depressed

Cracked edge reveals white gills

White stem, no ring

Domed when young

The name is well earned, since the cap and stalk base of this mushroom have a rosy hue and the white patches on the cap turn reddish in older specimens. Even more characteristic is the white flesh, which goes pink on cutting. It is indigestible or even poisonous when raw, but can be eaten cooked. However, it is best avoided, because it is easily confused with the panther cap, which is very poisonous but whose flesh does not turn pink on cutting.

Common in deciduous woods.

This colourful and unmistakable fungus causes hallucinations and intoxication, and often violent gastric upsets, but it is rarely fatal. Its common name comes from the medieval practice of using it to repel flies. Apart from the brilliant scarlet cap with its white scales – a common sight in birch woods – it has the telltale amanita collar on the stem, though the volva is reduced in this case to a few warty rings. The white scales can wash off in rain.

Common in birch and pine woods.

Often called the false blusher, or simply the panther, this poisonous species can be confused with the blusher, which is sometimes eaten. Both have flaky scales on the cap – though the panther's are pure white – and both grow in similar mixed woodland. But only the blusher's flesh turns pink when cut and its cap is a more rosy-brown. The panther cap shows a series of rings at the base of the stalk – the remnants of the volva. The scales are crowded at the 'button' stage.

Occasional in deciduous or coniferous woods.

This brightly coloured fungus is called the sickener because its flesh has a burning hot taste and will cause vomiting if eaten in quantity. It lives on damp soil under pines, though a very similar but smaller species, *Russula mairei*, inhabits beech woods. Brightly coloured russulas – there are many species – usually have the white flesh showing where little bits of the cap are missing or they have cracked with age. When the cap is fully expanded, the centre may be depressed.

Common in pine woods.

| J | F | M | A | M | J | **J** | **A** | **S** | **O** | N | D |

| J | F | M | A | M | J | **J** | **A** | **S** | **O** | N | D |

| J | F | M | A | M | J | **J** | **A** | **S** | **O** | N | D |

| J | F | M | A | M | J | **J** | **A** | **S** | **O** | N | D |

189

Mushrooms and toadstools

Cep

Boletus edulis

2-10in (5-25cm) tall

White flesh flushed reddish or straw in cap; no change when cut

Brown bun-like cap to 8in (20cm) across

Mass of pores beneath cap; no gills

Thick bulbous stem, white netting

The most sought-after of all edible fungi, the cep can be distinguished from the many other similar-shaped *Boletus* species by its brown cap and sturdy stem with fine netting. Its relations – some of them poisonous – have different-coloured caps and stalks, and flesh that may change colour a few minutes after cutting. The cep's alternative name of 'penny bun' is very apt, since the cap is the colour and size of an English bun.

Fairly common in deciduous woods, especially in southern England; sometimes in pine woods.

Devil's boletus

Boletus satanas

3-4in (7.5-10cm) tall

Flesh pale blue when cut

Greyish or buff-white domed cap 3-10in (7.5-25cm) across

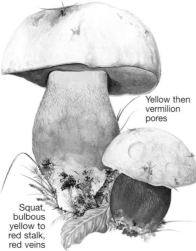

Yellow then vermilion pores

Squat, bulbous yellow to red stalk, red veins

It is obvious from its shape that the devil's boletus is related to the prized cep or 'penny bun', but there is no mistaking the lurid colours of this poisonous species. This is particularly true when it is cut, for the flesh turns a beautiful sky blue. It also smells unpleasant. It is usually found under mature beeches in southern England. There are a number of similar *Boletus* species, some regarded as good eating, but with brown or yellow caps, and without the carrion smell of this species.

Rare; in woodland on chalky soils.

Morel

Morchella esculenta

2½-8in (6.5-20cm) tall

Mop-like brown head

White or creamy hollow stem

False morel
Gyromitra esculenta

Yellowish sponge-like cap darkens with age

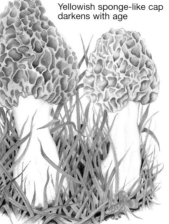

Stem base swollen and furrowed

A relative of the cep, this is another gourmet's species. The morel may be found in long grass in damp meadows in spring. It also lives in woodland, but only where there is plenty of light and rich green grass. All morels have characteristic sponge-like cups all over the cap. At first this is pale yellowish-brown, but it changes to a dark brown after a few days.

The false morel, also found in spring in pine woods, can be deadly poisonous raw.

Damp meadows and woodlands; infrequent.

Stinkhorn

Phallus impudicus

4-8in (10-20cm) tall

Gelatinous layer under skin

Black honeycombed cap; green slimy spore mass

White mycelial 'root'

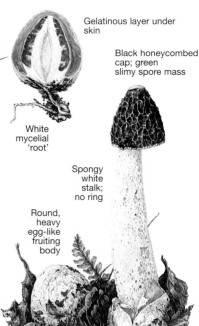

Spongy white stalk; no ring

Round, heavy egg-like fruiting body

The phallic shape of the mature stinkhorn develops in only a few hours from a gelatinous 'egg' up to 2in (5cm) across that sits in the leaf-litter of woodland or gardens. The delicate spongy stalk has a conical honeycombed cap covered in a dark green slimy mass of spores. The slime smells foul, attracting flies from afar which spread the spores. The 'egg' stage is said to be edible. A common related species is the narrower orange-tipped dog stinkhorn (*Mutinus caninus*).

Common on rich soil in woodland and gardens.

J F M A M J **J A S O** N D

J F M A M J **J A S O** N D

J F M A **M** J **J A S** O N D

J F M A M J **J A S O** N D

Dead man's fingers

Xylaria polymorpha
1-3in (25-75mm) tall

Flesh white; spore sacs at edges

Irregular club-shaped black 'fingers'

Grow upright from stumps

Narrow stalk at base

The stiff finger-like projections, slightly smaller than human fingers, are often found growing on old tree stumps, especially those of beech. The base of each black 'finger' is narrow and it twists slightly towards the broader top; each has a unique shape. When broken the fingers reveal a pure white tough flesh which has rows of tiny sacs all around the edge where spores are formed. It is not an edible species.

Common, especially in beech woods.

J F M A M J J A S O N D

Sulphur tuft

Hypholoma fasciculare
2-4in (5-10cm) tall

Flesh yellow, brownish towards base

Domed cap 1-3in (25-75mm) across; bright yellow, centre tinged orange-brown

Yellow gills with black spores

Honey-coloured stems, growing in tufts

Bright yellow clumps of fungi sprouting from old tree stumps, or even covering the ground in a forestry plantation where side-branches have been left to rot, are probably the sulphur tuft. It could be mistaken for honey fungus, but is generally smaller and brighter coloured, the domed cap maturing from sulphur-yellow to a pale honey colour. The gills are yellowish but bear black spores, giving an overall greenish tinge. Sulphur tufts are inedible.

Very common in woodland.

J F M A M J J A S O N D

Fairy-ring champignon

Marasmius oreades
1-4in (2.5-10cm) tall

Mature cap concave

Whitish to flesh-pink, wrinkled

Gills crowded

Clitocybe rivulosa

Whitish flesh; fresh sawdust smell

Cap up to 2in (5cm) across; central cone

Cap buff to tan, tinged pink

Tough stem, whitish to pale buff

Broad paler gills spaced widely

The tough stalk and dry, ochre-coloured cap help to distinguish this edible species. But beware the poisonous small white or pinkish-white *Clitocybe rivulosa*, which can also form rings in short grass. The 'fairy ring' is sometimes a strip or cluster rather than a ring, which is formed when the mycelia ('roots') spread out equally in all directions from a central cluster. The mycelia enrich the soil with nitrogen, and the grass grows greener at the ring's edge.

Common on village greens, meadows, parks and gardens.

J F M A M J J A S O N D

Honey fungus

Armillaria mellea
2½-5in (6.5-12.5cm) tall

White flesh, strong smell

Stem base tapered or swollen

Rough cap 1-4in (2.5-10cm) or more across

Cap variable honey-brown, dark scales

Ring often flecked yellow

Stem honey-coloured below ring

This serious pest kills trees in gardens, parkland and forestry plantations. It may grow directly from the bark of the tree on the trunk or branches, from the roots around its base, or in clusters from an old stump. Its colour varies widely from a pale to a dark honey shade; a stump may have several different clumps, each looking like a different fungus species. It spreads as a web of black 'bootlaces' in the soil and under the tree's bark. The caps are edible when cooked, with a sweet flavour.

Abundant everywhere on live and dead wood.

J F M A M J J A S O N D

191

Funnel and cup-shaped fungi

Chanterelle
Cantharellus cibarius
1½-3in (38-75mm) tall

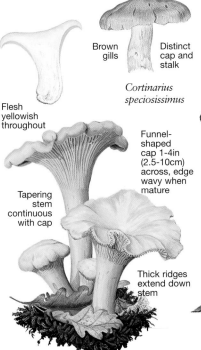

Brown gills

Distinct cap and stalk

Cortinarius speciosissimus

Flesh yellowish throughout

Funnel-shaped cap 1-4in (2.5-10cm) across, edge wavy when mature

Tapering stem continuous with cap

Thick ridges extend down stem

Horn of plenty
Craterellus cornucopioides
2½-3½in (65-90mm) tall

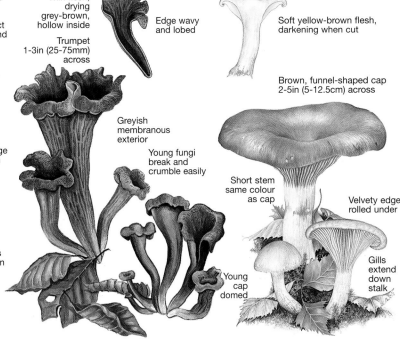

Thin flesh; drying grey-brown, hollow inside

Edge wavy and lobed

Trumpet 1-3in (25-75mm) across

Greyish membranous exterior

Young fungi break and crumble easily

Brown roll-rim
Paxillus involutus
2-5in (5-12.5cm) tall

Soft yellow-brown flesh, darkening when cut

Brown, funnel-shaped cap 2-5in (5-12.5cm) across

Short stem same colour as cap

Velvety edge rolled under

Young cap domed

Gills extend down stalk

Orange-peel fungus
Peziza aurantia
½-4in (1-10cm) across

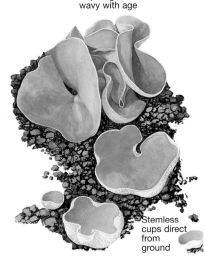

Bright orange inside

Outside whitish with fine down

Cup edge wavy with age

Stemless cups direct from ground

A small funnel-shaped fungus the colour of egg yolk growing amongst the leaf-litter of a deciduous or coniferous wood is likely to be the much-coveted chanterelle. The faint apricot smell distinguishes it from the worthless orange-coloured false chanterelle (*Hygrophoropsis aurantiaca*), which may live in the same habitat. More dangerously, the chanterelle has been confused with the similar-coloured but deadly poisonous *Cortinarius speciosissimus*.

Frequent, especially in birch, beech and pine woods.

In autumn, connoisseurs seek out the horn of plenty, or black trumpet, in the damp leaf litter below beech or sometimes oak trees. Its horn-like body is hollow right down to the base, which tapers. The edge is always very wavy. There are no gills or folds on the underside, but the outside of the thin trumpet has membranous streaks where the white spores are produced. When dried the fungus shrivels like a leaf and can be used in stews whole or powdered.

Widespread in deciduous woods, especially beech.

This common woodland species starts with a mushroom-shaped domed cap, but the margin remains inrolled so the cap forms a deep depression as it matures. It becomes sticky when wet. The whole fungus is a yellow-brown colour, darkening somewhat with age. The gills, which extend part-way down the stem, bruise dark brown and stain the fingers. They are easily detached by pushing a fingernail up the stalk. The stem often has rusty-looking patches. The brown roll-rim is poisonous.

Very common in woodland, especially birch.

The orange-peel fungus is one of the most attractive edible species. Some people eat it raw as a dessert, seasoned with sugar and kirsch. It grows on bare earth on well-worn footpaths, frequently in woods, its little cups of different ages and sizes springing from the ground after rain. The inside of the cup, where the spores are produced, is brightly coloured; the underside is whitish and downy. With age the round cups become irregular as they abut onto their neighbours.

Common along paths in woods and gardens.

| J | F | M | A | M | J | J | A | S | O | N | D |

| J | F | M | A | M | J | J | A | S | O | N | D |

| J | F | M | A | M | J | J | A | S | O | N | D |

| J | F | M | A | M | J | J | A | S | O | N | D |

Ball-shaped fungi

Giant puffball
Langermannia gigantea
3-32in (7.5-80cm) across

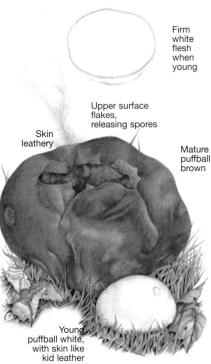

Firm white flesh when young

Upper surface flakes, releasing spores

Skin leathery

Mature puffball brown

Young puffball white, with skin like kid leather

Of all the British puffballs, most of them edible, the giant species probably makes the best eating. However, it must be gathered soon after it emerges from the ground so that the firm pure-white flesh can be sliced like bread and fried. When mature, it becomes a brown mass, with millions of dusty spores enclosed in the outer skin. It can be found in grassy orchards, meadows and farmland with plenty of organic material in the soil. Do not confuse them with the hard earthballs.

Occasional, in meadows and light wood.

Summer truffle
Tuber aestivum
1-3in (25-75mm) across

Flesh whitish to grey-brown, often tinged violet

Dark brownish to purplish-black warty skin

This underground fungus is a relative of the highly prized (and much more richly flavoured) black Perigord truffle (*Tuber melanosporum*) of France, which is traditionally sniffed out by pigs. The British species, a hard, round, warty tubercle, is nevertheless eagerly sought for its nutty flavour. It grows widely in mature southern beech woods, but is rarely found since it lives attached to tree roots just below ground. It occurs from spring to autumn but is commonest in high summer.

Under beech trees in chalky districts.

Common earthball
Scleroderma citrinum
1-3in (25-75mm) across

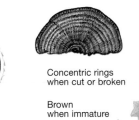

Central puffball with opening for spores

Outer skin segments form 'legs'

Collared earth star
Geastrum triplex

Mature ball contains blackish spore mass

Tough yellow skin with scales and irregular cracks

The earthball grows on bare earth in woodland. It is more rubbery than a puffball and has a purple-black interior containing spores that emerge from cracks in the wrinkled yellowish skin. Earthballs have sometimes been passed off in restaurants as truffles, but have a sour smell and are normally regarded as inedible. The related earth stars grow in woods. They look like an onion bulb at first, but the outer skin opens in a star shape and curls back to lift the ball off the ground.

Common on bare earth in woods.

Crampball
Daldinia concentrica
1-3in (25-75mm) across

Concentric rings when cut or broken

Brown when immature

Hard black balls on dead ash wood

These curious brown or black inedible hard balls grow almost exclusively on ash stumps. The name comes from an old belief that they prevent night cramp if placed in your bed; another name, King Alfred's cakes, is more descriptive since they look just like burnt cakes. They live a long time and release their spores at night after a year's growth. They are very difficult to dislodge from the tree. If they shatter, concentric internal rings – convolutions for spore production – can be seen.

Widespread on ash stumps.

Bracket fungi

Oyster fungus
Pleurotus ostreatus
Cap 2-6in (5-15cm) across

White flesh

Edge often wavy, lobed or split

Greyish, oyster-shaped brackets, waxy when young

Deep white to yellowish gills extend down stem

Stem short, indistinct

Although confused with true bracket fungi, this is a mushroom with gills, not pores. It grows on dead branches – especially beech but sometimes willow, poplar or birch. Most of the fungus lives within the wood. The brackets are the reproductive bodies, carrying the spores. As the brackets age they lose their grey colour and fade to light brown. The white to cream gills run down the stem. The oyster fungus is widely eaten, and is cultivated.

Widespread, especially on beech trees.

J F M A M J J A S O N D

Jew's ear
Auricularia auricula-judae
1-2½in (25-65mm) across

Velvety outer surface

Liver-brown ear-like brackets, cupped on underside

Translucent flabby flesh

Ear-like convolutions inside

This bracket fungus really does look like a human ear. The name comes from the tradition that Judas Iscariot hanged himself from an elder tree – the species' main host. Although found all year, it is commonest in late autumn. It is moist and rubbery when the tiny ½in (13mm) 'buds' break through the wood – always facing downwards – but they harden and go black when dry. Despite the texture, related species are widely used in oriental cuisine – both cooked and raw in salads.

Widespread and abundant, mainly on elder.

J F M A M J J A S O N D

Beefsteak fungus
Fistulina hepatica
4-10in (10-25cm) across

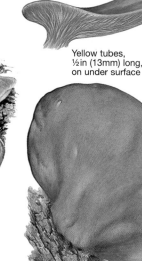

Cut flesh meat-like, with paler veins

Yellow tubes, ½in (13mm) long, on under surface

Sticky, orange-red upper surface

Grows on oak or chestnut trees

Growing like a tongue attached to a wound on a tree-trunk, the beefsteak fungus not only has rich red meat-like flesh but also juice that flows as bright as blood. It may take up to two weeks to reach full size, by which time it has become leathery in texture. The upper surface is sticky – and it moves like skin. It is edible but not very good to eat. It is usually found living on oak and sweet chestnut trees. Wood invaded by its 'roots' (mycelia) turns a rich brown colour and was once prized in cabinet-making.

Widespread in parkland, especially on oak.

J F M A M J J A S O N D

Sulphur polypore
Laetiporus sulphureus
6-16in (15-40cm) across

Cluster of densely packed lobes growing from bark

Pale yellow flesh; rather sour smell

Yellow or orange-yellow, bright yellow edges

Like many bracket fungi, the sulphur polypore is a parasite. Its spores enter the tree via a wound, and the spreading 'roots' (mycelia) rot the heartwood, turning it red in the process. The fungus's fresh yellow or orange-yellow fans shine as bright spots in the woodland scene. It usually grows on an oak but occasionally on a chestnut, cherry or conifer. As it ages the colour fades and the flesh becomes less firm. The sulphur polypore is sometimes called the 'chicken of the woods'; it is eaten but can cause upsets and is best avoided.

Widespread, especially in oak woods.

J F M A M J J A S O N D

Many of these are familiar to woodland ramblers, growing from the trunk or branches of a tree, or from a dead stump. They are often attractive and some make good eating, but sooner or later most spell death to any living tree they attack. The oyster and cauliflower fungi are not true bracket fungi but inhabit similar situations.

Razor-strop fungus

Piptoporus betulinus
4-8in (10-20cm) across

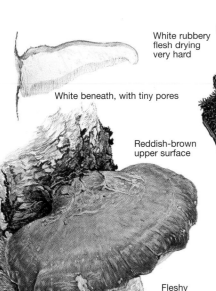

White rubbery flesh drying very hard

White beneath, with tiny pores

Reddish-brown upper surface

Fleshy bracket to 2½in (65mm) thick; wavy edge

The razor-strop is also called the birch polypore, for it lives exclusively on dead or dying birch trees. The brackets start as small white 'buds' breaking through the bark, usually in autumn. When mature they are light brown on top and purest white beneath – and very obvious on the trunk of a tree. The flesh of the fungus becomes leathery when dry and was once used for sharpening razors – hence the common name – or was cut into slices as cork, or made into pin-cushions. It is not edible.

Common in birch woods.

Artist's fungus

Ganoderma applanatum
4-16in (10-40cm) across

More or less flat bracket growing from tree

Greyish to brown dull, knobbly upper surface

Whitish lower surface

Tough red-brown flesh; strong smell

Thick lower tube layer

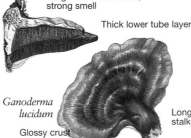

Ganoderma lucidum

Glossy crust

Long stalk

The white under surface of the artist's fungus can be scratched to expose the brown flesh below – hence the name. It grows on living trees, especially beech, which it slowly hollows out by digesting the dead heartwood. It is unusual in having perennial brackets, adding an extra layer each year; they become so woody that it is possible to stand on them. The closely related *Ganoderma lucidum* has a glossy brown upper surface and grows on stumps. Both species are inedible and produce clouds of brown spores, like spilt cocoa.

Widespread, especially on mature beech trees.

Many-zoned fungus

Coriolus versicolor
1-2½in (25-65mm) across

Tough white flesh

Under surface whitish, with pores

Top surface velvety; multicoloured bands

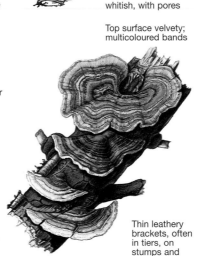

Thin leathery brackets, often in tiers, on stumps and old timber

The colour of this inedible species, a familiar sight on woodland walks, is extremely variable. The zones on the little brackets may be various shades of brown, black, even green, purplish or blue, with intervening light bands. The fungus attacks all sorts of dead wood – mainly stumps and fallen branches. When old it shrivels to a hard black bracket.
A related species, *Coriolus hirsutus*, is covered in short fine hairs.

Widespread and abundant on dead wood.

Cauliflower fungus

Sparassis crispa
6-18in (15-45cm) across

Cut flesh whitish

Hundreds of curly flattened lobes

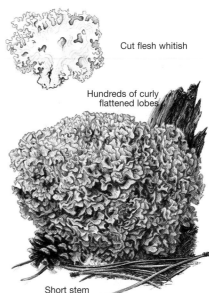

Short stem arising from pine needles

The aptly named cauliflower fungus is instantly recognisable by its yellowish to buff mass of flat branches with their lobed and crisped ends. It grows almost exclusively at the foot of pine trees – sometimes larch or spruce. It is good to eat when young, fresh and sweet-smelling, but removing the dirt, insects and pine needles that inevitably are found among its tightly packed lobes is a laborious job. Some specimens weigh up to 6½lb (3kg) and keep well in cool storage.

Occasional under mature pines in woods and parks.

J F M A M J J A S O N D   J F M A M J J A S O N D   J F M A M J J A S O N D   J F M A M J J A S O N D

195

Ferns

Maidenhair spleenwort
Asplenium trichomanes
Fronds 2-8in (5-20cm) long

Oblong to rounded bright green pinnae

Pinnae sometimes toothed; sori (spore-cases) in rows

Wiry glossy-black midrib

Midrib persists after pinnae drop

Grows on walls

Sori cover under surface from June

Irregular evergreen fronds 1-4in (2.5-10cm) long

Fan-shaped, dull dark green pinnae

Wall-rue
Asplenium ruta-muraria

Thirty or more fronds of different lengths may arise from the central rosette of the maidenhair spleenwort. Each has a distinctive black midrib, with oblong or rounded pinnae ('leaflets') unlike those of the true maidenhair fern, which are fan-shaped. It grows on the vertical surface of rocks and walls, often with rustyback and wall-rue, especially in wet limestone regions. The smaller wall-rue has irregular fronds with dull green fan-shaped pinnae.

Commonest in wetter west and north.

J F M A M J J A S O N D

Rustyback fern
Ceterach officinarum
Fronds 2-6in (5-15cm) long

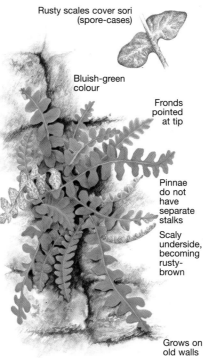

Rusty scales cover sori (spore-cases)

Bluish-green colour

Fronds pointed at tip

Pinnae do not have separate stalks

Scaly underside, becoming rusty-brown

Grows on old walls

The rusty colour of the mature scales on the underside of the fronds identify – and name – this fern of rocks and walls. The scales, which hide the spore-producing sori, are silvery when they first form. The dark green fronds, which persist well into winter, are pointed at the tip and have wider pinnae than those of the maidenhair spleenwort, a close relative. The rustyback is found particularly in limestone crevices and the mortar of old walls.

Commonest in south and west.

J F M A M J J A S O N D

Hard fern
Blechnum spicant
Sterile fronds 4-20in (10-50cm) long

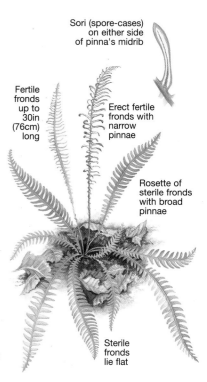

Sori (spore-cases) on either side of pinna's midrib

Fertile fronds up to 30in (76cm) long

Erect fertile fronds with narrow pinnae

Rosette of sterile fronds with broad pinnae

Sterile fronds lie flat

The hard fern has two distinct types of frond. Ladder-like fertile fronds, which bear the reproductive sori, stand erect in the centre, and the more leafy sterile fronds radiate on all sides. The fertile fronds die back, but the non-reproductive fronds remain green all winter. The hard fern is often the only green plant to be seen on the woodland floor in winter, especially in a coppice wood. It may also be found on mossy banks, along lanes and on moors.

Woods and hedgerows, especially in west.

J F M A M J J A S O N D

Ferns are primitive plants, some or all of whose leaves, or fronds, double as reproductive parts; their undersides have spore-bearing bodies called sori. When the spores fall on moist soil, they develop into small scale-like structures called prothalli. These in turn, after a second reproductive stage, give rise to the graceful adult fern.

Polypody
Polypodium vulgare
Fronds 2-24in (5-60cm) long

Bracken
Pteridium aquilinum
Fronds 12-72in (30-180cm) tall

Broad buckler fern
Dryopteris dilatata
Fronds 12-48in (30-120cm) tall

Hart's-tongue fern
Phyllitis scolopendrium
Fronds 6-24in (15-60cm) long

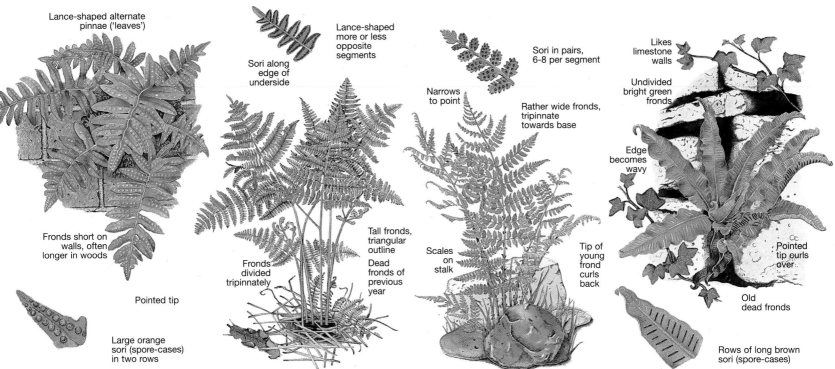

Lance-shaped alternate pinnae ('leaves')

Fronds short on walls, often longer in woods

Pointed tip

Large orange sori (spore-cases) in two rows

Sori along edge of underside

Lance-shaped more or less opposite segments

Fronds divided tripinnately

Tall fronds, triangular outline

Dead fronds of previous year

Sori in pairs, 6-8 per segment

Narrows to point

Rather wide fronds, tripinnate towards base

Scales on stalk

Tip of young frond curls back

Likes limestone walls

Undivided bright green fronds

Edge becomes wavy

Old dead fronds

Pointed tip curls over

Rows of long brown sori (spore-cases)

This common evergreen fern is recognised by the fronds which vary considerably in length and taper to a point. They arise in two rows from the top of a rhizome that grows along or just below the soil surface, or on a wall, rock or tree. The pinnae, or segments, are lance-shaped and the upper ones have distinctive large round sori in rows beneath; these start yellowish but mature bright orange in late autumn. Other polypodies have broader fronds.

Woodland, walls and banks, especially in west.

Bracken, one of our commonest ferns, is described by botanists as tripinnate. This means that each frond is divided into frond-like branches, these in turn branch, and the branches and sub-branches themselves are divided into individual segments. The fern grows from spreading underground rhizomes, and is a pest to farmers as it is difficult to eradicate. It can poison livestock, but they usually avoid it. New fronds arise as curled hairy 'croziers' in spring.

Common everywhere except on shallow chalk.

Pale brown pointed scales with a dark central stripe cover the lower part of the stalks, and are the most characteristic distinguishing feature of this bushy fern. Its fronds, which narrow to a sharp point, are arranged rather like a basket, the young ones curling back at the tip. As with bracken, the fronds are tripinnate, the individual segments being joined together towards the top of each frond.

Widespread except in parts of eastern England and central Ireland; dislikes shallow limestone.

The wavy-edged tongue-like fronds – undivided and more like conventional leaves than normal fern fronds – make this species instantly recognisable. Thanks to the beautiful bright green colour it is a garden favourite, but it does need a moist spot. It often grows on old walls or moss-covered rocks in damp woods. The young fronds unwind from a curled-over tip in spring. The long narrow sori – either side of the midrib – are also distinctive.

Widespread in lowlands, especially in west.

J F M A M J J A S O N D   J F M A M J J A S O N D   J F M A M J J A S O N D   J F M A M J J A S O N D

Ferns and horsetail

Male fern
Dryopteris filix-mas
Fronds 12-48in (30-120cm) long

Pinna segments toothed
Round sori (spore-cases) with flap

Brown scales on stalks

'Shuttlecock' habit

Broad, tapering evergreen fronds, bipinnate

Hook-shaped sori

Lady fern
Athyrium filix-femina

Delicate, feathery fronds
Deeply toothed segments

The name is misleading, since all ferns are bisexual, but it does emphasise the male fern's bolder, coarser and more upright habit as compared with the lady fern. The two can be confused. Both live in woodland, hedgerows and mountains, and both have tapering, pointed fronds that grow from a stout rootstock; both have scales on the lower stalk. But the lady fern's fronds are more feathery and spreading, with more deeply toothed pinnae (frond segments); they die down in winter. It is also slightly less common than the evergreen male fern (one of our commonest species) and is not found in limestone areas.

Common and widespread almost everywhere.

Royal fern
Osmunda regalis
Fronds 24-84in (60-210cm) long or more

Flower-like erect fertile fronds

Broad sterile fronds in tufts

Copious brown spore-bearing organs

Blunt-tipped pale green segments

This fern deserves its title since it is a magnificent bushy plant, resplendent in autumn with its strident orange-brown fronds. Planted in ideal conditions it can reach a height of 10ft (3m) – some specimens live hundreds of years – but on an exposed cliff it may be less than one-quarter as big. It likes wet places: woods, fens, river banks. The broad green outer fronds are sterile and do not produce spores. The fertile spore-bearing fronds grow upright in the middle, and when loaded with spores in summer look rather like flower spikes. Its distinctiveness earns the royal fern a place in a separate fern family.

Common only in western Scotland and western Ireland.

Common horsetail
Equisetum arvense
To 36in (90cm) tall

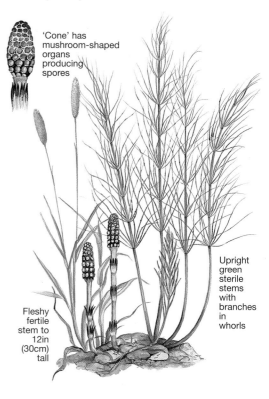

'Cone' has mushroom-shaped organs producing spores

Upright green sterile stems with branches in whorls

Fleshy fertile stem to 12in (30cm) tall

Perhaps best known as a troublesome weed, the common or field horsetail is one of the few survivors of a very primitive group allied to the ferns. Their ancestors included tall trees that formed huge coal deposits 80 million years ago. It grows from a deep, spreading rhizome that is difficult to eradicate. This sends up unbranched brownish fertile stems in April; they are tipped by cone-like spore-bearing structures and die after a few weeks. Then the taller green sterile stems appear, with the narrow branches in layered whorls, giving them a tail-like appearance. These die down in autumn.

Damp places everywhere except northern Scotland.

J F M A M J J A S O N D

J F M A M J J A S O N D

J F M A M J J A S O N D

Mosses

Mosses are difficult to identify without a magnifying glass, but when enlarged you can spot differences in the shape of the leaves and spore capsules (which grow on stalks). They are not so dependent on moisture as the closely related liverworts and may be seen in exposed places.

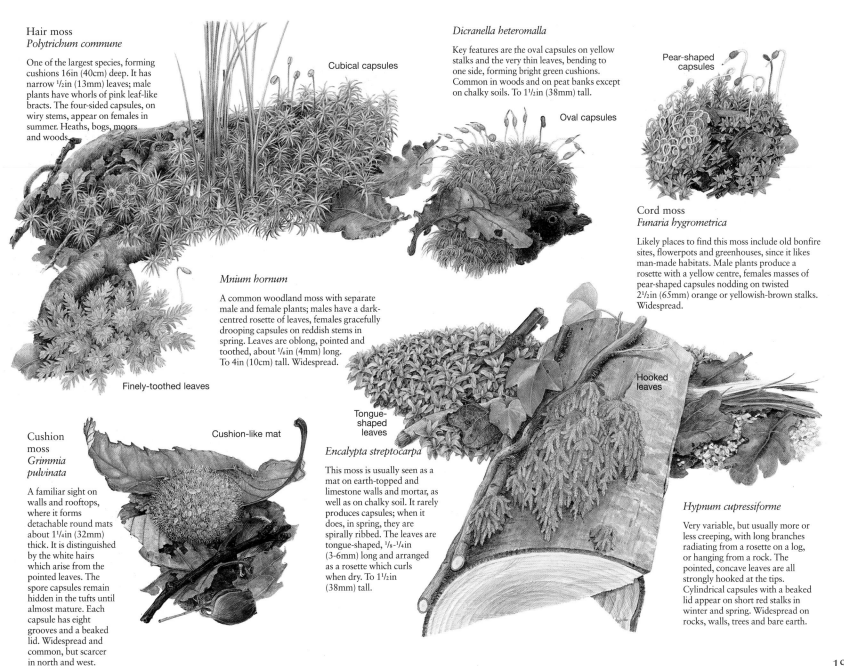

Hair moss
Polytrichum commune

One of the largest species, forming cushions 16in (40cm) deep. It has narrow ¹⁄₂in (13mm) leaves; male plants have whorls of pink leaf-like bracts. The four-sided capsules, on wiry stems, appear on females in summer. Heaths, bogs, moors and woods.

Cubical capsules

Dicranella heteromalla

Key features are the oval capsules on yellow stalks and the very thin leaves, bending to one side, forming bright green cushions. Common in woods and on peat banks except on chalky soils. To 1¹⁄₂in (38mm) tall.

Oval capsules

Pear-shaped capsules

Cord moss
Funaria hygrometrica

Likely places to find this moss include old bonfire sites, flowerpots and greenhouses, since it likes man-made habitats. Male plants produce a rosette with a yellow centre, females masses of pear-shaped capsules nodding on twisted 2¹⁄₂in (65mm) orange or yellowish-brown stalks. Widespread.

Mnium hornum

A common woodland moss with separate male and female plants; males have a dark-centred rosette of leaves, females gracefully drooping capsules on reddish stems in spring. Leaves are oblong, pointed and toothed, about ¹⁄₆in (4mm) long. To 4in (10cm) tall. Widespread.

Finely-toothed leaves

Hooked leaves

Tongue-shaped leaves

Cushion moss
Grimmia pulvinata

A familiar sight on walls and rooftops, where it forms detachable round mats about 1¹⁄₄in (32mm) thick. It is distinguished by the white hairs which arise from the pointed leaves. The spore capsules remain hidden in the tufts until almost mature. Each capsule has eight grooves and a beaked lid. Widespread and common, but scarcer in north and west.

Cushion-like mat

Encalypta streptocarpa

This moss is usually seen as a mat on earth-topped and limestone walls and mortar, as well as on chalky soil. It rarely produces capsules; when it does, in spring, they are spirally ribbed. The leaves are tongue-shaped, ¹⁄₈-¹⁄₄in (3-6mm) long and arranged as a rosette which curls when dry. To 1¹⁄₂in (38mm) tall.

Hypnum cupressiforme

Very variable, but usually more or less creeping, with long branches radiating from a rosette on a log, or hanging from a rock. The pointed, concave leaves are all strongly hooked at the tips. Cylindrical capsules with a beaked lid appear on short red stalks in winter and spring. Widespread on rocks, walls, trees and bare earth.

199

Mosses

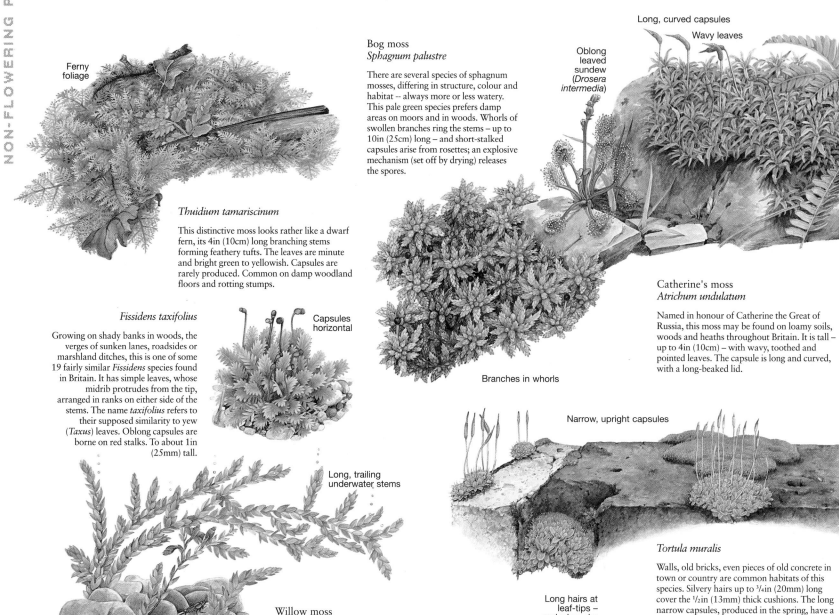

Ferny foliage

Thuidium tamariscinum

This distinctive moss looks rather like a dwarf fern, its 4in (10cm) long branching stems forming feathery tufts. The leaves are minute and bright green to yellowish. Capsules are rarely produced. Common on damp woodland floors and rotting stumps.

Fissidens taxifolius

Growing on shady banks in woods, the verges of sunken lanes, roadsides or marshland ditches, this is one of some 19 fairly similar *Fissidens* species found in Britain. It has simple leaves, whose midrib protrudes from the tip, arranged in ranks on either side of the stems. The name *taxifolius* refers to their supposed similarity to yew (*Taxus*) leaves. Oblong capsules are borne on red stalks. To about 1in (25mm) tall.

Capsules horizontal

Long, trailing underwater stems

Willow moss
Fontinalis antipyretica

This underwater species is the largest moss in Britain; its long, willowy, branched stems often grow up to 30in (76cm) long. It is usually attached to rocks in clear rivers and streams.

Bog moss
Sphagnum palustre

There are several species of sphagnum mosses, differing in structure, colour and habitat -- always more or less watery. This pale green species prefers damp areas on moors and in woods. Whorls of swollen branches ring the stems – up to 10in (25cm) long – and short-stalked capsules arise from rosettes; an explosive mechanism (set off by drying) releases the spores.

Branches in whorls

Oblong leaved sundew (*Drosera intermedia*)

Long, curved capsules

Wavy leaves

Catherine's moss
Atrichum undulatum

Named in honour of Catherine the Great of Russia, this moss may be found on loamy soils, woods and heaths throughout Britain. It is tall – up to 4in (10cm) – with wavy, toothed and pointed leaves. The capsule is long and curved, with a long-beaked lid.

Narrow, upright capsules

Long hairs at leaf-tips – curl when dry

Tortula muralis

Walls, old bricks, even pieces of old concrete in town or country are common habitats of this species. Silvery hairs up to ³/₄in (20mm) long cover the ¹/₂in (13mm) thick cushions. The long narrow capsules, produced in the spring, have a pointed lid. If this is removed the 32 twisted hair-like teeth are exposed.

Liverworts and lichens

Liverworts are closely related to mosses but have flat green fleshy lobes. Some lichens are also leaf-like but others form mere crusts on a rock or tree; all are a unique combination of a fungus and an alga living as one unit.

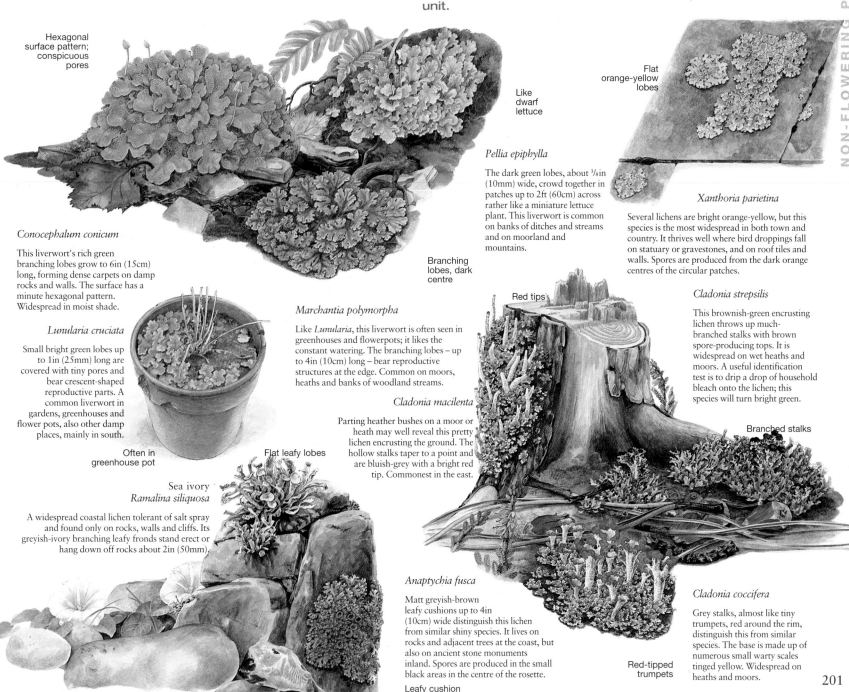

Hexagonal surface pattern; conspicuous pores

Conocephalum conicum

This liverwort's rich green branching lobes grow to 6in (15cm) long, forming dense carpets on damp rocks and walls. The surface has a minute hexagonal pattern. Widespread in moist shade.

Lunularia cruciata

Small bright green lobes up to 1in (25mm) long are covered with tiny pores and bear crescent-shaped reproductive parts. A common liverwort in gardens, greenhouses and flower pots, also other damp places, mainly in south.

Often in greenhouse pot

Flat leafy lobes

Sea ivory
Ramalina siliquosa

A widespread coastal lichen tolerant of salt spray and found only on rocks, walls and cliffs. Its greyish-ivory branching leafy fronds stand erect or hang down off rocks about 2in (50mm).

Like dwarf lettuce

Pellia epiphylla

The dark green lobes, about $\frac{3}{8}$in (10mm) wide, crowd together in patches up to 2ft (60cm) across rather like a miniature lettuce plant. This liverwort is common on banks of ditches and streams and on moorland and mountains.

Branching lobes, dark centre

Marchantia polymorpha

Like *Lunularia*, this liverwort is often seen in greenhouses and flowerpots; it likes the constant watering. The branching lobes – up to 4in (10cm) long – bear reproductive structures at the edge. Common on moors, heaths and banks of woodland streams.

Cladonia macilenta

Parting heather bushes on a moor or heath may well reveal this pretty lichen encrusting the ground. The hollow stalks taper to a point and are bluish-grey with a bright red tip. Commonest in the east.

Red tips

Anaptychia fusca

Matt greyish-brown leafy cushions up to 4in (10cm) wide distinguish this lichen from similar shiny species. It lives on rocks and adjacent trees at the coast, but also on ancient stone monuments inland. Spores are produced in the small black areas in the centre of the rosette.

Leafy cushion

Flat orange-yellow lobes

Xanthoria parietina

Several lichens are bright orange-yellow, but this species is the most widespread in both town and country. It thrives well where bird droppings fall on statuary or gravestones, and on roof tiles and walls. Spores are produced from the dark orange centres of the circular patches.

Cladonia strepsilis

This brownish-green encrusting lichen throws up much-branched stalks with brown spore-producing tops. It is widespread on wet heaths and moors. A useful identification test is to drip a drop of household bleach onto the lichen; this species will turn bright green.

Branched stalks

Cladonia coccifera

Grey stalks, almost like tiny trumpets, red around the rim, distinguish this from similar species. The base is made up of numerous small warty scales tinged yellow. Widespread on heaths and moors.

Red-tipped trumpets

201

birds

Around 550 species of bird have been recorded in the British Isles out of a worldwide total of some 9000, but many of them have been seen only on a few rare occasions after being swept our way by storms or other freak weather conditions. Only about 130 species are resident all year, but there are many other regular visitors, either coming here to breed after wintering in warm places, or flying in from far northern breeding grounds for our relatively mild winters. Others occur regularly as so-called passage migrants in spring and/or autumn, stopping off in the British Isles on their migratory journeys between their breeding and wintering grounds. The 244 species and two subspecies covered in these pages include both residents and regular visitors.

On the whole, birds live rather brief lives. Their mobility and wildness make them fascinating to watch but also difficult to identify. They cannot be studied at leisure, nor is there a simple way to help you narrow down the possibilities. For this reason, the birds are grouped here in their families, and practice is needed in quickly observing the key features needed to make a positive identification. Get to know the general characteristic of each group, and look out for these important points:

What size is the bird? What is the overall colour and patterning of the plumage? (This often differs between the sexes and with maturity and the time of year.) Are there any conspicuous 'fieldmarks'? What is its stance and how does it behave, in flight or on the ground? The song or call may be another clue. Finally, take into account where and when you observe the bird: the habitat as well as geographical area, the time of day as well as the date, are all useful clues. But remember that there will always be occasions when such 'rules' are broken.

Divers and grebes

Red-throated diver
Gavia stellata
21-23in (53-58cm)

Deep wing-beats

Summer

Uptilted bill

Red throat

Winter

Throat colouring absent; division between cap and throat blurred

Black-throated diver
Gavia arctica
23-27in (58-68cm)

Black and white neck and back

Shallow wing-beats

Summer

Winter

Darker than red-throated diver, with sharper demarcation between dark and light areas

Great northern diver
Gavia immer
30-33in (76-84cm)

Slow, powerful wing-beats

Summer

Spotted back

Neck patches

Heavy, straight bill, held horizontally

Winter

White-billed diver
Gavia adamsii

Paler, uptilted bill

In summer dress, the smallest and commonest of our divers sports a blood-red throat patch and a pearl-grey head, which make it unmistakable. Identification in winter is complicated by the fact that the face and red throat are moulted to white, making confusion with other divers a possibility. At this time, however, it may be distinguished by the distinctly upturned bill, the blurred division between the white throat and darker cap, and the white speckled back. During the breeding season the red-throated diver inhabits small, shallow lochs in the Scottish Highlands.

Breeds on Highland lochs; seen all around coasts in winter.

One of the most haunting sounds to be heard in the Highlands of north-western Scotland is the wailing cry of the black-throated diver as it asserts ownership of its territory. Unlike the slightly smaller red-throated diver it prefers bigger stretches of water on which to breed, but moves to the coast in the winter. In summer the adults have large white patches on the black back, and a black throat. After breeding, however, this plumage is lost, leaving only the straighter bill, and sharper demarcation between the dark and light areas of the head and neck, to distinguish this species from its red-throated cousin.

Breeds in Highlands; on most coasts in winter.

Many cinemagoers will have heard the wailing call of this species, for it is so eerie that recordings are often dubbed into scenes in thrillers to heighten suspense. In Britain this breeding call is unlikely to be encountered in the wild, for the species is mainly a winter visitor from Iceland. The striking greenish and black and white plumage of the breeding season is lost in winter, but the bird is still identifiable by its large size, bulky dagger-like bill and steep forehead.
The white-billed diver, a vagrant from arctic Russia, is even larger, with paler colouring.

Coasts of north-west Britain and western Ireland.

J F M A M J J A S O N D   J F M A M J J A S O N D   J F M A M J J A S O N D

Divers are large, streamlined swimming and diving birds, with stout necks, short tails and sharply pointed bills. Grebes are smaller, with longer, thinner necks. Both families are clumsy on land but expert in the water, diving for fish and other food. Flight is fast and direct.

Little grebe

Tachybaptus ruficollis
10-12in (25-30cm)

No wing-patches

Often in flight, low over water

Pale patch near base of bill

Chestnut throat

Summer

Dull brown above

Winter

Buffish below

Alarmed bird submerges body

During its busy search for small fish or water insects on reedy lakes and rivers, the little grebe resembles a small fluffy ball, diving frequently and bobbing up again. It is often called the dabchick, and is not much larger than a duckling. Its nest is a floating platform of water plants, often close to the bank.

Widespread on lowland fresh water.

Great crested grebe

Podiceps cristatus
19in (48cm)

Winter

White wing-patches

White extends over eye

Double crest

No white above eye

Dusky neck

Ruff

Chicks carried on back

Red-necked grebe
Podiceps grisegena

Before the breeding season, both sexes of great crested grebe acquire conspicuous dark head plumes which are erected during their elaborate courtship display. This involves head-shaking, diving, fluffing out the plumes and presenting each other with water plants while rising from the water breast to breast. During the autumn the plumes are lost, but the bird can still be distinguished by its very white face and neck.

The red-necked grebe ventures to Britain's eastern coasts chiefly when in its drab winter garb. It is smaller and darker than the great crested grebe.

Large lakes, except in far north; also on coasts in winter.

Black-necked grebe

Podiceps nigricollis
12in (30cm)

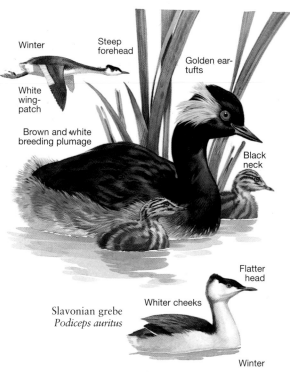

Winter

Steep forehead

Golden ear-tufts

White wing-patch

Brown and white breeding plumage

Black neck

Flatter head

Slavonian grebe
Podiceps auritus

Whiter cheeks

Winter

This attractive small grebe prefers ponds with a rich growth of water plants for breeding. Only a few such sites exist in Britain, and the bird is scarce as a breeding species. Outside the breeding season it makes for estuaries, sea channels and large lakes.

The scarce Slavonian grebe breeds on the remoter lochs of the Scottish Highlands and winters on coastal waters. Its summer plumage is chestnut and black with long golden ear-tufts; in winter, the flatter head, straight bill and white neck distinguish it from the black-necked grebe.

Breeds very locally in England and Scotland; winters mainly on south coast.

Gannet and cormorants

Gannets are large seabirds, often seen diving from the air for fish. Cormorants dive from the surface, chasing fish under water, and often sit on a rock with their wings half open, drying out their feathers.

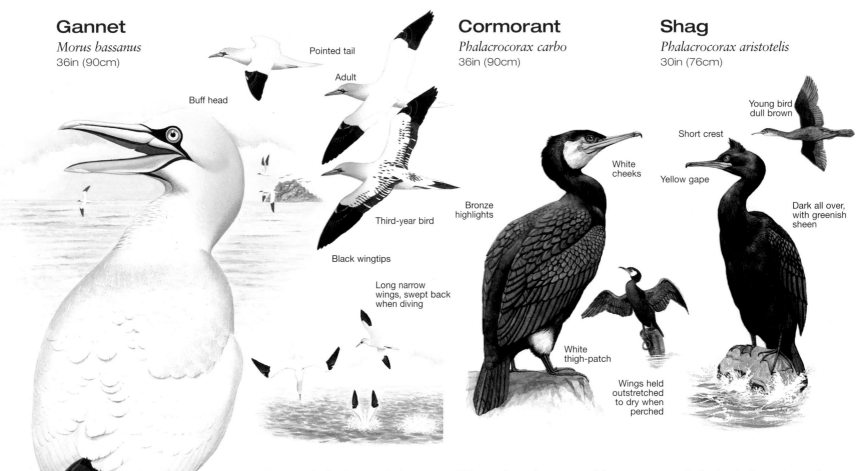

Gannet
Morus bassanus
36in (90cm)

Pointed tail

Adult

Buff head

Third-year bird

Black wingtips

Long narrow wings, swept back when diving

Cormorant
Phalacrocorax carbo
36in (90cm)

White cheeks

Bronze highlights

White thigh-patch

Wings held outstretched to dry when perched

Shag
Phalacrocorax aristotelis
30in (76cm)

Young bird dull brown

Short crest

Yellow gape

Dark all over, with greenish sheen

When crowds of gannets nest in dense colonies they become very aggressive. Although each nest is set beyond pecking distance of neighbouring birds, squabbles often break out between adults as they stray into each other's territory. Gannets plummet into the sea from as high as 100 ft (30m) to catch herring-sized prey. Nest sites include sea-cliffs and adjacent slopes.

Coastal; breeds mainly on islands and sea-cliffs in north and west.

Fishermen have often persecuted the cormorant in the belief that its fishing skill and enormous appetite deplete stocks of fish such as trout and salmon. In fact the bird's diet consists mainly of flat fish and eels, and it takes only a small proportion of game fish. Cormorants nest in colonies which sometimes number hundreds of pairs. They build large nests on cliff tops and rocky islets, or in trees beside lakes.

On coasts, and inland on lakes and rivers.

Over the last hundred years the shag has generally increased around British coasts, and it is still doing so in some areas. In summer, the shag's glossy, greenish-black plumage, short crest on the front of the head and smaller size distinguish it from the related cormorant. Its bulky nest of sticks and seaweed, lined with grass, is placed on inaccessible rocky ledges and in sea caves.

Rocky coasts, breeding mainly in west and north.

| J | F | M | A | M | J | J | A | S | O | N | D |

| J | F | M | A | M | J | J | A | S | O | N | D |

| J | F | M | A | M | J | J | A | S | O | N | D |

Shearwaters and petrels

The ocean-going shearwaters have long, narrow wings and a short tail, and often glide close to the water. Petrels are also ocean birds, but may appear inland after gales. They flutter over the water searching for food.

Fulmar

Fulmarus glacialis
18½in (47cm)

Manx shearwater

Puffinus puffinus
14in (36cm)

Storm petrel

Hydrobates pelagicus
6in (15cm)

Long narrow wings, held stiffly outstretched

White underparts catch light as birds bank and turn

White head

Grey back

Heavy bill with tube-shaped nostrils

Courting pairs squat on cliff ledges

White underparts

Huge flocks gather at dusk

Black upper parts

Long, narrow, stiff wings

Sits high in water

Sooty-brown plumage

White patches under wings

Pale wing-bar

White rump; square tail

The population of this species in Britain and Ireland has increased astonishingly since the first pioneers arrived in the Shetlands in 1878. There are now over 500,000 breeding pairs around our coastline. This expansion was probably aided by the growth of the fishing industry, and the large quantities of offal discarded by fishermen after cleaning their catch. Intruders into the fulmar's nesting areas may be sprayed with a stinking oil if they approach too close to the rocky ledge where a bird has laid its single egg. This fluid comes from the bird's stomach and is shot out through its nostrils to a distance of 4ft (1.2m).

On most cliffbound coasts.

Powerful wing-beats, interspersed with long glides which appear to 'shear' the waves, make the Manx shearwater one of nature's longest and most accomplished travellers. One bird ringed on the Welsh island of Skokholm reached Australia during its oceanic wanderings, while another taken to Massachusetts, more than 3000 miles across the Atlantic, had returned to its chick within 12 days. The Manx shearwater is usually seen in flight, or sitting buoyantly on the sea. It comes to land only during the breeding season, and then only during darkness.

Breeds on isolated islands in western Britain, Ireland and on Shetland.

Mariners once believed that the sight of petrels following a ship was a warning of an impending storm. In fact the birds – the smallest European seabirds – were feeding on the marine life disturbed by the ship's wake. The storm petrel spends most of its life on the open sea and has the habit of pattering – apparently 'walking' – on the surface of the water. When breeding, it comes ashore at night, to remote islands off western Britain. Colonies vary in size from a few pairs to many thousands. The favourite nesting sites are on beaches, but stone walls are also used as shelter for the single egg.

Breeds on islands off western Britain and on coast of Ireland.

J F M A M J J A S O N D

J F M A M J J A S O N D

J F M A M J J A S O N D

Herons

This family of large, long-legged wading birds has only two resident British members – the familiar grey heron and the much shier bittern. Both hold their necks retracted in flight and often also on the ground. In the air, their legs stretch out behind.

Grey heron

Ardea cinerea
36in (90cm)

Neck drawn back in flight

Courting male stretches neck up then lowers it over back

Black crown

Bill brighter colour in breeding season

Purplish-brown fore-wing and belly

White crown

Black crest

Long sharp bill

Purple heron
Ardea purpurea

Streaked neck

Long legs

Bittern

Botaurus stellaris
30in (76cm)

Compact and owl-like in flight

Birds climb reeds by grasping them in clumps

Plumage brown with black streaks

Stockier build than heron

Green legs

Its large size, long legs, and long sinuous neck make the heron unmistakable. Poised alert and motionless in or beside shallow water, the watchful bird waits patiently for a fish or small animal to come within range of its long, dagger-like bill. Then it stabs the prey and swallows it whole. Herons nest in colonies, usually in tall trees, but sometimes on cliffs and in reed-beds and bushes. The purple heron is an uncommon visitor from mainland Europe to reed pools in southern England.

Widespread on rivers, lakes, estuaries and seashores.

Once fairly common in marshy areas of Britain, the bittern was exterminated as a breeding species in 1868 by man's shooting. Forty years later migrants from Europe managed to recolonise in Britain, but after a peak in the 1950s they again declined with the drainage of many reed-beds. Bitterns rarely take to the air, preferring to skulk among the reeds where their brown and black streaked plumage makes them difficult to see. They have a resonant 'boom' call in the breeding season.

Reed-beds, mainly in eastern England.

J F M A M J J A S O N D

Swans

Swans are the largest of Britain's wildfowl, graceful in the water and powerful in the air. The orange bill distinguishes the resident mute swan from the two winter visitors, which also hold their necks straighter. The migrant species are also much noisier.

Bewick's swan

Cygnus columbianus
48in (120cm)

Neck shorter than in other swans

Rounded head

Small, rounded yellow patch

Grey-brown

Juvenile

Adult

Mute swan

Cygnus olor
60in (152cm)

Neck outstretched

Wings produce throbbing 'hum'

Orange bill with black knob at base

Curved neck

Whooper swan

Cygnus cygnus
60in (152cm)

Slow, silent wing-beats

Large angular yellow patch

Long thin neck

Family groups graze in fields in winter

Juvenile paler brown than Bewick's swan

High-pitched honking and crooning sounds are heard as V-shaped skeins of Bewick's swans wing across Britain's winter skies. These migrants from Siberia appear on marshland pools, where they feed on seeds and water plants. Bewick's swan is very goose-like, with a short neck and rounded head. These features, together with the rounded yellow bill patch, help to distinguish it from the larger whooper swan. The bill patches, like human fingerprints, are peculiar to individuals.

Wet meadows in eastern and western Britain and Ireland.

In powerful flight, mute swans present a graceful spectacle and make an exciting, unmistakable sound – a throbbing 'wing music'. Britain's only resident swans, they are quieter than the migrant species but not entirely mute: they hiss and snort when angry, and can even trumpet weakly. The mute swan is the world's second heaviest flying bird – only slightly behind the Kori bustard of Africa – and can weigh 40lb (18kg). A serene appearance belies its aggressiveness when breeding.

Widespread in Britain and Ireland, except extreme north.

Unlike the mute swan, the whooper is particularly noisy, producing the loud trumpeting call which accounts for its name. In flight, on the other hand, whoopers are relatively quiet, their wing-beats making only a swishing sound. Although the odd pair may nest in Scotland, they are mainly winter visitors from Iceland. The long thin neck and long bill distinguish the whooper from the smaller Bewick's swan.

Open waters and farmland in Ireland and Britain; occasionally breeds in Scotland.

J F M A M J J A S O N D
J F M A M J J A S O N D
J F M A M J J A S O N D

209

Geese

Bean goose
Anser fabalis
28-35in (70-89cm)

Long dark head and neck visible in flight

Flocks forage in fields

Juvenile bird has duller legs

Black and yellow bill

Dark head and neck

Orange legs and feet

Pink-footed goose
Anser brachyrhynchus
24-30in (60-76cm)

Pale forewing visible in flight

Often fly in large, noisy flocks

Grey back

Pink on bill

Juvenile browner, with duller legs and feet

Pink legs and feet

Flocks pick over stubble after harvest

White-fronted goose
Anser albifrons
26-30in (66-76cm)

Dark forewing visible in flight

Flocks may also feed at night

White forehead

Black bars

Orange legs

On marshy grassland in East Anglia or central Scotland in winter, a scattered flock of large geese may sometimes appear. At first they resemble greyish-brown farmyard geese, but closer inspection may reveal the browner plumage, long necks and black and yellow bills of bean geese, scarce visitors from northern Scandinavia and Russia. The bean goose is unique among our geese in frequently nesting amongst the birch and pine trees of northern Europe instead of in open country. The nest is concealed under the trees.

Wet meadows in eastern England and central Scotland.

To the wildfowl enthusiast, the sound of an approaching flock of wild and wary pink-footed geese is thrilling music. The calls of individuals vary widely in pitch between 'ang-ang' and 'wink-wink', producing a chorus that has led some experts to call them the most musical of grey geese. Numbers visiting Britain have increased in recent decades, probably because of a reduction in shooting and better protection of their roosts. The visitors to Britain come from Greenland and Iceland.

Farmland in Scotland and England, close to secure estuarine or open water roosts.

A white blaze on the forehead, black bars across the belly, and orange legs and feet make the white-fronted goose the most distinctive of the grey geese. Young birds, however, lack these characteristic body markings and can be confused with other species. White-fronted geese flock in from their Arctic breeding grounds in Greenland and Russia in late September or early October to winter on the marshlands of Britain and Ireland.

Marshes and meadows, mainly on western and eastern coasts.

Immature bird lacks white forehead

J F M A M J J A S O N D   J F M A M J J A S O N D   J F M A M J J A S O N D

Geese are large, heavily built birds, with a long neck and short legs. They fly fast and powerfully, frequently in large flocks that maintain formation. Although they are good swimmers they feed mainly on land, and are usually seen on marshes and fields.

Greylag goose

Anser anser
30-35in (76-89cm)

Canada goose

Branta canadensis
36-40in (90-100cm)

Barnacle goose

Branta leucopsis
23-27in (58-68cm)

Brent goose

Branta bernicla
22-24in (55-60cm)

Wild geese often fly in V-shaped 'skein'

Paired adults re-enact courtship ritual

Orange bill

Pink legs

White chin-patch

Black head and neck

Long neck and deep wing-beats

Birds stay in safe waters after breeding

Light belly

Wings flicked in mating display

White face

Black neck

Grey back with white-edged black bars

Dark plumage and white stern

Black head and neck

White neck-patch

Young bird lacks white neck-patch

Ashy grey belly

White stern

The greylag was once the only goose that bred in Britain. It was the ancestor of the farmyard fowl, and its cackles and 'aang-ang-ang' honk in flight are similar to those of the domestic bird. The greylag was driven back to the remoter parts of Scotland when agricultural development destroyed its breeding grounds. In recent times, however, it has been re-introduced to many of its old areas.

Breeds on wetlands throughout Britain.

The first specimens of this very large goose were brought to Britain from Canada in the 17th century as decorative birds for parkland lakes. Later, attempts were made to increase their numbers as game birds; but the Canada goose is too tame and flies too low to make a sporting target. It has now spread out of its parkland homes, and its numbers are still increasing. The nest consists of plant material placed close to water, and is defended aggressively.

Grassland by lakes, mainly in England.

In the air or on the ground, family groups of barnacle geese bicker amongst themselves with a noisy dog-like yapping. Rarely silent for long, they produce the loudest clamour of all when taking flight. Coastal grass is their favourite food, but they will also graze on pasture land. Wintering flocks come to Britain from Spitzbergen or Greenland. Introduced birds now nest locally in England.

Coastal marshes, mainly in western Scotland and Ireland.

Small dark Brent geese begin to arrive in eastern and southern England in large numbers during October from their breeding grounds in the Arctic tundra. The young fly south when only three months old. The almost black plumage is broken only by a small, white neck-patch and striking white undertail coverts. In flight Brent geese form long wavering lines, usually low above water or ground.

Tidal flats and estuaries in eastern and southern England, and coastal Ireland.

J F M A M J J A S O N D

J F M A M J J A S O N D

J F M A M J J A S O N D

J F M A M J J A S O N D

Ducks

Shelduck
Tadorna tadorna
24in (60cm)

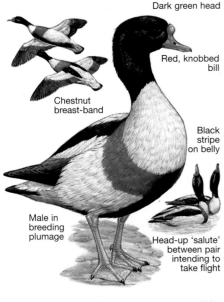

Dark green head

Red, knobbed bill

Chestnut breast-band

Black stripe on belly

Male in breeding plumage

Head-up 'salute' between pair intending to take flight

Female similar, lacks bill knob

Wigeon
Anas penelope
18in (45cm)

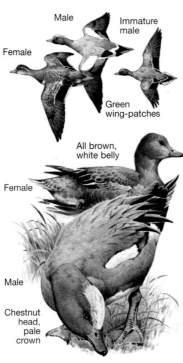

Male

Immature male

Female

Green wing-patches

All brown, white belly

Female

Male

Chestnut head, pale crown

Teal
Anas crecca
14in (36cm)

Female

Male

Chestnut head with green eye-patch

Male

Black and yellow under-tail feathers

Female dark brown, with green wing-patch

Male in eclipse

Garganey
Anas querquedula
15in (38cm)

Male

Blue-grey forewing

Female

Male

Chocolate-brown head

Pale eye-stripe

Grey-brown wings

Female has stripey face

Large numbers of these colourful waterfowl leave Britain each summer to moult in the tidal estuaries of the Heligoland Bight off the north German coast. They return in autumn, and after three to six months their dull eclipse plumage gives way to handsome breeding dress. The nest is often placed in a rabbit hole or similar location. On hatching, the young are led to water and form crèches under the eye of a few resident adults.

Estuaries all around coast; sometimes inland.

Although they feed in water and occasionally 'up-end' in typical duck fashion, wigeon are unusual in that they also often graze on grass like geese. They may be seen flying in long irregular lines as they pass by on migration or to winter feeding grounds, when the sweet 'whee-ooo' call of the drake can be heard. The majority of birds wintering in Britain are migrants from northern Europe.

Freshwater pools and estuaries; breeds in north.

With their variegated colouring, male teal are attractive little ducks, but because they are a favourite quarry of wildfowlers they are often too wary to allow close views. Teal fly fast, and spring vertically into the air when alarmed. They are found on small ponds, marshland drains and mud-flats, where they dabble in the shallows for plants and small animals. Only about 1500 pairs breed in the British Isles, but winter migrants increase the numbers.

Ponds and marshes; commoner in winter.

Garganey are particularly timid birds, and the most frequent sighting of them may be of a pair springing from a pool in alarm, the drake showing a blue-grey forewing. On the water the male is readily distinguished by a white crescent above and behind the eye. In their post-breeding plumage, drakes resemble ducks with their mottled brown colouring. Both display green patches in the hindwing.

Reedy pools; breeds mainly in eastern England.

J F M A M J J A S O N D   J F M A M J J A S O N D   J F M A M J J A S O N D   J F M A M J J A S O N D

In nearly all species of ducks, males in breeding plumage are more brightly coloured than females. Like other wildfowl, ducks fly fast and powerfully. They are mainly to be seen around the coast, on inland waters such as ponds and lakes, and on marshes. Only the wigeon commonly feeds on dry land like geese.

Gadwall
Anas strepera
20in (50cm)

Male

Female

Sharp-pointed wings; white rear wing-patch

Pale brown head and black bill

Female

Male

Black tail coverts

Red-brown, black and white wing-patches

Orange legs

Mallard
Anas platyrhynchos
23in (58cm)

Female

Green head

Male

Yellow bill

White collar

Rich brown breast

Violet-blue wing-patch

Orange-yellow bill

Moulting male resembles female but bill yellow

Female

Shoveler
Anas clypeata
20in (50cm)

Blue-grey forewing

Male

Green head

Shovel-shaped bill

White breast; chestnut sides

Brown head and body; blue-grey forewing

Female

Pintail
Anas acuta
22in (55cm) excluding drake's tail

Male

Long slender neck

Female

Chocolate head, white neck-stripe

Female pale brown

Male

Long tail

Black under-tail coverts

Before 1850 this duck was known only as a winter immigrant from its homelands of central and western Asia and North America. Today, a few breed in Scotland and, after a huge increase, many hundreds of pairs now nest in England, the majority in East Anglia. Eggs are laid in May in a ground hollow. As with all ducks, the nest is lined with down feathers pulled from the female's breast.

Freshwater wetlands throughout lowland Britain.

In town and country, this is the most familiar duck in the British Isles. It is as much at home on a park lake or city canal as it is on a quiet country backwater or remote reservoir. Mallards in towns are very tame, but those in rural areas are wary, for they are much sought after by wildfowlers. Mallards are typical of dabbling ducks in that they feed on the water surface, or by up-ending, and spring straight up into the air from the water.

Resident near water in all areas.

The long spade-like bill that gives the shoveler its name is used, in the typical manner of dabbling ducks, for sifting large volumes of water to filter out small aquatic plants and animals. The shoveler's patchy distribution is governed partly by the availability of marshy areas with muddy shallows rich in food. The ducklings start to develop the large bill when very young.

Shallow waters in most lowland parts of Britain and Ireland.

Both on the ground and in the air, the pintail is the most elegant of the British ducks. Its long slender neck and wings and its long tail, which can add an extra 8in (20cm) to the male's length, make it easy to recognise and attractive to watch. Most pintails spend only the winter in Britain, and the breeding population is no more than 50 pairs, in scattered sites in Scotland and eastern England. Their nests are less camouflaged than those of other ducks.

Estuaries, flooded grassland and large lakes.

J F M A M J J A S O N D

J F M A M J J A S O N D

J F M A M J J A S O N D

J F M A M J J A S O N D

213

Ducks

Pochard

Aythya ferina
18in (45cm)

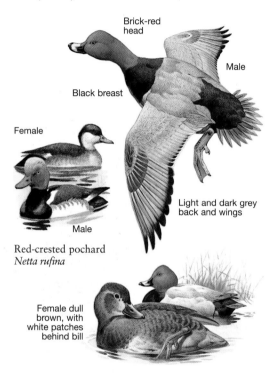

Brick-red head

Male

Black breast

Female

Light and dark grey back and wings

Male

Red-crested pochard
Netta rufina

Female dull brown, with white patches behind bill

The brick-red head, black breast and grey beak of the pochard drake are distinctive; the female is a dowdy greyish-brown with lighter streaks. Pochards nest in reed-beds and other vegetation bordering fresh water. In winter, sites such as gravel pits and reservoirs are favoured; there the birds dive for molluscs, other animals and plants. The nest may be on the ground, but sometimes a platform is made of water plants built up from the bottom in shallow water.
The attractive red-crested pochard, when seen wild in Britain, is probably an escape from a wildfowl collection.

Lowland waters, mostly in eastern Britain; winter immigrants widespread.

Tufted duck

Aythya fuligula
17in (43cm)

Long, pointed wings

White wing-bar

Purplish head and tuft

White sides

Male

Female brown

Male strikes 'bill down' pose after mating

A stranger to Britain before 1849, the tufted duck is now the country's commonest diving species. These ducks, which have become very tame, have been helped by the development of lakes from disused gravel pits and by the spread of reservoirs. Another encouragement was the introduction of the zebra mussel, a favourite source of food, from Russia in the last century. These ducks also eat small fish and insects. Males are black and white with a delicate drooping crest; females are brown, with pale sides. In after-breeding plumage drakes become browner. Both sexes show a white wing-bar in flight.

Widespread on fresh water; immigrants in winter.

Scaup

Aythya marila
19in (48cm)

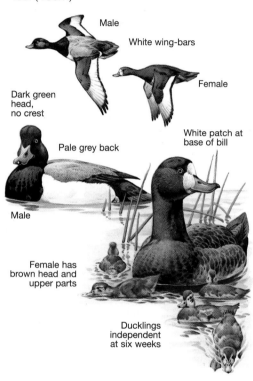

Male

White wing-bars

Dark green head, no crest

Female

Pale grey back

White patch at base of bill

Male

Female has brown head and upper parts

Ducklings independent at six weeks

This duck's name may have come from its habit of feeding on broken shells, called scaup; its diet is largely made up of mussels, which it obtains by diving under water. In winter it is mainly a visitor from northern Europe, gathering in bays and estuaries; but rivers and lakes are preferred by the very few that have bred in Britain. These are protected by law. Scaup often nest in loose colonies on islands, where a hollow is selected and a nest made from grass and feathers.

Estuaries and coasts, rarely breeds.

Scaup rest on sand-banks

J F M A M J J A S O N D   J F M A M J J A S O N D   J F M A M J J A S O N D

Eider

Somateria mollissima
23in (58cm)

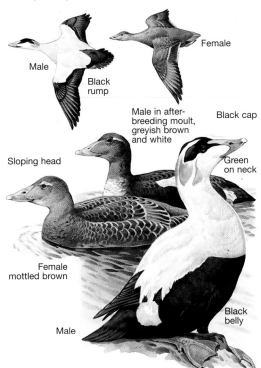

Male

Black rump

Female

Male in after-breeding moult, greyish brown and white

Black cap

Sloping head

Green on neck

Female mottled brown

Black belly

Male

The soft breast feathers of the eider duck have long been prized by man as a filling for the warm bedcover to which they have given their name – the eiderdown. In nature, the duck plucks the down from her breast to line her nest and protect the clutch of three to ten eggs. The down is still collected from the nests, in strictly controlled quantities, for commercial use. The nest is on the ground, often in an exposed site where the duck's drab colouring acts as camouflage. The male, by contrast, has striking black and white plumage, with tints of pink and green. Eiders feed mainly on crabs and molluscs.

Rocky coasts, breeding in Scotland and northern Ireland.

Goldeneye

Bucephala clangula
18in (45cm)

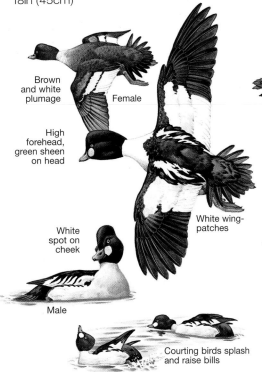

Brown and white plumage

Female

High forehead, green sheen on head

White wing-patches

White spot on cheek

Male

Courting birds splash and raise bills

This striking bird is normally a winter visitor from northern Scandinavia and Asia, but it also nests in Scotland, usually in tree holes close to water. The male has black and white plumage and a large angular head with a green sheen; there is a white spot on each cheek. The female is mainly grey except for a chocolate head. In winter, goldeneyes inhabit coasts and inland waters, diving for water animals and shellfish. Their wings make a loud whistling noise.

Inland waters and coasts, except in south-west.

May nest in holes in trees

Common scoter

Melanitta nigra
19in (48cm)

Female

Male

All-dark wings

Yellow patch on bill

Male

Pale face, dark cap

Female

White patches on face and wings

Velvet scoter
Melanitta fusca

In contrast to the multi-coloured finery of many drakes, the male common scoter is almost black, with only an orange-yellow patch on its bill. Females are all brown, with a pale face. The 100 or so British breeding pairs select lochs in mountainous or moorland country, building a nest in a ground hollow near the shore. At this time adults feed on vegetable matter, insects and a few molluscs. When they move to the coast they feed almost exclusively on mussels. In winter the British population is joined by birds from Arctic and sub-Arctic regions. The larger velvet scoter has white patches on the wings. It is mainly a winter visitor to northern and eastern coasts.

Around the coast, breeding inland in Ireland and Scotland.

Ducks

Red-breasted merganser
Mergus serrator
23in (58cm)

Female

Dark green head

Larger white wing-patches

Double crest

Male

Ginger head

Red-brown breast

White chin and throat

Female

Male

Goosander
Mergus merganser
26in (66cm)

Male

Female

Large white wing-patch

White wing-patch

Chestnut head

Green head

White chin, dark neck

Female

Male

Pinkish-white body

Smew
Mergellus albellus
16in (40cm)

Female

Dense flocks fish together

White wing-flashes stand out

Male

Courting male uses head gestures

White crest

Female

Black eye-patches

Chestnut cap; white cheeks

Black lines on breast

Male

These ducks have a bad reputation among fishermen because of their taste for trout and salmon fry. Their defenders, however, say that they take as many predatory coarse fish, thus helping the game fish to survive. Like the goosander and smew, the red-breasted merganser has tooth-like projections on its bill, which help it to grasp slippery fish. Since about 1950, the merganser has spread from Scotland into England and Wales as a breeding species. It nests in dense vegetation by rivers.

Breeds near fresh water except in lowland England; winter visitor to coasts.

Although the goosander suffers persecution by man because of its liking for small salmon and trout, it has managed to spread from Scotland into England, Wales and Ireland. Goosanders nest in holes in banks, among boulders, or in holes in trees. When breeding they stay close to rivers, but in winter they move to larger stretches of water, mainly inland. The female has a brown foreneck, unlike the similar female merganser.

Nests in hollow tree or rock crevice

Upland rivers, wintering also on lowland lakes.

From Russia and northern Scandinavia these small handsome ducks come to winter in our south-eastern estuaries and freshwater lakes. Although striking in appearance, the smew is not easily or often seen, for it is a shy and elusive bird that flies fast. Smews sometimes eat vegetable matter, but their main source of food is fish, molluscs and crustaceans which they catch under water with the aid of a saw-edged bill. They breed in crevices and holes in trees in coniferous forests close to shallow lakes and narrow meandering rivers.

Lakes and reservoirs in eastern and southern England.

J F M A M J J A S O N D

J F M A M J J A S O N D

J F M A M J J A S O N D

Long-tailed duck

Clangula hyemalis
16in (40cm) excluding drake's tail

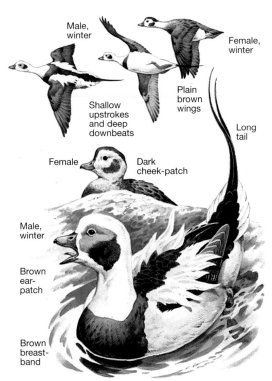

Male, winter

Female, winter

Plain brown wings

Shallow upstrokes and deep downbeats

Long tail

Female

Dark cheek-patch

Male, winter

Brown ear-patch

Brown breast-band

Ruddy duck

Oxyura jamaicensis
16in (40cm)

Male

Plain wings

Brown cap

Female

Stripe on face

Male

Black cap and blue bill; chestnut body and white face

Mandarin duck

Aix galericulata
17in (43cm)

Male

Female

Green wing-patches

White streak on head

Female

White 'spectacles'

Golden ruff

Chestnut-orange wing-fans

Male

The period from the end of September to the end of October sees the arrival in British waters of the wintering population of long-tailed ducks from their northern breeding grounds. The voice of the male is melodious, resonant and far-carrying, and the sound of several males calling together has been likened to the distant skirl of bagpipes. The nest, a mere scrape in the ground, sparsely lined with plant material and down, is usually sited in thick vegetation not far from water. The elongated tail feathers extending to 5in (12.5cm), which give this duck its name, are borne only by the male.

Off coasts of northern Ireland, Scotland and eastern England.

This species is a member of the stiff-tailed duck family, so called because of their habit of holding the tail erect. It is a native of North America which was introduced into ornamental wildfowl collections and subsequently escaped into the wild. So successful have these escapees been that ruddy ducks are now a British breeding species, and still spreading. They inhabit ponds, lakes and reservoirs, especially those with reedy edges and other vegetation. Nests are woven baskets, attached to reeds.

Breeds on lakes chiefly in central and southern England.

This wonderfully plumaged little duck is not a native of Europe, for it is an introduced species from eastern Asia that has managed to escape from captivity and establish itself as a breeding species. In this country mandarin ducks are to be found on large parkland lakes with well vegetated islands, and undisturbed tree-lined river banks. The nest is placed in holes in trees, or in nest-boxes. The male is unmistakable with its golden ruff and erectile, chestnut-orange wing-fans. Females are drab greyish-brown with whitish 'spectacles'.

Breeds on lakes and rivers, mainly in southern England.

J F M A M J J A S O N D

J F M A M J J A S O N D

J F M A M J J A S O N D

217

Birds of prey

Marsh harrier
Circus aeruginosus
20-23in (50-58cm)

Black wingtips

Grey plumage, black wing-bar

Male

Grey tail

Pale head

Brown upper parts

Reddish underparts

Female

Buff shoulders and head

Young

Chocolate body

Female is largest of all harriers

Reduced to a handful of pairs in the 1960s, this powerful hunter has staged a remarkable recovery. Helped by protection, around 150 pairs now nest in Britain, chiefly in eastern England. Most migrate, but a few remain near their breeding grounds in winter. At the approach of spring they somersault and dive in mid-air in a courtship display, the male often passing food to the female. Marsh harriers feed on small mammals and birds.

Reed-beds and marshes, mainly in eastern England.

Montagu's harrier
Circus pygargus
16-18in (40-45cm)

Female

Wings raised in shallow V

Male

Brown streaks on whitish underparts

Female brown

Narrow white rump-patch

Reduced to probably fewer than ten breeding pairs in southern England, Montagu's harrier is one of Britain's rarest breeding birds. Unlike its close relative the hen harrier, this species is only a summer visitor, arriving during late April from its wintering grounds in Africa to breed in southern England before returning south in September. The harriers nest on the ground amongst vegetation in many kinds of open country from farmland to sand-dunes. Their flight is very buoyant.

Marshes, farmland and moors in southern England.

Hen harrier
Circus cyaneus
17-20in (43-50cm)

Black wing-tips

Male

Pale grey above, white below

White rump

Male drops prey to female who rolls over to catch it

Female brown

Female shows white rump in flight

Barred tail

Almost any creature up to the size of a hare or duck is game for the powerful talons of this moorland marauder. Centuries ago, when the bird was more widespread, it preyed on poultry, and so obtained its name. Earlier this century hen harriers declined in number, and ceased to breed on the mainland; but thanks to protection some have today recolonised old haunts.

Breeds on moors in northern Britain, north Wales and Ireland; winter visitor elsewhere.

J F M A M J J A S O N D

J F M A M J J A S O N D

J F M A M J J A S O N D

All birds of prey have sharp, hooked bills and strong, curved talons, for holding and tearing meat. They are masters of flight: some soar, some hover, some stoop or dive on their prey, some fly their prey down. In most species, females are larger than males.

Red kite
Milvus milvus
24in (60cm)

Pale head

Long wings, often held angled; white wing-patches

Male; female is duller

Long, reddish, forked tail

Buzzard
Buteo buteo
20-22in (50-55cm)

Wings pointed when gliding

Wing feathers extended when soaring

Rounded wings, with dark patches

White tail, black band at tip

Pale underparts

Short neck

Fan-like tail, with many narrow bars

Rough-legged buzzard
Buteo lagopus

Honey buzzard
Pernis apivorus
20-23in (50-58cm)

Long tail

Slender body and wings

Narrow head

Distinct bars on tail, broad band at tip

The child's toy kite was named after this bird because of its habit of hovering over rural hillsides. By about the 1900s gamekeepers – wrongly believing them to be a threat to their birds – had shot and poisoned them to the verge of extinction in Britain. Although strictly protected for several decades their recovery was slow, and birds were confined to central Wales. Aided by reintroductions in England and Scotland, numbers have now soared to more than 300 pairs. Apart from carrion, the kite eats small mammals, birds and insects.

Wooded hills in Wales, Scotland and central England.

A familiar sound in hilly country in western and northern Britain is the mewing 'kiew' of a buzzard as it sails, apparently without effort, over a wooded hillside. The keen-sighted bird meanwhile scans the ground for small mammals, in particular rabbits. Buzzards prefer open hillsides with wooded valleys and mountain ledges on which to breed. They build a large nest of sticks which is decorated with greenery.
The rough-legged buzzard, an occasional visitor to eastern Britain from Scandinavia, has paler underparts and an almost white tail with a dark band at its end.

Hills and wooded valleys in western and northern Britain, now spreading into central and eastern England.

The main diet of this buzzard is unusual for a bird of prey, consisting mainly of wild bees and their honey, and other insects. It may, however, be supplemented by more usual prey such as small mammals, nestling birds, lizards and frogs. The honey buzzard is one of our rarest breeders, for no more than 50 or so pairs nest each year, scattered throughout Britain.

Summer visitor to woodlands in various parts of Britain.

J F M A M J J A S O N D

J F M A M J J A S O N D

J F M A M J J A S O N D

Birds of prey

Sparrowhawk
Accipiter nisus
12-15in (30-38cm)

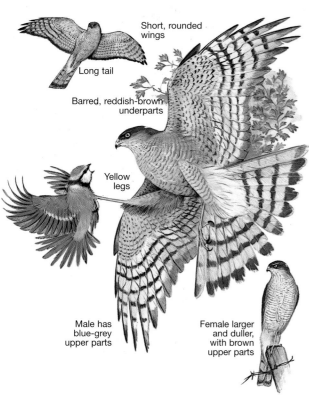

Short, rounded wings

Long tail

Barred, reddish-brown underparts

Yellow legs

Male has blue-grey upper parts

Female larger and duller, with brown upper parts

A watcher must be alert to spot the quick flurry and chorus of frantic alarm calls as this yellow-eyed predator darts down a woodland ride or along a hedgerow, scattering terrified birds. Although most prey is captured with the advantage of surprise, the sparrowhawk is capable of overtaking a quarry by its sheer speed and agility. The nest is a flattened bulky platform of sticks, sometimes based on an old nest of another species. Males are much smaller than the females.

Woodland, farmland and large gardens throughout Britain and Ireland.

Goshawk
Accipiter gentilis
20-24in (50-60cm)

Short, rounded wings

White eye-stripe

Long tail, white under-tail coverts

Barred under-parts

This dashing hawk is a very efficient killer. Swift but controlled, it swoops through the trees to take its prey completely unawares with its powerful talons. The goshawk is much larger than the commoner and similarly plumaged sparrowhawk, and can take wood-pigeons, crows, game birds, rats and hares. Until the 1950s the goshawk was a rare breeding species, but since then its numbers have increased considerably, aided by escapes from falconry training and by deliberate introductions. The nest of sticks is made high in a tree.

Well-wooded areas in various parts of Britain.

Osprey
Pandion haliaetus
20-23in (50-58cm)

Long wings angled at wrist

Dark brown upper parts

Dark mark through eye

White crown and underparts

An osprey making a kill is a spectacular sight. Fish make up the bulk of its diet, and a hunting bird flies over water at a considerable height, with alternate spells of flapping and gliding, until it spots a fish near the surface. It pauses, sometimes hovering momentarily, before turning and plunging feet first with partly closed wings. In the 1950s a pair of ospreys bred near Loch Garten in Scotland after an absence of 50 years. Since then their numbers have increased to more than 100 pairs. They build bulky stick nests in tall trees, close to large expanses of water.

By lakes and rivers in Scotland; now spreading to England.

J F M A M J J A S O N D

J F M A M J J A S O N D

J F M A M J J A S O N D

Golden eagle

Aquila chrysaetos
30-35in (76-89cm)

White-tailed eagle

Haliaeetus albicilla
28-36in (70-90cm)

Kestrel

Falco tinnunculus
13½in (34cm)

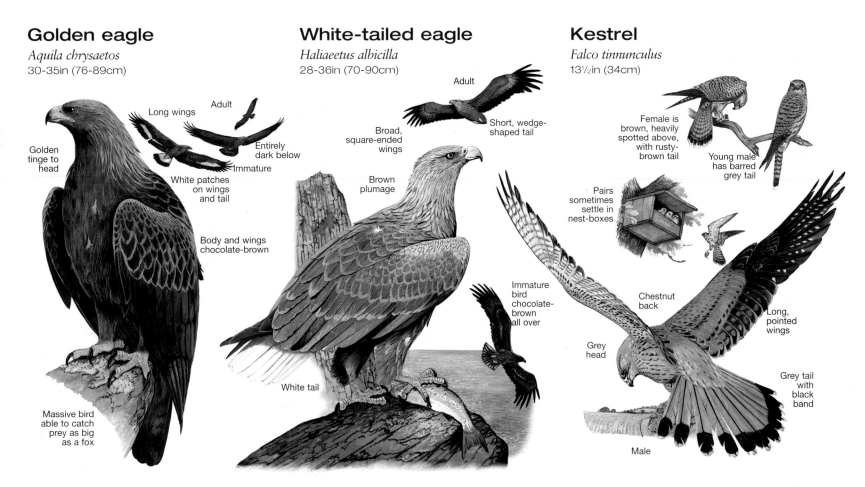

Golden eagle labels: Long wings · Adult · Entirely dark below · Immature · White patches on wings and tail · Golden tinge to head · Body and wings chocolate-brown · Massive bird able to catch prey as big as a fox

White-tailed eagle labels: Adult · Short, wedge-shaped tail · Broad, square-ended wings · Brown plumage · Immature bird chocolate-brown all over · White tail

Kestrel labels: Female is brown, heavily spotted above, with rusty-brown tail · Young male has barred grey tail · Pairs sometimes settle in nest-boxes · Chestnut back · Grey head · Long, pointed wings · Grey tail with black band · Male

In flight the 'king of birds' gives the impression of unequalled power and control, with its broad wings stretched to a 90in (230cm) wingspan and the wingtip feathers spread as though feeling for the currents of rising air which will bear it up to its mountain-top stronghold. Formerly more widespread, the golden eagle is now restricted mainly to remote mountain areas in Scotland, though one or two pairs have moved south. the bird builds a large stick nest on a rocky ledge or isolated tree; two eggs are laid, but normally only one eaglet survives.

Mountains and islands in Scotland and borders.

In 1907 these magnificent birds ceased to breed in the British Isles, after long years of persecution and disturbance by man. Until recently, the only sightings were of birds that strayed from their breeding grounds in Scandinavia. A reintroduction programme has now resulted in up to 20 breeding pairs becoming established in Scotland. Nests are usually a massive stick construction on a sea-cliff. The white-tailed eagle is more of a scavenger than the golden eagle and takes rotting fish stranded on the shore as well as live prey.

Western Isles and Highlands; rare winter migrant to east England.

A medium-sized, brownish falcon hovering above a roadside verge often catches the eye of the passing motorist – and usually means death for some small mammal below. Lift-like, the bird drops by stages, finally pouncing and grasping with its talons. Kestrels are today Britain's commonest birds of prey. They favour open country, but also flourish in urban areas with parks. They nest in holes in trees, tall buildings or nest-boxes. The adult male sports a russet back and wings, offset by a dove-grey crown and tail. Females are brown with dark spots.

All types of country throughout Britain and Ireland.

J F M A M J J A S O N D   J F M A M J J A S O N D   J F M A M J J A S O N D

Birds of prey

Merlin
Falco columbarius
10½-13in (27-33cm)

Female dull brown above, paler below

Long, pointed wings

Male has blue-grey upper parts

Shortish tail

Male

Streaked, reddish-brown underparts

Black tail-band

Hobby
Falco subbuteo
14in (36cm)

Long, scythe-shaped wings

Slate-coloured back

Red thighs

Shortish tail

Dark crown, black 'moustache'

Swift

Underparts streaked with black

Peregrine falcon
Falco peregrinus
15-19in (38-48cm)

Dark crown

White beneath, barred with black

Cheek 'moustache'

Grey upper parts

Young bird brown above, streaked below

Broad wings

Breeds on high rock ledges

Flying low and fast with quick, shallow wing-beats, the male merlin, little larger than a mistle thrush, quarters the moors in search of its prey – mainly small birds. On sighting a quarry, the merlin rises above it, then drops to sink its talons into its victim. Unfortunately, this beautiful small falcon is declining in numbers.

Breeds on moorlands in Scotland, northern England, Wales and Ireland.

Merlin often takes prey in level flight

For speed, grace and agility in flight the hobby has few rivals, even among its fellow falcons. Whether delicately picking a dragonfly out of the sky, or swooping down to seize a swallow in full flight, it presents a breathtaking spectacle. Hobbies are great travellers, for they spend the winter in Africa before migrating north to breed in central and southern England. They build a stick nest, usually high in a coniferous tree. The hobby's plumage is striking, and in flight its long wings, angled back, give it the appearance of a giant swift.

Farmland and heath in central and southern England.

The peregrine is a mere speck in the sky as it keeps a lonely, circling watch for prey. When it sights its victim it suddenly snaps back its wings and dives towards it in a rapid 'stoop' estimated as reaching 180mph (290km/h). If contact is made the quarry – typically a pigeon or duck – is killed instantly.

Breeds on cliffs and crags; now also on buildings, too.

Courting male loops the loop after mock dive at female

Game birds

These heavy-bodied land birds have stout, short bills, stubby, rounded wings and strong legs and feet adapted to scratching the ground for food. Game birds often run in preference to flying. Their flight is often fast and laboured, with very rapid wing-beats.

Red grouse

Lagopus lagopus
15in (38cm)

No wing-bar

Red wattle over eye

Reddish-brown plumage

Male; female is smaller and paler

Rounded tail

White legs

Exploding into chattering flight from the moorland heather almost beneath one's feet, a plump and handsome red grouse is an exhilarating sight. On the 'Glorious Twelfth' of August, beaters drive them towards the waiting guns and thus to restaurant tables. Stocks are maintained by managed breeding that includes the systematic burning of patches of heather to produce young shoots upon which the young grouse feed. They nest on the ground, and lay six to a dozen eggs. The cock is a rich reddish-brown, while the smaller female is a paler, duller brown.

Heather moors in northern Britain, Wales and Ireland.

Ptarmigan

Lagopus mutus
14in (36cm)

Female in winter lacks black face-patch

Red wattle over eye

Swift wing-beats alternate with glides

Male in breeding plumage

Greyish-brown body

White belly and wings

Rounded tail

Female paler than male

The ptarmigan is the only British bird that changes colour with the seasons to camouflage itself against predators. It changes from brown and white in summer to almost pure white in winter. A bird of northern climes, this grouse inhabits the rocky slopes of Scottish mountains, where bilberry, crowberry and heather provide food. The melting snow in May and June sees the start of the breeding season and skirmishes between cocks.

Winter plumage white except for black tail and male's black face-patch

Scottish mountain-tops.

Black grouse

Lyrurus tetrix
21in (53cm)

Male

Female

Female brown, barred plumage

Distinctive tail; white wing-bar

Notched tail

Black plumage

Red wattle over eye

Lyre-shaped tail

Male

In the mating season, black grouse gather for a communal courtship display known as a 'lek'. Each male holds a small area on which he stands with tail fanned and erect, wings spread and drooped. He faces a rival male and utters a prolonged bubbling sound which is sometimes interrupted by a loud, scraping 'tcheway'. Fights take place, with the victorious bird taking on another rival. Females strut nonchalently between the combatants before selecting their mates from the victors. The nest is well hidden in grass or heather.

Moorland fringes in northern Britain and Wales.

J F M A M J J A S O N D
J F M A M J J A S O N D
J F M A M J J A S O N D

223

Game birds

Capercaillie

Tetrao urogallus
34in (86cm)

Rounded tail

Bushy
throat
feathers

Male

Female

Large
fanned
tail

Male

Black
and brown
plumage

Felling of coniferous forests and shooting had eliminated the capercaillie from Britain by 1785. In 1837 it was reintroduced into Scotland, but attempts in England and Ireland have failed. Like the black grouse, the male is noted for its courtship display and extraordinary vocal accompaniment. The call starts with a slow series of clicks which speed up into a rattle; this is followed by a 'klop' then a final hissing.

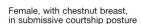

Conifer woods in north-central Scotland.

Female, with chestnut breast, in submissive courtship posture

Red-legged partridge

Alectoris rufa
13½in (34cm)

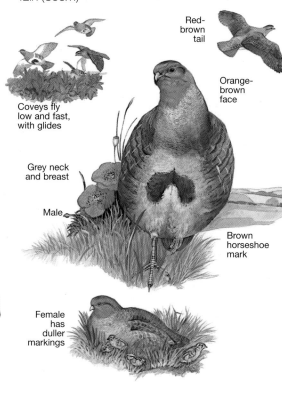

Red
bill

Red
tail

Black
throat
band

Red
legs

White
cheeks

Birds
seek safety
in cover,
not flight

Thousands of chicks and eggs imported from France in 1790 introduced this bird to Britain. Further imports occurred in the 19th and 20th centuries, and today the red-legged partridge outnumbers our native species. Although introduced for shooting purposes, the red-leg is not a particular favourite of hunters because of its reluctance to fly and preference to remain in cover, where its plumage helps it to escape detection. The nest is placed in cover on the ground. Sometimes two clutches are produced, and the male has to care for one of the nests.

Heaths and farmland in central and southern England.

J F M A M J J A S O N D

Common partridge

Perdix perdix
12in (30cm)

Red-
brown
tail

Orange-
brown
face

Coveys fly
low and fast,
with glides

Grey neck
and breast

Male

Brown
horseshoe
mark

Female
has
duller
markings

One of Britain's most popular game birds, the partridge often has its population boosted by imports from the Continent to supplement the native breeding population for the shooting season. Shoots take place in early September, when the birds have gathered together in family coveys. By February the surviving birds have paired and taken up territories which the cocks defend vigorously. Courtship often takes the form of a running chase, the cock and hen taking turns to pursue each other. The nest is on the ground. Occasionally two hens will lay in one nest clutches of up to 40 eggs.

Farmland in lowland Britain and Ireland; declining rapidly.

J F M A M J J A S O N D

Rails

Quail

Coturnix coturnix
7in (18cm)

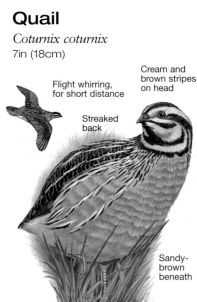

Flight whirring, for short distance

Cream and brown stripes on head

Streaked back

Sandy-brown beneath

Female drabber brown

The distinctive liquid 'whit-whit-whit' call of the quail rises from southern cornfields and hayfields in spring and early summer. The bird resembles a tiny partridge; unlike the partridge, however, it is not a resident but a migrant from Africa that breeds in Britain. Its nest, on the ground, is well hidden amongst tall vegetation. The male bird sometimes has more than one mate, sharing a single nest.

Lowland farmland throughout British Isles; numbers fluctuate.

Pheasant

Phasianus colchicus
21-35in (53-89cm)

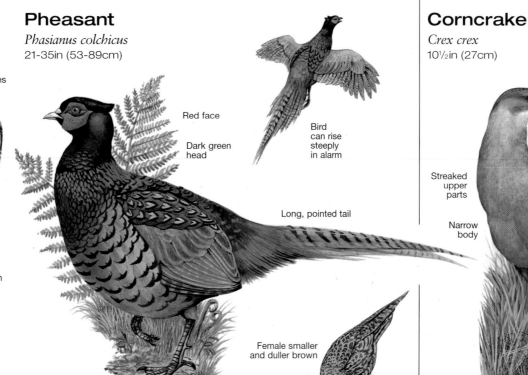

Red face

Dark green head

Bird can rise steeply in alarm

Long, pointed tail

Male

Female smaller and duller brown

Though this colourful and handsome creature is Britain's most widespread game bird it is not a native, for it was introduced from Asia in the Middle Ages. The rearing of pheasants for shooting has encouraged estate owners to maintain wooded habitats, so bringing benefit to other forms of wildlife. However, many foxes, crows and birds of prey have been killed to protect the pheasant from predators. The males are distinctive with their long tail, iridescent green head and metallic sheen on their mainly brown body.

Woods and farms throughout Britain and Ireland, except north-west Scotland.

Some males dark green; some have white neck-ring

Corncrake

Crex crex
10½in (27cm)

Red-brown wing-patches

Legs dangle

Streaked upper parts

Narrow body

Although fields of corn are sometimes used by the corncrake as breeding grounds, its common name is less appropriate than the alternative, land rail. Formerly the bird nested in large numbers throughout the British Isles, and its grating call, like a stick drawn across a notched piece of wood, was a feature of summer days and nights in country areas. Today the corncrake is a rarity, and its numbers are still decreasing.

Summer visitor to grasslands in western Scotland and Ireland.

Rails

These medium-sized or small long-legged birds usually live in or near water. They are frequently secretive, preferring running or swimming to flying, despite the lack of webbing between their toes. Their flight is laboured, with legs trailing behind.

Water rail
Rallus aquaticus
11in (28cm)

Moorhen
Gallinula chloropus
13in (33cm)

Coot
Fulica atra
15in (38cm)

Water rail: Male · Brown wings · Trailing legs · Brown back · Brown back · Red bill · Long toes · Striped flanks · Female has duller colouring; chicks black · Short bill · White spots; buff under-tail · Spotted crake *Porzana porzana*

Moorhen: Long pattering run to take-off · Dark brown and black plumage · White under-tail · Red bill-base · Green legs · Juvenile has dull brown body and pale face · Chicks have bare blue crown

Coot: Long legs trail in flight · White bill and forehead · Black body · Lobed feet · Juvenile has pale throat and breast · Chicks have brightly coloured bare head

The call of the water rail from dense waterside vegetation sounds like a pig squealing with fear. A glimpse, however, is enough to identify the bird, for its long, red bill, brown upper parts and black and white striped flank markings on slate-grey underparts are distinctive. The rail's long legs and toes are adapted for walking on floating vegetation, and its narrow body helps it to slip between close-growing stems.

The rare spotted crake, a summer visitor to a few marshes in Britain, differs in having a shorter bill, white spots on its upper parts and buff under its tail.

Reed-beds and marshes in most of Britain and Ireland, except north-west Scotland.

In spite of its name this is not a moorland species; the name comes from the Anglo-Saxon *more*, meaning mere or bog. It eats water plants, insects, spiders, worms and other invertebrates which it picks from the surface or by up-ending. Like its cousin the coot it is aggressive in defence of its territory, attacking any encroaching neighbour with feet and bill. The feet are exceedingly long, but have no webbing, so its swimming action is jerky and laboured. If alarmed, the bird will dive and stay submerged with only its bill above the water. The nest is usually a woven structure, anchored to vegetation.

Watersides throughout Britain and Ireland, except extreme north-west Scotland.

In striking contrast to their black plumage, coots of both sexes have an area of bare skin on their forehead that matches their shiny white bill. Males squabble frequently over territory, with the white shield playing an important part in their aggressive displays. It is held forward, low on the water, with wings and body fluffed up behind, presenting a menacing impression. As two birds approach each other they produce an unmusical ringing call – rather like a hammer striking a steel plate. The coot's diet consists mainly of plant material for which it dives to bring to the surface to eat.

Open lakes and pools throughout Britain and Ireland, except north-west Scotland.

Waders

Waders are generally plump shorebirds, with long legs and bills and pointed wings. In most species males and females are alike. Flight is strong and swift. Most waders are highly migratory, travelling together in huge flocks sometimes many thousands strong.

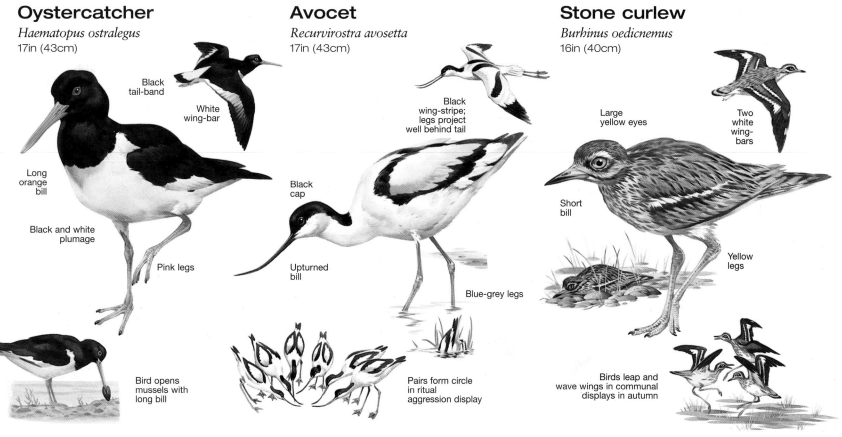

Oystercatcher

Haematopus ostralegus
17in (43cm)

Black tail-band

White wing-bar

Long orange bill

Black and white plumage

Pink legs

Bird opens mussels with long bill

Avocet

Recurvirostra avosetta
17in (43cm)

Black wing-stripe; legs project well behind tail

Black cap

Upturned bill

Blue-grey legs

Pairs form circle in ritual aggression display

Stone curlew

Burhinus oedicnemus
16in (40cm)

Large yellow eyes

Two white wing-bars

Short bill

Yellow legs

Birds leap and wave wings in communal displays in autumn

With its immaculate black and white plumage, orange chisel-like bill and pink legs, the oystercatcher is one of our most handsome shorebirds. From late summer through to spring, wintering flocks grace our sandy shores and rocky beaches wherever food in the form of shellfish and other invertebrates is plentiful. Breeding takes place after the flocks have broken up in spring. The courtship display usually consists of several birds walking around agitatedly whilst uttering a noisy, piping chorus of 'kleep-kleep-kleep'. The birds nest on the ground in coastal fields and on beaches.

Coasts and estuaries, and inland beside rivers and lakes.

This striking black and white wader owes its presence in Britain to the Second World War. It had ceased to breed amongst the brackish pools and low islets of the east coast when the land was reclaimed. During the war, however, access to the coast was restricted and avocets returned. Now strictly protected, they thrive in many colonies in east and southeast England. Avocets feed on tiny invertebrates which they sift from the water by sweeping movements of their curved bills. In a ritual aggressive display, known as 'grouping', pairs of birds join in groups, often in a circle with lowered heads. Fighting sometimes occurs.

Breeds mainly on coastal lagoons in east and south-east England; winters also in south-west England.

The plaintive 'coooeee' cry of this wader hangs hauntingly over the chalk downs and sandy heaths of south and east England. Snails, slugs and insects are its main food, although it may take larger prey such as frogs and field mice. Stone curlews are early migrants from warmer regions that come to breed in Britain. Their numbers have declined drastically, however, and only 200 or so pairs now breed. They nest on bare ground.

Downs and heaths in south and east.

Courting birds face opposite ways

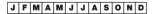

Waders

Dotterel
Charadrius morinellus
8½in (22cm)

Summer

Female

White breast-band

Winter plumage mainly grey

Male, duller than female, guards and feeds young

Chestnut underparts

Unusually in the bird world, the male dotterel has duller plumage than the female. This gives it camouflage as it incubates the eggs – for the female, after laying, loses interest in her offspring and leaves all parental duties to the male. In addition to visiting birds, a few dotterels breed on barren mountain-tops above 2500ft in the north of Britain. They feed on insects, other invertebrates and seeds.

Scarce breeder on northern mountains.

Male feigns broken wing to distract predators from nest

Ringed plover
Charadrius hiaticula
7½in (19cm)

White wing-bar visible in flight

Dark bill

No wing-bar

Little ringed plover
Charadrius dubius

Black breast-band

Black-tipped orange bill

Orange legs

Mottled plumage disguises chicks on sand or shingle

When feeding, ringed plovers walk quickly along the seashore for a few paces, pause as if lost in thought, then suddenly seize food from the mud with a swift dip of their beak. While halted they patter with their feet to disturb prey such as shellfish and other invertebrates. The adult is brown and white, distinctly marked about the head and breast with black; the base of the bill and legs are bright orange. Juveniles have indistinct black markings, incomplete breast-bands and pale legs.
The little ringed plover is smaller, with a yellow eye-ring and a white line on the forehead. It is a summer visitor to inland waters in England.

Coastal areas all around Britain and Ireland.

Golden plover
Pluvialis apricaria
11in (28cm)

Winter

Summer

Winter

Faint wing-bar

Pale underparts

Gold speckles

Winter

Summer

Winter

Black under-parts

Black spot

Mottled grey and white, faint wing-bar

Grey plover
Pluvialis squatarola

Black belly in summer, more extensive in male

Grey plover hunched

Golden plover upright

The plaintive 'klew-ee' call of the golden plover is a characteristic sound of the high moors where it breeds in summer. During late summer, the adult's black belly moults to white, and some birds migrate south as far as Spain to spend the winter. More northerly nesting birds, from Iceland and Scandinavia, move south to Britain to winter, and form large flocks on ploughed fields or marshland sites, where they feed on grubs and insects. The grey plover is mottled grey and white when it visits Britain in autumn and winter from its breeding grounds in the Arctic tundra. A black patch under the base of each wing is conspicuous in flight.

Northern moors in summer; widespread in winter.

J F M A M J J A S O N D

J F M A M J J A S O N D

Turnstone

Arenaria interpres
9in (23cm)

Double white wing-bar

Chestnut and black back

White head

Summer

Orange legs

Winter

Dark brown breast and back

As it walks across the sands and rocks at low tide in search of food, this colourful wader quickly reveals the origin of its name. Not only stones, but also seaweed, shells and driftwood are diligently lifted by the bird's probing bill in the hope of finding sandhoppers and other shore-life. In Britain, the turnstone is a passage migrant, on its way to and from its breeding grounds in the Arctic. It frequently winters in coastal areas, and some birds stay for the summer. Young birds resemble winter adults, with dark brown and white plumage.

Rocks and pebbly shores all round Britain and Ireland.

Knot

Calidris canutus
10in (25cm)

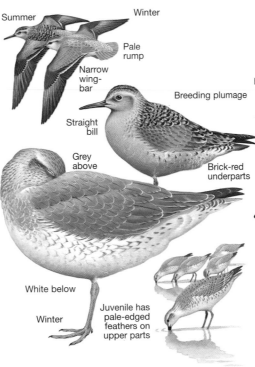

Summer

Winter

Pale rump

Narrow wing-bar

Breeding plumage

Straight bill

Grey above

Brick-red underparts

White below

Winter

Juvenile has pale-edged feathers on upper parts

The knot standing at the tide's edge has been likened to a tiny King Canute standing on the shore and ordering the tide to retreat. Its common name, however, is probably derived from the bird's call, a low-pitched 'knut'. It is a winter visitor to coastal mud-flats, where thousands often gather to probe for food that consists of marine invertebrates. Early arrivals may still bear traces of the brick-red breeding plumage, lost in winter.

Coastal mud-flats around most of the British Isles.

Lapwing

Vanellus vanellus
12in (30cm)

Male

Broad, rounded wings

Long crest

Black breast-band

Black tail-bar

White wing-tips

Female

Metallic-green back

Male

Orange under-tail

White underparts

The lapwing is sometimes known as the green plover because of its iridescent back plumage, or the peewit because of its distinctive call. At one time, it was a common sight on ploughed fields, but numbers have fallen due to modern farming methods. The male's territorial display – involving a twisting, rolling dive – is a delight to watch.

Open country throughout most of Britain and Ireland.

Female whiter on tail and throat

Waders

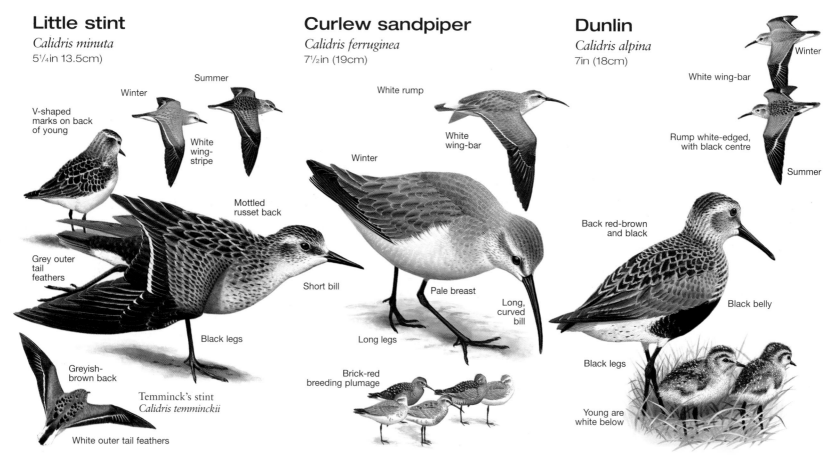

Little stint
Calidris minuta
5¼in 13.5cm)

Summer

Winter

V-shaped marks on back of young

White wing-stripe

Mottled russet back

Grey outer tail feathers

Greyish-brown back

Temminck's stint
Calidris temminckii

White outer tail feathers

Black legs

Curlew sandpiper
Calidris ferruginea
7½in (19cm)

White rump

White wing-bar

Winter

Short bill

Pale breast

Long, curved bill

Long legs

Brick-red breeding plumage

Dunlin
Calidris alpina
7in (18cm)

Winter

White wing-bar

Rump white-edged, with black centre

Summer

Back red-brown and black

Black belly

Black legs

Young are white below

The smallest wader in Britain, the little stint is a short-stay autumn visitor, passing through on its way from its breeding grounds in northern Russia and Siberia to winter in Africa. Most little stints seen in Britain are young birds, distinguished by two pronounced V-shapes on the back. Adults at this season are grey above and white below, with faint streaks on the breast.
Temminck's stint, a rare passage visitor, is distinguished by its white outer tail feathers, finer bill and yellow-green legs.

Marshes, freshwater margins and coastal mud-flats, especially on east coast.

Many curlew sandpipers pass through eastern Britain in autumn on their way back from their breeding grounds in Siberia to their winter quarters in equatorial Africa. They can be seen foraging for small aquatic animals on mud-flats beside inland waters. Young birds, with a pinkish-buff breast, usually outnumber adults. In summer, adults are brick-red below and rusty-brown above. After the autumn moult they are grey and white, and best distinguished from the dunlin by their long, curved bill, and in flight by their white rump.

Coasts and inland waters, especially in eastern England.

These little birds blend so well into the background of Pennine and Scottish moorland that they are not easily seen, though they breed there in large numbers. The nest is a neat little cup, hidden in a grass tussock. After breeding, dunlins migrate in autumn to spend the winter in southern Europe and northern Africa. They are replaced in Britain by more northerly breeders which come to spend the winter here. In their grey and white winter plumage they can be seen in their thousands feeding on coastal mud-flats on flies, molluscs, crustaceans and worms.

Moorlands in summer; widespread on coasts in winter.

J F M A M J J A S O N D

J F M A M J J A S O N D

J F M A M J J A S O N D

Woodcock

Scolopax rusticola
13½in (34cm)

Eyes set high for wide vision

Broad wings

Russet plumage

Short legs

Bars on head and breast

Long bill

Snipe

Gallinago gallinago
10½in (27cm)

Tail feathers produce bleating sound

Shorter bill

No white on tail

Jack snipe
Lymnocryptes minimus

Creamy-striped head and back

Reddish-brown tail fringed with white

Long bill

Ruff

Philomachus pugnax
11½in (29cm)

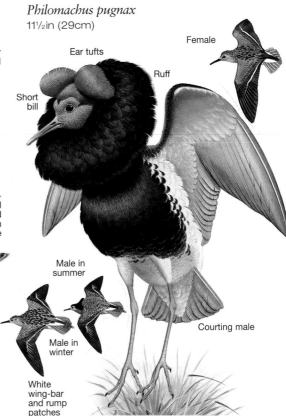

Female

Ear tufts

Ruff

Short bill

Male in summer

Courting male

Male in winter

White wing-bar and rump patches

It is difficult to believe that the woodcock is related to the waders, so firmly established is it as a bird of damp woodland with plenty of bracken and bramble for cover. Woodcocks need soft ground in which to feed, probing with their bills for worms, insects and their larvae, centipedes and spiders. The male's territorial display is very distinctive. It flies over its territory at dusk, on slow-beating wings, uttering two calls – a throaty 'og-og-og' and a 'chee-wick'. When disturbed, the woodcock's flight is fast and twisting as it manoeuvres deftly between the trees.

Widespread in damp woodlands.

Endowed with a bill which is about a quarter of its total length, the snipe is easy to identify. The tip of the bill is flexible and highly sensitive, allowing the bird to detect the worms and invertebrates on which it mainly feeds, and then dig deep into the mud for them. The snipe has a spectacular courtship display in spring. The jack snipe is smaller, and is only a winter visitor to Britain.

Boggy areas with good cover throughout Britain and Ireland.

Nest in tussock of grass; chicks dark

Courting male ruffs present an extraordinary sight. They gather together on display grounds called 'leks' in plumage as extravagant as that of an Elizabethan dandy to overawe their rivals and dazzle the females. The neck ruffs can be black, red-brown, purple, white or creamy, or else striped, barred or spotted. Most fly south to Africa in autumn.

Freshwater marshes in southern Britain, mainly autumn migrant; a few breed occasionally.

Female

Mottled buff, with pale bill and legs

Waders

Greenshank

Tringa nebularia
12in (30cm)

White on back and rump

Uniform wing colour

Upturned bill

Winter plumage less strongly barred

Greenish legs

With its long green legs and sleek body, the greenshank is one of Britain's most elegant waders. It is generally to be found on estuaries and pools and lagoons in freshwater marshes. Most birds migrate to Africa in the autumn, but some spend the winter in Britain. In the breeding season they move to the wild remote bogs of the Scottish Highlands and the Hebrides. The nest is a hollow in the ground lined with plant debris.

Breeds in Scottish Highlands and Hebrides; winters in Ireland and south-west Britain.

Parent bird fiercely defends young

Redshank

Tringa totanus
11in (28cm)

Summer

Red-based bill

White on back and rump

White wing-patches

Orange-red legs

Long bill, black on upper side

Grey and white winter plumage

Spotted redshank
Tringa erythropus

A volley of harsh, piping notes issues from the so-called 'sentinel of the marsh' as soon as any intruder approaches. This ear-piercing cry contrasts sharply with the redshank's more musical and liquid call of 'tew-ew-ew'. The bird is a common breeder in most forms of marsh, where it makes a root-lined nest hidden in a grass tussock. It feeds on all sorts of invertebrates, small fish and frogs, seeds, buds and berries. In winter, redshanks often form large flocks on estuary mud-flats.
The spotted redshank is larger and greyer, with a longer bill and finely barred and speckled upper parts. It is mainly an autumn passage migrant.

Breeds widely inland; frequents coasts in winter.

Sanderling

Calidris alba
8in (20cm)

Winter

White wing-bar

Summer

Grey back

Dark shoulder

Winter

White underparts

Black legs

Summer

Rusty head and body

With frenetic bursts of energy, flocks of sanderlings scurry along the seashore, their heads down in pursuit of retreating waves. They snatch a few morsels of food, then race back in advance of the next wave to avoid getting washed off their feet. If disturbed they rise with a chorus of liquid 'twick, twick, twick' calls and move along the beach to resume their feeding. In winter, sanderlings can be found on most sandy shores in Britain, but in summer they are birds of the high Arctic where they breed. In spring the plumage changes from almost white to reddish, marked with black on the back, head and breast.

Sandy coasts around most of Britain and Ireland.

Black-tailed godwit

Limosa limosa
16in (40cm)

Winter

Summer

Black and white tail

White wing-bar

Straight bill

Chestnut head, neck and breast

Summer

Winter

Grey plumage

Black and white tail; white under-tail

Re-established since the 1950s as a breeding bird in Britain, the black-tailed godwit now has a small, stable population. Adults in summer have chestnut and white plumage; but in winter, when large flocks can be seen around the coast, the plumage is rather grey. At all times the prominent black and white tail and broad, white wing-bar distinguish this species from its cousin, the bar-tailed godwit.

Marshes and estuaries.

Often breeds on inland grasslands

Bar-tailed godwit

Limosa lapponica
15in (38cm)

Winter

Barred tail

White rump

Flocks plunge down in spirals

Plumage brownish, with streaked upper parts

Barred tail

Chestnut underparts

Bill slightly upturned

Summer

Winter

Unlike the black-tailed godwit, the bar-tailed godwit does not breed in Britain, but large flocks can be seen in spring on their way to Scandinavia and Russia. Returning here in autumn, many stay to winter, feeding in estuaries and marshes. The adult in summer has chestnut head, neck and underparts, while the winter plumage is mainly a buffy brown, with more streaking than in the black-tailed godwit. The rump is white and the tail barred all year.

Estuaries around all coasts.

Bill probes deeply into sand

Curlew

Numenius arquata
22in (55cm)

White rump

Uniform head colouring

Long, down-curved bill

Striped crown

Short, down-curved bill

Whimbrel
Numenius phaeopus

Europe's largest wader, the curlew is distinguished by its long, down-curved bill and its haunting, plaintive 'coor-li' call. On the marshes, moors and heaths where curlews breed, males establish their territories by flying in wide circles delivering a rising, bubbling song. In winter the birds move to coastal estuaries.

The related whimbrel is one of Britain's rarest breeding waders, confined to the north of Scotland, the Orkneys and Shetlands. Its shorter bill, distinctive crown stripe and whistling 'pee-pee-pee-pee-pee-pee-pee' call distinguish it.

Breeds in uplands over most of Britain and Ireland; winters on estuaries.

233

Waders

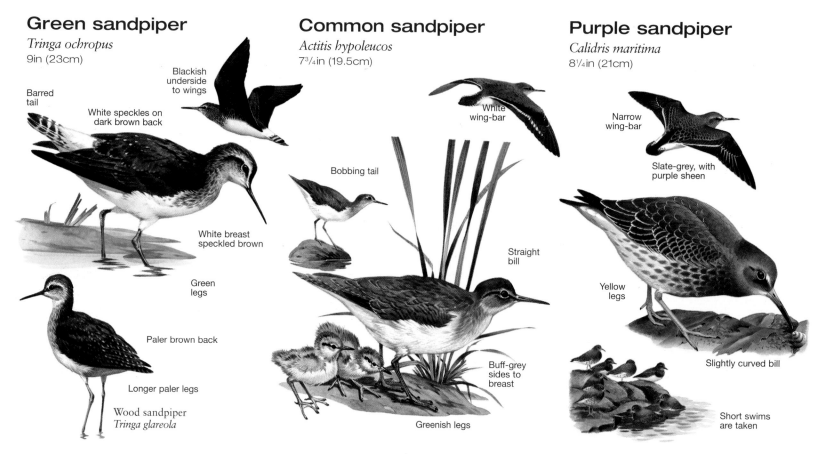

Green sandpiper
Tringa ochropus
9in (23cm)

Barred tail

White speckles on dark brown back

Blackish underside to wings

White breast speckled brown

Green legs

Paler brown back

Longer paler legs

Wood sandpiper
Tringa glareola

Common sandpiper
Actitis hypoleucos
7³⁄₄in (19.5cm)

White wing-bar

Bobbing tail

Straight bill

Buff-grey sides to breast

Greenish legs

Purple sandpiper
Calidris maritima
8¹⁄₄in (21cm)

Narrow wing-bar

Slate-grey, with purple sheen

Yellow legs

Slightly curved bill

Short swims are taken

Usually seen only when flushed, as it rises into the air calling 'klee-weet-tweet', the green sandpiper looks black and white – only its legs are green. This passage migrant is seen mainly in spring and autumn. The nest is unusual for a wader: normally the old nest of a blackbird or jay in a tree. The wood sandpiper is smaller. A few pairs breed in Scotland.

Almost vertical rise when flushed

Marshes and ditches in Britain and south and east Ireland.

The tail, bobbing or wagging, is a distinctive feature of the common sandpiper. Its flight is also characteristic: shallow, rigid wing-beats are followed by glides on bowed wings, often accompanied by the loud, musical 'twi-wi-wi-wee' call. The bird feeds on insects, small molluscs and crustaceans. Its upper parts are brownish and the underside is white; a dark line runs through the eye. The nest is a sparsely lined scrape in the ground; both adults share in incubating the eggs for 20-23 days. Some birds winter in Britain, but most migrate to Africa.

Breeds on lakes, ponds and streams in northern and western Britain and Ireland; common migrant elsewhere.

Frequently seen on piers, groynes and slipways at high tide, the purple sandpiper is a winter visitor favouring rocky shores and shingle beaches, feeding on insects, molluscs, worms and crustaceans. It is often seen with turnstones. The slate-grey upper parts have a purplish sheen; the short legs are a dull yellow, and the bill is black with a dull yellow base. The call is a weak 'wheet-wheet', with sometimes a short trill. This sandpiper, a very rare breeder in Britain, lays its eggs in a hollow in the ground. Adults have more variegated plumage in summer.

Rocky coasts all around Britain and Ireland in winter; a few pairs now nest on Scottish mountains.

J F M A M J J A S O N D   J F M A M J J A S O N D   J F M A M J J A S O N D

Skuas

Skuas, looking like dark gulls, are the pirates of the bird world. They fly down other seabirds and force them to disgorge their last catch – which they then eat – as well as robbing nests of eggs and young.

Red-necked phalarope

Phalaropus lobatus
8in (20cm)

Black eye-patch

Pale grey back

Stouter bill

Long white wing-bar

Pale stripes on back

Winter

Dark line through eye

Grey phalarope
Phalaropus fulicarius

Dark head

Reddish-brown neck

Female in summer

Male paler in colour

Needle-like bill

Unusual in the bird world, the female phalarope has much brighter plumage than the male. Indeed, in both red-necked and grey phalaropes the usual sexual roles are reversed, the males incubating the eggs and caring for the young: the females even court the males. Again, unusually for a wader, phalaropes have lobed toes like coots and spend much of their time swimming. The bird is mainly a rare summer visitor, breeding in shallow pools in the Outer Hebrides, Shetland and Ireland.

The grey phalarope, a rare autumn visitor, is slightly heavier in appearance and lighter in colour, with a shorter, stouter bill.

Rare. Breeds in Scotland; migrant in England.

Great skua

Catharacta skua
23in (58cm)

White wing-patch conspicuous

Flight heavy and purposeful

Bird swoops on intruders head-on

White wing-patch

Brown underparts

The largest of the skuas – the bonxie as it is known in the Shetlands where it nests – is a sturdily built, aggressive bird. It feeds partly by stealing other seabirds' most recent meals, grasping a wingtip in its bill and making the victim plunge into the sea where it disgorges its food. Intruders to the nesting territory are attacked by 'dive bombing'.

Passage migrant on coasts; some breed in far north.

Kittiwake chicks are frequent victims

Arctic skua

Stercorarius parasiticus
18in (45cm)

Immature birds brownish

Projecting tail feathers

Pale underparts

Dark cap

Adult feigns injury to protect young

Smaller than the great skua, and with less prominent wing-patches, the Arctic skua is nevertheless a fearless, piratical bird from which no other bird's nest is safe. Eggs and nestlings are often plundered by a pair of birds working together, one fighting off the parents while the other robs the nest. Passing Arctic terns are also preyed upon and pursued relentlessly until a bird disgorges its food. Although most often seen as a passage migrant, the Arctic skua breeds on windswept moors in the far north. Its nest is a shallow depression in the ground.

Passage migrant on coasts; some breed in far north.

235

Gulls

Black-headed gull
Larus ridibundus
14-15in (36-38cm)

Winter

Chocolate-brown head

Dark spot behind eye

Dark red bill

Summer

Flocks follow plough

Young birds have brown in plumage

Little gull
Larus minutus
11in (28cm)

Black head

Winter

Wings light grey above, dark below

Young bird

Bright red legs

Summer

Winter

Light under-wing

Mediterranean gull
Larus melanocephalus

Common gull
Larus canus
16in (40cm)

Head streaked grey-brown; dark eye

Winter

Yellow-green bill

Yellow-green legs

Other birds may be pursued and robbed of food

Black and white wing-tips

Head white

Summer

The name is misleading as this gull has a chocolate-brown head. And it hardly deserves the name seagull as three-quarters of those in Britain nest inland around reservoirs, gravel pits or sewage farms – and especially on boggy areas near northern lakes. Only in the south does it breed mainly on the coast, often on salt-marshes or in sand-dunes. Flocks of black-headed gulls can be seen following the plough, picking up insects and worms. They are also a common sight on refuse tips in winter. The calls are a repeated 'cuck-cuck' and a rasping 'kee-arr', which sounds overpowering when a colony is in full cry. Both parents incubate the eggs, laid in a cup of vegetation.

Widespread in Britain and Ireland.

Unlike its cousin the black-headed gull, this smallest of the gulls really does have a black head rather than a brown one. It is tern-like and has a very buoyant flight, dipping to the surface to pick up small crustaceans, fish or insects. The pale grey upper parts, black head and the bright red bill and legs are distinctive; outside the breeding season, however, the black head colouring is replaced by a dark smudge on the crown and behind the eye. This bird is mostly seen on passage. The Mediterranean gull, which has whiter wings than the black-headed gull now breeds locally in England.

Fairly widespread on passage; scarcer, mainly coastal, in winter.

The common gull is, in fact, not nearly as common as the name suggests, except in north-west Ireland and Scotland. In the south it is most often seen in winter. Its varied diet includes insects and worms, water creatures and even coastal refuse. Sometimes other seabirds are pursued to be robbed of food or eggs, and the young of other species are taken. The adult has a grey back, white head and yellow-green bill and legs; the wing-tips are black, with white spots or 'mirrors'. Its voice is a shrill 'keeeyar'. The common gull nests in small colonies on rocks, islets or boggy areas of grass and moorland, in small hollows lined with plant material.

Breeds chiefly in Scotland and Ireland; widespread in winter.

J F M A M J J A S O N D   J F M A M J J A S O N D   J F M A M J J A S O N D

Though common seabirds, gulls often scavenge inland. They have webbed feet, pointed wings and sturdy tails. Plumage is usually grey, white and black. Males and females are alike; young birds are flecked with brown. They are gregarious birds, nesting in large, noisy colonies.

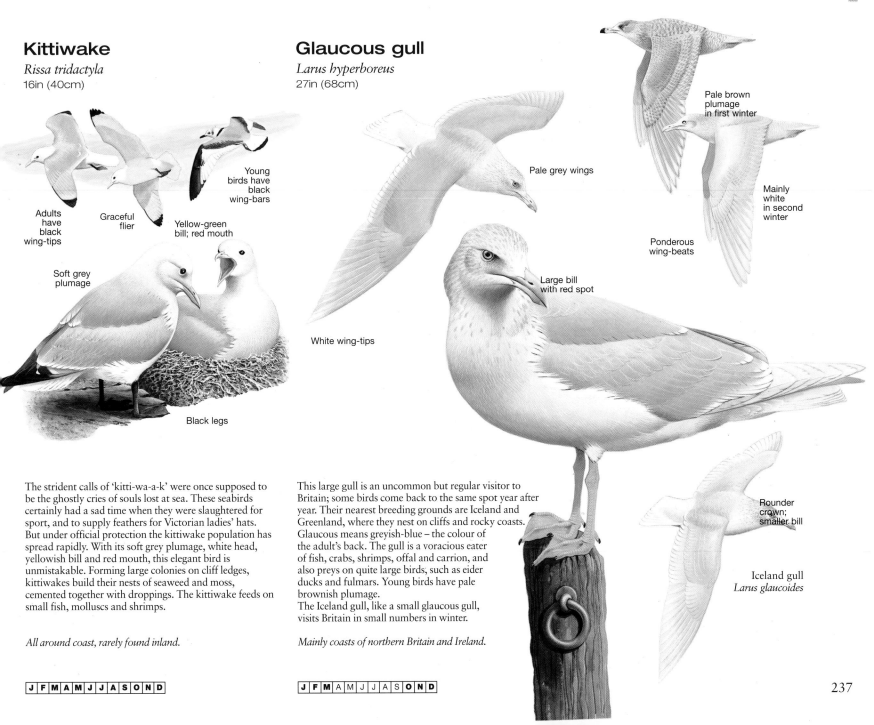

Kittiwake

Rissa tridactyla

16in (40cm)

Young birds have black wing-bars

Adults have black wing-tips

Graceful flier

Yellow-green bill; red mouth

Soft grey plumage

Black legs

Glaucous gull

Larus hyperboreus

27in (68cm)

Pale grey wings

Pale brown plumage in first winter

Mainly white in second winter

Ponderous wing-beats

Large bill with red spot

White wing-tips

Rounder crown; smaller bill

Iceland gull
Larus glaucoides

The strident calls of 'kitti-wa-a-k' were once supposed to be the ghostly cries of souls lost at sea. These seabirds certainly had a sad time when they were slaughtered for sport, and to supply feathers for Victorian ladies' hats. But under official protection the kittiwake population has spread rapidly. With its soft grey plumage, white head, yellowish bill and red mouth, this elegant bird is unmistakable. Forming large colonies on cliff ledges, kittiwakes build their nests of seaweed and moss, cemented together with droppings. The kittiwake feeds on small fish, molluscs and shrimps.

All around coast, rarely found inland.

This large gull is an uncommon but regular visitor to Britain; some birds come back to the same spot year after year. Their nearest breeding grounds are Iceland and Greenland, where they nest on cliffs and rocky coasts. Glaucous means greyish-blue – the colour of the adult's back. The gull is a voracious eater of fish, crabs, shrimps, offal and carrion, and also preys on quite large birds, such as eider ducks and fulmars. Young birds have pale brownish plumage.

The Iceland gull, like a small glaucous gull, visits Britain in small numbers in winter.

Mainly coasts of northern Britain and Ireland.

J F M A M J J A S O N D

J F M A M J J A S O N D

Gulls

Herring gull

Larus argentatus
22-26in (55-66cm)

Hard-shelled prey is dropped to be smashed

Young bird brown; tail speckled, with black band

Red spot on bill

Young take several years to get full adult plumage

Grey upper parts and black wing-tips

Pink legs

Lesser black-backed gull

Larus fuscus
21-22in (53-55cm)

Slender build

First-year bird is darker than herring gull

Dark grey upper parts

Yellow legs

Great black-backed gull

Larus marinus
25-31in (64-79cm)

Heavy hooked bill with red spot

Young bird contrasts with elders in flight

Black upper parts

White on tail, with black band

Pink legs

Young has white on head and underparts

The harsh, loud 'kee-owk-kyowk-kyowk-kyowk' call of the commonest of our larger gulls evokes the very spirit of the seashore. It is heard at close quarters in coastal towns where the birds are increasingly breeding on rooftops. Although herrings may form part of its diet, it also eats shellfish, small mammals and birds and, like many other gulls, refuse from rubbish tips. Hard-shelled prey like mussels are dropped from a height to smash them The yellow-legged gull, with its darker grey back, has become a regular visitor and now nests in southern England.

Coasts all around Britain and Ireland; widespread inland in winter.

This handsome migratory bird is smaller and rather paler-looking than its great black-backed relative. It is similar in size to the herring gull – and similar, too, in its habits, feeding on virtually anything that comes its way, from the chicks of other species to waste from dumps and ships. Moors, bogs and inland waters are its favourite breeding places. The nest is built in a shallow hollow in the ground.

Large winter roosts form on inland waters

Breeds most commonly near western and northern coasts.

A fearsome butcher of a bird, the great black-backed gull is the largest of Britain's native gulls. Its numbers have been steadily growing, due to the gradual warming of the North Atlantic and to the increase in edible refuse thrown away by man. Large and powerful, with a heavy hook-tipped bill, it eats refuse, carrion, crabs and fish. But it also preys on adult birds and chicks of other species, including puffins, shearwaters and kittiwakes. The gull breeds singly or in small colonies, usually on high rocky stacks, building a large nest of sticks and seaweed.

Rocky coasts around most of Britain and Ireland; fewer breed in east.

J F M A M J J A S O N D   J F M A M J J A S O N D   J F M A M J J A S O N D

Terns

Terns are slim, delicate relatives of the gulls with narrow wings and forked tails; their bills are more slender and pointed than those of gulls. They fly lightly over the water, sometimes hovering, before diving for food or picking it from the surface.

Roseate tern

Sterna dougallii
15in (38cm)

Winter

Summer

Long tail streamers

Dark bill, red at base

Summer

Rose tint on breast

In courting, wings droop and tail and neck point upwards

Common tern

Sterna hirundo
14in (36cm)

Summer

Winter

White forehead

Translucent patch in wing

Summer

Birds hover and plunge to catch fish

Black forehead

Black-tipped bill

Summer

Sandwich tern

Sterna sandvicensis
16in (41cm)

Winter

Summer

White forehead

Black bill with yellow tip

Shaggy crest

Summer

Chicks have spiky plumage

Courting birds hold out wings

Perhaps the most graceful and attractive of all the terns that visit Britain, the roseate tern gets its name from the soft rosy flush on the breast which appears briefly in the breeding season. In flight the roseate tern shows the longest tail streamers of any tern; it also has a whiter overall appearance, and a more slender build. Like other terns it dives into the water to catch small fish. Eggs are laid in June in a shallow depression.

Rare summer visitor to coasts of Irish Sea and north-east England.

Downy young have spiky plumage

Although not Britain's most numerous tern, this is the most widely distributed. It differs from the Arctic tern in having a black tip to its red bill, longer legs and shorter tail streamers; it has a translucent patch in the wing. The common tern eats small fish, crustaceans, marine worms and molluscs. Breeding takes place in coastal and inland colonies which sometimes number many hundreds of pairs. The nest is a scrape in the ground. Incubation and care of the young are mainly undertaken by the female.

Breeds locally on most coasts, also inland in central and eastern England.

A harsh 'kirrick, kirrick' announces the presence of the Sandwich tern. This handsome summer visitor has a large, shaggy black crest and forehead and a long, yellow-tipped black bill. It is the largest breeding tern in Britain. At the start of the breeding season a pair go on a 'fish-flight': the male offers fish to the female in mid-air as part of the courtship ritual. On the ground they face each other with their heads pointing upwards and their folded wings held out. Breeding is in colonies and the eggs are laid in a bare scrape. Sandwich terns dive to catch marine worms, small fish and sand eels.

Sand-dunes, shingle and marshes in scattered coastal sites.

J F M A M J J A S O N D

J F M A M J J A S O N D

J F M A M J J A S O N D

239

Terns

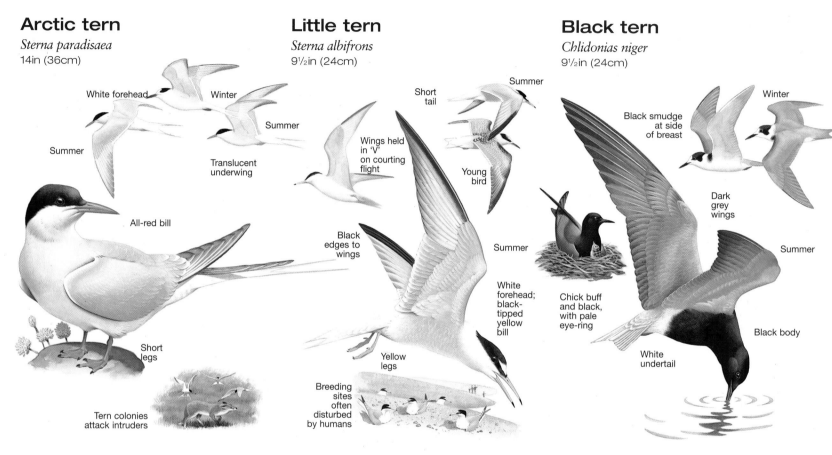

Arctic tern
Sterna paradisaea
14in (36cm)

White forehead

Winter

Summer

Summer

Translucent underwing

All-red bill

Short legs

Tern colonies attack intruders

Little tern
Sterna albifrons
9½in (24cm)

Short tail

Summer

Wings held in 'V' on courting flight

Young bird

Black edges to wings

Summer

White forehead; black-tipped yellow bill

Yellow legs

Breeding sites often disturbed by humans

Chick buff and black, with pale eye-ring

Black tern
Chlidonias niger
9½in (24cm)

Summer

Black smudge at side of breast

Winter

Dark grey wings

Summer

Black body

White undertail

The Arctic species is the most numerous breeding tern in Britain, with some 40-50,000 nesting pairs. Like the common tern, this summer visitor may breed in mixed colonies. It is an aggressive bird and will attack any intruders, including humans, by striking with its bill. Some other species of birds, especially eider ducks, breed in colonies of Arctic terns for protection. In flight, the Arctic tern appears darker on the underparts than the common tern, and its whole underwing is translucent. Breeding begins in May, when eggs are laid in a bare scrape and incubated by both parents.

Mainly Scottish islands and coast, and some south to coasts of north Wales and Ireland.

After wintering in Africa, the little tern arrives in Britain between late April and the end of May. The smallest of Britain's breeding terns, it breeds on sandy beaches and shingle banks, though invasions of holidaymakers have severely reduced its numbers. Unlike other British terns, the little tern has a white forehead in summer. The narrow wings flicker rapidly in fast flight as it dives for small fish, shrimps and marine worms.

Around coasts; rare inland.

Birds share incubation and catch fish for each other

Once a regular breeder in south-eastern Britain, the black tern now occurs only as a spring and autumn migrant. Its decline is probably due to the drainage of the fens, swamps and marshes it favours. A black body, dark grey wings and tail and white under-tail make the black tern easily recognisable in summer. In autumn the main distinguishing feature is the black smudge at the side of the breast. Black terns usually swoop and dip over the water, taking insects from the surface. The nest is a floating platform of vegetation.

Marshes and coasts mainly in southern and eastern England.

J F M **A M J J A S** O N D

J F M **A M J J A S** O N D

J F M **A M J J** A S O N D

Auks

These stout, black and white seabirds have a short tail and short, pointed wings. On land their stance is upright; in the water they are expert swimmers and divers, chasing fish below the surface. The biggest of their family, the great flightless auk, has been extinct since 1844.

Razorbill

Alca torda
16in (40cm)

Summer

Rapid wing-beats

Heavy bill with vertical white stripe

Black above and white below

Summer

Summer

Cocked tail

Winter

White throat and breast

Guillemot

Uria aalge
16½in (42cm)

White throat and breast, with black line behind eye

Slim pointed bill

Winter

Summer

Dark brown above and white below

Summer

Black guillemot
Cepphus grylle

Black body

White wing-patch

Red feet

Winter

Upper parts barred black and white

Puffin

Fratercula arctica
12in (30cm)

White face

Rapid wing-beats

Colourful triangular bill

Summer

Webbed feet spread as 'brakes' when landing

Orange legs

Brown chick fed for 40 days, then left to find own way to sea

A sharp, hooked upper mandible, well suited to grasping fish and marine invertebrates, also enables the razorbill to defend itself against predators. The stout bill is the razorbill's main distinction from the guillemot, but it also has blacker upper parts and a thicker neck. In winter the black throat changes to white. Razorbills are often seen floating in the sea in great 'rafts', when their dumpy silhouettes and upward-pointing tails are distinctive. They breed in the same areas as guillemots, but usually in more protected rock crevices. Both parents share the incubation.

Breeds on sea-cliffs, mainly in north and west; all around coast in winter.

When standing upright with its black-brown upper parts and pure white underparts fully visible, the guillemot presents the nearest approximation in Britain to the penguin of Antarctic waters. Breeding in dense colonies on precipitous cliff ledges, the guillemots pack themselves together, often with just room to stand. The single egg, laid on the bare ledge, is tapered so that it rolls in a circle without falling off. Both sexes incubate the egg, balancing it on their feet and covering it with their belly plumage. The black guillemot has black breeding plumage, broken only by a white wing-patch. It is rare except in Ireland and north-west Britain.

Breeds on sea-cliffs in north and west; all coasts in winter.

Unmistakable in summer with its huge red, blue and yellow triangular bill, clown-like face and orange feet, the puffin is one of Britain's best known birds. The distinctive bill has given rise to such popular names as 'sea parrot' and 'bottlenose'. In winter the face is darker and the bill becomes smaller and less colourful. The puffin is a marine species and catches fish, a breeding adult being capable of carrying up to ten at a time crosswise in its bill. The birds breed in large colonies and lay their single egg in an old rabbit burrow, or else in a hole excavated from the soft turf on cliff tops.

Breeds on cliffs and islands in north and west; scarce visitor elsewhere.

J F M A M J J A S O N D   J F M A M J J A S O N D   J F M A M J J A S O N D

241

Pigeons and cuckoo

Turtle dove
Streptopelia turtur
11in (28cm)

Rounded white tip to tail

Male bows in courtship ritual

Chestnut and black above

Black and white neck-patch

Pink breast

Stock dove
Columba oenas
13in (33cm)

Short black wing-bars

Grey rump

No white on neck or wings

Longer wing-bars

White rump

Rock dove
Columbia livia

Collared dove
Streptopelia decaocto
12½in (32cm)

Dark wingtips

White tail-band

Black half-collar on nape of neck

Buffish-grey above

Buffish-grey body

Flocks gather around grain stores

The throbbing, purring coo of the turtle dove is a bird call of high summer that mingles well with the sound of cricket on the village green. This dove visits Britain in summer from its winter quarters in sub-Saharan Africa. Adults of both sexes sport a heavily chequered black and chestnut back and a black and white half-collar, while the underparts have a pinkish tinge. In flight the undertail shows a rounded white tip. Juveniles are duller. The male and female turtle doves incubate the eggs alternately.

Frail nest of sticks often built low amongst brambles

Woodlands and hedges, mainly in southern England.

This plump dove is basically all grey with two black bars on the wing, and a greenish neck-patch. It nests in holes in trees and rock faces and, very occasionally, in disused rabbit burrows and isolated farm buildings. A similarly plumaged species breeding in remote coastal areas of Ireland and northern Scotland is called the rock dove. This is most readily distinguished from the farmland-dwelling stock dove by its white rump. Most pigeons found in urban areas are descended from a domesticated form of the rock dove, even though few resemble the wild bird in plumage. They roost on buildings instead of cliffs, and are often known as feral or London pigeons.

Woods and cliffs, except extreme north and north-west.

Before the 1930s the range of the collared dove in Europe was restricted to parts of the Balkans. Since then, in an amazing population explosion, it has colonised much of Europe as far north as Iceland. By 1955 it was nesting in Britain, and it has now reached almost plague proportions in some areas. The new arrivals seemed to find an ecological niche that no other bird was filling, and so were able to establish themselves without competition. Favourite habitats are in the vicinity of farms, chicken runs, corn mills and docks, where grain and other animal feed is often spilled. The birds nest mainly in conifers, and can produce five broods – each of one or two young – between March and September.

Widespread over most of Britain and Ireland.

J F M A M J J A S O N D   J F M A M J J A S O N D   J F M A M J J A S O N D

Pigeons and doves are stout, rather heavy birds with small heads and broad, pointed wings, angled at the wrist. They often feed in flocks on the ground. Flight is rapid and powerful, with occasional gliding. The birds live on rocky ledges, in woods and in urban areas. The cuckoo is unrelated.

Wood-pigeon

Columba palumbus
16in (40cm)

White neck-patch

White wing-patches

Pinkish breast

Black band on tail

To many farmers, the largest of Britain's pigeons is 'public enemy number one'. The wood-pigeon does immense damage to crops, particularly in winter when the population is joined by continental immigrants, and huge flocks feed on rape, turnips and clover – plentiful alternatives to their traditional foods of ivy berries, acorns and weed seeds. At other times of the year they feed on cereals, potatoes, beans, peas and greens. The wood-pigeon is distinguished from other pigeons by its white neck-patch and wing-patches, which are conspicuous in flight. The adult's voice is a soft cooing. In display flight the bird climbs steeply, noisily claps its wings together and then glides down.

Widespread, residents joined by immigrants in winter.

Cuckoo

Cuculus canorus
13in (33cm)

The familiar 'cuc-coo, cuc-coo' call of the male bird in April is a sure sign that summer is on its way. Cuckoos are notorious for their parasitic breeding habits. The female selects a territory and then finds a suitable nest, built by a much smaller bird, in which to lay her eggs. One egg is deposited in each nest, after the cuckoo has carefully removed one of the host's eggs. The nests of meadow pipits, dunnocks and reed warblers are mostly used. When hatched the young cuckoo ejects the remaining eggs or nestlings of the host bird.

Widespread throughout Britain and Ireland.

J F M A M J J A S O N D

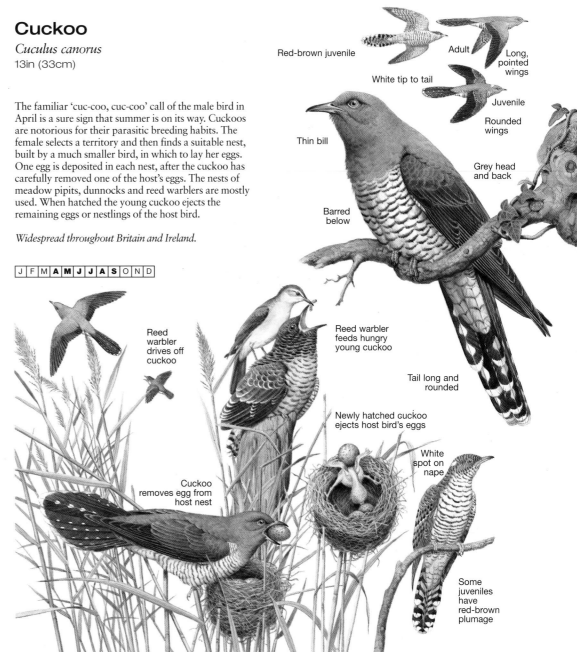

Red-brown juvenile

Adult

Long, pointed wings

White tip to tail

Juvenile

Rounded wings

Thin bill

Grey head and back

Barred below

Reed warbler drives off cuckoo

Reed warbler feeds hungry young cuckoo

Tail long and rounded

Newly hatched cuckoo ejects host bird's eggs

Cuckoo removes egg from host nest

White spot on nape

Some juveniles have red-brown plumage

Owls

Barn owl

Tyto alba
13½in (34cm)

White face

Golden-buff upper parts

All-white appearance in flight

White underparts

Long legs

A white shape silently leaving a barn or quartering a field is all an observer usually sees of the barn owl. The rare closer view reveals golden-buff upper parts lightly mottled with grey, a white heart-shaped face with dark eyes, white underparts and long feathered legs. The barn owl feeds mostly on rats, mice and voles, which are caught mainly at night. It breeds in barns, ruins and church towers, and also in natural sites such as cliff holes and hollow trees.

Widespread in agricultural country, except in far north.

Wings spread out on barn floor to shelter nestlings

Snowy owl

Nyctea scandiaca
21-24in (53-60cm)

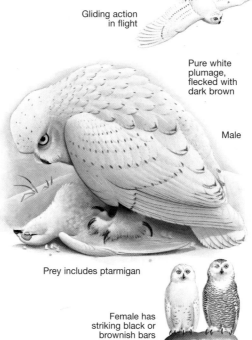

Gliding action in flight

Pure white plumage, flecked with dark brown

Male

Prey includes ptarmigan

Female has striking black or brownish bars

Until 1967 the snowy owl was considered no more than a rare visitor to northern Britain. That year, however, and for the next eight years until 1975 a pair nested in the Shetlands, and 23 young fledged during that period. The male has almost pure white plumage, adapted to the Arctic region where it normally breeds, while the larger female is heavily barred with brown. The snowy owl's status has now reverted to that of a rare winter visitor to parts of Scotland and Ireland. It lives on rabbits, mice, voles and birds.

Rare. Irregular winter visitor to northern Scotland and Ireland.

Little owl

Athene noctua
8½in (22cm)

Flight low, swift and undulating

Plumage grey-brown and white

'Frowning' expression

Yellow eyes

Fence post used as day-time perch

The smallest of our breeding owls was introduced to this country from the Continent in the late 19th century. Although mainly nocturnal, it can be seen in daylight, especially on warm summer evenings. The birds breed in holes and crevices, and frequent agricultural land, parks, orchards, quarries and sea-cliffs. Insects, small birds and mammals form the main part of the little owl's diet. Its call is a plaintive 'kiew', interspersed with a yelping 'werrow'.

Widespread in England and Wales.

Fledglings bob heads when anxious

Mainly night-hunting birds of prey, owls have round heads, short, hooked beaks and large forward-facing eyes. They live in buildings, marshes and woods, fly silently, hunting for prey, and are most often seen at dusk.

Tawny owl

Strix aluco
15in (38cm)

Rounded face

Brown body with dark streaks below

Dark eyes

Owl watches for prey from tree perch, then pounces without warning

The long, quavering, 'hoo-hoo-hoo-hooooo', often described as 'to-whit-to-wooo', is one of the best known of British bird calls. In woodland, its favoured habitat, the tawny owl's presence is often given away by the alarm calls of small birds mobbing an adult dozing on the branch of a tree. Its wholly dark eyes distinguish the tawny owl (sometimes called the brown owl) from other owls. Soft plumage and specially adapted wing feathers make it silent in flight.

Woodlands throughout Britain; absent from Ireland.

Fledglings have barred downy underparts

Short-eared owl

Asio flammeus
15in (38cm)

Dark 'wrist' patches

Short 'ear' tufts

Yellow eyes

Dark front, lighter at rear

Bold tail bars

Rounded wings

Of all the owls, the short-eared owl is the one most often seen hunting in daylight. It looks like a huge moth as it hunts low over moorland, estuaries and marshes. In display the bird makes a loud clapping noise as the wingtips meet below the body. The short ear-like tufts are often difficult to see, but the bright yellow eyes stare from the round face. It preys mainly on voles.

Open country, mainly in northern England and Scotland; more widespread in winter.

Nest is on ground

Long-eared owl

Asio otus
13½in (34cm)

Long 'ear' tufts

Tiny bars on tail

Faint 'wrist' patches

Orange eyes

Uniformly dark underparts

Slender posture adopted when alarmed

The apparent 'ears' are in fact no more than tufts of feathers – the actual ears are under the feathers on the sides of the head. The prominent, bright orange eyes in the facial disc, combined with the 'ear' tufts, distinguish this species from all other owls. Strictly nocturnal, roosting in dense tree cover during the day, the long-eared owl's preferred habitat is coniferous woodland. Communal roosts often form in winter. The owls breed usually in the old nest of another bird, or in a squirrel's drey. The call is a low, drawn out 'oooo' 'oooo' 'oooo', and the young birds make a noise like a creaking gate.

Widespread but uncommon in woodland; continental visitors in winter.

Kingfisher and nightjar

The brightly coloured kingfisher has a large head, short tail and long sharp bill used for catching fish. In marked contrast, the nightjar is a nocturnal insect-eater, with long wings and tail and a short bill.

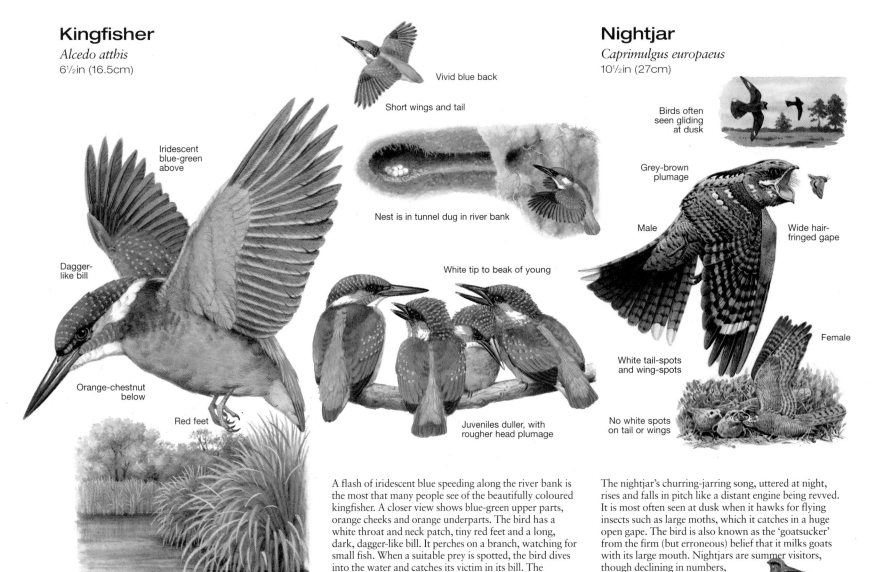

Kingfisher

Alcedo atthis
6½in (16.5cm)

Vivid blue back

Short wings and tail

Iridescent blue-green above

Nest is in tunnel dug in river bank

Dagger-like bill

White tip to beak of young

Orange-chestnut below

Red feet

Juveniles duller, with rougher head plumage

Nightjar

Caprimulgus europaeus
10½in (27cm)

Birds often seen gliding at dusk

Grey-brown plumage

Male

Wide hair-fringed gape

Female

White tail-spots and wing-spots

No white spots on tail or wings

A flash of iridescent blue speeding along the river bank is the most that many people see of the beautifully coloured kingfisher. A closer view shows blue-green upper parts, orange cheeks and orange underparts. The bird has a white throat and neck patch, tiny red feet and a long, dark, dagger-like bill. It perches on a branch, watching for small fish. When a suitable prey is spotted, the bird dives into the water and catches its victim in its bill. The kingfisher returns to its perch, beats the fish against the branch to kill it, then swallows it head first. Both adults dig a long tunnel in the river bank to make their nest.

Along slow-moving rivers; scarce in Scotland.

The nightjar's churring-jarring song, uttered at night, rises and falls in pitch like a distant engine being revved. It is most often seen at dusk when it hawks for flying insects such as large moths, which it catches in a huge open gape. The bird is also known as the 'goatsucker' from the firm (but erroneous) belief that it milks goats with its large mouth. Nightjars are summer visitors, though declining in numbers, and nest in a scrape in the ground among dead leaves.

Heaths, moors and woods, mainly in southern Britain.

Bird perches along branch, not across it

J F M A M J J A S O N D

J F M A M J J A S O N D

Hoopoe, oriole and swift

After wintering in Africa, a few golden orioles and hoopoes bring a dash of exotic colour to southern England. By contrast, swifts arrive from Africa in spring in huge flocks, and breed over a large area.

Hoopoe

Upupa epops
11in (28cm)

Barred wings and tail

Crest flattened into hammer shape

Black-edged crest raised in alarm or when alighting

Pink-brown body

Long down-curved bill; feeds on insects and grubs

Golden oriole

Oriolus oriolus
9½in (24cm)

Male

Bill pinkish-red

Black wings with yellow patches

Black tail with yellow tips

Male

Yellow body

Female

Yellow-green plumage

Swift

Apus apus
6½in (16.5cm)

Screaming mobs dash around rooftops

Tiny bill

Flocks circle higher at dusk

Dark underparts

Sickle-shaped wings

Nests often in gaps in stonework under roofs

When erected, the long, black-edged crest of the hoopoe is unmistakable. This is a bird that spends the winter in Africa and migrates to southern and central Europe in summer. Some birds extend their journey northwards to Britain, and a few may stay to breed. Surprisingly for such a boldly marked bird it can be difficult to see when perched or when feeding on the ground. In flight, which is undulating with lazy flaps on broad, rounded wings, the striking appearance is unmistakable. The song is a low 'poo-poo-poo', often repeated and far-carrying.

Rare. Wood edges and coastal grassland in southern England; has occasionally bred.

Despite the brilliant yellow and black plumage of the male golden oriole it is seldom visible in its favoured woodland habitat. Usually the only clue to its presence is a fluty, melodious whistle 'weela-wheco'. Like the hoopoe it is a rare summer visitor, a few pairs staying to breed. Female and juveniles are yellow-green, streaked on the breast, and can be confused with the green woodpecker. Some old adult females can resemble adult males. The female builds the nest, a suspended cup, in the fork of a branch. The eggs are incubated by both parents.

A few pairs breed in poplar woods in East Anglia; elsewhere, rare spring migrant.

Few birds spend more of their lives in the air than the swift. They collect all their food and nesting material in flight, and drink and bathe without alighting. Swifts even mate on the wing, and at dusk parties of them can be seen circling higher and higher into the sky to spend the night 'cat-napping' while airborne. Insects are funnelled into the large gape with the help of stiff bristles around the mouth. Food for the young is stored in a throat pouch which can often be seen bulging with gorged insects. Alighting only to nest and feed the young, the swift breeds in holes and crevices in cliffs and buildings.

Widespread visitor, except in north-west Scotland.

J F M A M J J A S O N D   J F M A M J J A S O N D   J F M A M J J A S O N D

Swallows and martins

These small birds have long, pointed wings, a forked tail, short legs and small feet. They are fast and graceful in flight, and use their wide mouth to catch insects while on the wing.

Swallow

Hirundo rustica
7½in (19cm)

Russet throat and forehead in both sexes

Graceful flight with constant swooping

Upper parts blue-black in both sexes

Female

Shorter tail feathers

Male

Long tail feathers, white marks

Although swallows are regarded as harbingers of summer, the first birds often arrive from their African wintering grounds as early as March. Swallows rarely land on the ground except to gather nest material. They drink on the wing, skimming over the water, and most of their insect food is captured in low-level flight. Their pleasant, twittering song is often heard as they perch on telephone wires. The swallow's russet throat and the long tail streamers of the male are unmistakable. Swallows frequently return to the same locality, even the exact site, to breed. The nest is a cup of mud and straw.

Widespread in open country, near buildings and water.

House martin

Delichon urbica
5in (12.5cm)

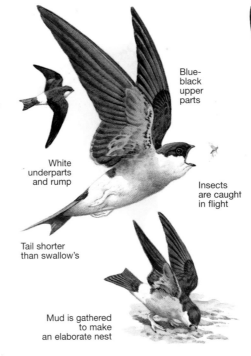

Blue-black upper parts

White underparts and rump

Insects are caught in flight

Tail shorter than swallow's

Mud is gathered to make an elaborate nest

A short tail and a white rump distinguish the house martin from the swallow. Traditionally it is a cliff-nesting species, but it has adapted to living under house eaves. Its nest of mud and plant fibres is flask-shaped, with only a small entrance at the top. Arriving from Africa in April or May, house martins start breeding almost immediately and raise two or even three broods each year. The birds swoop and wheel to catch flying insects for food. Nests built by house martins that arrive early are often re-used by later migrants.

Widespread, except in far north.

Sand martin

Riparia riparia
4¾in (12cm)

White underparts with brown breast-band

Tail less forked than house martin's

All-brown upper parts

Birds roost in reed-beds

The sand martin population is prone to fluctuations caused by the weather in its African winter quarters. Periodic droughts drastically reduce the numbers of aerial insects on which the birds feed. Smaller than the house martin, the sand martin takes its name from its nesting habits: it digs out a tunnel in a sand-bank, leading to a nesting chamber lined with plant material and feathers. Both parents share incubation.

Widespread, except in northern Scotland.

Feet are used to dig out nest tunnel in sand

[J F M A M J J A S O N D]

[J F M A M J J A S O N D]

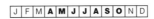

Woodpeckers

The strong, sharp bill of these colourful birds is adapted for chipping and boring into tree-trunks, both to create a nesting chamber and to catch insects for food. The short stiff tail is used for support when climbing trees. The flight is undulating.

Wryneck

Jynx torquilla
6½in (16.5cm)

Grey-brown plumage has camouflage pattern

Barred underparts and tail

Flight slow and hesitant

This bird's habit of twisting and turning its neck when startled is responsible for its name. Camouflaged with mottled grey and brown plumage, this relative of the woodpecker spends much of its time on the ground, feeding on ants. Holes in trees are nesting places. Once common in England as a breeding bird, the species had virtually vanished by the early 1980s. But birds, apparently from Scandinavia, have started breeding in the Scottish Highlands.

Rare migrant on east and south coasts.

Green woodpecker

Picus viridis
12½in (32cm)

Red crown

Red 'moustache'; female's is black

Male

Green back

Yellow rump, conspicuous in flight

An attractive call of 'kew-kew-kew' sounds like laughter and gives this bird the local country name of 'yaffle'. It is the largest British woodpecker, brightly and distinctively coloured. Both sexes have the black face-patch; the female has a black 'moustache', which on the male is red and black-edged. The green woodpecker seldom 'drums' like related species. It also feeds more on the ground.

Widespread in woods in England and Wales, spreading in Scotland; not in Ireland.

Lesser spotted woodpecker

Dendrocopos minor
5¾in (14.5cm)

Red cap

Male

White bars on black back

Black and white pattern noticeable in flight

Female has buff cap

Young bird brownish on head

Scarcely larger than a sparrow, the lesser spotted woodpecker can be elusive among the top branches of trees where it likes to live. Its other name, the barred woodpecker, describes its plumage; the upper parts are black with white barring. The male's cap or crown is red, the female's buff. The bird drums on wood to establish territory; the sound usually lasts longer than that of the great spotted woodpecker.

Woods in England and Wales; not in Scotland or Ireland.

Great spotted woodpecker

Dendrocopos major
9in (23cm)

Female

Shoulder patches visible in flight

Red patch

White shoulder patches

Male

Red under-tail

Rapid blows with its bill on a dead branch produce the characteristic drumming sound of a great spotted woodpecker establishing territory. It is also known as the pied woodpecker because of its mainly black and white plumage. The large shoulder patches are easily recognised. Both sexes have bright red under-tail patches, but only the male has a red patch on the nape. Young birds have a red cap. The nest is a chamber in a tree-trunk.

Widespread in woods, but not in Ireland.

J	F	M	A	M	J	J	A	S	O	N	D

J	F	M	A	M	J	J	A	S	O	N	D

J	F	M	A	M	J	J	A	S	O	N	D

J	F	M	A	M	J	J	A	S	O	N	D

Larks

These streaky brown birds of coasts, marshes, heaths and fields nest and feed on the ground; their bill is adapted to a diet of both insects and seeds. Their song is usually delivered in flight. Larks often gather in flocks when the breeding season is over.

Skylark

Alauda arvensis
7in (18cm)

Migrant skylarks seen on open sand-dunes

Yellow and black head with black 'horns'

Shore lark
Eremophila alpestris

Song sustained while in flight, often when hovering

Uniform brown plumage

Crest

Long tail, with white outer feathers

As it flies above the open fields and downs, giving voice loudly, the skylark is a difficult bird to ignore. It breeds more widely in Britain than any other bird, and is seen over farmland, grassland, meadows, commons and sand-dunes. It is larger than the woodlark and has a bigger crest. Rising almost vertically in a hovering flight, often to several hundred feet, the skylark sustains its clear warbling song for several minutes, before sinking gradually to the ground, still singing.
The shore lark, a rare winter visitor to the east coast, is distinguished by its boldly patterned head.

Widespread; joined by migrants from northern Europe in winter.

Woodlark

Lullula arborea
6in (15cm)

Buff eye-stripes

Small crest

White wing marks

Buff nape band

Seeds supplement insect diet in autumn

The woodlark's generic name *Lullula* is derived from its highly distinctive song, a rich medley of various phrases often repeated and interspersed with an occasional 'loo-loo-loo'. The song is delivered in a circular song flight, or from a tree or bush. Similar in appearance to the skylark, the woodlark has a shorter tail, buffish eye-stripes meeting across the nape, a black and white patch on the coverts and a much smaller crest. Woodlarks once bred in nearly every county in England and Wales, but were hit hard by cold winters in the 1960s. Mild winters and helpful woodland management have brought about a resurgence.

Wooded areas, mainly in southern England and East Anglia.

| J | F | M | A | M | J | J | A | S | O | N | D |

Pipits

Its name gives the clue to the type of country in which each species of pipit is likely to be seen. All are delicate, slender, long-tailed birds, with a fine pointed bill adapted for catching insects. They are all brownish, and sing in flight like larks.

Tree pipit
Anthus trivialis
6in (15cm)

Stout bill

Yellowish breast

Pink legs; short hind claws

Male 'parachutes' down in song flight

Birds feed on ground insects

Meadow pipit
Anthus pratensis
5¾in (14.5cm)

Long tail, white-edged

Slender shape

Long hind claws

Streaked breast

Brownish legs

Rock pipit
Anthus petrosus
6½in (16.5cm)

Long bill

Olive-brown back

Whitish eye-stripe

Grey tail edges

White outer tail feathers

Bird feeds on seaweed insects

Greyer plumage

Water pipit
Anthus spinoletta

From mid-April onwards the tree pipit arrives in Britain to breed, after a winter spent in central Africa. Fluttering up steeply from its perch high in a tree, the bird delivers a loud, far-carrying trill ending in 'zeea-zeea-zeea'. The bird continues to sing as it floats down, wings and tail held out like a parachute. The tree pipit is plumper than the meadow pipit, and has a stouter bill, shorter hind claws and pinker legs. It breeds in heather, open woodland and parkland, the nest being well hidden in vegetation on the ground.

Widespread in mainland Britain in open woodland; only sporadic breeder in Ireland.

Present all year, although many migrate south to France or Spain for the winter, the meadow pipit is a bird of open country and one of Britain's commonest song-birds. Its long hind claws and more densely streaked breast, and its thin 'eest' or 'tissip' call, distinguish this bird from the tree pipit. The conspicuous song flight by the male starts as the bird flies up to 100ft (30m) from the ground, uttering an accelerating series of 'pheet' notes. These reach a climax and are replaced by slower and more liquid notes as the bird 'parachutes' down.

Widespread on heaths and grassland; coasts in winter.

Longer billed than other breeding pipits, the rock pipit inhabits rocky shores and cliffs. It is identified by its heavily streaked breast and olive-brown upper parts, and grey outer tail feathers. Its alarm call, a thin 'phist', is also distinctive. The nest of grass, moss and fine seaweed, built by the female, is usually concealed in a hole in a crevice or cliff.

The water pipit, a continental species, visits Britain in winter and may be seen in marshes and watercress beds. It is identified by its white outer tail feathers, clear whitish eye-stripe and finely streaked breast. It breeds in mountainous areas of central and southern Europe.

Rocky coasts, all round Britain and Ireland.

J F M A M J J A S O N D   J F M A M J J A S O N D   J F M A M J J A S O N D

Wagtails

Like the pipits, to which they are closely related, wagtails are slender birds. They have a long narrow tail which they wag up and down. They feed mostly on the ground, using their long thin legs for running or walking. The plumage is more colourful in the males than in the females.

Pied wagtail

Motacilla alba yarrellii
7in (18cm)

Young bird brownish-grey, with black breast-band

Black cap and bib merge in one

Long wagging tail, with white outer feathers

Greyer upper body

White wagtail
Motacilla alba alba

Named for its pied plumage and tail-wagging habit, this bird is to be found in towns and gardens, on farmland and near water. All such places provide an abundance of the flies and other insects which form its diet. The pied wagtail is black above and white below, with a white face-patch and black chin and bib; the outer tail feathers are white. Cavities in cliffs, stream banks, walls and trees provide sites for its nest, an accumulation of grass, leaves and moss lined with hair, wool or feathers.

A European form of the bird, seen in Britain on migration and known as the white wagtail, has pale grey upper parts and rump.

Widespread. Some migrate to Continent in winter.

Yellow wagtail

Motacilla flava
6½in (16.5cm)

Yellow head, greenish crown

Flying insects are caught

Young birds buff below, with black throat

White edges to tail

The slender yellow wagtail, a summer visitor, can often be seen sitting on a fence or tall plant calling with a loud, prolonged 'tsee-eep'. It lives in flooded territory, marshes, water-meadows and fields and is extremely wary, making close approach difficult. It feeds on insects, and is frequently found near fly-bothered cattle. The male bird has a bright yellow face and underparts; the crown, ear coverts and back are greenish. The female is slightly duller. Migrant birds from the Continent with varying head colours appear from time to time; a blue-headed wagtail is one. The yellow wagtail's cup-shaped nest is built in a hollow on the ground.

Marshes and wet meadows in most of England.

Grey wagtail

Motacilla cinerea
7in (18cm)

Birds roost together in winter

Young bird yellow only under tail

Grey above

Female duller

Male

Black throat

Yellow below

A walk beside one of the rushing, tumbling streams in Britain's hill country may afford a glimpse of this elegant little bird. It spends much of its time walking along the water's edge or perching on a boulder, twitching its long tail as it watches for insects such as flies, midges, dragonflies and water beetles. Its yellow underparts are as bright as those of the yellow wagtail; but unlike the migrant yellow wagtail the grey wagtail is present all year. The nest is always by fast-flowing water and usually in a crevice or hollow.

Widespread, but less common in eastern England.

J F M A M J J A S O N D

J F M A M J J A S O N D

J F M A M J J A S O N D

Shrikes and waxwing

The hooked bills of shrikes are suitable for catching small birds and rodents. Their tails are long and rounded and their wings broad. The waxwing is a short-tailed fruit-eater with a distinctive crest on its head.

Great grey shrike

Lanius excubitor
9½in (24cm)

Hooked bill

Black band through eye

Black, grey and white plumage

White edge to tail

Small birds are chased and killed

Red-backed shrike

Lanius collurio
6¾in (17cm)

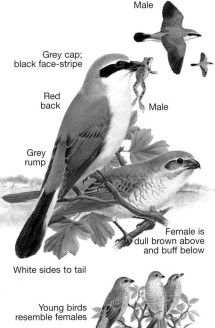

Male

Grey cap; black face-stripe

Red back

Male

Grey rump

Female is dull brown above and buff below

White sides to tail

Young birds resemble females

Waxwing

Bombycilla garrulus
7in (18cm)

Crest

Birds gather in groups to eat berries

Black on throat and around eyes

Waxy red tips on wing feathers

Multi-coloured wings

Yellow tip to tail

Birds fly swiftly and directly in flocks

Ponds and puddles are drinking places

Beginning to arrive in September from northern Europe, the great grey shrike winters in Britain. Like other shrikes it is a lonely hunter and 'butcher bird' – so called because victims are often impaled on thorns or spikes and stored as if in a larder. Its prey consists of insects, small mammals and birds. Finches and buntings are favourite prey, though larger birds such as blackbirds and thrushes are sometimes taken.

Winter visitor, sparsely scattered throughout Britain.

The red-backed shrike population has decreased drastically in Britain and Europe. Owing to colder, wetter summers and consequent food scarcity, and to the destruction of the birds' habitat, none of these summer visitors now breed in England. Shrikes like to perch on a bush, fence or telegraph wire, watching for prey. Like its great grey relative, the red-backed shrike frequently stores its victims by impaling them on thorns.

Scarce migrant on east and south coasts; occasionally nests in Scotland.

Waxwings are winter visitors and do not breed in Britain; every few years large numbers come from their Arctic and sub-Arctic breeding grounds looking for berries – the rowan is a favourite. Adults are pinkish-brown, darker on the back, with black eye areas and throat, grey rump and yellow-tipped black tail. Black, yellow and white appear in the wings with the red waxy-looking blobs that give the bird its name.

Shrubs and trees, mainly in east; numerous only in occasional years.

J F M A M J J A S O N D

J F M A M J J A S O N D

J F M A M J J A S O N D

253

Starling and crows

Starling
Sturnus vulgaris
8½in (22cm)

Glossy, iridescent plumage

Yellow bill

Juveniles mouse-brown, becoming blackish with white spots

This brash, aggressive bully of a bird is as familiar in towns as it is in the country. Starlings' droppings foul buildings and pavements, and their voracious appetite can strip fields of corn. On the good side, however, they eat leatherjackets, wireworms and other pests. An expert mimic, the starling imitates other bird-calls and songs, its own song being only a jumble of squeaks and whistles. Hundreds of thousands of starlings are present in Britain all year, and in winter their numbers are swollen by millions more that arrive from the Continent to take advantage of our milder winter climate.

Widespread throughout Britain and Ireland.

J F M A M J J A S O N D

Chough
Pyrrhocorax pyrrhocorax
15½in (39cm)

Curved red bill

Glossy purple-black plumage

Wing-ends upturned in rolling flight

Red legs

Master of the air, performing aerobatics as it glides, dives, rolls and soars above the cliffs, the chough is a bird of mountain crag and sea-cliff. Its glossy, purple-black plumage, long red curved bill and red legs are its most striking features. Once fairly widespread, the population has declined for reasons unknown. Choughs feed on worms, caterpillars and other insects, and also eat small shellfish. They are often seen in small flocks, and their wild, excited 'keeaar' call is distinctive. The nest is built on a ledge or crevice in a cave, or on a cliff.

Coasts of Isle of Man, Inner Hebrides, Wales and Ireland.

J F M A M J J A S O N D

Jackdaw
Corvus monedula
13in (33cm)

Dark band on under-wing

Grey nape

Grey eyes

Birds flock together on derelict buildings

Black back

Dark grey underparts

The most notorious robber in the crow family, the jackdaw is well known for its thieving habits. Apart from having a reputation for snatching and hiding inedible objects, it also hides food, especially cereals, potatoes, fruit and berries. Other items in the bird's diet are insects, mice and worms. The jackdaw also steals the eggs and nestlings of other birds. It is easily identified by its grey nape, which contrasts strikingly with its darker grey and black plumage. Jackdaws nest in tree holes, old nests of other birds, rabbit burrows or chimney pots. Their call is a loud, explosive 'tchack' or a shrill 'keeya'.

Cliffs and old woodlands throughout Britain and Ireland.

J F M A M J J A S O N D

Dark, speckled plumage identifies the starling in urban areas, woods and pastures. Crows, the largest perching birds, have broad wings and strong bills, legs and feet. With notable exceptions, their plumage is mostly black. They walk or hop on the ground, often feeding in flocks.

Magpie

Pica pica
18in (45cm)

Once heavily persecuted and considered a pest, the magpie is now increasing in numbers, especially in suburban areas where it was once unknown. Its black and white plumage and long narrow tail, coloured a dark iridescent green, make it one of the easiest of all birds to identify. During the nesting season magpies often stay hidden in overgrown hedgerows and thickets, their presence only revealed by their call, a hoarse, laughing chatter 'chacha-chacha-chak'. The birds build a large, domed nest of sticks. Young birds, which have shorter tails, leave the nest after three to four weeks.

Widespread, except in northern Scotland.

J F M A M J J A S O N D

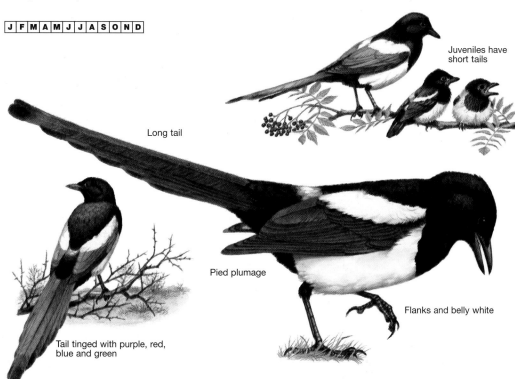

Wedge-shaped tail

White wing-patches

Magpies steal eggs and sometimes nestlings of other birds

Juveniles have short tails

Long tail

Tail tinged with purple, red, blue and green

Pied plumage

Flanks and belly white

Jay

Garrulus glandarius
13½in (34cm)

White rump and wing-patches

Streaked crest

Jay buries acorns and other food

Pinkish-brown back

Blue wing-patch

The jay is heavily dependent on trees – especially oaks – and is more often heard than seen. It greets woodland intruders with a raucous, scolding 'caaarg-caarg'. When seen, the pinkish-buff plumage, streaked crown, white rump and blue wing-patch are distinctive. Jays like acorns, which are often collected and buried among fallen leaves and twigs to be eaten in the winter. Beech-nuts, peas, fruit and berries are similarly stored. Jays also eat small mammals, insects and worms, and will sometimes raid the nests of other birds.

Woodlands in most of British Isles, except northern Scotland and western Ireland.

J F M A M J J A S O N D

Crows

Raven

Corvus corax
25in (64cm)

Stout, heavy bill

Shaggy throat feathers

Diamond-shaped tail visible in high, soaring flight

All-black plumage

Carrion crow

Corvus corone corone
18½in (47cm)

Heavy, rounded bill

Square tail

Birds feed on carrion

Grey body; black wings

Glossy, all-black plumage

Square-ended tail

Hooded crow
Corvus corone cornix

Rook

Corvus frugilegus
18in (45cm)

Throat pouch extended when carrying food

Black plumage with purple gloss

Bare white face-patch

Long thigh feathers

In the past the raven, the largest member of the crow family, was regarded as a bird of ill-omen and harbinger of death. It probably acquired its reputation because of its black plumage and its habit of feeding on corpses hanging on the gibbet. Once common throughout Britain, it has been forced by man to withdraw to remote sea-cliffs, mountains, quarries, moors and windswept hills. Like most crows, the raven's main method of acquiring food is by scavenging. But it will also kill birds or small mammals, and forage for eggs, reptiles, insects and seeds. Its stick nest is usually placed on a rock ledge or in a tree.

Upland and coasts in north and west British Isles.

Smaller than the raven but far more common, the carrion crow has few friends. Its scavenging habits and harsh croaking call have never endeared it to man, and farmers and gamekeepers have persecuted it because of its liking for grain and root crops and its thieving of the eggs and chicks of game birds. Birds south of the Scottish Highlands are usually all black, while the more northerly birds have grey bodies and are known as hooded crows. In between, the two forms often interbreed, producing black plumage with some grey markings.

Widespread. Hooded crow in Scottish Highlands, Ireland, Isle of Man; carrion crow elsewhere.

As many as 6000 rooks have been counted in one raucous rookery, their nests in tall trees standing out in springtime against a network of bare branches. But their numbers have declined locally, possibly because of the ploughing-up of permanent pasture in favour of temporary crops of grass and clover, where the soil does not harbour so many of the bird's favourite insect foods – leatherjackets and wireworms. A bare, white face-patch, elongated thigh feathers and highly glossy plumage distinguish the adult rook from the similar but more solitary carrion crow. Juveniles are all black.

Widespread, especially in farmland with tall trees.

J F M A M J J A S O N D

J F M A M J J A S O N D

J F M A M J J A S O N D

Dipper, wren and dunnock

The plump dipper is a bird of mountain and moorland streams, while the tiny wren, with its upturned tail, appears in many a garden. The dunnock sings a wren-like warble in woodland.

Dipper
Cinclus cinclus
7in (18cm)

Fast flight on short wings

White throat and breast

Short tail

Chestnut waistband

Young bird has duller plumage

Bird walks underwater with back slanted and head down

Wren
Troglodytes troglodytes
3¾in (9.5cm)

Whirring flight on short rounded wings

Cocked tail

Eats small insects

Reddish-brown above; buff plumage below

Call is often made from a rock or wall

Dunnock
Prunella modularis
5¾in (14.5cm)

Grey head

Thin bill

Streaked flank

Young bird is spotted like young robin

Dunnock's song similar to wren's

Unlike any other bird, the dipper walks underwater on the bed of fast-flowing streams seeking food – mainly aquatic insects and their larvae. As it walks upstream with its head down, the force of the current against its slanting back holds it on the bottom. It also swims underwater and on the surface, propelling itself with its wings. Looking like a large wren with a white throat and breast, it is often seen on a rock, bobbing up and down and flexing its legs before plunging in. The domed nest is built under a bridge, behind a waterfall or beneath the overhang of a river bank.

Wet places, mainly in north and west Britain and in Ireland.

Tiny and inconspicuous, the wren can be located by sounds. An explosive 'tit-tit-tit' is the call; the song is a shrill, rattling warble. The wren is reddish-brown above, buff below, and its short tail is held in a cocked position. The male builds a number of domed nests, one of which the female chooses and lines with feathers. It is a prolific breeder, but is hard hit by severe winters.

In most habitats throughout Britain and Ireland.

Bird briefly glimpsed flitting between thickets

The dunnock uses its wings in a curious display. A pair or even a small party of birds will perch in the open and wave their wings at each other in a sort of semaphore. The dunnock has long been called the hedge sparrow, but it is not a true sparrow. It is identified by its grey head and underparts and thin bill. Its song is a jingling warble like the wren's but less aggressive in tone. A cup-shaped nest is lined with hair and feathers.

Widespread in gardens and woodland.

Often rears cuckoo from egg laid in nest

J F M A M J J A S O N D

J F M A M J J A S O N D

J F M A M J J A S O N D

Warblers

Grasshopper warbler

Locustella naevia
5in (12.5cm)

Heavily streaked brown upper parts

Buff under-parts

Long, rounded tail

Reddish-brown upper parts, unstreaked

Broad, rounded tail

Savi's warbler
Locustella luscinioides

Sedge warbler

Acrocephalus schoenobaenus
5in (12.5cm)

Spread tail

Streaked back; reddish rump

Creamy eye-stripe

Marsh warbler

Acrocephalus palustris
. 5in (12.5cm)

Flat head

Olive-brown upper parts

Pink legs

Reed warbler

Acrocephalus scirpaceus
5in (12.5cm)

Rounded head

Warm brown upper parts

Dark legs

Dark, reddish-brown upper parts

Cetti's warbler
Cettia cetti

Pale underparts

The distinctive trilling song or 'reel', like the sound of a cricket, is usually the only clue to this warbler's presence. When disturbed, the bird usually creeps away through the grass, and only rarely is it seen flitting between bushes or tussocks. Savi's warbler is probably our rarest breeding warbler, confined to south-east England and East Anglia. Its song is rather like the grasshopper warbler's, but is louder, faster and pitched lower.

Widespread in undergrowth, except far north.

A summer visitor from Africa, the sedge warbler favours damp areas such as reed and osier beds, and ditches and bushes near water. But it will also breed in standing crops and young forestry plantations. The song, a continuous and hurried series of harsh, chattering notes, is usually delivered from the top of a reed or bush. In courtship display, the male flies vertically upwards, singing, then descends on spread wings and tail.

Water edges throughout Britain and Ireland.

Breeding only in a few areas of southern England, the marsh warbler is similar in appearance to the reed warbler, but with more olive upper parts, whiter underparts and pinkish legs. Its most striking feature, however, is its song. The marsh warbler is a supreme mimic and has been known to imitate more than 50 other species. It is usually found near water, among willows and osiers. The cup-shaped nest is slung from plant stems by 'basket handles'.

Rare breeding visitor to southern England.

A common victim of the cuckoo, this summer visitor from Africa likes reed-beds, osiers and other dense vegetation near or over water. The harsh repetitive song, 'jag-jag-jag . . . chirruc-chirruc-chirruc', is similar to that of the sedge warbler, but harsher and more even. Cetti's warbler, a rare marshland resident, is a secretive bird, resembling a small nightingale but usually located only by its loud, repetitive song.

Reed-beds in England and Wales.

J F M A M J J A S O N D   J F M A M J J A S O N D   J F M A M J J A S O N D   J F M A M J J A S O N D

All warblers are slim, active, mainly insect-eating birds, mostly with rather dull brown plumage. Many species are shy and secretive, more likely to be heard than seen. Their different habitats and wide variety of songs help to distinguish the different species.

Garden warbler

Sylvia borin
5½in (14cm)

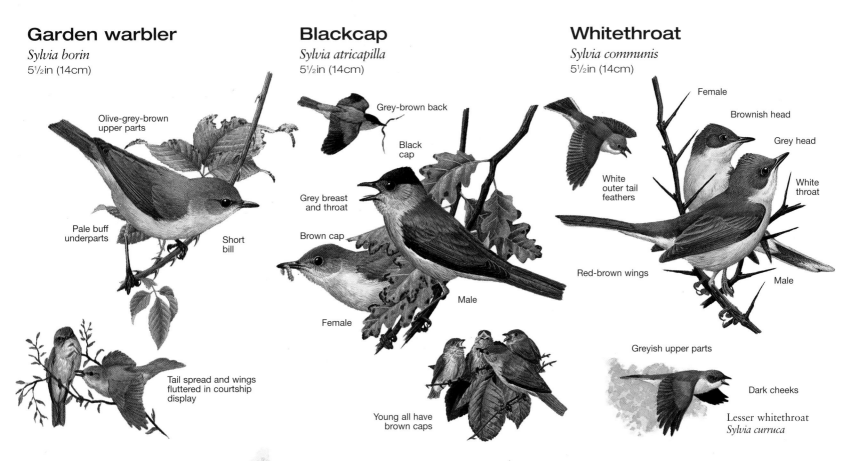

Olive-grey-brown upper parts

Pale buff underparts

Short bill

Tail spread and wings fluttered in courtship display

Blackcap

Sylvia atricapilla
5½in (14cm)

Grey-brown back

Black cap

Grey breast and throat

Brown cap

Male

Female

Young all have brown caps

Whitethroat

Sylvia communis
5½in (14cm)

Female

Brownish head

Grey head

White outer tail feathers

White throat

Red-brown wings

Male

Greyish upper parts

Dark cheeks

Lesser whitethroat
Sylvia curruca

Despite its name, the garden warbler is more a bird of open woodland and copses, where it announces its presence by its melodic song or a harsh 'tacc-tacc' alarm call. In appearance it is a rather heavy-looking, short-billed bird, with no distinctive features. Before migrating back to Africa in winter it fattens up on berries.

Woods and bushy areas, except northern Scotland and Ireland.

Nest built low, often in brambles

In recent years a few blackcaps have started braving the English winter, visiting bird-tables and gardens for food, but they are normally summer visitors from the Mediterranean. They have a rich, warbling song, shorter and more variable than that of the garden warbler. The male's black cap distinguishes it from all other British warblers. Insects are its main food.

Woodlands, hedgerows and scrub, except in northern Scotland.

Courting male raises feathers, droops wings and spreads tail

Once the commonest warbler to be found in Britain, the whitethroat suffered a severe fall in numbers between the autumn of 1968 and the spring of 1969, because of a severe drought south of the Sahara, where the birds winter. Since then the breeding population has fluctuated at around 1 million pairs. The whitethroat's song, a short, scratchy warble, is often uttered by the male in a brief dancing display flight.
The lesser whitethroat is a slightly smaller bird, with a shorter tail and without the red-brown wings. It lives in woods and thorny scrub, mainly in England.

Widespread in scrub and hedges, except in Scottish Highlands.

J F M A M J J A S O N D   J F M A M J J A S O N D   J F M A M J J A S O N D

Warblers

Dartford warbler
Sylvia undata
5in (12.5cm)

Long tail

Bushy grey head

Brown back

White spots on throat

Male; female is less colourful

Reddish-brown below

Birds flit furtively between bushes

A secretive bird, the Dartford warbler is usually seen only when flitting between bushes. It is unusual among British warblers in that it does not migrate for the winter. In periods of severe frost its numbers may decline substantially, as it cannot find the insects which form its diet. Despite the loss of its favoured heathland habitat, a succession of mild winters has enabled it to increase and spread. When perched, the long tail is held cocked. The grey head contrasts with a dark brownish back and reddish-brown underparts. Females are generally paler than males.

Heaths with gorse in southern England.

Goldcrest
Regulus regulus
3½in (9cm)

Nest usually built in conifer

Yellow crest

Female

Whitish-buff underparts

Dull green upper parts

Short tail

Orange crest

Male

Double wing-bars

With the firecrest, this is one of the smallest birds in Britain. Although often allowing close approach, it spends most of its time flitting from branch to branch in the top of coniferous trees, seeking the spiders, insects and larvae which form its diet. The orange centre to the crown of the male is visible only at close range; the female lacks the orange feathering and has only a yellow crest. A thin, blackish 'moustache' at the sides of the bill tends to give the goldcrest a mournful appearance. Juveniles lack the coloured crest altogether, being dull greenish above and pale below.

Coniferous and other woodland in all areas.

Firecrest
Regulus ignicapillus
3½in (9cm)

Courting male raises crest

Black and white eye-stripe

Juvenile lacks crest

Orange crest

Bronze shoulder-patches

Green upper parts

White underparts

Male

Although similar in appearance to the goldcrest – and just as tiny – the firecrest is much less commonly seen, being mainly a passage migrant in autumn and spring. It can be distinguished from the goldcrest by the black and white eye-stripes in both adults and juveniles and, in the adult male, more orange in the black-bordered crest. Some females also have a little orange in the crown. Adult firecrests are a brighter green on the back and whiter below than goldcrests, and have bronze-coloured patches on the shoulders, especially in the male.

Scarce breeder in woods and passage migrant in southern Britain; some winter in scrub in south-west England.

J F M A M J J A S O N D

J F M A M J J A S O N D

Willow warbler
Phylloscopus trochilus
4½in (11.5cm)

Chiffchaff
Phylloscopus collybita
4½in (11.5cm)

Wood warbler
Phylloscopus sibilatrix
5in (12.5cm)

Wingtips fanned out in flight

Pale streak above eye

Greenish or olive above

Yellowish below

Legs usually pale

Young birds more yellow

Courting male 'floats' down on spread wings

Greyish-green above

Pale yellow below

Dark legs

Parent alights above nest and slips down through foliage

Some prey taken in flight

Yellow eye-streak

Yellow-green back

White belly

Lemon-yellow breast

Bird sings a trill while flitting from tree to tree

Courting male fans out tail and makes wings quiver

By far the commonest of all British warblers, the willow warbler is a summer visitor from Africa. Some 2 million pairs inhabit open woodland and bushy places – for the bird is not confined to willows. Its song, a cadence of soft, liquid notes, can be heard all summer as the bird darts restlessly among the foliage, feeding on its insect prey. Unusually, the willow warbler moults completely, replacing all its plumage twice a year.

Widespread in woods and scrub.

The chiffchaff is very similar in appearance to the willow warbler and can most readily be identified by its distinctive song, after which it is named. Two quite separate notes are uttered, 'chiff' or 'tslip' and 'chaff' or 'tsalp', and repeated several times. Its call is a clear 'hooeet'. Closely observed, the chiffchaff is dumpier and more rounded than the willow warbler, and more greyish-green in colour above and paler yellow below, with less-pronounced streaks above the eyes. The dark legs are another distinction. Young chiffchaffs are much yellower than adults and even more like willow warblers.

Mature woodlands, except northern Scotland; some winter.

This handsome bird is named after its favoured habitat of oak, beech or birch woods. Foraging for insects high in the tree canopy, the bird can be elusive and would be easily overlooked were it not for its distinctive song. This is usually divided into two parts: first a series of 'stip' notes accelerating into a trilling 'sweeee', which is repeated regularly and then interspersed with a plaintive 'pew-pew-pew'. Larger, more brightly coloured than the willow warbler and chiffchaff, it is greener above, clear lemon yellow on the breast and above the eyes, with a pure white belly.

Widespread in Britain, especially in west; rare in Ireland.

J F M **A M J J A S** O N D   J F M **A M J J A S** O N D   J F M **A M J J A S** O N D   261

Flycatchers and thrushes

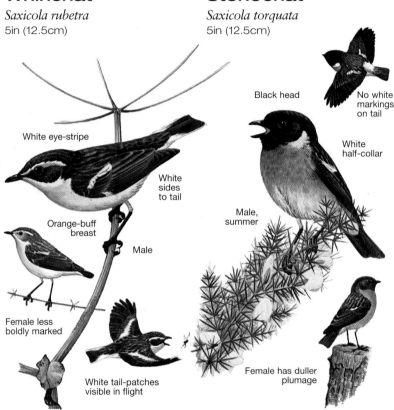

Spotted flycatcher
Muscicapa striata
5½in (14cm)

Broad-based bill makes audible snap

Streaks on breast and head

Grey-brown upper parts, pale below

Juveniles prominently speckled

Pied flycatcher
Ficedula hypoleuca
5in (12.5cm)

White forehead

Black above

White wing-patch

White breast

Male

Female brown above, without white forehead

Whinchat
Saxicola rubetra
5in (12.5cm)

White eye-stripe

White sides to tail

Orange-buff breast

Male

Female less boldly marked

White tail-patches visible in flight

Stonechat
Saxicola torquata
5in (12.5cm)

Black head

No white markings on tail

White half-collar

Male, summer

Female has duller plumage

Sitting on a low branch or other perch, the spotted flycatcher watches for its insect prey. Periodically it darts out and catches flies with an audible snap of its broad-based bill. The bird's plumage is mainly grey-brown above, pale below. Arriving from Africa in May, the birds seek out woodland edges, gardens, parks and heathland. Both birds build a cup-shaped nest of dried grass and lichen on a ledge or in a hole in a tree or an open-fronted nest-box.

Widespread, open wooded areas; decreasing.

Normally breeding in holes in trees, these migrants from Africa have spread in Britain with the encouragement of nest-boxes deliberately placed for them in woodland. The male is black above and white below, with a large white wing-patch and white forehead; the female lacks the white forehead and is brown above. The birds catch insects in flight and snatch prey such as caterpillars from leaves.

Woodlands of western and northern Britain; elsewhere passage migrant.

The attractive little whinchat, a breeding visitor, has disappeared from many areas in recent years. Heathland with plenty of gorse and bracken is the likeliest place to see breeding birds – 'whin' is another name for gorse. The male is noticeable for its white eye-stripe and orange-buff breast; the female is less boldly marked. The whinchat builds its nest in thick ground cover. Prominent perches are used for fly-hunting and spotting prey on the ground.

Mainly upland areas of Britain and Ireland.

Perching on a high vantage point, the stonechat scolds intruders with its alarm call, 'wee-tac-tac', like two pebbles being knocked together – hence the bird's name. It is a resident, and feeds on insects, worms, spiders and, occasionally, seeds. The male has a black head, white half-collar, white patches on the wings and rump, and an orange-red breast. A nest of moss, grass and hair is well concealed, low in a bush.

Widespread on heaths and grassland, scarce in eastern England.

J F M A M J J A S O N D

J F M A M J J A S O N D

J F M A M J J A S O N D

J F M A M J J A S O N D

Flycatchers perch upright watching for insects, then dart to catch them in mid-air in their flat, pointed bills. In addition to the plump, typical thrushes, the thrush family includes warbler-like birds such as the chats. They feed mainly on the ground, eating worms, insects and berries.

Black redstart

Phoenicurus ochruros
5½in (14cm)

Male

Sooty black upper body

White wing-patch

Wing-patches show as bird hovers

Female

No wing-patch

Rusty red tail

Migrant birds seen on rocky beaches

The black redstart population in Britain began to grow in the Second World War, when bombed and derelict buildings made ideal nesting sites. But only about 50 pairs now breed in Britain, on sites that include factories, power stations and railway yards. Males are sooty black on top, females browner. The song is a loud, reedy warble. The birds usually nest on a ledge or crevice.

Buildings, cliffs and quarries, mainly in southern England; a few winter.

Juvenile is speckled, otherwise like female

Redstart

Phoenicurus phoenicurus
5½in (14cm)

Female in flight

Female brown above and paler below

Black mask

Grey back

Reddish belly

Male

Rusty tail

Chick is speckled

A flash of bright, rusty red, low down in a tree or bush, reveals the redstart as it 'shimmers' its tail. This flickering motion plays an important part in courtship. A summer visitor from northern Africa, the male has a grey back, black face and throat, white forehead, reddish belly and rusty rump and tail. The female is brown above and paler below, but has a rusty rump and tail like the male. The call is loud 'hooeeet-tac'; the song is a brief warble ending with a jingling rattle. The nest is in a crevice or hole in a wall or tree.

Widespread in British woods and parkland; rare in Ireland.

Wheatear

Oenanthe oenanthe
6in (15cm)

Insects are often chased in vertical flight

Bold black 'T' on tail

Female is browner

White rump

Black mask

Male

Birds hunt insects in ploughed fields

Among the earliest summer migrants, the wheatear can be seen in southern England from the second week of March. It often nests under stones, in holes, in drystone walls or even in rabbit burrows. The male has grey upper parts, black mask and wings, buff breast and white rump. The black tail with white sides shows a bold 'T' pattern. The call is a distinctive 'wee-chat-chat' and the song a brief warble, mingled with rattles, squeaks and whistles.

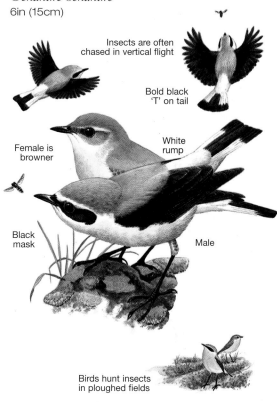

Upland, heathland and coasts.

Young bird is brown and heavily speckled

Thrushes

Robin
Erithacus rubecula
5½in (14cm)

Red breast, with pale grey border

Brown back and tail

Male defends territory aggressively

Nest built in any convenient container such as an old teapot

Nightingale
Luscinia megarhynchos
6½in (16.5cm)

Brown back

Buff underparts

Chestnut tail, conspicuous in flight

Song thrush
Turdus philomelos
9in (23cm)

Brown above

Yellow under-wing distinguishes song thrush from redwing

White, speckled belly

Buff breast

The robin's association with Christmas is appropriate, for it is during the winter that its colours are most marked, with its breast at its reddest and its back a rich brown, both contrasting brilliantly with whitish underparts. Young birds have speckled plumage and look like juvenile nightingales. The bird is tame in town and city gardens, and often accompanies gardeners to search for insects and larvae as the ground is dug over. Away from habitation, however, it is shy and retiring, inhabiting woodland and hedges. Males are aggressive and guard their territory possessively. The song is a high, pleasant warble and the loud alarm call a penetrating 'tic-tic'.

Widespread throughout Britain and Ireland.

The most famous songster of all, the nightingale is a shy bird more often heard than seen. Renowned for singing at night, and especially at dusk, it can also be heard by day. The rich song has short phrases, single notes and harsh trills. Nightingales live in thickets of thorn bushes, or copses and woodland with plenty of cover, but with open areas for feeding. They eat insects, worms and berries.

Thickets or open woods in south-east.

Juveniles heavily spotted

Broken snail shells littering the ground around a large stone indicate the presence of the song thrush. Tapping noises coming from cover may be the sound made by the bird as it smashes open the snails on a favourite 'anvil' to get at the contents. Aptly named, the bird is noted for its loud, rich song, which consists of repeated musical phrases together lasting for five minutes or more. This impressive aria is usually delivered from a high perch. The bird's call is a thin 'sipp' or a 'tchuck-tchuck' alarm note. The song thrush is warm brown above, with a buffish breast and a white belly speckled with small, neat spots. It feeds on snails, worms and berries.

Widespread throughout Britain and Ireland.

J F M A M J J A S O N D

J F M A M J J A S O N D

J F M A M J J A S O N D

Mistle thrush

Turdus viscivorus
10½in (27cm)

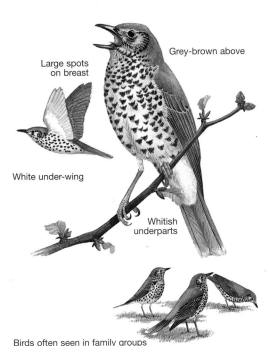

Large spots on breast

Grey-brown above

White under-wing

Whitish underparts

Birds often seen in family groups

Fieldfare

Turdus pilaris
10in (25cm)

Black tail contrasts with grey rump in flight

Pale beneath

Grey rump

Chestnut back

Grey head

Buff breast heavily speckled

Redwing

Turdus iliacus
8¼in (21cm)

Red under-wing patches

Eye-stripe

Brown back

Red flanks

Redwings feed with other thrushes

The largest of the British thrushes is also the most boldly marked, with bigger and blacker spots than those of the song thrush. In flight, its white under-wing and pale tips to the tail help to distinguish it from other thrushes. The call is a harsh, rattling chatter and the song is a prolonged series of short, fluty notes of varying pitch and considerable carrying power. This is usually sung from the top of high trees, and can be heard throughout the winter as well as in spring and summer. Mistle thrushes feed on berries (including mistletoe), slugs, snails, worms and insects. After the breeding season small family groups often feed together.

Widespread in woods and gardens, except in far north.

Autumn and winter see large flocks of fieldfares arriving in the British Isles from Scandinavia and more northerly regions. The bird's 'chack-chack' chattering call can be heard as it flies onto fields to feed on worms and insects. Noisy scuffles are common among flocks of fieldfares as they feed, and berry-bearing bushes are defended against all-comers.

Hedges and copses; a few breed in Scotland and northern England, but mainly winter visitor.

Noisy scuffles common as birds feed

Leaving its northern European breeding grounds in autumn, the redwing is a passage migrant and winter visitor to the British Isles, although some breed regularly in Scotland. In September and October flocks can be heard flying over, uttering their thin 'seeip' contact call. The adults are smaller and darker than the song thrush, brown above, with a pale streak above the eye and a long moustache. Below they are white with dark speckles, and reddish flanks and under-wing patches. Their favoured food is hawthorn, holly and other berries. Worms, snails and insects are also eaten.

Fields, scrub, woodland everywhere; breeds in Scotland.

Thrushes

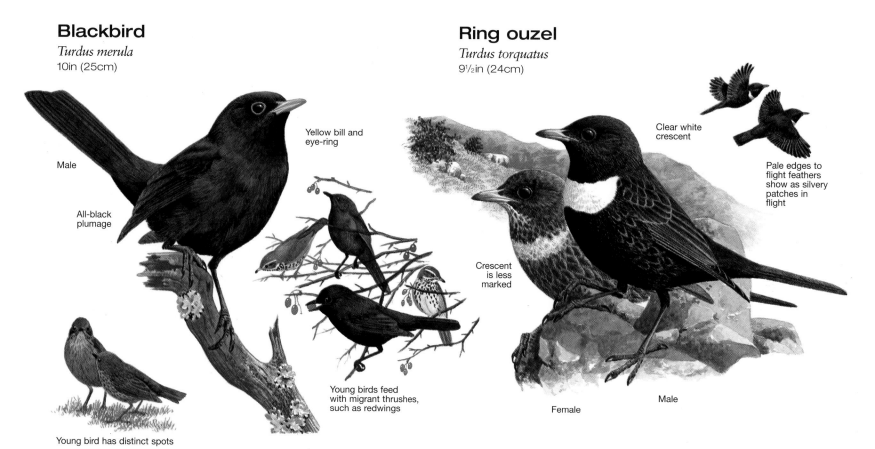

Blackbird
Turdus merula
10in (25cm)

Male

Yellow bill and
eye-ring

All-black
plumage

Young birds feed
with migrant thrushes,
such as redwings

Young bird has distinct spots

Ring ouzel
Turdus torquatus
9½in (24cm)

Clear white
crescent

Pale edges to
flight feathers
show as silvery
patches in
flight

Crescent
is less
marked

Female

Male

Female is brown,
with blurred spots on underparts

The blackbird's rich, fluty, warbling song, punctuated by pauses, is a herald of spring, and its alarm call of 'pink-pink-pink' is also a familiar sound. The male is distinguished by its all-black plumage, yellow bill and eye-ring, while females are brown, with mottled underparts. The blackbird's diet consists of insects, berries and worms, and it can often be seen standing with head cocked to one side, listening for worms before pulling them out of the ground.

Widespread in Britain and Ireland; continental visitors in winter.

Departing
migrants
feed in groups

Similar to the blackbird in appearance, the male ring ouzel is distinguished by the white crescent or gorget across its breast and the pale patches on its wings. Females are brownish-grey, with a less clearly marked crescent. These summer visitors from the Mediterranean are normally shy and nervous, but they can be aggressive at nesting time. The harsh 'chak-chak-chak' alarm call is more metallic than the blackbird's.

Mountains and moors, mainly in west and north Britain.

J F M A M J J A S O N D

J F M A M J J A S O N D

Nuthatch, treecreeper and reedling

Nuthatch

Sitta europaea
5½in (14cm)

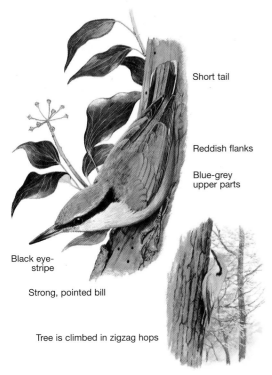

Short tail

Reddish flanks

Blue-grey upper parts

Black eye-stripe

Strong, pointed bill

Tree is climbed in zigzag hops

The nuthatch is unique in its ability to hop down a tree-trunk as easily as it hops up. It picks up insects from the bark, and also feeds on hazel and beech-nuts, acorns and seeds. The nuts are placed into a crack in the bark, and the bird hammers them open to reach the kernel. The nuthatch's call is a loud 'chwit' or a rapid, trilling 'chiririri'. Nuthatches breed in tree-holes or nest-boxes, making the entrance smaller by filling it with mud which hardens in the sun.

Mud is used to adjust size of nesting hole

Wooded areas of England and Wales.

Treecreeper

Certhia familiaris
5in (12.5cm)

Down-curved bill

White eye-stripe

Pointed tips to tail

Large claws for climbing

Long, stiff tail

Thanks to its large claws and stiff, fairly long tail, the treecreeper can progress jerkily up tree-trunks in its search for bark-dwelling insects. It climbs spirally up one tree and then – because it cannot hop down like the nuthatch – flies down to the base of another and starts its upward journey again. The treecreeper's song is thin and high-pitched, its call a shrill 'tseee'. In winter, treecreepers and nuthatches associate with flocks of tits as they search for weevils, beetles, earwigs, small moths, woodlice and spiders. The nest is usually behind loose bark or ivy on an old tree. The young climb well, but are poor fliers at first.

Woodlands in most parts of Britain and Ireland.

Bearded reedling

Panurus biarmicus
6½in (16.5cm)

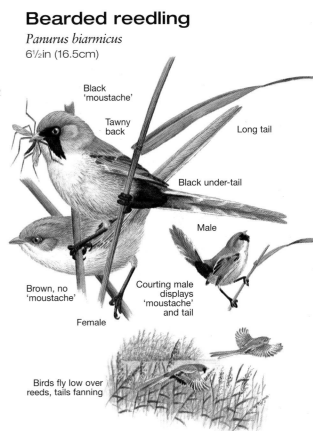

Black 'moustache'

Tawny back

Long tail

Black under-tail

Male

Courting male displays 'moustache' and tail

Brown, no 'moustache'

Female

Birds fly low over reeds, tails fanning

A twanging 'ping-ping' call-note announces a flock of tit-like bearded reedlings flying low over their territory. These birds inhabit dense beds of reeds. They are also known as bearded tits, but despite their tit-like appearance, they belong to a different family. They are called 'bearded' because of the black face markings on the male bird; these are lacking on the female. Bearded reedlings live mainly on the seeds of the *Phragmites* reeds they inhabit. The nest, built by both adults, is a deep cup of dead reeds and sedges, placed low down amongst the reed stems and lined with the feathery flower-heads of reeds. Two broods are often raised in a year.

Reed-beds in East Anglia and parts of southern England.

J F M A M J J A S O N D   J F M A M J J A S O N D   J F M A M J J A S O N D

Tits

Long-tailed tit

Aegithalos caudatus
5½in (14cm)

Birds gather in small groups

Long tail conspicuous in flight

White crown

Black head band

Tiny bill

Pink flanks

Crested tit

Parus cristatus
4½in (11.5cm)

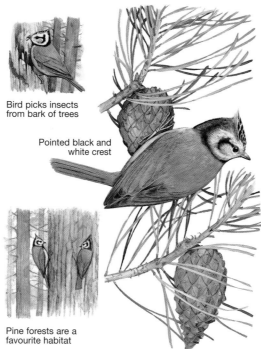

Bird picks insects from bark of trees

Pointed black and white crest

Pine forests are a favourite habitat

Willow tit

Parus montanus
4½in (11.5cm)

Matt black crown

Pale wing-patch

Bird excavates nest-hole in soft tree-trunk

The tail of this tiny tit is more than half the bird's total length and very conspicuous, especially in flight. The bird is always on the move in woods, commons, wasteland and hedges, feeding on insects, spiders, seeds and buds. Severe winters take their toll, and in some years reduce the bird's population by 80 per cent. The oval nest is a masterpiece in moss, bound together with cobwebs and hair and lined with hundreds of feathers.

Juvenile duller, with shorter tail

Widespread except in far north of Scotland and western Ireland.

A common bird on the Continent of Europe, from the Mediterranean northwards to Scandinavia, the crested tit in Britain is found only in a small area in Scotland. Perhaps this is because it is largely a sedentary species and has not yet discovered the other large areas of planted pine forests which are its favoured habitat. The bird is easy to identify by its pointed black and white crest. Like the coal tit and treecreeper, it moves along the tree-trunks picking insects and their larvae from the bark, as well as eating pine seeds and berries.

Scottish pine woods, mainly in Spey valley.

For years the willow tit was mistaken for the marsh tit, and it was not recognised as a separate species until 1897. Gradually it was realised that the willow tit is almost as common as the marsh tit. For unknown reasons, both species have declined in recent years. It is distinguished from the marsh tit by its matt black crown, more diffuse black bib and pale panel on the wing. It can also be identified by its voice. The call is a loud, harsh 'tchay' and a high pitched 'zee-zee-zee'; the song is a melodious warbling, unusual among tits.

Damp woods and hedges in England, Wales and south-west Scotland.

J F M A M J J A S O N D   J F M A M J J A S O N D   J F M A M J J A S O N D

Tits are small, very lively and acrobatic birds, many of them brightly coloured. They appear in woods, hedgerows and reed-beds; some are familiar garden visitors, feeding – often upside-down – from bird-tables, bags of nuts and milk bottles. In winter, tits often fly and feed in mixed flocks.

Marsh tit

Parus palustris
4½in (11.5cm)

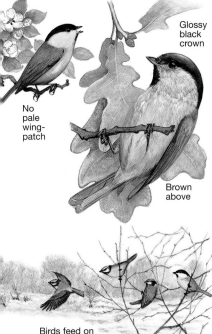

Glossy black crown

No pale wing-patch

Brown above

Birds feed on lower vegetation

In spite of its name, the marsh tit is rarely found in marshland but favours woods, heaths and hedges. It is almost indistinguishable from the willow tit at a distance except by its call and song. The call is a distinctive 'pitchew' or a scolding 'chickabee-bee-bee-bee'. The song is a single, repeated 'chip, chip, chip'. Closely observed, its black crown is glossy and it lacks the pale wing-patch of the willow tit.

Deciduous woods in England and Wales; rare in Scotland and absent from Ireland.

J F M A M J J A S O N D

Great tit

Parus major
5½in (14cm)

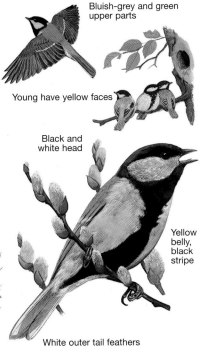

Bluish-grey and green upper parts

Young have yellow faces

Black and white head

Yellow belly, black stripe

White outer tail feathers

The great tit is the biggest, brightest and noisiest member of the tit family. Experts have identified no less than 57 distinct calls, some of them uncomfortably shrill to the human ear.. It is also extremely adept at finding ways to reach tempting morsels of food, such as a nut at the end of a piece of string. Its nest is usually in a tree or a wall, but it is not averse to a nest-box, letter-box or drainpipe.

Woods and gardens in all areas except Orkney and Shetland.

J F M A M J J A S O N D

Blue tit

Parus caeruleus
4½in (11.5cm)

Young birds have yellow faces

Blue crown

White face

Yellow underparts

Bluish upper parts

Birds readily use nest-boxes with small entrance hole

For many people the blue tit is the star performer in the garden bird show, combining the talents of acrobat, conjurer and songster. It is small and clever: it was the blue tit that cracked the problem of how to pierce milk-bottle tops. Originally a woodland bird, the blue tit will nest wherever there is a suitable small hole in a tree or a nesting-box. It has a large vocabulary of calls and a song consisting of two or three notes followed by a rapid trill.

Woods and gardens in all areas.

J F M A M J J A S O N D

Coal tit

Parus ater
4½in (11.5cm)

Young have yellow head patches

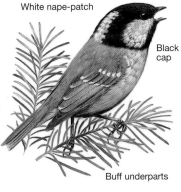

Birds eat insects under tree bark

White nape-patch

Black cap

Buff underparts

The smallest British tit is also one of the most elusive. The coal tit's favourite habitat is coniferous woodland, where its high-pitched 'tsee' call can be heard. While less bold than other tits, it will nevertheless forsake the tall pines to forage in gardens, especially for nuts, meat and suet scraps left on bird-tables in winter. It survives prolonged periods of snow by feeding on insects beneath the bark of trees.

Woodlands in all areas, except Orkney and Shetland.

J F M A M J J A S O N D

269

Sparrows

These town and country cousins are small and sturdy birds, with stout, strong bills for cracking seeds. They hop along the ground when feeding, and are highly gregarious.

Finches

House sparrow

Passer domesticus
5¾in (14.5cm)

Female's plumage is duller, without grey crown or black bib

Female   Male

Grey crown

Large black bib

Streaked brown back

Male

House sparrows mingle with finches in winter feeding

Tree sparrow

Passer montanus
5½in (14cm)

All-chestnut crown

Black patch on cheek

Black bib

Courting birds bow and run at each other

Twite

Acanthis flavirostris
5¼in (13.5cm)

Bill grey in summer, yellow in winter

Male

Notched tail

Streaky brown plumage

Pink rump

Twites mix with other finches in winter

Possibly the most familiar of British birds, the house sparrow is, surprisingly, less numerous than the chaffinch, blackbird or wren. It is intelligent, has adapted to major changes in the environment, and is largely dependent on man for its food and nesting places. A rather persistent 'chee-ip' is the commonest call, and there is endless twittering. The sparrow's nest is a rather untidy affair in a hole or under the eaves of a building. Where there are no suitable man-made structures it will build a domed nest in a hedge, bush or tree.

Towns, villages and farmland throughout Britain and Ireland, but recently in much reduced numbers.

This country cousin of the house sparrow is distinguished by a chestnut crown, neater bib and black cheek spot. Males, females and young are similarly marked. Populations fluctuate, and the bird's distribution is to some extent dependent on the availability of suitable nesting sites. Breeding is in colonies; both sexes build the domed nest in a suitable hole or cavity in a tree, wall or cliff face or in a nest-box, using dried grass or straw. In flight, tree sparrows give a high, distinctive 'teck-teck' call. After the breeding season birds often mix with flocks of finches and house sparrows in the search for food.

Locally widespread, but declining everywhere.

Although it breeds in moorland and upland areas, the twite extends its range in winter to salt-marshes, stubble fields and coastal areas, where the birds can be seen in small flocks, often feeding with other finches. Resembling the linnet, the male twite has streaky brown plumage, with a buff throat and a pink rump which is noticeable in flight. The female has a greyish rump and is drabber. In both sexes the bill is grey in summer and yellow in winter. Food is mainly seeds, but insects are also eaten. The call sounds like 'twa-it'. The nest, a cup on the ground, is built by the female.

Moors and hills in northern Britain and Ireland.

J F M A M J J A S O N D

J F M A M J J A S O N D

J F M A M J J A S O N D

Finches are small, seed-eating song-birds with stout bills that are specially adapted to each species' particular diet. Males are usually more brightly coloured than females, with distinctive wing patterns. In courting, male and female touch bills in a 'kiss'. Different species often mingle.

Crossbill
Loxia curvirostra
6½in (16.5cm)

Female

Green plumage

Crossed bill

Male

Forked tail

Young have uncrossed bill

Red plumage

Chaffinch
Fringilla coelebs
6in (15cm)

Slate-blue crown and neck

Chestnut back

Male

White wing-bar and shoulder patch

Pink-brown below

Females less colourful but have same wing pattern

Brambling
Fringilla montifringilla
5¾in (14.5cm)

Black head and beak

Bill yellow in winter

Orange-buff leading edge on wing

Orange-buff breast

Male in breeding plumage

Female duller, without black markings

Feeding mainly on pine seeds, this bird has developed a peculiar bill, one mandible of which crosses over the other. This enables it to prise open the cones to extract the seeds. Although conifers are preferred, the crossbill also eats the seeds of rowan, ivy, hawthorn and thistles; some insects are taken too. The male has brick-red or orange-red plumage, while the female is yellowish-green. Young birds are greenish-grey with dark streaks. Breeding is usually from February to July; the nest of twigs and grass is built high in a conifer.

Widespread in coniferous woodlands; rare in Ireland.

One of Britain's commonest birds – there may be 7 million pairs – the chaffinch nests in hedges and tree-forks but roams widely in large flocks by day seeking seeds and insects in gardens, parks and farmyards. There is no mistaking the colouring of the male chaffinch when at rest, while in flight it shows its white shoulder patches and wing-bar. The nest is a cup of grass, moss and lichens, lined with hair. The loud jangling song starts slowly, accelerates down the scale and ends with an exuberant flourish; it is repeated up to five or ten times a minute. The alarm call is a 'pink, pink, pink'.

Widespread, in all areas.

Although mainly a winter visitor, the brambling does occasionally breed in Britain. But the male's splendid breeding plumage is a rare sight, as most birds leave before attaining the full black head and back. In winter plumage, the male has buff mottling, greyish sides to the neck and a bright orange breast. Females are paler and lack the black on the head. Both sexes have a prominent white rump. The brambling's favourite food is beech mast, but it also eats weed seeds, berries and grain. The nest, similar to that of the chaffinch, is usually in the fork of a tree. The call note is a hoarse 'tsweek'.

Woodlands, except in far north and west.

J F M A M J J A S O N D

J F M A M J J A S O N D

J F M A M J J A S O N D

Finches

Goldfinch
Carduelis carduelis
4³/₄in (12cm)

Black on head

Red face

Broad yellow
wing-bar

Young birds lack
red and black
head markings

Black and yellow wings flutter above a patch of thistles or groundsel in summer and early autumn as goldfinches cling to the heads of the plants and pluck out the seeds with their delicate bills. They use the thistledown, too, when building their nests in small trees in farmlands or orchards. The liquid, twittering song which once made the goldfinch a popular cage bird is heard as 'charms', or flocks, dance from plant to plant.

Farmlands, gardens and open areas, except northern Scotland.

Greenfinch
Carduelis chloris
5³/₄in (14.5cm)

Wing and tail
patterns distinctive
in flight

Pink bill

Yellow on wing
and tail

Male

Female duller,
with slightly
streaked
underparts

Bright yellow wing and tail flashes help to identify the greenfinch against a background of woodland and bushes. It is a frequent visitor to town gardens, especially when there is water for bathing or peanuts on the bird-table. Greenfinches build their nests in bushes, using twigs, moss or roots. Their call is a loud rapid twittering on one note, followed by four or five musical notes; after a pause there is sometimes a single 'greee'. In song flight the male often flits and weaves erratically.

Near human settlements in most areas.

Lesser redpoll
Acanthis cabaret
4¹/₂-6in (11.5-15cm)

Red forehead

Black
chin

Pink breast
and rump –
rarely
seen on
female

Buff
wing-
bars

Male

Birds feed on
tree and
plant seeds

Dull brown with pink patches, like the linnet, the lesser redpoll's main distinguishing mark is its black chin. It nests in a variety of sites, from low gorse bushes and alder thickets to the high branches of silver birches and conifers. The nest is a cup of twigs, grass and plant stems. The redpoll's flight call is a rattling bell-like 'ching, ching, ching'. Flocks of redpolls often rise from the treetops and wheel in the air a few times before settling again.

Widespread in woods; winter visitors include the larger, paler mealy redpoll.

Linnet
Acanthis cannabina
5¹/₄in (13.5cm)

White edges to wing
and tail feathers
seen in flight

Red forehead

Chestnut
back

Red
breast

Male

Female
duller, more
streaked

In Victorian and Edwardian times linnets were often kept in cages for their musical song, a varied twittering heard between late March and late July. Now a protected bird, it faces a new enemy in the increased use of weedkillers, which are depleting the chickweeds and dandelions whose seeds it eats. The linnet is largely dull brown, though the male has a red forehead and breast. Linnets build their nests a few feet off the ground, often in gorse or bramble.

Thickets, hedges and open country, except northern Scotland.

J F M A M J J A S O N D

J F M A M J J A S O N D

J F M A M J J A S O N D

J F M A M J J A S O N D

Bullfinch

Pyrrhula pyrrhula
6in (15cm)

Black cap; stubby bill

Red underparts

Male

White wing-bar and white rump conspicuous in flight

Female dull salmon-pink below

Siskin

Carduelis spinus
4½in (11.5cm)

Black crown and chin

Yellow-green rump; yellow tail-patches

Yellow wing-bars

Siskins often feed with redpolls

Male

Female less yellow, more streaked

Hawfinch

Coccothraustes coccothraustes
7in (18cm)

Short tail, white band; white wing-bars

Large head

Chestnut body and head

Large, strong bill

Male

Young bird lacks black bib

Female plumage duller than male's

Male

Despite its distinctive colouring, the bullfinch is not often seen; it is a shy bird, keeping to cover in hedgerows and bushes. Often the only clue to its presence is the soft piping 'dew' of its call note. The male has red underparts, black cap, grey upper parts and white rump. The female is similar, except for dull salmon-pink underparts. Bullfinches often raid orchards to strip fruit trees of their young buds, the stubby bill being specially adapted for this diet. The birds breed mainly in woodlands but also in parks and hedges. The nest of twigs, moss and lichens varies from a shallow platform to a bulky cup.

Widespread except in extreme north and north-west.

Conifer forests are the main home of the siskin, which feeds largely on the seeds of pine and spruce. It used to live only in the pine forests of the Scottish Highlands; though it now breeds more widely, most birds seen in southern Britain are winter visitors. Siskins like to flock together, often in alder trees during winter. Yellow-green in colouring, like the greenfinch, it is a smaller bird and the male has a black crown and chin. Siskins build their nests high in conifer trees, and towards the tips of branches. The nest is usually made of small twigs bound with grass, moss, plant fibres and wool.

Scotland and forests in England; joined by winter migrants.

The largest of Britain's finches also has the most powerful bill. It uses this to crack open the stones of fruit such as cherries and sloes to get at the edible kernels which form its principal diet. It also feeds on beech mast, haws and holly berries. Hawfinches are elusive birds, keeping mainly to the shelter of the forests, orchards and well-wooded parks. An abrupt 'tik' occasionally indicates the bird's presence overhead. The nest, often built high in a fruit tree, is a bulky construction of twigs and moss lined with roots and grass.

Wooded areas, especially in south-east. Rare in Scotland and Wales; absent from Ireland.

J F M A M J J A S O N D   J F M A M J J A S O N D   J F M A M J J A S O N D   273

Buntings

Lapland bunting
Calcarius lapponicus
6in (15cm)

Chestnut nape

Female

Male

Black and white head

Yellow bill

Male, winter

Duller plumage; female similar, without bold black markings

Male, summer

Birds join skylarks and finches to seek food

A male Lapland bunting's handsome summer plumage is not often seen in Britain, as it is mainly a winter visitor to England's east coast, feeding in stubble fields and on salt-marshes. Its hind toe has a long claw – probably an adaptation to living mainly on the ground. In its winter plumage the Lapland bunting is similar to a female reed bunting, but it has a yellowish bill, pale crown stripe and, more markedly in the male, a chestnut nape. The bird can be identified by its call of 'ticky-tick-teu'. Its nest is a cup on the ground, made of grass, moss and roots and lined with hair and feathers.

Open country and coast, mainly in eastern England.

Corn bunting
Miliaria calandra
7in (18cm)

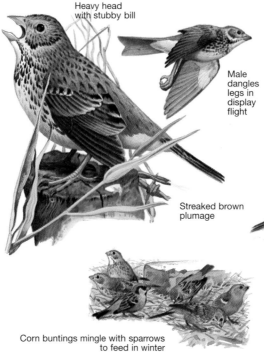

Heavy head with stubby bill

Male dangles legs in display flight

Streaked brown plumage

Corn buntings mingle with sparrows to feed in winter

The largest of the British buntings inhabits cornfields and other lowland arable areas, as well as heaths and downlands. Adult birds of both sexes have streaked brown plumage and a heavy head with a yellowish, stubby bill. The song – usually delivered from the top of a bush, fence post or telegraph wire – consists of a rapidly repeated note speeding up into a flourish. It sounds like a bunch of jangling keys. The fluttering flight is sparrow-like, and the male often leaves his legs dangling during a display flight. A loose cup of grasses and plant stems, lined with hair, forms the nest, which is built on the ground or in a bush or hedge.

Arable farmland; rare in west; declining everywhere.

Ortolan bunting
Emberiza hortulana
6½in (16.5cm)

Brown rump

Eye-ring

Grey-green head and breast; yellow throat

Females and young birds are paler

In spring and especially in autumn, small numbers of ortolan buntings reach Britain, blown off course as they migrate between their breeding grounds in Europe and their winter quarters in Africa. The male is mainly pinkish-brown, with a grey-green head and breast, yellow throat, pink bill and clearly defined eye-ring. Females and young birds are paler, with streaked underparts. The birds, which feed on seeds and insects, are usually seen in coastal grass or stubble fields, and their presence is often indicated by the call – a soft 'chip' or 'chew'.

Scarce passage migrant, seen mainly in east and south.

Looking rather like the finches, buntings are seed-eaters with stout bills. They inhabit open country, generally keeping away from towns; they live and usually nest on the ground or only a few feet above it, and fly with an undulating motion. The male's plumage is generally more colourful than the female's.

Reed bunting
Emberiza schoeniclus
6in (15cm)

White on nape and tail

Black head and throat

Male

Brown head

Female

White 'moustache'

Streaked brown upper parts

Male, summer

White outer tail feathers

Snow bunting
Plectrophenax nivalis
6½in (16.5cm)

White head

Largely white wings

Black back

Male, summer

Female, like male in winter, is brown and white

Yellowhammer
Emberiza citrinella
6½in (16.5cm)

Bright yellow head

Yellow breast

Streaked brown and black upper parts

Chestnut rump

White tail outer feathers

Cirl bunting
Emberiza cirlus
6½in (16.5cm)

Olive-green rump

Black and yellow head

Male

Greenish breast-band

Male

Birds join yellowhammers to feed in winter

Once confined to the reed-beds of marshes, fens and rivers, the reed bunting has spread to other types of habitat. In summer the male's black head and throat are distinctive, and both sexes have a white moustache-like streak. Reed buntings often mingle with yellowhammers, house sparrows and finches in winter when they search for food. The song is a repetitive 'cheep-cheep-cheep-chizzup'. The nest, a cup of grass and moss lined with hair, is built on or close to the ground.

Widespread near water in Britain and Ireland.

Although snow buntings normally breed in the Arctic, a few birds have nested in Highland parts of Scotland for 100 years. As winter visitors they are often seen in small flocks on the shoreline. In summer, the male is mainly white with black on the wings, back and tail. The female is brown and white, as is the male in winter. The song is a brief fluting 'turee-turee-turee-turiwee'. The nest is usually in a cranny between rocks and is made of moss and grasses, lined with wool or feathers.

Scottish Highlands; near coasts elsewhere.

'A little bit of bread and no cheese' is a popular version of the yellowhammer's song. Really it is a repeated series of notes ending in 'zeee' or 'chwee' – with some variations. The bird inhabits hedged fields, scrubby heathland and commons, feeding on seeds, berries, grain and insects. The male has a bright yellow head and breast, and a chestnut back streaked with black; its chestnut rump distinguishes it from the cirl bunting. Females are duller and more streaked. The nest is a neat cup of grasses.

Widespread in open areas with scrub or hedges.

Once spread across southern England, cirl buntings have declined in numbers and are now more localised. The male's black and yellow head, black chin, grey-green breast-band and olive-green rump distinguish it from the male yellowhammer. Females lack the head pattern and breast markings and are buffish. The song is a rattling, repeated single note. The cirl bunting feeds on seeds, berries and some insects. The nest is made of grass, moss and rootlets.

Arable farms with many hedges and trees; now found only in Devon.

J F M A M J J A S O N D

J F M A M J J A S O N D

J F M A M J J A S O N D

J F M A M J J A S O N D

275

butterflies, moths & other small creatures

The world of 'creepy-crawlies' is a fascinating and extraordinarily diverse one and, despite some people's feelings of revulsion for some of its inhabitants, the majority of them are harmless or positively beneficial.

This section of the book deals with all invertebrates except those that live predominantly under water (see pp. 376-9 and 386-402). The emphasis is on insects, which are by far the most successful members of the animal kingdom in terms of numbers of species – about a million worldwide, more than all other animal species put together. Some 22,500 species live in the British Isles, the best known of which are the butterflies and moths. The division between these two groups has no real scientific basis but depends on a few simple characteristics. Butterflies generally fly by day, fold their wings together vertically when at rest and their antennae have club-shaped ends. Moths, on the other hand, are generally duller coloured, fold their wings flat or tent-wise over the body, do not have clubbed antennae and most of them fly by night.

Other familiar creatures – including dragonflies and damselflies, grasshoppers and crickets, bees, wasps, ants and beetles, as well as snails, slugs and spiders – are grouped by family. Further sections cover a representative selection of common predators and scavengers, soil creatures and insects that live on or next to water.

Skippers

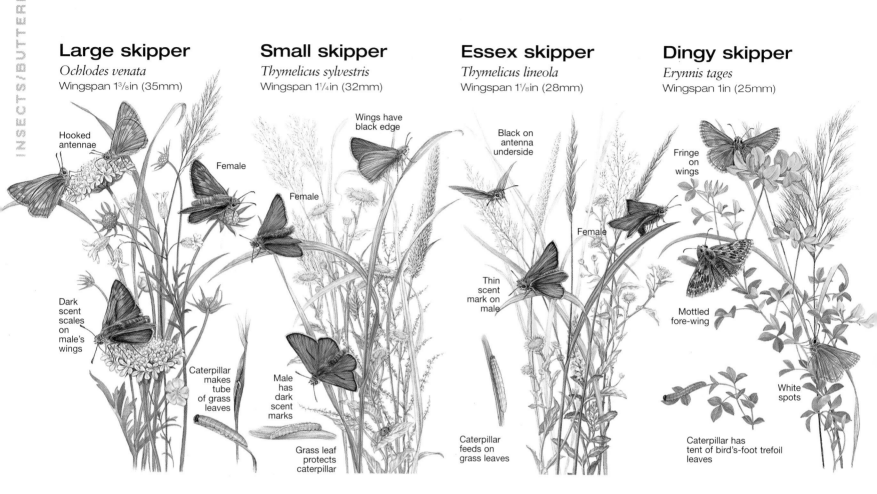

Large skipper
Ochlodes venata
Wingspan 1³/₈in (35mm)

Hooked antennae

Female

Dark scent scales on male's wings

Caterpillar makes tube of grass leaves

Small skipper
Thymelicus sylvestris
Wingspan 1¹/₄in (32mm)

Wings have black edge

Female

Female

Male has dark scent marks

Grass leaf protects caterpillar

Essex skipper
Thymelicus lineola
Wingspan 1¹/₈in (28mm)

Black on antenna underside

Female

Thin scent mark on male

Caterpillar feeds on grass leaves

Dingy skipper
Erynnis tages
Wingspan 1in (25mm)

Fringe on wings

Mottled fore-wing

White spots

Caterpillar has tent of bird's-foot trefoil leaves

Often found in sunny glades or along woodland paths, the largest British skipper likes to have a perching point, such as a leaf, from which to guard its territory. It is pugnacious like other skippers and will fly off to intercept other insects, including butterflies, that come too near. Eggs are laid on cock's-foot and other grasses. The young caterpillar feeds inside a tube it spins from grass leaves. To avoid the attention of predators it ejects its droppings far away.

Most of England and Wales; southern Scotland.

This alert and active flier is the commonest of the grassland skippers. In the air, it darts about at great speed. At rest, it sometimes holds its hind-wings slightly backwards – like a swing-wing aircraft – and the fore-wings may be slightly raised; this attitude may help it to absorb the sun's warmth. The small skipper favours long grass and such wild flowers as scabious, mayweeds, dandelions, thistles and knapweeds. Eggs are laid on a variety of grasses.

Widespread in southern England and Wales.

The Essex skipper is frequently confused with the small skipper, but can be distinguished by the black tip on the underside of each antenna and by the thinner scent mark on the male's wings. A more fundamental distinction is that the Essex skipper overwinters as an egg and not as a caterpillar; this helped the species to survive the calamitous 1953 floods in East Anglia. It is not confined to Essex, but was discovered there in 1888.

East Anglia and parts of southern England.

A grey mottled fore-wing with a darker hind-wing justify this butterfly's name and make it easy to identify. The margin of grey hairs around the wings is also a feature. It rests in a different way from other skippers – with its wings spread out, rather like some moths. It enjoys nectar from wayside flowers. The dingy skipper winters as a caterpillar on bird's-foot trefoil, its food plant, turning into a chrysalis in spring.

Patchy distribution, mostly in chalk and limestone areas of Britain and Ireland.

J F M A M J J A S O N D

J F M A M J J A S O N D

J F M A M J J A S O N D

J F M A M J J A S O N D

Skippers get their name from their style of flight, characterised by a fast wing-beat and a darting motion. They are small and compact – rather like moths in appearance and often folding their wings in moth fashion. They are especially active on warm, sunny days, and are often aggressive in defending their small territory.

Lulworth skipper
Thymelicus acteon
Wingspan 1in (25mm)

Chequered skipper
Carterocephalus palaemon
Wingspan 1¼in (32mm)

Silver spotted skipper
Hesperia comma
Wingspan 1⅜in (35mm)

Grizzled skipper
Pyrgus malvae
Wingspan 1⅛in (28mm)

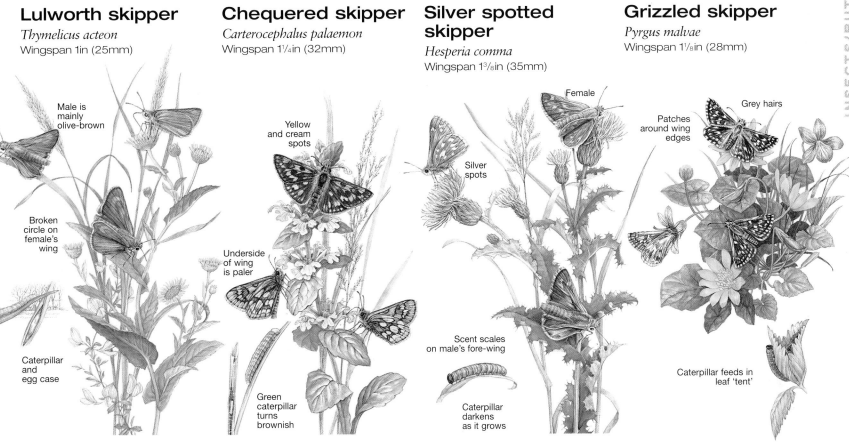

Male is mainly olive-brown

Broken circle on female's wing

Caterpillar and egg case

Yellow and cream spots

Underside of wing is paler

Green caterpillar turns brownish

Female

Silver spots

Scent scales on male's fore-wing

Caterpillar darkens as it grows

Grey hairs

Patches around wing edges

Caterpillar feeds in leaf 'tent'

A pale brown broken circle on the fore-wing of the female identifies the Lulworth skipper, which could be confused with the large skipper. Lulworth Cove, in Dorset, remains its main location, though there are populations elsewhere along the Dorset coast. It prefers warm, south-facing slopes. The female butterfly lays her eggs on the curled underside of a grass leaf: the caterpillars feed on tor grass.

Dorset coast.

A chequered pattern of yellow and cream spots on the wings makes this the most strongly marked of the British skippers. It became extinct in England in 1976, and is now a protected species. There have been re-establishment trials. The loss of woods and copses and the spread of conifer plantations have destroyed the breeding grounds where its favourite grasses grew. The caterpillar, pale green at first, becomes yellowish-brown with pinkish stripes.

Western Scotland only.

Both sexes of this active little butterfly have silver spots on the underside of the hind-wing; the female also has cream spots on the fore-wing. Once found in a broad belt following the chalky areas of the Downs and Chilterns, its population is declining and confined to isolated areas. Agricultural development has eliminated the caterpillar's food plant – sheep's-fescue. The butterflies take nectar from a variety of wild plants.

In south-east and south-west England.

Abundant grey hairs give this skipper its grizzled appearance. It is the only British skipper to overwinter as a chrysalis, emerging in spring to feed on the nectar of such flowers as hawkbits. It is often found in chalky areas such as the Downs, where its caterpillar's food plants – mainly wild potentillas – grow. Grizzled skippers rest with their wings open in the sun but closed on dull days and at night. In warm years there may be two generations.

Southern England and Wales.

J F M A M J **J A** S O N D

J F M A M **J** J A S O N D

J F M A M J **J A** S O N D

J F M A M **J J** A S O N D

279

Whites and yellows

Large white

Pieris brassicae
Wingspan 2½in (64mm)

Gardeners know this familiar butterfly, easily spotted by the bold black markings on its white wings, as the cabbage white. Summer specimens have darker marks than those that appear in spring. Before 1940 large whites were a serious cabbage pest, but chemical pesticides have helped to restrict numbers and they were also hit by a virus in 1955. Tens of thousands of immigrant large whites top up the British population each year. Caterpillars, which are mottled green and black, can feed on 60 wild members of the cabbage family, as well as garden brassicas and nasturtiums.

Widespread as resident and immigrant.

J F M **A M J J A S** O N D

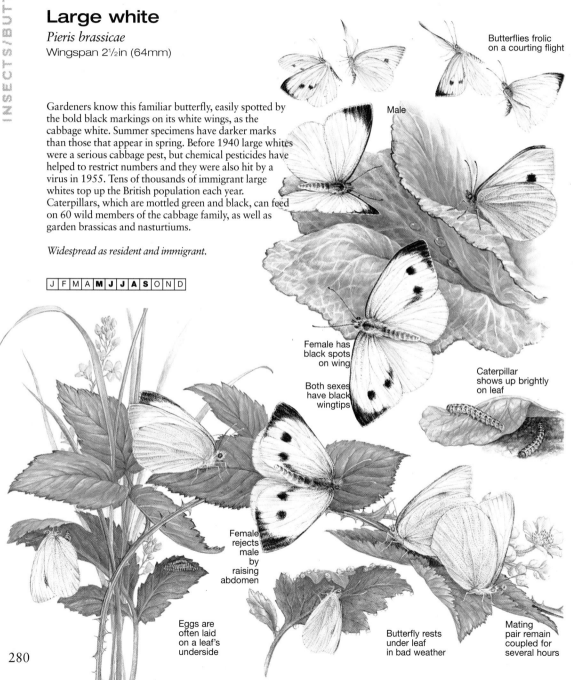

Butterflies frolic on a courting flight

Male

Female has black spots on wing

Both sexes have black wingtips

Caterpillar shows up brightly on leaf

Female rejects male by raising abdomen

Eggs are often laid on a leaf's underside

Butterfly rests under leaf in bad weather

Mating pair remain coupled for several hours

Small white

Pieris rapae
Wingspan 1⅞in (48mm)

Male has one black spot on wing

Grey on wingtips

Frail body; wingspan 1⅝in (42mm)

Wood white
Leptidea sinapis

Female – two black spots

Caterpillar feeds on cabbage heart, then outer leaves

The small white is distinguished from the large by its size and its less-pronounced black markings. It flies in the same places, shares the same food plants and is also a pest with 'cabbage white' as an alternative name. The light green caterpillar feeds deep in the heart of a cabbage before moving to its outer leaves. Populations can be augmented with immigrants.

The wood white, found locally in woodland clearings, is the smallest of the whites. It is extinct in northern England and Scotland. The caterpillar feeds on vetches.

Widespread as resident and immigrant.

J F M **A M J J A S** O N D

White and yellow butterflies belong to the Pieridae family. There is some seasonal colour variation among them, those born in summer tending to be more heavily marked than the spring generation. The caterpillars feed mostly on plants of the cabbage and pea families; some are serious pests.

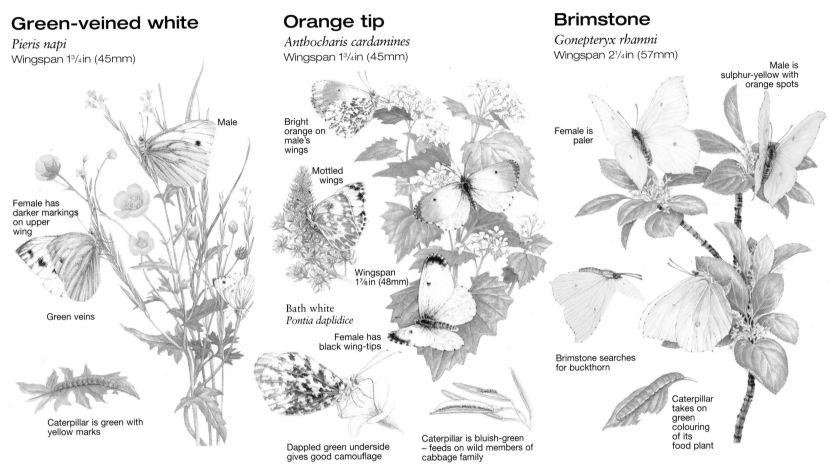

Green-veined white

Pieris napi
Wingspan 1¾in (45mm)

Male

Female has darker markings on upper wing

Green veins

Caterpillar is green with yellow marks

Orange tip

Anthocharis cardamines
Wingspan 1¾in (45mm)

Bright orange on male's wings

Mottled wings

Wingspan 1⅞in (48mm)

Bath white
Pontia daplidice

Female has black wing-tips

Dappled green underside gives good camouflage

Caterpillar is bluish-green – feeds on wild members of cabbage family

Brimstone

Gonepteryx rhamni
Wingspan 2¼in (57mm)

Male is sulphur-yellow with orange spots

Female is paler

Brimstone searches for buckthorn

Caterpillar takes on green colouring of its food plant

The green-veined white is sometimes confused with the female orange tip or, in flight, with the small white. The green veins on the underside of the hind-wings are best seen when the butterfly is at rest. It does no harm in the garden; the caterpillar feeds on wild members of the cabbage family such as watercress and garlic mustard. It is found in damp meadows and bogs, and along forestry rides rather than in gardens, which the large and small whites frequent. The green-veined white winters as a chrysalis; in most years two generations are produced.

Widespread except in Orkney and Shetlands.

Colourful symbols of springtime, orange tip males, which hatch first, can always be recognised by their distinctive orange patches. Females lack the orange colour and look like small whites or green-veined whites. They are mostly found along roadsides and in damp, flowery meadows. Males can be seen patrolling their territories while the females seek places to lay their eggs.
The Bath white is a migrant deriving its name from the city where it was depicted in an 18th-century embroidery. It likes clover and lucerne.

Widespread in England and Wales; patchy elsewhere.

The butter-coloured brimstone is often the first and last butterfly seen each year, since it hibernates late in November and flies on fine February days. The word 'butterfly' probably refers to the male's bright colour, while 'brimstone' is an old name for sulphur. Living for about ten months, brimstones hibernate among ivy leaves, whose shape their wings resemble. The incidence of brimstones depends entirely on the distribution of buckthorns, their food plant. Primroses are pollinated by the long tongues of brimstones fresh out of hibernation.

England, Wales and parts of Ireland.

J F M A M J J A S O N D   J F M A M J J A S O N D   J F M A M J J A S O N D

Yellows

Pale clouded yellow

Colias hyale
Wingspan 1⁷⁄₈in (48mm)

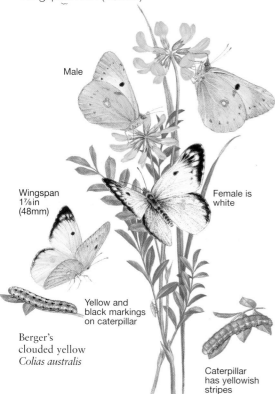

Male

Wingspan
1⁷⁄₈in
(48mm)

Female is
white

Yellow and
black markings
on caterpillar

Berger's
clouded yellow
Colias australis

Caterpillar
has yellowish
stripes

Only the male of this rare species is a true pale yellow. The female is white, with diffuse black margins and two yellow spots on each hind-wing. Both sexes have white spots on the black border of the fore-wings. Pale clouded yellows seen in Britain are immigrants from the Continent and are not seen every year. A few early arrivals may lay eggs, but the damp winter kills the caterpillars, which are pale green and velvety, with a yellowish stripe each side. Berger's clouded yellow is similar to the pale, but the male has a deeper lemon shade.

Mostly in southern England and southern Ireland.

J F M A M **J J A S O** N D

Clouded yellow

Colias croceus
Wingspan 2in (50mm)

Male –
no
spots
on wing
border

Female –
wing border
is yellow-
spotted

Caterpillar has
white hair
and yellow stripes

The strong, fast-flying clouded yellow feeds and lays its eggs on various species of clover – less common in Britain today. Spring migrants from southern France may give rise to autumn butterflies raised in Britain, but none survives the winter in any stage. There are big clouded yellow years such as 1983 and 1996, when tens of thousands migrated across the country from late May; otherwise they may not be seen for a decade or more. The deep yellow colour – 'clouded' by a broad dark border – makes the butterfly obvious from afar.

Occasional, anywhere in Britain.

J F M A M **J J A S O** N D

Aristocrats

Camberwell beauty

Nymphalis antiopa
Wingspan 2¹⁄₂in (64mm)

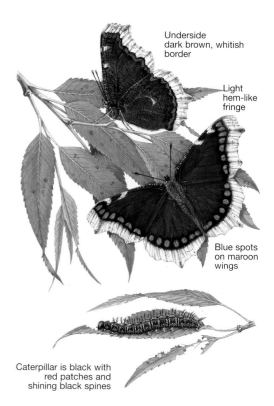

Underside
dark brown, whitish
border

Light
hem-like
fringe

Blue spots
on maroon
wings

Caterpillar is black with
red patches and
shining black spines

This majestic butterfly was first seen in Britain in 1784 at Camberwell, then a village 2 miles south of London. Because of the pale edge to the wings it was first given the name of 'white petticoat'. It is native to much of the Continent and some Camberwell beauties have evidently come from Scandinavia as hibernating adults on timber ships. But they are powerful fliers and could reach England or Scotland across the North Sea. There is no record of this species breeding in Britain, but the caterpillar may be seen on butterfly farms.

In eastern Britain, but very rare.

J F **M A M J** J A S O N D

The biggest and most splendid-looking butterflies were dubbed aristocrats by early entomologists, who gave them grand-sounding names. With the fritillaries they make up a family called nymphalids, or brush-footed butterflies; their non-functional front legs bear long hairs. Most hibernate as adults, but some die of cold.

Large tortoiseshell

Nymphalis polychloros
Wingspan 2³/₈in (60mm)

No white patches on wing

Caterpillars are black and yellow

White patches on wing

Blue half-moons

Caterpillars spin a web on leaves

Small tortoiseshell

Aglais urticae
Wingspan 2in (50mm)

Comma

Polygonia c-album
Wingspan 1⁷/₈in (48mm)

Female has darker underside
Wings have ragged edges

Male

White 'comma' on wing underside

Young caterpillar resembles bird dropping

Painted lady

Cynthia cardui
Wingspan 2¹/₄in (57mm)

Caterpillar has yellow stripes

White markings on fore-wings

With the loss of elm trees by disease, the large tortoiseshell has declined dramatically since the late 1940s, when it was abundant. Eggs are laid high in elm trees where the caterpillars live gregariously in a silk web, dropping to the ground to pupate in June. The large tortoiseshell may survive on other food plants; caterpillars have been seen on willows, birches and pears.

Rare; now thought to be extinct.

One of the commonest British butterflies, the small tortoiseshell is easily recognised by its brightly speckled orange and black wings. It feeds as readily on thistles and nettles in the countryside as on buddleias and ice-plants in the garden. However, it always lays its eggs on stinging-nettle leaves. The black and yellow caterpillars turn into chrysalises, which often hang under window sills or the eaves of sheds.

Seen almost everywhere.

White, comma-shaped marks on the underside of the hind-wings give this resident butterfly its name. The ragged edges to its wings camouflage the butterfly among dead leaves when it hibernates. Commas that have hibernated produce offspring in July. These are lighter coloured than their parents and die in the autumn, after producing a darker second generation which will live through the winter.

Southern England and Wales.

A journey of hundreds of miles lies behind the painted lady butterflies that appear in Britain every May and June. They have come from North Africa and the Mediterranean coast of France, and although some of them breed in Britain, the offspring produced in September and October die when cold weather arrives. The painted lady takes nectar from a variety of flowers, including buddleia.

Regular migrant; seen in most areas.

| J | F | M | A | M | J | J | A | S | O | N | D |

| J | F | M | A | M | J | J | A | S | O | N | D |

| J | F | M | A | M | J | J | A | S | O | N | D |

| J | F | M | A | M | J | J | A | S | O | N | D |

Aristocrats

Purple emperor

Apatura iris
Wingspan 2⁷⁄₈in (73mm)

Female is browner

Underside has false eyes

Only the male has purple sheen on wings

Caterpillar grows 'horns'

Peacock

Inachis io
Wingspan 2³⁄₈in (60mm)

Underside is almost black

'Eyes' are used to scare birds

Hairy caterpillars feed on nettles

Red admiral

Vanessa atalanta
Wingspan 2¹⁄₂in (64mm)

White markings on fore-wings

Caterpillar rests in leaf 'tent'

Red bands

Mottled brown wing underside

Patient watching in a New Forest glade on a July or August morning may afford a glimpse of the purple emperor as it sucks moisture and salts from a puddle, animal dung or a small dead animal. But by noon the butterflies will have flown to the top of the oak trees. The iridescent sheen which gives the butterfly its name is seen only on the male – when the light strikes the upper surface of its wings. The female is a duller brown. The purple emperor lays its eggs on sallow or goat willow. The green caterpillars hibernate on sallow twigs.

Only in central southern England.

'Eyes' resembling those on a peacock's tail feathers make the peacock butterfly easy to recognise. They are a means of defence; when the butterfly is disturbed by a predatory bird it opens and closes its wings rapidly, displaying all four eyes to frighten off the intruder. Peacocks feed on garden flowers such as buddleia, and are attracted to rotting fruit. Eggs are laid on nettle leaves, where the hairy black caterpillars feed before turning into chrysalises which hang from the plant stem by a silk pad. In winter peacocks hibernate in sheds and outhouses.

Widespread except northern Scotland.

Most red admirals seen in Britain arrive from the Continent in May and June. The early arrivals lay eggs, and in the summer a new resident generation visits gardens and orchards to feed on Michaelmas daisies, buddleia and rotting fruit. In the wild they feed on the teasel, scabious and clover; they drink from puddles and the sap exuding from trees. The name 'admiral' is a corruption of 'admirable', referring to its bright colours. Some may fly back south in the autumn and a few hibernate, but most perish when cold weather comes.

Regular migrant, seen in most areas.

Fritillaries

White admiral
Ladoga camilla
Wingspan 2³⁄₈in (60mm)

Patterned underside

White wing-band

Caterpillar has reddish spines

The white admiral is a butterfly of secluded woodlands with rough undergrowth. The white wing-bands are distinctive as the butterflies flit across a woodland clearing in July, the males swooping low over their territory and the females looking for honeysuckle leaves on which to lay their eggs. Brambles are their favourite source of nectar. The caterpillar hibernates in autumn and in the following June forms a chrysalis that resembles a shrivelled leaf. The butterfly emerges after 16 days.

Southern and central England only.

J F M A M J J A S O N D

Queen of Spain fritillary
Argynnis lathonia
Wingspan 1³⁄₄in (45mm)

Eggs laid on violet leaves

Large silver patches on wing

Caterpillar has six rows of spines

The large silver spots on the underside, glistening in the summer sun, distinguish this butterfly; its visual glory also inspired its regal name. The sexes are similarly patterned with the typical spottedness of fritillaries, and it could be confused at first glance with some of the commoner British woodland species. Unfortunately this fritillary is not often seen here; it is an occasional immigrant from the Continent. The caterpillars feed on violet leaves and the butterfly is an avid feeder on flower-heads such as those of scabious and knapweed.

Seen occasionally in the south.

J F M A M J J A S O N D

High brown fritillary
Argynnis adippe
Wingspan 2³⁄₈in (60mm)

Red and silver spots

Bramble flowers are favoured

Caterpillar is reddish-brown, light or dark

Eggs laid on violet leaves

A row of small red spots on the underside of the hind-wings are the best identification mark of the high brown fritillary. The sexes are difficult to tell apart, though females are usually bigger than males and have rounded instead of pointed fore-wing tips. This butterfly competes with the dark green fritillary (p. 286), vying for the same bramble blossom. Eggs are laid on leaves of dog violet and sweet violet. The high brown has vanished from many areas since 1970 due to the disappearance of many of its habitats – downland scrub, woods and copses.

Stronger colonies in south-west England and Wales than in east.

J F M A M J J A S O N D

285

Fritillaries

Dark green fritillary

Argynnis aglaja
Wingspan 2¼in (57mm)

Green wing underside with silver spots

Black spiny caterpillar has red spots

Spots on wing have varied shapes

Chrysalis forms at base of violet plant

Unlike the other two big fritillaries – the silver-washed and the high brown (p. 285) – this resident butterfly prefers open country to woodlands. The dark green fritillary gets its name from the green background on the underside of its hind-wings; on this background silver spots are ranged in at least two bands. This fritillary is a fast flier, skimming and soaring between flowers; thistles are a favourite source of nectar. A Scottish sub-species is considerably darker. Another form is *charlotta*, once named the Queen of England fritillary, in which the silver spots at the wing base are fused together.

Mainly western Britain and Ireland.

Silver-washed fritillary

Argynnis paphia
Wingspan 2¾in (70mm)

Caterpillar has reddish-brown spines and yellow stripes

Male has dark scent bars

Chrysalis has gold patches

Eggs are laid on tree-trunks or walls close to violets

Silver wash marks

Certainly the largest British fritillary, the silver-washed is often considered the handsomest, too. The male has distinctive dark scent bars across the upper side of the fore-wings. A delicate washing of silver across the underside of the hind-wing in both sexes gives the butterfly its name. A true woodland species, it differs from its relatives in laying its eggs on a tree-trunk, close to violet clumps, instead of on the leaves of violets. The caterpillars then have to find their food plant. Silver-washed fritillaries often chase each other playfully across sunny clearings, resting on leaves and tree-trunks.

South-west England, Wales and Ireland.

J F M A M J **J A** S O N D

J F M A M J **J A** S O N D

Fritillary butterflies get their name from the spotted flowers that they resemble. They are predominantly orange-brown, with dark spots in greatly varied patterns. Most British fritillaries live in open areas of woodland; the clearing of woods and other changes in their habitat have dramatically reduced their populations.

Heath fritillary

Mellicta athalia
Wingspan 1¾in (45mm)

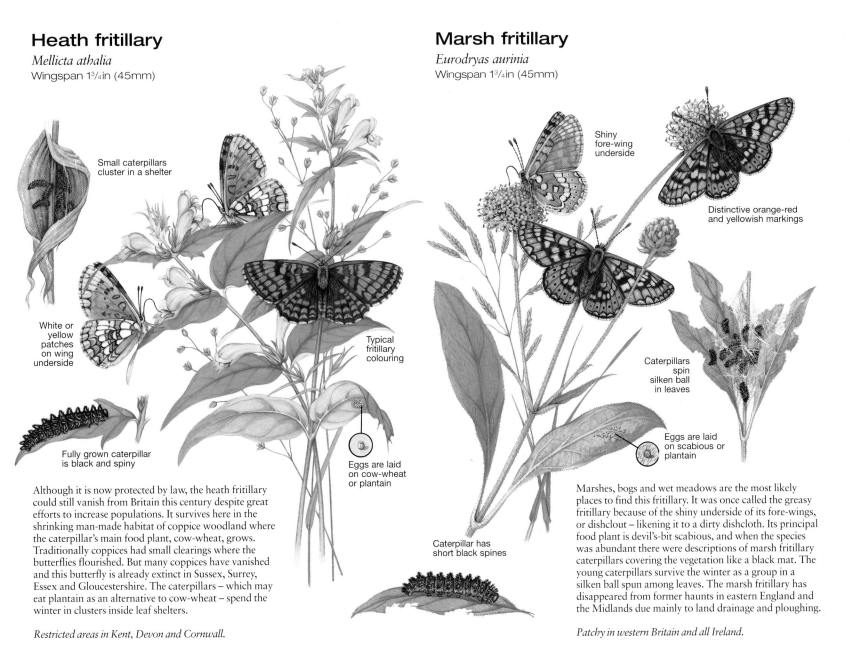

Small caterpillars cluster in a shelter

White or yellow patches on wing underside

Typical fritillary colouring

Fully grown caterpillar is black and spiny

Eggs are laid on cow-wheat or plantain

Although it is now protected by law, the heath fritillary could still vanish from Britain this century despite great efforts to increase populations. It survives here in the shrinking man-made habitat of coppice woodland where the caterpillar's main food plant, cow-wheat, grows. Traditionally coppices had small clearings where the butterflies flourished. But many coppices have vanished and this butterfly is already extinct in Sussex, Surrey, Essex and Gloucestershire. The caterpillars – which may eat plantain as an alternative to cow-wheat – spend the winter in clusters inside leaf shelters.

Restricted areas in Kent, Devon and Cornwall.

Marsh fritillary

Eurodryas aurinia
Wingspan 1¾in (45mm)

Shiny fore-wing underside

Distinctive orange-red and yellowish markings

Caterpillars spin silken ball in leaves

Eggs are laid on scabious or plantain

Caterpillar has short black spines

Marshes, bogs and wet meadows are the most likely places to find this fritillary. It was once called the greasy fritillary because of the shiny underside of its fore-wings, or dishclout – likening it to a dirty dishcloth. Its principal food plant is devil's-bit scabious, and when the species was abundant there were descriptions of marsh fritillary caterpillars covering the vegetation like a black mat. The young caterpillars survive the winter as a group in a silken ball spun among leaves. The marsh fritillary has disappeared from former haunts in eastern England and the Midlands due mainly to land drainage and ploughing.

Patchy in western Britain and all Ireland.

J F M A M **J** **J** A S O N D

Fritillaries

Pearl-bordered fritillary

Boloria euphrosyne
Wingspan 1³/₄in (45mm)

Seven 'pearls' on hind-wing underside

Caterpillar has yellow spines

Small pearl-bordered fritillary

Boloria selene
Wingspan 1⁵/₈in (42mm)

Caterpillar has spiny 'horns'

Extra white patches and black markings on wing

Row of 'pearls'

Glanville fritillary

Melitaea cinxia
Wingspan 1¹/₂in (38mm)

Beige and orange bands on underside

Black dots in orange spots

Young caterpillars hibernate in web

Seven spots, with some resemblance to a string of pearls, adorn the underside of the fritillary's hind-wing, along the rear edge. It shares this feature with its similar, but slightly more decorated, smaller relative. It was once called the April fritillary, but that referred to the pre-1752 calendar, and it usually appears in May. The caterpillars of the pearl-bordered fritillary, like those of its close relatives, feed on violet leaves growing along shady lanes and in woods. The butterflies sip nectar from a variety of wild flowers. Grubbing-out of woods and hedgerows has eliminated much of their habitat.

South and west Britain and west Ireland.

Flying slightly later in the year than the similar pearl-bordered fritillary, the smaller species is also found in some different places such as mountain slopes and sea-cliffs. Both species may, however, be found in woodland clearings. Only violet leaves sustain the caterpillars, and loss of its habitat has driven this butterfly from many of its former haunts in eastern England. The string of seven 'pearls' on the underside of the hind-wings is accompanied in this species by more silver patches than its relative bears. The caterpillar has orange tubercles on its body and a large pair over the head.

Mostly western Britain, but not Ireland.

Named after the 18th-century amateur naturalist Eleanor Glanville, this butterfly is now restricted entirely to the Isle of Wight. The species could be confused with the marsh fritillary but for the row of orange spots, with black dots, on the hind-wing. Young caterpillars are easy to identify; they are gregarious and sunbathe on the silk web they spin on the leaves of ribwort plantain, their food plant. (The butterfly's alternative name, plantain fritillary, refers to this.) The caterpillars overwinter in their web and resume feeding in the spring. Adults fly in colonies of several hundred.

South-western Isle of Wight.

Swallowtail and monarch

Duke of Burgundy fritillary

Hamearis lucina
Wingspan 1¼in (32mm)

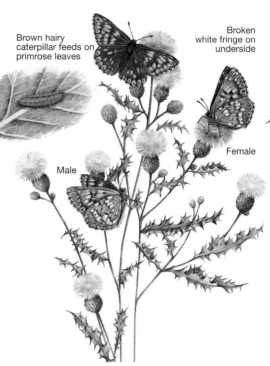

Brown hairy caterpillar feeds on primrose leaves

Broken white fringe on underside

Male

Female

Known as the smallest British fritillary, this butterfly actually belongs to another family whose members live mostly in South America. In the 18th century it was called Mr Vernon's small fritillary, and the spotted wings give it the characteristic fritillary look. Primroses and cowslips are the caterpillar's food plants, and meadows and woodland glades where these grow are likely places to see the butterflies. They are active fliers, visiting only a few flowers, such as thistles, for nectar. They live in colonies which become more isolated as grassland and primrose-rich downland vanish.

Patchy between Yorkshire and southern England.

JFMA**MJ**JASOND

Swallowtail

Papilio machaon
Wingspan 3¾in (95mm)

Distinctive yellow and black markings

Prominent 'tail'; red and blue 'eyes'

Fully grown caterpillar is green, black and orange

The largest and one of the loveliest of British native species, the swallowtail is conserved in marshy areas of the Norfolk Broads where its food plant, milk parsley, grows. This butterfly cannot be confused with any other. The long 'tail' extensions and false eyes give the impression of a head and antennae to confuse predators. When young, the camouflaged caterpillars look like bird droppings; fully grown, they are attractively coloured and display orange 'horns' for defence. Immigrant swallowtails may be found in southern England, but the English form is slightly smaller and darker.

Resident only in parts of Norfolk.

JFMA**MJ**JASOND

Monarch

Danaus plexippus
Wingspan 4in (10cm)

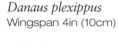

Black veins in orange wings

Largest wingspan seen in Britain

Caterpillar has black 'horns' at each end

Regularly blown 3500 miles across the Atlantic from America, this species has been known in Britain since 1876. It is also called the milkweed after the North American food plant of its caterpillar. The butterflies migrate southward in the autumn, and those reaching Britain appear to have been blown by very strong winds, with influxes in 1995 and 1999. There are no wild-growing food plants for them in Europe. The monarch is poisonous to predators. The poison is taken from the milkweed by the caterpillars and carried through the chrysalis and adult stages.

Migrant to the western seaboard.

JFMAMJ**JASO**ND

289

Blues

Chalk-hill blue

Lysandra coridon
Wingspan 1³⁄₈in (35mm)

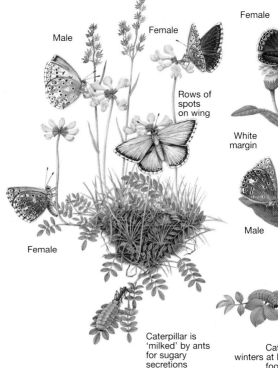

Male

Female

Rows of spots on wing

Female

Caterpillar is 'milked' by ants for sugary secretions

The silvery-blue of the male's upper wings immediately distinguishes this butterfly from the common blue. The female is brown on the upper side. Once this butterfly had a wider distribution in East Anglia and the Salisbury Plain; it is now restricted to the chalky area of the Chilterns, Cotswolds and Downs, and is actively conserved. It prefers steep hillsides with plenty of its caterpillar food plant, horseshoe vetch.

Chalky areas in the south.

J F M A M J **J A S** O N D

Common blue

Polyommatus icarus
Wingspan 1³⁄₈in (35mm)

Female

Female normally brown, may be blue

Female

White margin

Male

Male

Male

Caterpillar winters at base of food plant

Glinting in the summer sun, the iridescent colour of the common blue is distinctive. Only the male always has this bright colour; the female is usually brown. Common blues normally fly in large numbers, often in company with other blues, and they feed for hours on plants such as fleabane. Their caterpillar food plants include bird's-foot trefoil and clovers, which grow mostly on downland. Common blues rest head-down on stems.

Widespread in Britain and Ireland.

J F M A M J **J** J **A S** O N D

Silver-studded blue

Plebejus argus
Wingspan 1³⁄₈in (35mm)

Fully grown caterpillar has black and white stripes

Silver studding

Male

Female brown with orange marking

Female

A silver-blue row of spots on the margin of the hind-wing underside is the identity key to this species – but they are not always pronounced. The male has a dark margin on the upper wing surface and the female has dark brown upper wings with a row of orange half-moons. Scores of butterflies may be seen feeding on hot July days, especially in heathland and chalky areas. Caterpillars feed on gorse and rock-rose, and secrete a fluid that is drunk by ants.

South and south-west, East Anglia, Wales.

J F M A M J **J A S** O N D

Adonis blue

Lysandra bellargus
Wingspan 1¼in (32mm)

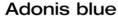
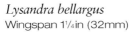

Male is vivid blue

Caterpillar has yellow stripes

Chequered wing edges

Female

The males have the brightest colour of any English blues, but the most reliable identification keys to the Adonis are the alternate black and white bars to the edge of all wings. These are seen on the male and the brown female. The butterfly is confined to chalky areas where it finds its food plant, horseshoe vetch. Adonis blues congregate in suitable places, preferring grassland slopes. As with many blues, the caterpillar is 'milked' by ants.

Chalky parts of southern England.

J F M A M **J** J **A S** O N D

The family Lycaenidae embraces the blues, the hairstreaks and Britain's two copper butterflies. All are small and most fly swiftly. Most female 'blues' are in fact brown, and there are some all-brown species. Some caterpillars of this family have a secretion 'milked' by ants, which in turn carry the caterpillars to food plants.

Short-tailed blue
Everes argiades
Wingspan 1in (25mm)

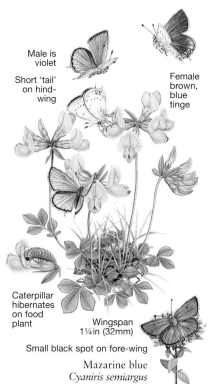

Male is violet

Short 'tail' on hind-wing

Female brown, blue tinge

Caterpillar hibernates on food plant

Wingspan 1¼in (32mm)

Small black spot on fore-wing

Mazarine blue
Cyaniris semiargus

This rare immigrant to Britain has never been known to breed here. Only a handful have been recorded in Britain since 1885. The 'tails' on the hind-wings are tiny and covered in scales. Together with the small, false eye-spots at the base of the hind-wing the tails mimic the head and antennae, thus confusing predators.

Mazarine blues are now extinct in Britain, but some fly in from the Continent to the south coast in summer.

Rare migrant in southern England.

Long-tailed blue
Lampides boeticus
Wingspan 1⅜in (35mm)

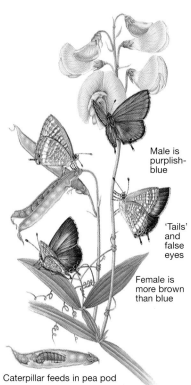

Male is purplish-blue

'Tails' and false eyes

Female is more brown than blue

Caterpillar feeds in pea pod

Although this is one of the commonest butterflies in the world, it is very rare in Britain. There are false eyes at the base of the long 'tails', giving a head-and-antennae impression more pronounced than in the short-tailed blue. The underside of the hind-wing has a pattern of thin wavy lines and light bands. Caterpillars feed inside a developing pea pod; hence the old name of pea-pod argus.

Rare migrant in southern England.

Small blue
Cupido minimus
Wingspan ⅞in (22mm)

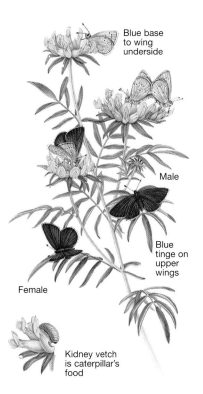

Blue base to wing underside

Male

Blue tinge on upper wings

Female

Kidney vetch is caterpillar's food

Small it certainly is – the tiniest butterfly in Britain – but the blue is only a faint dusting over the upper surface of the wings and at the base of their underside; otherwise both sexes are predominantly brown. Though frail, these butterflies are vigorous fliers. The small blue is dependent on kidney vetch as a food plant. Typical habitats are disused railway cuttings, sandhills, coastal cliffs and chalk and limestone areas.

Mainly in the south; scattered elsewhere.

Large blue
Maculinea arion
Wingspan 1½in (38mm)

Black spots on wing

Male has smaller spots

Female

Ant 'milks' caterpillar

Caterpillar feeds on thyme

Extinct in Britain since 1979, the large blue may make a come-back. Attempts have been made to introduce continental forms of the butterfly to its old localities in the West Country. The large blue has a much darker hue to its wings than other blues, and there are distinctive black spots on the fore-wings. Unusually, both male and female are blue. Caterpillars, like those of some other blues, secrete a fluid on which ants feed, and are even carried to the ants' nest.

Once extinct in Britain, but re-establishing.

| J | F | M | A | M | **J** | **J** | A | S | O | N | D |

| J | F | M | A | M | **J** | **J** | **A** | S | O | N | D |

| J | F | M | A | **M** | **J** | J | **A** | S | O | N | D |

| J | F | M | A | M | **J** | **J** | **A** | S | O | N | D |

Blues and coppers

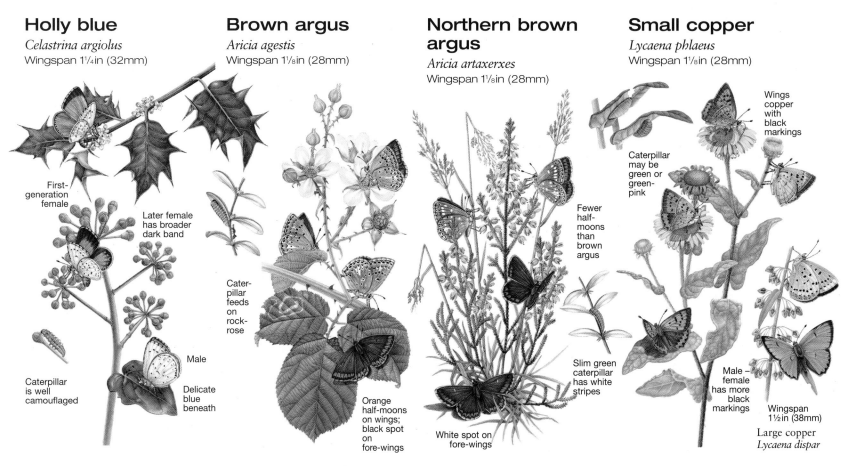

Holly blue
Celastrina argiolus
Wingspan 1¼in (32mm)

First-generation female

Later female has broader dark band

Caterpillar is well camouflaged

Male

Delicate blue beneath

Brown argus
Aricia agestis
Wingspan 1⅛in (28mm)

Caterpillar feeds on rock-rose

Orange half-moons on wings; black spot on fore-wings

Northern brown argus
Aricia artaxerxes
Wingspan 1⅛in (28mm)

Fewer half-moons than brown argus

Slim green caterpillar has white stripes

White spot on fore-wings

Small copper
Lycaena phlaeus
Wingspan 1⅛in (28mm)

Wings copper with black markings

Caterpillar may be green or green-pink

Male – female has more black markings

Wingspan 1½in (38mm)

Large copper
Lycaena dispar

Unlike other British butterflies, the holly blue alternates its caterpillar food plants. First-generation caterpillars, in May and June, feed mainly on holly buds; the second, in August and September, usually on ivy. Second-generation females have a broader dark band at the edge of the fore-wing than those of the first. The delicate blue underside of both sexes is unique. Holly blues like hedgerows, gardens, sandy hillsides and woodland clearings.

Common in England, Wales and Ireland.

Although it is a member of the blue family, this species has only a suggestion of blue at the base of its wing undersides. It is chocolate-brown on the upper surface with a band of orange half-moons at the edge of the wings. The male is very slightly darker and smaller than the female. The undersides have a dark background colour covered in a pattern of spots. There are two generations of butterflies each year.

Southern England; north and south Wales.

This close relative of the brown argus, found only in northern Britain, is distinguished mainly by a white spot on the fore-wing. It is the same colour as the brown argus and its caterpillars feed on the same plant, rock-rose. But the two species do not overlap their territory. The sexes are indistinguishable. These butterflies may be plentiful in habitats such as open grassland, cliffs and mountain scree slopes.

Northern England and Scotland.

The distinctive small copper may be seen in gardens as well as the countryside. It has bright coppery wings with black spots and borders. Some small coppers have a row of blue marks around the edge of the hind-wing. A quick flier, it visits many different plants for nectar; its caterpillars feed on docks and sorrels.
The large copper became extinct in Britain in 1865, but a colony has since been established in Cambridgeshire.

Abundant except on high uplands.

J F M **A** M J **J A** S O N D   J F M A M **J J** A S O N D   J F M A M **J J** A S O N D   J F M A M **J J** A S O N D

Hairstreaks

Members of the same family as the blues and coppers, the hairstreaks get their name from the thin white line running across the underside of the wings. Most of them are less flamboyantly coloured than the blues and inhabit mainly woodland.

Black hairstreak

Strymonidia pruni
Wingspan 1¼in (32mm)

Male has scent patch on fore-wing

Female

Black dots, orange band and hairstreak

'Tails' on hind-wings

Caterpillar feeds on blackthorn flowers, buds and leaves

This rarity is predominantly brown; the 'black' in its name refers to the row of spots on the underside of each wing, between the white hairstreak and an orange band near the edge. The upper surface of the wings is a darker brown than the lower, with another margin of orange marks – more prominent on the female – and a subtle bluish scent patch on the male's fore-wings. Woodland clearances have driven it from many of its old haunts.

Woods in the east Midlands.

Brown hairstreak

Thecla betulae
Wingspan 1⅜in (35mm)

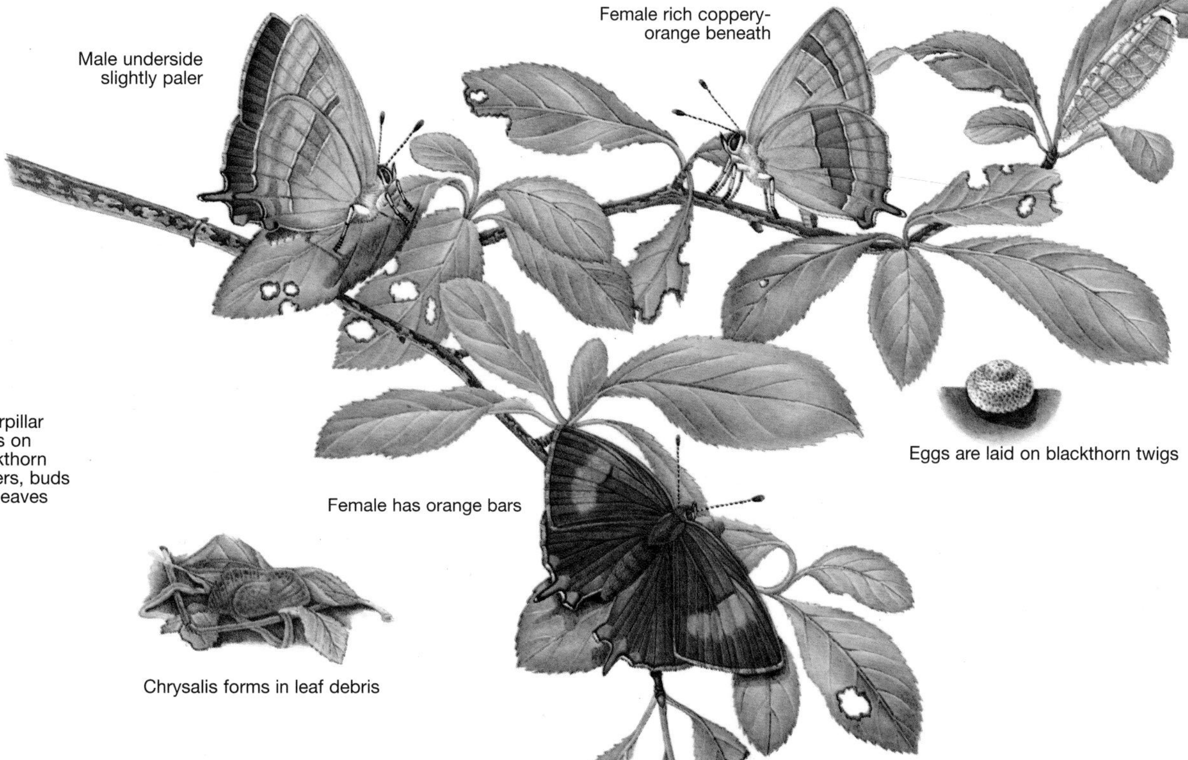

Male underside slightly paler

Female rich coppery-orange beneath

Young blackthorn leaves are caterpillar's food

Eggs are laid on blackthorn twigs

Female has orange bars

Chrysalis forms in leaf debris

Male

'Tails' on hind-wings

Colonies of brown hairstreaks do not overlap with the black species, but their caterpillars share the same food plant – blackthorn. This species is rich brown. The female has a bar of orange on each fore-wing. The 'tails' are quite distinctive and there is an obvious white band on the underside. Seen close up, the brown hairstreak cannot be confused with anything else. It is a secretive butterfly of hedgerows and wood edges, best seen basking in the late afternoon. It spends much time feeding on honeydew, the sugary waste of aphids. There is one generation a year.

Western England and Wales; isolated colonies elsewhere.

J F M A M **J J** A S O N D   J F M A M J **J A S** O N D

Hairstreaks

Purple hairstreak

Quercusia quercus
Wingspan 1⁵/₁₆in (34mm)

The most likely place to see these butterflies
is high up in an oak wood, where clouds of
them may swirl around the treetops in July
and August. Purple hairstreaks feed on the
sugar-rich honeydew deposited by aphids
and jealously guard their lofty territory –
even sometimes tackling wasps. They
rarely descend to the ground. The purple is
seen more in the male than the female. The
silvery-grey underside, with its 'tails',
shows a watermark appearance behind the
hairstreak line.

Abundant in the south; isolated elsewhere.

J F M A M J **J A** S O N D

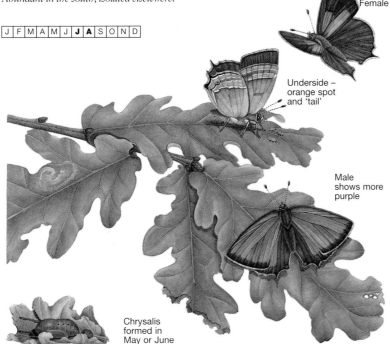

Caterpillar
emerges
in spring

Eggs are
laid on oaks
in July

Caterpillar
has slug-like shape;
feeds on oak leaves
and buds

Female

Underside –
orange spot
and 'tail'

Male
shows more
purple

Chrysalis
formed in
May or June

Green hairstreak

Callophrys rubi
Wingspan 1¹/₈in (28mm)

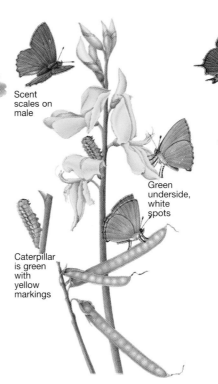

Scent
scales on
male

Green
underside,
white
spots

Caterpillar
is green
with
yellow
markings

This butterfly's vivid green is only on the
underside; the upper surface is dark brown.
The sexes are indistinguishable except for
the tiny bluish scent mark on the male's
fore-wings. The butterflies delight in warm
south-facing slopes where there are plenty
of scrubby bushes and open meadows with
their food plants such as broom, gorse and
bramble. They are extremely territorial and
gregarious, spending much time basking.

Common in south; widespread elsewhere.

J F M A **M J** J A S O N D

White-letter hairstreak

Strymonidia w-album
Wingspan 1¹/₄in (32mm)

Female

Male has
scent scales

Caterpillar
emerges
in March

White
letter
W

White-
tipped
'tail'

A mark like the letter W forming part of
the hairstreak line helps to distinguish this
species from the black hairstreak. Behind
its 'tails' – longer on the female – there is a
row of orange half-moons. The male has a
small scent mark on the top of each fore-
wing, but the upper surfaces are revealed
only in flight. Populations declined in the
1970s with the death of many elms, the
caterpillars' only food plant.

England and Wales; not in far north.

J F M A M J **J A** S O N D

Browns

Brown butterflies, forming the Satyridae family, all have false eyes on the upper or lower surface of the wings, or on both – marks which confuse predators. The marbled white is the one member of the group that is not brown. Almost all species spend the winter as grass-eating caterpillars.

Meadow brown
Maniola jurtina
Wingspan 2in (50mm)

Gatekeeper
Pyronia tithonus
Wingspan 1½in (38mm)

Small heath
Coenonympha pamphilus
Wingspan 1⅛in (28mm)

Female's underside is paler than male's

Male

Male is smaller and darker

Female has prominent 'eyes' on orange patches

Caterpillar takes 8-9 months to develop

Male is smaller and richer in colour, with scent scales

Usually two white spots in fore-wing 'eye'

Female

White spots on hind-wing underside

Caterpillar is brown with white stripes

Faint dark spot at wing-tip

Caterpillar from May egg is butterfly by August

Wingspan 1⅜in (35mm)

False eye

Southern form (*davus*)

Scottish form (*scotica*)

Territorial subspecies vary in spotting on wings

Large heath
Coenonympha tullia

Females of the meadow brown, perhaps Britain's commonest butterfly, are unusual in being more distinctive than the males. The female's 'eyes' stand out strongly against the large orange patches on the upper side; the male is smaller and darker, with less prominent 'eyes'. Like other browns, they have only four walking legs. On sunny days meadow browns are always on the wing, and they will also fly in dull or even wet weather. They are seen in lowland and upland meadows and often on roadside verges. There is one generation a year.

Abundant in most of Britain and Ireland.

True to its name, the gatekeeper patrols hedgerows and corners of fields day after day. These butterflies, also known as hedge browns, are very territorial and spend much of their lives in one place, basking in the sun or drinking nectar from bramble blossom, fleabane and many other nectar-rich flowers. Colonies in orchards and along overgrown hedgerows may number several hundreds. Key identification features are the dark bands across the orange wings of the male and the white spots on the hind-wing underside of both sexes.

Widespread in southern Britain and south of Ireland.

This is one of the most widespread British butterflies, looking like a miniature meadow brown, with a false 'eye' spot on the underside of each fore-wing, dark hind-wings with a lighter patch, and light brown upper surfaces. There are two generations each year. Roadside verges and ditches can be teeming with them in late August.
The large heath, an upland butterfly, is confined mainly to northern Britain and Ireland, but shows regional variations even within this area. It feeds on sedges, fescues and cottongrass.

Abundant throughout Britain and Ireland.

J F M A M **J J A S** O N D   J F M A M J **J A S** O N D   J F M A M **J J A S** O N D

Browns

Wall
Lasiommata megera
Wingspan 2in (50mm)

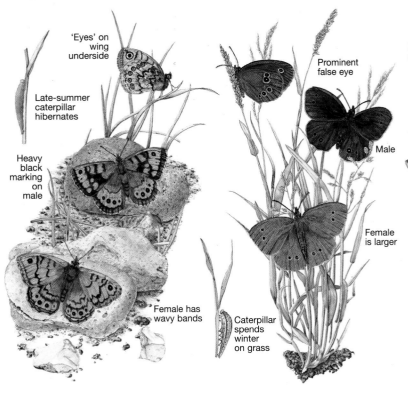

'Eyes' on wing underside

Late-summer caterpillar hibernates

Heavy black marking on male

Female has wavy bands

Walkers often accidentally disturb this butterfly basking in the sun on tracks or roads, and by the edges of woods. It is also known as the wall brown. The males are territorial and will spend most of their time perching and patrolling. Patterns on the wings are distinctive – noticeable bars across the fore-wings of the male, wavy bands on the female and 'eyes' on the fore-wing underside in both sexes. There are two generations a year.

Common; only local in Scotland.

Ringlet
Aphantopus hyperantus
Wingspan 1⁷⁄₈in (48mm)

Prominent false eye

Male

Female is larger

Caterpillar spends winter on grass

This sombre-looking butterfly can be plentiful in its favourite grassy habitats such as road verges and rich meadows. Its name comes from the large false eyes on the underside of all four wings. These have a white 'pupil' and pale outer ring. The female lays her eggs while flying, so it is not certain that they will land on grasses that the caterpillars eat. Young caterpillars hibernate through the winter and resume feeding high on grass stems in the spring.

Especially southern Britain and Ireland.

Scotch argus
Erebia aethiops
Wingspan 1³⁄₄in (45mm)

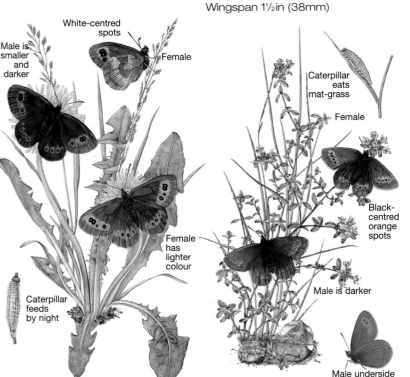

White-centred spots

Male is smaller and darker

Female

Female has lighter colour

Caterpillar feeds by night

Scotland is the stronghold of this butterfly, but it is just surviving in two areas of the English Lake District. Grassy hillsides, sheltered valleys and moors are its haunts. It can live in colonies 10,000 strong. The key identification feature is the orange-red band, containing false eyes, on the upper surface of the wings. The caterpillar hibernates and resumes feeding in the spring; it lives about ten months, compared to one week or so as an adult.

Mostly in Scotland, below 1600ft (500m).

Small mountain ringlet
Erebia epiphron
Wingspan 1¹⁄₂in (38mm)

Caterpillar eats mat-grass

Female

Black-centred orange spots

Male is darker

Male underside

Looking like a smaller version of the Scotch argus, this little butterfly is usually found at altitudes of 1600ft (500m) and above. Males are usually smaller than females and have less prominent markings. The orange spots on the fore-wing have a black centre and, on the female, there is a light band across the underside of the wings. Large colonies of these butterflies may be found in areas where their only food plant, mat-grass, grows.

Western Scotland and Lake District.

J F M A **M J J A S** O N D

J F M A M J **J A S** O N D

J F M A M J **J A S** O N D

J F M A M J **J A** S O N D

Marbled white
Melanargia galathea
Wingspan 2⅛in (54mm)

Caterpillar hibernates autumn to February

Male is smaller

Broader fore-wings

Female

In spite of its colouring – black and grey with white patches – this butterfly belongs to the brown family. Females are usually larger than males, with slightly broader fore-wings. Some colonies of marbled whites number several thousand; they are found on unimproved pastures and along woodland rides or disused railway lines. Eggs are laid while the female flies; they should land among such grasses as red fescue. The caterpillar hibernates.

Mostly south-west England and Wales.

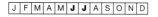

Grayling
Hipparchia semele
Wingspan 2in (50mm)

Wings are kept closed when at rest

Zigzag pattern

White-centred spots

Caterpillar hibernates, feeding in mild spells

Retracts fore-wing at rest

Natural grassland once supported large numbers of graylings throughout southern Britain, but numbers have declined with increased ploughing. It is now most likely to be seen flying swiftly over dunes, heaths and coastal downland. It rarely feeds from flowers but deftly alights on bare ground or rock. It has the knack of leaning sideways to avoid casting a large tell-tale shadow and rarely opens its wings when at rest. Both sexes look similar.

Mainly coastal areas of south and west.

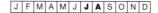

Speckled wood
Pararge aegeria
Wingspan 1¾in (45mm)

Males defend their territory

Butterflies rest on low tree branches

Male markings are more blurred

Autumn caterpillar grows through the winter

Female

'Eyes' with white centres on fore-wings

These butterflies, which can be seen dancing in the shafts of sunlight that sparkle through an oak-wood canopy, are strongly territorial; males will defend their 'patch' against other males. There is little to separate the sexes, except that males are slightly smaller, with rounded wingtips and less distinct markings than females. Grasses such as couch and cock's-foot are the caterpillar's food. There can be several generations a year, often overlapping.

Widespread in southern Britain and Ireland.

Hawk-moths

Eyed hawk-moth
Smerinthus ocellata
Wingspan 3½in (90mm)

Poplar hawk-moth
Laothoe populi
Wingspan 3½in (90mm)

Lime hawk-moth
Mimas tiliae
Wingspan 3⅛in (80mm)

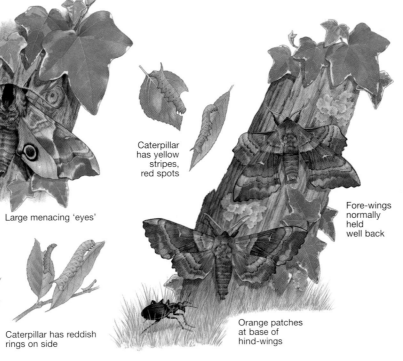

Mottled brown fore-wings

Large menacing 'eyes'

Caterpillar has reddish rings on side

Caterpillar has yellow stripes, red spots

Fore-wings normally held well back

Orange patches at base of hind-wings

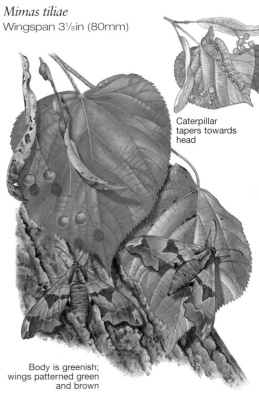

Caterpillar tapers towards head

Body is greenish; wings patterned green and brown

When alarmed by a small mammal or bird, this hawk-moth moves its fore-wings forward, thus revealing the menacing 'eyes' that pattern its rear wings. When the moth is at rest, only the mottled brown fore-wings are visible. The caterpillars, yellowish-green with white stripes and reddish rings on the sides, and a blue tail 'horn', can be seen from June to September. They are night-feeders, eating the leaves of young willows, aspens and poplars in damp places, and also feeding on apple, pear and plum trees. The moths also feed at night, on nectar from flowers. By day they rest on tree-trunks.

Locally common; absent from Scotland.

This night-flying moth gets its name from one of the main food plants of its caterpillars, and is often found in parks and recreation grounds where poplars are planted. During the day it rests on the bark of trees, where its grey-brown, sculptured wings help to conceal it. When alarmed, it brings its fore-wings forward, revealing not an 'eye' but a startling orange patch on each hind-wing, which may frighten a predator. It sometimes breeds twice a year – unlike other hawk-moths, which have only one generation. The caterpillars are green with yellow stripes and red spots, and have a typical 'horn' at the hind end.

England and Wales; patchy in Scotland and Ireland.

This slim-winged and delicately coloured moth lays its eggs on the leaves of lime and other trees, and is common in urban areas where lime trees have been planted in parks and squares. In May and June the moths can be seen resting in the daytime on walls, fences and tree-trunks, while at night they take to the wing and often blunder against lighted windows. The caterpillars, found from July to September, are yellowish-green with oblique stripes; the 'horn' at the tail is black, becoming green. They feed at night, resting on the underside of leaves by day. The chrysalis usually spends the winter in the soil.

Commonest in southern England; not in Scotland and Ireland.

J F M A **M J J A S** O N D

J F M A **M J J A S** O N D

J F M A **M J** O N D

Among the most spectacular of British moths, the hawk-moths have stout bodies, tapering to the rear, and comparatively narrow wings. They get their name from their ability to hover in the air like kestrels. Nearly all their caterpillars have a curved 'horn' or spine on their last segment.

Privet hawk-moth

Sphinx ligustri
Wingspan 4¹⁄₂in (11.5cm)

Elephant hawk-moth

Deilephila elpenor
Wingspan 2³⁄₄in (70mm)

Death's-head hawk-moth

Acherontia atropos
Wingspan 4¹⁄₂in (11.5cm)

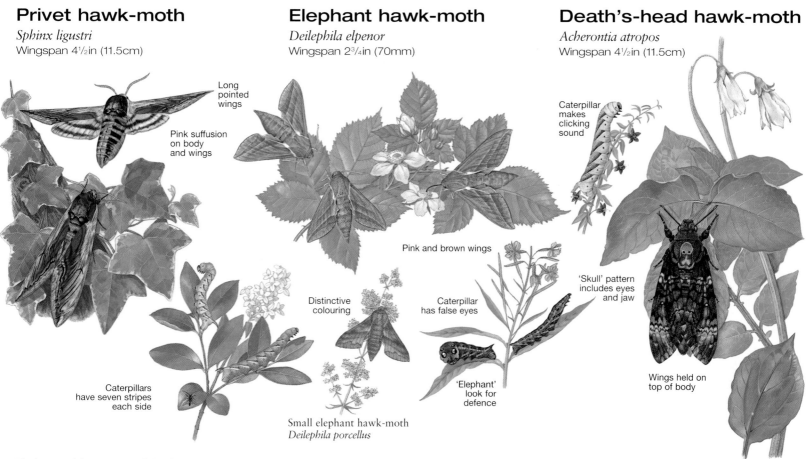

Long pointed wings

Pink suffusion on body and wings

Caterpillars have seven stripes each side

Distinctive colouring

Small elephant hawk-moth
Deilephila porcellus

Pink and brown wings

Caterpillar has false eyes

'Elephant' look for defence

Caterpillar makes clicking sound

'Skull' pattern includes eyes and jaw

Wings held on top of body

The largest of the native British hawk-moths, this striking species has been helped to spread by the popularity of garden privet hedges, on which its caterpillars feed. It gets its Latin name *Sphinx* from the winged monster of Greek mythology, which was usually represented with its wings folded over its body. The moth's pink-striped body is sometimes confused with that of the convolvulus hawk-moth (p. 300), but on the privet the coloration spreads over the wings. The large green caterpillars – seen in July and August – have zigzag stripes down their sides; the chrysalis may spend two winters buried underground.

Only in southern England.

Attractively coloured in shades of pink and light brown, this moth can be found in clearings, on waste ground or in any place where willowherbs – the caterpillar's main food plants – flourish. The adults may be seen in the evening feeding from the flowers of honeysuckle and petunias. The name comes from the browny-grey caterpillar – seen in July and August – which when threatened retracts its head into its body to produce an elephant's-head effect.

The small elephant hawk-moth, with a wingspan of 2¹⁄₈in (54mm), is seen in chalky areas from May to July.

Common in England and Wales; scattered elsewhere.

This large and distinctive migrant from Africa can be seen from spring to autumn. Early arrivals may breed here. The extraordinary skull-like markings from which the moth gets its name are on the top of the thorax. The caterpillars, which devour potato and tomato leaves, can grow to 5in (12.5cm) long; they have striking diagonal blue or purple stripes along the sides. The moth can utter a high-pitched squeak, said to sound like the 'piping' note of a queen bee, and this is emitted when the death's-head raids a hive for honey, its favourite food. The caterpillar produces a clicking noise if disturbed.

Found throughout British Isles.

| J | F | M | A | **M** | **J** | **J** | A | S | O | N | D |

| J | F | M | A | **M** | **J** | **J** | A | S | O | N | D |

| J | F | M | A | **M** | **J** | **J** | A | S | O | N | D |

Hawk-moths

Convolvulus hawk-moth

Agrius convolvuli
Wingspan 5in (12.5cm)

Pine hawk-moth

Hyloicus pinastri
Wingspan 2³⁄₄in (70mm)

Humming-bird hawk-moth

Macroglossum stellatarum
Wingspan 2¹⁄₄in (57mm)

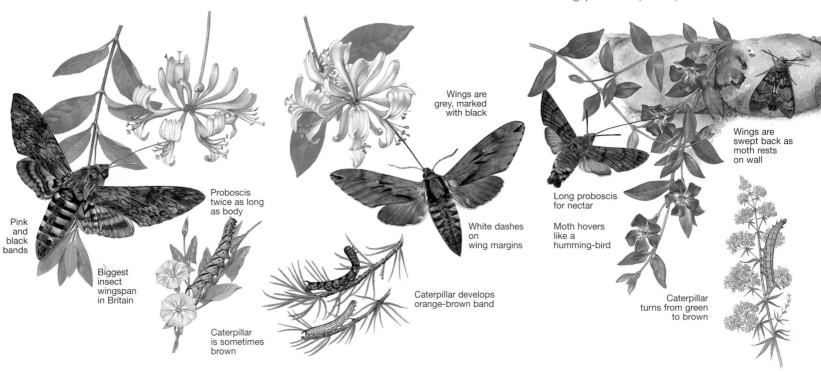

Pink and black bands

Biggest insect wingspan in Britain

Proboscis twice as long as body

Caterpillar is sometimes brown

Wings are grey, marked with black

White dashes on wing margins

Caterpillar develops orange-brown band

Long proboscis for nectar

Moth hovers like a humming-bird

Wings are swept back as moth rests on wall

Caterpillar turns from green to brown

This grey moth has the largest wingspan of any insect found in Britain. It is a powerful flier, migrating from the Mediterranean region. Its proboscis, with which it sucks nectar from long-tubed wild flowers such as honeysuckle, is also the longest of any British insect, reaching 3¹⁄₂in (90mm) – twice as long as its body. When the moth is not feeding, the proboscis curls up under the head. The irregular patterning of the wings acts as camouflage when the moth is resting on a tree or fence. Its caterpillars – rarely seen in Britain – feed on wild convolvulus (bindweed).

Throughout British Isles; numbers vary.

Thanks to the spread of conifer plantations, the range of the pine hawk-moth has also extended, so that it can now be found south of a line from Plymouth to The Wash. The dull-looking moth – grey marked with black – feeds at dusk on flowers such as privet and honeysuckle; the day is spent resting on the trunks of conifers. It lays its eggs on the needles of pine and spruce trees. The colourful caterpillars begin life bright green with white stripes, developing a broad orange-brown band along the back as they grow. They can be seen in August and September, and are forest pests in some areas.

Southern and eastern England, Channel Islands.

When it hovers over a flower like a humming-bird, the wings of this moth beat so fast that they are hardly visible, and they make an audible, high-pitched humming sound. Unlike most moths, it flies by day. It is a summer visitor from the south of France and feeds on garden flowers such as honeysuckle, geraniums and petunias, drinking their nectar with its long proboscis. Its overall colour is grey. Eggs are normally laid on bedstraw or wild madder. The caterpillars are green, turning brown as they grow, with a bluish 'horn' at the back; they are seen in July and August, but then die of cold.

Mainly on south coast, but reaches far north.

J F M A M J J **A S O** N D

J F M A M J **J** **A** S O N D

J F M A M **J** **J** **A** S O N D

Tiger moths

The tiger moths get their name from their often brilliant colouring. Their 'woolly bear' caterpillars are protected by an irritant which discourages all birds except cuckoos from eating them. In some species the caterpillars hibernate, feeding again in spring; others overwinter as a chrysalis.

Garden tiger
Arctia caja
Wingspan 3in (76mm)

Fore-wings marked brown and white

Bright hind-wings – colours vary

Hind-wings exposed to startle predators

Hairy caterpillar is often seen in gardens

Cream-spot tiger
Arctia villica
Wingspan 2½in (64mm)

Cream spots on fore-wing

Pinky-orange, black-spotted hind-wings

Caterpillar has pink or red head

Plainer colour; wingspan 1½in (38mm)

Ruby tiger
Phragmatobia fuliginosa

Cinnabar
Tyria jacobaeae
Wingspan 1¾in (45mm)

Moth flies by day or night

Caterpillar has orange and black bands

Bright red markings

The markings and colouring of the spectacular garden tiger vary greatly from one moth to another; the hind-wings are usually orange with blue spots, but may be red, crimson or yellow. Its rather giraffe-like fore-wing patterning breaks up its outline when at rest, and the moth warns off predators by flashing its brilliant hind-wing colours, rubbing its wings together with a grating noise, or exuding a distasteful yellow fluid. The adult has no tongue and survives without feeding. The caterpillars are found in May and June, and again from September, on dead-nettles, lettuces and strawberry plants.

Common throughout British Isles.

The pale spots from which this moth gets its name vary greatly in size; on some specimens they almost cover the fore-wing and on others are completely absent, so that the wing is black. The moth scares off predators by revealing its orange black-spotted hind-wings and pinkish abdomen, signalling that it is poisonous. The dark caterpillar, covered in reddish-brown hair, feeds on weeds such as dock, dandelion and plantain from July, and also eats young gorse shoots.
The smaller and less colourful ruby tiger is common throughout Britain in May and June.

Southern coastal areas.

The elegant red and black cinnabar is another colourful member of the tiger moth family. It gets its name from the red patches on its fore-wings, the colour of vermilion or cinnabar pigment. The coloration warns predators that the moth is one of the most poisonous in Britain, inedible to most insect-eating creatures. The distastefulness begins early in its life. Cinnabar caterpillars – brightly banded in orange and black and active in July and August – feed on ragwort and groundsel, both of which contain poisons that do not harm the caterpillars but make them unpleasant to eat.

Common in southern England and Wales, locally elsewhere.

J F M A M J J A S O N D   J F M A M J J A S O N D   J F M A M J J A S O N D

Tiger moths

Common footman
Eilema lurideola
Wingspan 1½in (38mm)

Yellow border on grey-blue wing

Only one wing visible when at rest

Caterpillar has black and yellow hairs; feeds on lichens

Footman moths look very different from other members of the tiger moth family. At rest, the common footman shows one grey fore-wing, piped with yellow, overlapping its other wings and body. This gives it a smart, narrow outline – like a footman standing stiffly in his livery. In spite of its name, this moth is becoming less common. It lays eggs on the lichen-covered bark of oak, poplar and ash trees. But atmospheric pollution kills the lichens the caterpillars eat, so the moth has disappeared from many areas. The caterpillars hatch in September and then hibernate before pupating the following June.

In woodland, mainly in England and Wales.

White ermine
Spilosoma lubricipeda
Wingspan 1¾in (45mm)

Female has broader abdomen

Furry white thorax

Caterpillar is dark and very hairy

Male

White wings with black spots

Buff with black spots

Caterpillar hairy, striped red and yellowish

Buff ermine
Spilosoma lutea

This furry white moth would be inconspicuous on the ermine trimming of a coronation robe, but it stands out clearly when at rest on a leaf or tree-trunk. It is a night flier which sometimes gets into houses. Obviously ill-equipped for concealment, it plays dead when threatened by a predator; it falls to the ground and lies there, ejecting an unpleasant fluid from its thorax. The caterpillars, too, feign death when disturbed.

The buff ermine, a close relative, has no poison to defend itself but looks sufficiently like the white to deter predators. It is seen in June and July.

Throughout British Isles.

Scarlet tiger
Callimorpha dominula
Wingspan 2¼in (57mm)

White and orange spots

Metallic sheen on fore-wings

Caterpillar yellow and black, with white spots

Scarlet with black markings

White stripes on fore-wings

Wingspan 2⅜in (60mm)

Jersey tiger
Euplagia quadripunctaria

The colourful scarlet tiger is found mainly in damp meadows, clearings and fens, where it may be seen flying by day with a distinctive dipping flight pattern. Its black fore-wings, spotted with white and orange, have a metallic sheeen, while the scarlet hind-wings and abdomen, marked with black, scare off predators. The caterpillars feed mainly on weeds such as comfrey and stinging nettles, and hibernate when still young.

The Jersey tiger is less brightly coloured. It is found only in south Devon and the Channel Islands from June to September.

Southern England and south Wales.

J F M A M J J A S O N D

J F M A M J J A S O N D

J F M A M J J A S O N D

Woodland and hedgerow moths

Puss moth
Cerura vinula
Wingspan 3⅛in (80mm)

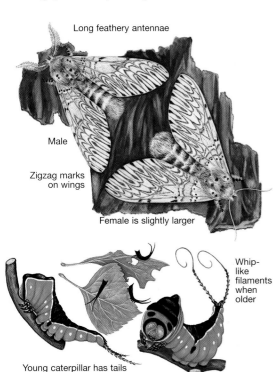

Long feathery antennae

Male

Zigzag marks on wings

Female is slightly larger

Whip-like filaments when older

Young caterpillar has tails

The name of this attractive grey to white moth comes from the fluffy hair that covers its body. The wings are covered in fine black zigzag markings. Puss moths favour damp areas in town and country where sallow and other willows grow; the moths lay their eggs on the leaves of these trees. The caterpillars hatch in July; as they grow they take on a spectacular appearance, coloured black and green, with red whip-like filaments or flagella protruding when danger threatens. They can also draw their head back into the thorax, showing a scarlet collar and two false eyes.

Common throughout British Isles.

Sallow kitten
Harpyia furcula
Wingspan 1¼in (32mm)

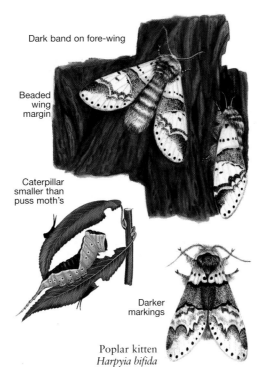

Dark band on fore-wing

Beaded wing margin

Caterpillar smaller than puss moth's

Darker markings

Poplar kitten
Harpyia bifida

The aptly named 'kittens' are smaller relatives of the puss moth, and generally similar both as caterpillars and as adults, though the wing markings are somewhat different. The caterpillar's main food plant is the grey willow, or grey sallow, one of the species that produce pussy-willow flowers in the spring. In southern England the sallow kitten has two generations a year, but only one farther north.
The poplar kitten is very similar to the sallow kitten, though somewhat larger – wingspan 1¾in (45mm). It lays its eggs on poplar leaves.

Common in English lowlands; scattered elsewhere.

Swallow prominent
Pheosia tremula
Wingspan 2⅜in (60mm)

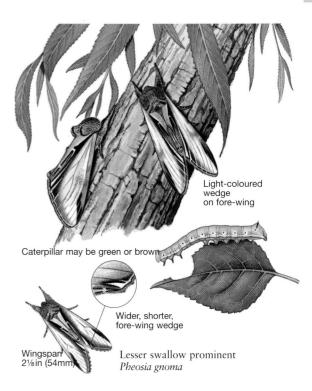

Light-coloured wedge on fore-wing

Caterpillar may be green or brown

Wider, shorter, fore-wing wedge

Wingspan 2⅛in (54mm)

Lesser swallow prominent
Pheosia gnoma

The so-called 'prominent' moths owe their name to wedge-shaped tufts of hair-like scales on the hind edge of the fore-wings; these stand up to form a small hump when the insect closes its wings. The swallow prominent is often attracted to light at night, and will fly in through open windows. In southern England it has two generations a year. It is mainly found in mixed woodland where its caterpillars feed on the leaves of sallow, willow, aspen and poplar.
The lesser swallow prominent is slightly smaller, with a wingspan of 2⅛in (54mm).

Widespread throughout British Isles.

303

Woodland and hedgerow moths

Coxcomb prominent

Ptilodon capucina
Wingspan 2in (50mm)

Faint
black lines
on wings

'Coxcomb'
projection
on fore-wing

Caterpillar bends
back head
when disturbed

A 'coxcomb' tuft of hairs on the trailing edge of the fore-wing gives this common woodland species its name; the tuft projects upwards when it is at rest. The moth has evolved along with such British native trees as birch, hazel and willow, and when it is resting on a tree the veined wings and projections give it the appearance of a dead leaf. Normally reddish, the coxcomb prominent can vary from yellow-brown to dark brown. Its caterpillar varies from pale yellow to dark green. If disturbed by a predator, it bends its head back over its body, breaking up the normal caterpillar outline.

Throughout England and Wales; scattered elsewhere.

Peach blossom

Thyatira batis
Wingspan 1½in (38mm)

Pink markings on
fore-wings

Camouflage
for life among
flowers

Caterpillar has
ridged back

The name of this pretty little woodland moth comes from the pale pink circular markings on its fore-wings. With this camouflage, it is hard to detect among the leaves of brambles and raspberries, where it lays its eggs. The caterpillar is as well camouflaged as the adult moth. Reddish-brown, and with a series of humps or ridges on its back, it looks like a twig. It pupates in a cocoon formed by drawing the edges of a bramble leaf together with silk. In warm years in the south there are two generations.

Throughout the south; locally elsewhere.

Red underwing

Catocala nupta
Wingspan 3⅜in (86mm)

Grey
caterpillar
feeds at
night

Drab fore-
wings are
camouflage

Brilliant
red and black
hind-wings

Wingspan 2¾in (70mm)
Hind-wings crimson-purple

Dark crimson underwing
Catocala sponsa

Drab, dappled fore-wings enable this large moth to conceal itself perfectly when at rest on the bark of a tree. The hind-wings, in contrast, are red and black – startling enough to scare predators when the moth raises its fore-wings to fly away. It flies mainly at night, and often feeds on the sap from a wound in the bark of a tree. The remarkable stick-like grey caterpillar hides by day in the crevices of a willow or poplar, coming out at night to feed on the leaves; it is seen from April to July.
The closely related dark crimson underwing is smaller and is probably restricted to the New Forest.

Common in southern and eastern England; rare elsewhere.

Many of the most familiar moths are woodland and hedgerow dwellers. Among them are those of the family Notodontidae, which includes the distinctive puss moth, the 'kitten' moths and the 'prominent' species which have a tuft of scales on the edge of the fore-wing. A rich variety of species inhabit oak woods.

Oak hook-tip
Drepana binaria
Wingspan 1in (25mm)

Oak beauty
Biston strataria
Wingspan to 2⅛in (54mm)

Oak eggar
Lasiocampa quercus
Wingspan to 3½in (90mm)

Green oak tortrix
Tortrix viridana
Wingspan ¾in (20mm)

Black dots on fore-wings

Hooked wingtips

Caterpillar raises both ends of body as aid to concealment

White speckled wings

Female

Male is smaller and has feathery antennae

Grey, stick-like caterpillar

Eggs are dropped in flight

Northern females

White spots on wings

Feathery antennae

Southern male

Caterpillar has orange hairs

White fringe on silvery-green wings; fold flat over body

Caterpillar curls oak leaf; when startled escapes by silk thread

Dark band on fore-wings

Wingspan 1½in (38mm)

Scalloped oak
Crocallis elinguaria

This small, unobtrusive yellowish-brown moth is one of five British hook-tip species, named from the shape of their fore-wing tips. The caterpillar has seven pairs of legs instead of the normal eight; it holds up both ends of its body to break up its outline and camouflage it from predators. The cocoon is formed inside two oak leaves drawn together with silk threads. The moth has no proper tongue, and takes no food during its two or three weeks of life.

Woods and hedgerows in England and Wales.

An early-flying moth, usually on the wing in March and April, the oak beauty is well camouflaged; at rest on the lichens and mosses on the bark of an oak tree, it is almost invisible. The male is slightly smaller than the female. The eggs, laid on the bark of oak, elm or birch, start to hatch in early May. The grey, stick-like caterpillars are hard to distinguish from twigs. They pupate by the middle of June, spending up to ten months as a chrysalis.

Widespread in Britain; rare in Ireland.

The various species of eggar moths may get their name from the egg-shaped cocoons their caterpillars spin. In Britain there are two sub-species of oak eggar, the northern being larger. The males, smaller and darker than the females, are fast fliers. Eggs are dropped at random while the female is in flight. Northern caterpillars feed mainly on heather; in the south they often eat oak and hawthorn leaves.

Hedges and waste ground in southern Britain, moors in north.

These dainty silvery-green moths rest by day, taking to the air when disturbed. The caterpillars, greyish-green with a black head, curl up oak leaves as they feed, using their own silk to bind them; they continue eating inside this shelter, unseen by predators. Large colonies can defoliate entire trees.
The scalloped oak, a pale buff moth, flies in July and August. Its eggs may also be laid on birch and other trees.

Widespread in southern England.

J F M A **M** J **J** A S O N D

J F **M** **A** M J J A S O N D

J F M A M J **J** **A** S O N D

J F M A M **J** **J** A S O N D

Woodland and hedgerow moths

Silk moth
Bombyx mori
Wingspan 1⅝in (42mm)

Only food plant mulberry – preferably white species

Silkworm can spin cocoon in one day

Cocoon has about ½ mile (800m) unbroken silk thread

Comb-like antennae

Short wings – unable to fly

Female lays hundreds of eggs on mulberry leaves or old cocoon

The silk moth was first domesticated in China about 5000 years ago, and no longer exists anywhere in the wild. The Chinese discovered how to spin the silk from which the moth's caterpillars, or silkworms, make their cocoons. After centuries of inbreeding, the moths can no longer fly and the 2in (50mm) long caterpillars can walk only short distances on the leaves of mulberry trees, their sole food. There is a silk farm in Dorset.

Only in captivity.

JFMAMJ**JA**SOND

Large emerald
Geometra papilionaria
Wingspan 1⅞in (48mm)

Caterpillar moves by looping its body

Caterpillar hatches brown, turns green with spring leaves

Caterpillar hibernates on twigs

Faint wavy lines on wing; green fades after death

Coloured like a green jewel, this moth flies at night in summer, and is attracted to lighted windows. The caterpillar feeds on the leaves of hazel, beech and other trees; it is a 'looper', moving its body by drawing it into a loop and then pushing the front end forward. Hatching in late summer, the caterpillars begin life brown and spend the winter hibernating on a twig. As the leaves turn green, so do they, thus preserving their camouflage.

All British Isles except northern Scotland.

JFMAMJ**JA**SOND

Yellow-tail tussock
Euproctis similis
Wingspan 1⅜in (35mm)

Caterpillar moves by looping its body

Grown caterpillar colourful

Male has marks on fore-wing

Yellow 'tail'

Female covers eggs with protective hairs

Brown-tail caterpillars striped; hairs very irritating to skin

This white moth is distinguished by its golden-yellow 'tail', or tussock, of hairs. Its caterpillars feed on a wide variety of trees and wild plants, including oak, hawthorn, apple and plum; they sometimes strip most of the leaves. They hatch in August and hibernate in a protective silk cover. The caterpillar's hairs can cause irritation to humans, but less than the brown-tail's. All tussocks are distasteful to most birds – but the cuckoo's gullet can deal with them.

Commonest in southern England.

JFMAMJ**JA**SOND

Hind-wings lighter than fore-wings

Caterpillar has yellow tufts

Pale tussock
Dasychira pudibunda

A pale grey moth with darker patterning over the wings and white hairs on the head. The caterpillars have tufts of hair distasteful to predators. Widespread in England and Wales, May and June. Wingspan up to 2¾in (70mm).

Brown tuft on abdomen

Brown-tail tussock
Euproctis chrysorrhoea

The caterpillar is a serious orchard pest, stripping leaves from trees. The adult flies at the same time as the similar yellow-tail. Mainly south and south-east. Wingspan 1½in (38mm).

Wingless female

Male has white spots

Caterpillar has tufts of hair

Vapourer moth
Orgyia antiqua

The female is wingless and spends all her life on her own cocoon, even laying her eggs there; the male's wingspan 1⅜in (35mm). Seen from July to October except in parts of Scotland and Ireland.

Brindled beauty
Lycia hirtaria
Wingspan 1½in (38mm)

Hairy head and thorax

Male has feathery antennae

Female

Grown caterpillar grey or reddish-brown – is a looper

Only male has wings – span 1¾in (45mm)

Pale brindled beauty
Apocheima polosaria

No longer common in city streets, the brindled beauty can still be found on silver birch in parks and on urban wasteland, as well as in the country. The name comes from the attractive mottled pattern on its wings, which can vary in colour. The grey or reddish-brown caterpillars feed on the leaves of most deciduous trees.
The pale brindled beauty is seen from January to March in most parts of Britain. The wingless female lives on tree-trunks.

Throughout British Isles, mostly in south.

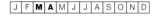

Lobster moth
Stauropus fagi
Wingspan 2¾in (70mm)

Caterpillar strikes 'lobster' pose for defence

Male has feathery antennae

Female

Hairy body

Fore-wings may be light or dark – inconspicuous against tree bark

When threatened by a predator, the bizarre-looking lobster moth caterpillar raises its head and rears up its claw-like tail section, giving itself a lobster-like appearance. As a second line of defence it can squirt formic acid. The moth is most commonly seen in deciduous woodland. There are two different forms – one with light fore-wings, the other with dark. Its caterpillars feed on the leaves of beech, birch, oak, hazel and other deciduous trees.

Southern Britain; south-west Ireland.

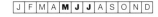

Peppered moth
Biston betularia
Wingspan to 2⅛in (54mm)

Female is longer

Lighter form, with dark speckles

Male

Twig-like caterpillar is green or brown

Male; wingspan 1⅝in (42mm)

Wingless female

Mottled umber
Erannis defoliaria

Named after the pepper-and-salt speckling on its wings, this moth occurs in two forms. The darker (*carbonaria*) evolved in the 19th century as a better camouflaged form in areas of industrial pollution. The caterpillars have warts like blemishes on a twig; they feed on deciduous tree leaves. The related mottled umber is common in hedgerows from October to March. The wingless female hatches on the ground and climbs up to a branch to mate.

Widespread throughout British Isles.

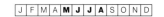

Kentish glory
Endromis versicolora
Wingspan to 3⅛in (80mm)

Caterpillar has cream stripes

Male's antennae are sensitive to scent

Female

Hairy body

In spite of its name, this large and striking moth is no longer found in Kent. Its caterpillars feed on the leaves of young silver birch, and as areas of suitable scrubland have been ploughed up, so its habitat has diminished. The males fly during the day; the larger females only at night. The male's feathery antennae enable it to detect a female's scent hundreds of yards away. The chrysalis may take as long as four years to develop.

Now only in northern Scotland.

307

Wayside moths

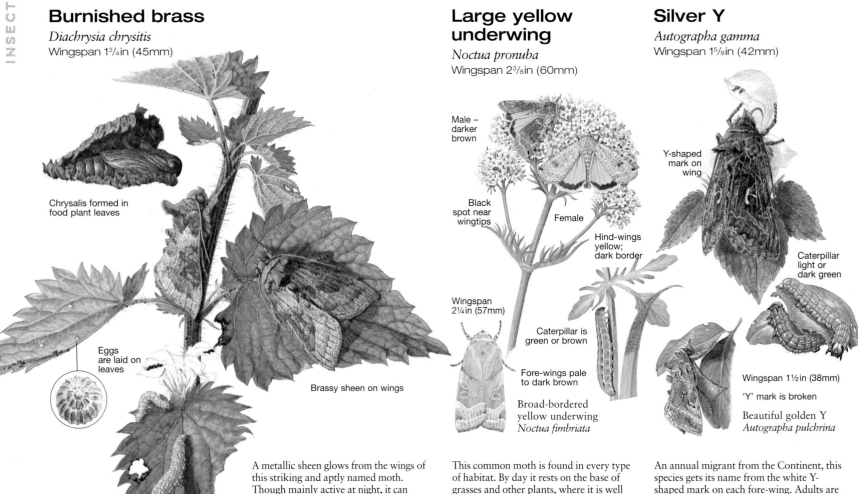

Burnished brass

Diachrysia chrysitis
Wingspan 1¾in (45mm)

Chrysalis formed in
food plant leaves

Eggs
are laid on
leaves

Brassy sheen on wings

Caterpillar
hibernates on
food plant

Large yellow
underwing

Noctua pronuba
Wingspan 2⅜in (60mm)

Male –
darker
brown

Black
spot near
wingtips

Female

Hind-wings
yellow;
dark border

Wingspan
2¼in (57mm)

Caterpillar is
green or brown

Fore-wings pale
to dark brown

Broad-bordered
yellow underwing
Noctua fimbriata

Silver Y

Autographa gamma
Wingspan 1⅝in (42mm)

Y-shaped
mark on
wing

Caterpillar
light or
dark green

Wingspan 1½in (38mm)

'Y' mark is broken

Beautiful golden Y
Autographa pulchrina

A metallic sheen glows from the wings of
this striking and aptly named moth.
Though mainly active at night, it can
sometimes be seen flying by day. Its main
food sources are stinging and dead-nettles
and other plants of ditches and wasteland.
The caterpillar is light green, marked with
dark green lines. When feeding it takes up a
humped posture. It hibernates in winter,
feeds again in spring and pupates in May.

*Widespread in England, Wales, parts of
Scotland.*

This common moth is found in every type
of habitat. By day it rests on the base of
grasses and other plants, where it is well
camouflaged; it flies at night to feed on the
nectar of flowers. The caterpillars, known
as cutworms, feed at night on low-growing
plants, including crops, gnawing through
stems at ground level.
The broad-bordered yellow underwing is
slightly smaller and lives mainly in woods.
The dark hind-wing border is wider.

Abundant throughout British Isles.

An annual migrant from the Continent, this
species gets its name from the white Y-
shaped mark on each fore-wing. Adults are
often seen in the day. The caterpillars feed
by night on low-growing plants, including
lettuces, and rest in crevices by day. There
are two generations a year.
The beautiful golden Y is found in similar
habitats and has the same food plants. Its
fore-wings are golden-orange, and the Y-
shaped mark is broken to form a V and dot.

Gardens and farmland everywhere.

J F M A M **J J A S** O N D

J F M A M J **J A S** O N D

J F M A M **J J A S** O N D

Along country lanes, urban roads and motorway embankments live moths which feed, from the caterpillar stage onwards, on wayside plants and grasses. Among the most spectacular are the burnished brass, with its metallic gleam, and the underwing moths, whose hind-wings flash brightly in flight.

Lappet
Gastropacha quercifolia
Wingspan 2¾in (70mm)

Blair's shoulder knot
Lithophane leautieri
Wingspan 1⅝in (42mm)

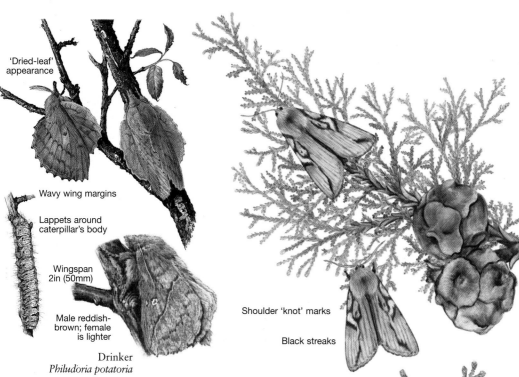

'Dried-leaf' appearance

Wavy wing margins

Lappets around caterpillar's body

Wingspan 2in (50mm)

Male reddish-brown; female is lighter

Drinker
Philudoria potatoria

Shoulder 'knot' marks

Black streaks

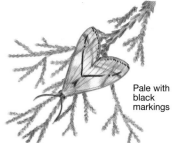

Pale with black markings

Cypress pug
Eupithecia phoeniceata

A Mediterranean moth, the cypress pug was first seen in Cornwall in 1959. Its caterpillar feeds on cypress shoots. It is seen in southern England over a very long flying season – from the end of May to October. Wingspan 1½in (38mm).

Varies from pale to deep yellow

Large thorn
Ennomos autumnaria

First sighted in Kent in 1855, this native of Europe and Asia is now found in various parts of south-east England and East Anglia, in September and October. The brown, twig-like caterpillars feed on various trees and shrubs. Wingspan to 2in (50mm).

The lappet's name comes from the flaps, or lappets, around the caterpillar's body; these hide its legs and give it a twig-like appearance. The moths are equally well camouflaged, with russet-coloured, crinkly edged wings, which when at rest look like dead leaves. The caterpillars feed on blackthorn, sloe, hawthorn and apple. The drinker is slightly smaller. It gets its name from the caterpillar's fondness for drinking dewdrops.

Southern England and Wales.

Several moth species have settled in Britain over the years – attracted by newly introduced trees or influenced by some change of climate. One of the most successfully established is Blair's shoulder knot, first seen in the Isle of Wight in 1951, having probably arrived from France. Its caterpillars feed on Monterey cypress, introduced from California, and juniper. The moth's name comes from markings like timber knots on its fore-wings.

In south, spreading northwards.

Caterpillars feed high in cypress trees

Varied coronet
Hadena compta

Dark brown, boldly patterned black and white

This boldly patterned moth was first recorded in Kent in 1948; it is also found in East Anglia. The yellow-brown caterpillar feeds mainly on sweet william, also the seeds of bladder campion. Adults fly in June and July. Wingspan 1⅛in (28mm).

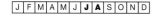

J F M A M J J A S O N D

J F M A M J J A S O N D

309

Garden moths

Heart markings on wing

Heart and dart
Agrotis exclamationis

This moth is found in gardens and fields and on waste ground. Its caterpillars feed on chickweed, cleavers (goosegrass) and other plants, and may attack turnips and potatoes. Common throughout the British Isles in May and June; a July-August second generation occurs in the south. Wingspan 1½in (38mm).

Elaborate wing pattern

Setaceous (bristly) wing edges

Setaceous Hebrew character
Xestia c-nigrum

A common moth throughout the British Isles, particularly in the south-east. Its fore-wings range from reddish-grey to purple-brown in colour. Caterpillars feed on many kinds of wayside weeds, including chickweed, dock and groundsel. Flies from May to July. Wingspan 1⅝in (42mm).

Brown markings on yellow wings

Brimstone moth
Opisthograptis luteolata

On the wing throughout the British Isles from April to October – having two generations a year – the brimstone is one of those moths that are attracted to lighted windows. Its twig-like caterpillars, found in both brown and green forms, feed on hawthorn and other shrubs. Wingspan 1⅜in (35mm).

'8' mark on fore-wing

Figure of eight
Diloba caeruleocephala

Named from the whitish marking on the fore-wing, this moth is common in southern and central England, favouring orchards, woods and hedges. The caterpillar – bluish-grey with black spots and yellow bands – feeds mainly on hawthorn and blackthorn. Flies October-November; wingspan 1½in (38mm).

Moth can look like dead leaf

Early thorn
Selenia dentaria

A widespread British moth, the early thorn has two generations a year. The caterpillars look like thorn twigs and feed on hawthorn and sloe. The moths, with wings folded, often resemble dead leaves. The two generations of adults fly in April and May and in July and August. Wingspan 1½in (38mm).

Broad butterfly-like wings

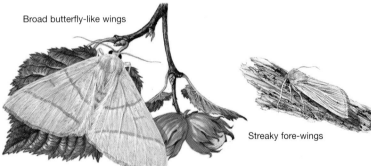

Swallow-tailed moth
Ourapteryx sambucaria

A large, broad-winged moth, it has a 'tail' on each hind-wing but this is normally obscured. It is common in gardens everywhere except northern Scotland, and is often seen around lights. The caterpillar feeds on hawthorn, ivy and privet. The adult flies in June and July. Wingspan 2⅜in (60mm).

Streaky fore-wings

Common wainscot
Mythimna pallens

All wainscot moths have plain-coloured, streaky fore-wings. This one is widespread and has two generations a year. The caterpillars feed at night on cock's-foot and other grasses. The adult moth is seen from June to October; a night flier, it is attracted to flowering grasses. Wingspan 1⅜in (35mm).

Wing pattern varies

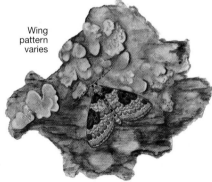

Garden carpet
Xanthorhoe fluctuata

This little moth, with variable and well-camouflaged wings, is common in gardens and at waysides throughout Britain. Its green caterpillars feed on cabbages, currant bushes, wallflowers and many other plants. There are two or more generations a year on the wing from April to October. Wingspan ⅞in (22mm).

Many delightful-looking moths come into the 'common or garden' category, but the feeding habits of their caterpillars are frequently less than delightful to the farmer or gardener. While some help by attacking weeds, others are a menace to fruit and vegetable crops. The codling moth, for instance, is a notorious apple pest.

Buff tip
Phalera bucephala
Wingspan 2½in (64mm)

Magpie moth
Abraxas grossulariata
Wingspan 1¾in (45mm)

Goat moth
Cossus cossus
Wingspan to 3½in (90mm)

Codling moth
Cydia pomonella
Wingspan ½in (13mm)

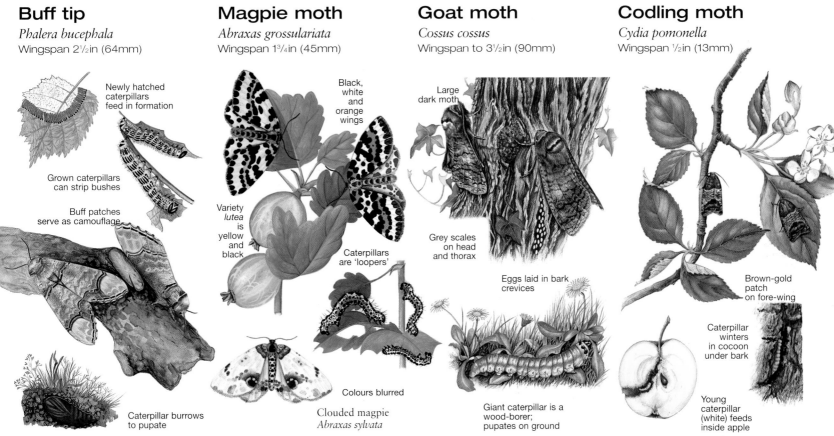

Buff tip — Newly hatched caterpillars feed in formation. Grown caterpillars can strip bushes. Buff patches serve as camouflage. Caterpillar burrows to pupate.

Magpie moth — Black, white and orange wings. Variety *lutea* is yellow and black. Caterpillars are 'loopers'. Colours blurred. Clouded magpie *Abraxas sylvata*.

Goat moth — Large dark moth. Grey scales on head and thorax. Eggs laid in bark crevices. Giant caterpillar is a wood-borer; pupates on ground.

Codling moth — Brown-gold patch on fore-wing. Caterpillar winters in cocoon under bark. Young caterpillar (white) feeds inside apple.

When the moth is at rest on the bark of a tree the mottled buff ends of its fore-wings look like the broken ends of twigs. The caterpillars cause serious damage to foliage. They attack limes and poplars in cities; oak, sallow, aspen, birch and hazel in woodland; and apple, cherry and cobnut in orchards. When newly hatched in July, the caterpillars eat together in formation; as they grow, they go on their own way, pupating below their food tree.

Common in England and Wales, local elsewhere.

This brightly coloured moth has been nicknamed the currant moth, as its caterpillars can be a serious pest of currant bushes. They will also attack gooseberry bushes and apricot trees, while in the countryside they live on hazel and other common hedgerow plants. The caterpillars match the orange, white and black colouring of the adults, and are 'loopers' – progressing by looping their bodies. The clouded magpie's wings bear a lighter, blurred version of the magpie's colouring.

Widespread throughout British Isles.

The adult goat moth has no tongue for feeding and lives for only about a month on the energy reserves built up in its large caterpillar stage. This has an unpleasant goat-like smell, from which the name derives. It is up to 4in (10cm) long and ½in (13mm) wide. One of the largest wood-boring insects in Britain, it burrows into oak, poplar and other trees. As many as 20 or 30 caterpillars live inside a single tree, where they may spend up to four years. They are a favourite food of woodpeckers.

Woodlands and orchards of British Isles.

Identifiable by brown-gold patches on the fore-wings, this little moth is inconspicuous in its adult form. But it is all too well known as a caterpillar for the way it burrows into apples, leaving an entry hole full of brown droppings. It gets its name from the codling or codlin, one of the oldest apple varieties. On leaving the fruit it spends the winter in a cocoon under the bark, where it changes into a chrysalis and hatches in early summer. As well as apples, it will attack quince and walnut.

Orchards throughout British Isles.

JFMAM**J****J**ASOND   JFMAM**J****J****A**SOND   JFMAM**J****J****A**SOND   JFMAM**J****J****A**SOND

Garden moths

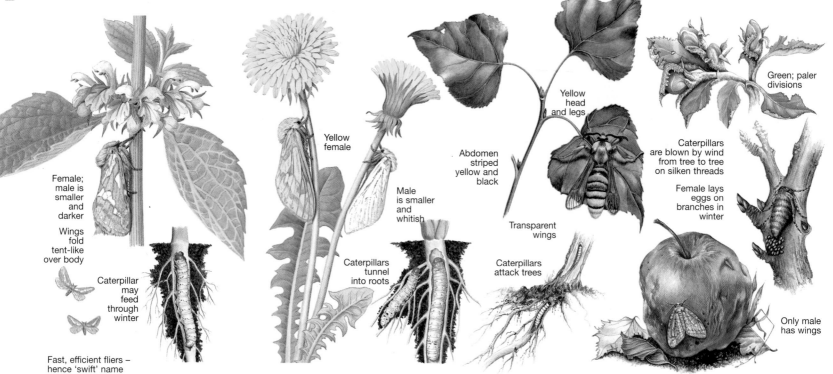

Common swift
Hepialus lupulinus
Wingspan to 1½in (38mm)

Female; male is smaller and darker

Wings fold tent-like over body

Caterpillar may feed through winter

Fast, efficient fliers – hence 'swift' name

Ghost swift
Hepialus humuli
Wingspan 2in (50mm)

Yellow female

Male is smaller and whitish

Caterpillars tunnel into roots

Hornet moth
Sesia apiformis
Wingspan 1⅝in (42mm)

Yellow head and legs

Abdomen striped yellow and black

Transparent wings

Caterpillars attack trees

Winter moth
Operophtera brumata
Male wingspan 1in (25mm)

Green; paler divisions

Caterpillars are blown by wind from tree to tree on silken threads

Female lays eggs on branches in winter

Only male has wings

Also known as the garden swift, this small mottled, brownish moth is no friend of growers. Its caterpillars attack the roots and stems of barley and wheat, eat lettuces and other market-garden crops, and burrow into garden bulbs. Male common swifts fly at dusk, swooping over pasture in search of females. The female, larger and paler than the male, lays her eggs in flight. Caterpillars hibernate in the roots of plants except in mild winters.

Farms and gardens throughout British Isles.

The ghost swift is similar in many respects to the closely related common swift – and is equally destructive; up to 50,000 caterpillars have been found on a single acre of land. Named from the male's habit of hovering over churchyards and damp meadows at dusk, it has an erratic, dipping flight. Females are yellow with reddish markings, while the male has thin, white-powdered wings. Reversing the usual roles, males release a scent to attract females.

All over British Isles, often in wet grass.

In spite of its ferocious, hornet-like appearance, this yellow and black day-flying moth is completely harmless. But its forbidding looks deter birds from attacking it. Unlike the true hornet (p. 325), it has no maroon on its abdomen. It can sometimes be seen resting on the bark of a tree. The caterpillar, yellowish-white with a reddish-brown head, bores into the bark (or sometimes deeper) of willows, aspens and poplars. Its burrow is very narrow.

Scarce; mainly eastern England.

Unlike most moths, which fly during the summer, the male winter moth is on the wing from October to January in gardens, orchards and scrubland. The females are wingless and climb up the trunks of trees to mate and lay their eggs. Before the use of chemical pesticides, the caterpillars were serious orchard pests; fruit-growers put grease bands around their trees to catch the climbing females. The caterpillars can defoliate oak trees and blackthorn bushes.

Throughout British Isles.

J F M A M J J A S O N D   J F M A M J J A S O N D   J F M A M J J A S O N D   J F M A M J J A S O N D

Day-flying moths

Often confused with butterflies because of their bright colours, moths that fly by day tend to rely on methods of defence other than camouflage. Some, like the burnets, are highly poisonous; others trust to speedy flight.

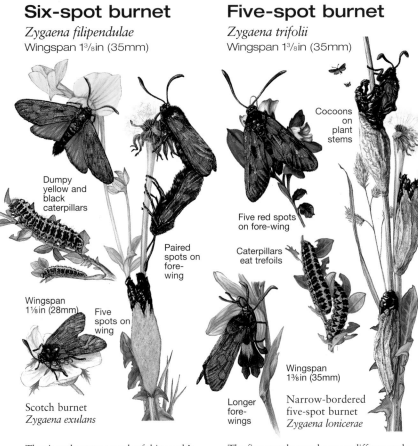

Six-spot burnet
Zygaena filipendulae
Wingspan 1³⁄₈in (35mm)

Dumpy yellow and black caterpillars

Paired spots on fore-wing

Wingspan 1¹⁄₈in (28mm)

Five spots on wing

Scotch burnet
Zygaena exulans

Five-spot burnet
Zygaena trifolii
Wingspan 1³⁄₈in (35mm)

Cocoons on plant stems

Five red spots on fore-wing

Caterpillars eat trefoils

Longer fore-wings

Wingspan 1³⁄₈in (35mm)

Narrow-bordered five-spot burnet
Zygaena lonicerae

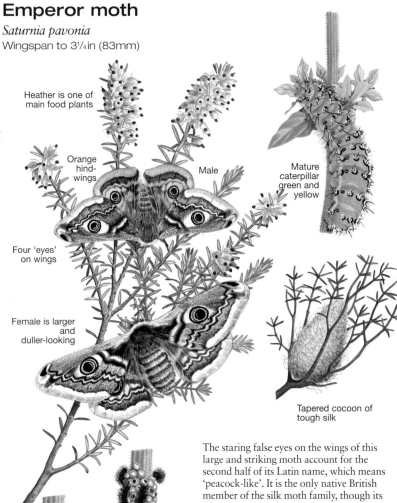

Emperor moth
Saturnia pavonia
Wingspan to 3¹⁄₄in (83mm)

Heather is one of main food plants

Orange hind-wings

Male

Four 'eyes' on wings

Female is larger and duller-looking

Mature caterpillar green and yellow

Tapered cocoon of tough silk

Eggs laid around plant stem

Caterpillars hatch black and hairy

The six red spots on each of this moth's greenish-black fore-wings are arranged in three distinct pairs. The main food plants of the caterpillars are trefoil and vetch, which contain cyanide derivatives. Passed on from the caterpillar stage to the mature moth, these poisons deter predatory birds, which learn to avoid the moths as food. The smaller Scotch (or mountain) burnet is rare and found only in the Aberdeen area. Its semi-transparent wings have five spots.

Meadows and downland everywhere.

The five-spot burnet has two different sub-species: *palustrella*, found on chalk downs in south-east England in May and June; and the rather larger *decreta*, on the wing in July and early August in marshy areas of the west and Wales. The five spots are arranged in two pairs, plus one at the back. A close relative, the narrow-bordered five-spot burnet, has longer fore-wings and more pointed hind-wings. It is found in June and July.

Southern England, Wales and Isle of Man.

The staring false eyes on the wings of this large and striking moth account for the second half of its Latin name, which means 'peacock-like'. It is the only native British member of the silk moth family, though its silk has no commercial value. The male can be seen on the wing by day, but the female flies only at night. The male has perhaps the most highly developed sense of smell of any moth – able to detect a female more than 6 miles (10km) away.

Heath and moorland of British Isles.

Day-flying moths

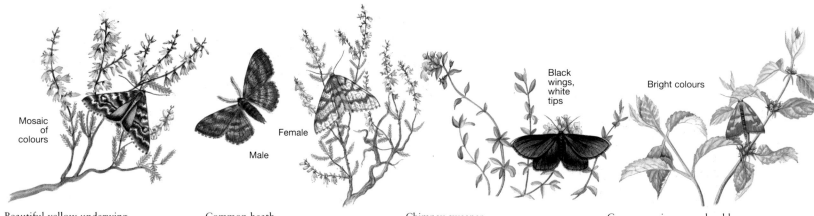

Mosaic of colours

Female

Male

Black wings, white tips

Bright colours

Beautiful yellow underwing
Anarta myrtilli

This pretty little moth feeds on the nectar of wild flowers and is on the wing from late April to August. Its caterpillar eats heather leaves, so it is found only on heathland and moors. Wingspan 1⅛in (28mm)

Common heath
Ematurga atomaria

Found on open habitats from April to June throughout most of Britain. Its wing pattern is variable. Caterpillars feed on heather, broom and clover. Wingspan 1⅛in (28mm).

Chimney sweeper
Odezia atrata

A distinctive little black moth with white wing-tips, seen mainly in meadows in June and July in central and northern England, Wales and Scotland, rarely in southern England. Caterpillars feed on pignut foliage. Wingspan 1in (25mm).

Common crimson and gold
Pyrausta purpuralis

The caterpillars of this cheerfully patterned little moth feed on mint and thyme. It has two generations a year, seen from May to August throughout Britain on hillsides and marshes. Wingspan ⅞in (22mm).

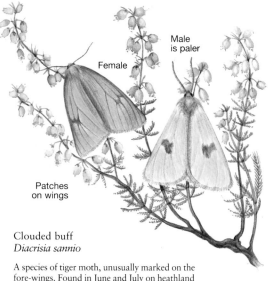

Female

Male is paler

Patches on wings

Clouded buff
Diacrisia sannio

A species of tiger moth, unusually marked on the fore-wings. Found in June and July on heathland in Surrey and Dorset and moors in western Scotland, it lays its eggs on bell heather. Wingspan 1⅝in (42mm).

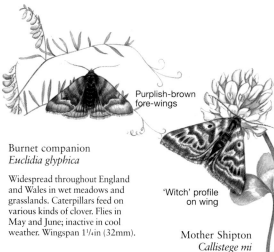

Purplish-brown fore-wings

Burnet companion
Euclidia glyphica

Widespread throughout England and Wales in wet meadows and grasslands. Caterpillars feed on various kinds of clover. Flies in May and June; inactive in cool weather. Wingspan 1¼in (32mm).

'Witch' profile on wing

Mother Shipton
Callistege mi

The wing pattern was thought to resemble the hook-nosed profile of the 16th-century Yorkshire witch. Frequents sunny banks, meadows and railway cuttings in May and June, often alongside the burnet companion. Caterpillars eat clover and melilot. Wingspan 1⅜in (35mm).

Dappled wings

Speckled yellow
Pseudopanthera macularia

Often seen flitting along woodland clearings in June and July, this dappled moth lays its eggs on wood sage. Found mainly in southern England, northern Wales and Ireland. Wingspan 1³⁄₁₆in (30mm).

Dragonflies

Dragonflies belong (with damselflies) to the Odonata order of insects. They have whirring transparent wings and brilliantly coloured bodies. There are two groups: the long-bodied 'hawkers' and the squatter 'darters'. All feed on smaller insects.

Emperor dragonfly

Anax imperator
Wingspan 4¹/₈in (10.5cm)

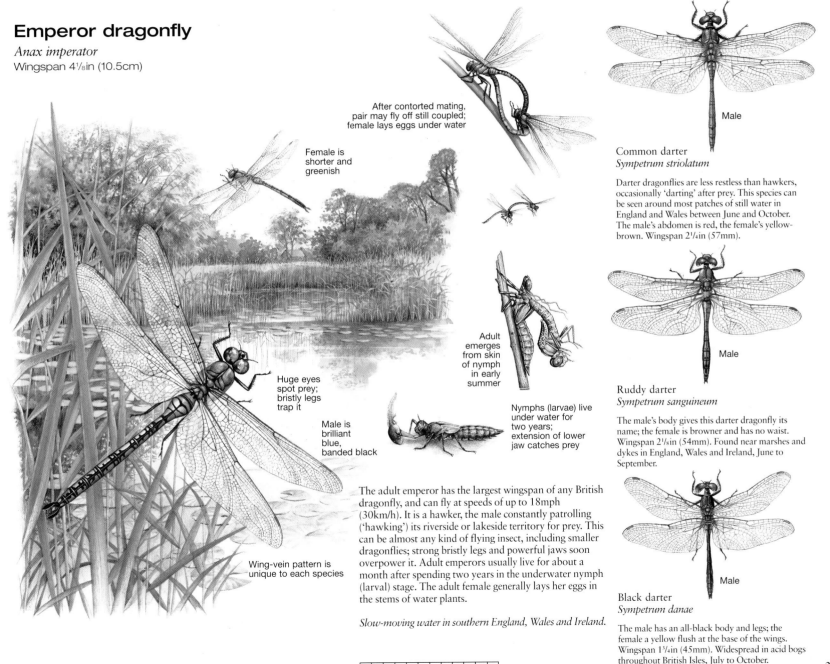

After contorted mating, pair may fly off still coupled; female lays eggs under water

Female is shorter and greenish

Adult emerges from skin of nymph in early summer

Huge eyes spot prey; bristly legs trap it

Male is brilliant blue, banded black

Nymphs (larvae) live under water for two years; extension of lower jaw catches prey

Wing-vein pattern is unique to each species

The adult emperor has the largest wingspan of any British dragonfly, and can fly at speeds of up to 18mph (30km/h). It is a hawker, the male constantly patrolling ('hawking') its riverside or lakeside territory for prey. This can be almost any kind of flying insect, including smaller dragonflies; strong bristly legs and powerful jaws soon overpower it. Adult emperors usually live for about a month after spending two years in the underwater nymph (larval) stage. The adult female generally lays her eggs in the stems of water plants.

Slow-moving water in southern England, Wales and Ireland.

Male

Common darter
Sympetrum striolatum

Darter dragonflies are less restless than hawkers, occasionally 'darting' after prey. This species can be seen around most patches of still water in England and Wales between June and October. The male's abdomen is red, the female's yellow-brown. Wingspan 2¹/₄in (57mm).

Male

Ruddy darter
Sympetrum sanguineum

The male's body gives this darter dragonfly its name; the female is browner and has no waist. Wingspan 2¹/₈in (54mm). Found near marshes and dykes in England, Wales and Ireland, June to September.

Male

Black darter
Sympetrum danae

The male has an all-black body and legs; the female a yellow flush at the base of the wings. Wingspan 1³/₄in (45mm). Widespread in acid bogs throughout British Isles, July to October.

| J | F | M | A | M | J | J | A | S | O | N | D |

315

Dragonflies

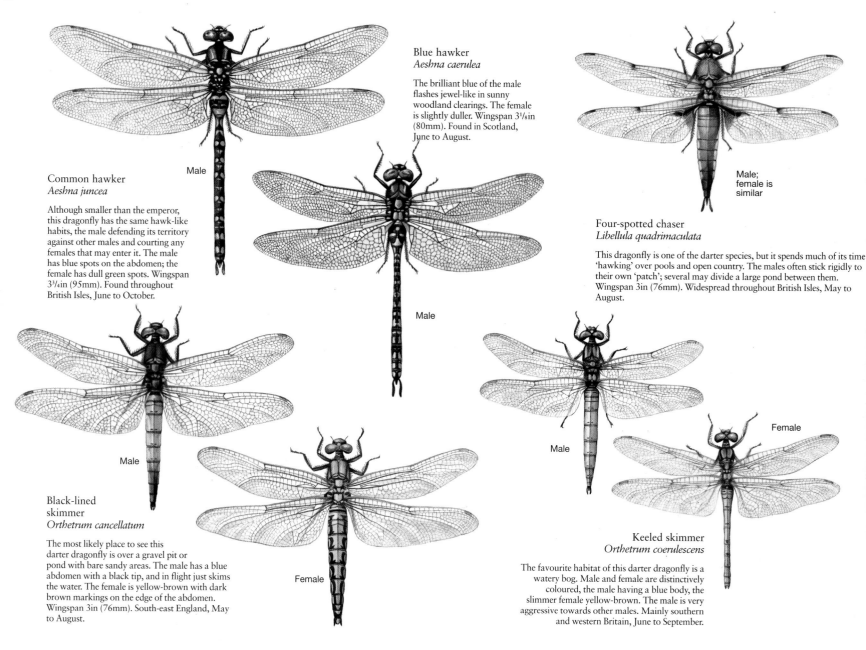

Blue hawker
Aeshna caerulea

The brilliant blue of the male flashes jewel-like in sunny woodland clearings. The female is slightly duller. Wingspan 3¹/₈in (80mm). Found in Scotland, June to August.

Male

Common hawker
Aeshna juncea

Although smaller than the emperor, this dragonfly has the same hawk-like habits, the male defending its territory against other males and courting any females that may enter it. The male has blue spots on the abdomen; the female has dull green spots. Wingspan 3³/₄in (95mm). Found throughout British Isles, June to October.

Male

Male; female is similar

Four-spotted chaser
Libellula quadrimaculata

This dragonfly is one of the darter species, but it spends much of its time 'hawking' over pools and open country. The males often stick rigidly to their own 'patch'; several may divide a large pond between them. Wingspan 3in (76mm). Widespread throughout British Isles, May to August.

Male

Female

Black-lined skimmer
Orthetrum cancellatum

The most likely place to see this darter dragonfly is over a gravel pit or pond with bare sandy areas. The male has a blue abdomen with a black tip, and in flight just skims the water. The female is yellow-brown with dark brown markings on the edge of the abdomen. Wingspan 3in (76mm). South-east England, May to August.

Female

Male

Keeled skimmer
Orthetrum coerulescens

The favourite habitat of this darter dragonfly is a watery bog. Male and female are distinctively coloured, the male having a blue body, the slimmer female yellow-brown. The male is very aggressive towards other males. Mainly southern and western Britain, June to September.

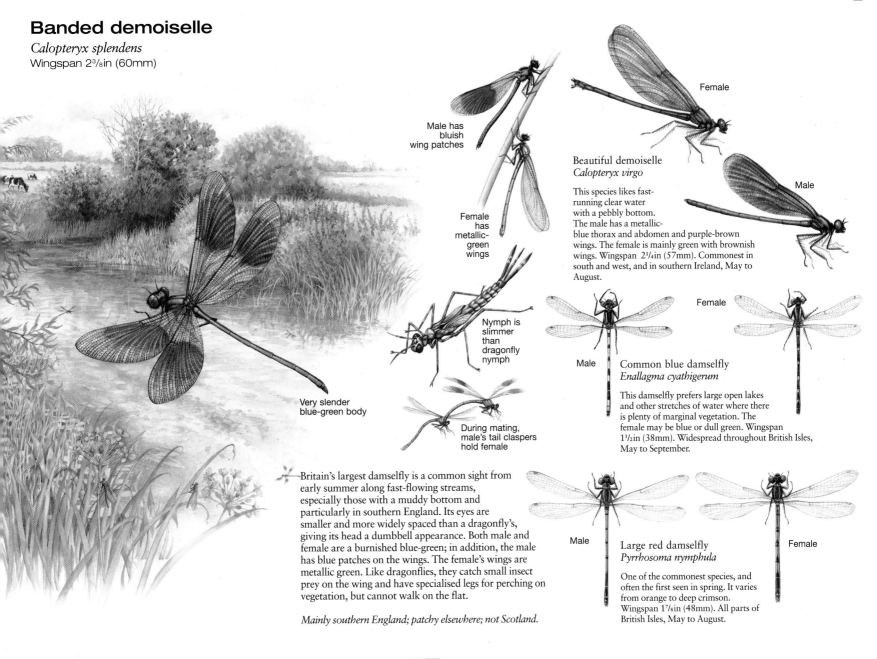

Damselflies

Damselflies are smaller and more delicate than dragonflies, with a weaker, more fluttering mode of flight. Their fore and hind-wings are more or less equal in size, whereas the dragonfly's hind-wings are wider. At rest, damselflies may hold their wings up; dragonflies always spread them.

Banded demoiselle

Calopteryx splendens
Wingspan 2³/₈in (60mm)

Male has bluish wing patches

Female has metallic-green wings

Nymph is slimmer than dragonfly nymph

Very slender blue-green body

During mating, male's tail claspers hold female

Britain's largest damselfly is a common sight from early summer along fast-flowing streams, especially those with a muddy bottom and particularly in southern England. Its eyes are smaller and more widely spaced than a dragonfly's, giving its head a dumbbell appearance. Both male and female are a burnished blue-green; in addition, the male has blue patches on the wings. The female's wings are metallic green. Like dragonflies, they catch small insect prey on the wing and have specialised legs for perching on vegetation, but cannot walk on the flat.

Mainly southern England; patchy elsewhere; not Scotland.

Female

Male

Beautiful demoiselle
Calopteryx virgo

This species likes fast-running clear water with a pebbly bottom. The male has a metallic-blue thorax and abdomen and purple-brown wings. The female is mainly green with brownish wings. Wingspan 2¹/₄in (57mm). Commonest in south and west, and in southern Ireland, May to August.

Female

Male

Common blue damselfly
Enallagma cyathigerum

This damselfly prefers large open lakes and other stretches of water where there is plenty of marginal vegetation. The female may be blue or dull green. Wingspan 1¹/₂in (38mm). Widespread throughout British Isles, May to September.

Male

Female

Large red damselfly
Pyrrhosoma nymphula

One of the commonest species, and often the first seen in spring. It varies from orange to deep crimson. Wingspan 1⁷/₈in (48mm). All parts of British Isles, May to August.

Aquatic insects

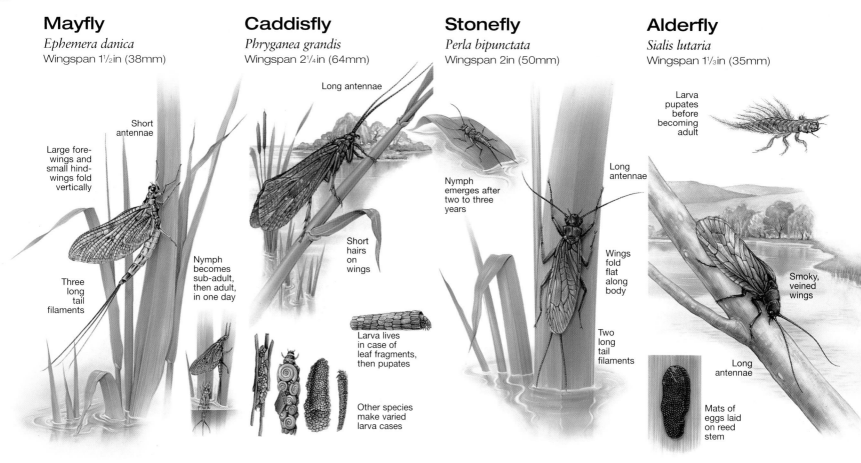

Mayfly
Ephemera danica
Wingspan 1½in (38mm)

Short antennae

Large fore-wings and small hind-wings fold vertically

Three long tail filaments

Nymph becomes sub-adult, then adult, in one day

Caddisfly
Phryganea grandis
Wingspan 2¼in (64mm)

Long antennae

Short hairs on wings

Larva lives in case of leaf fragments, then pupates

Other species make varied larva cases

Stonefly
Perla bipunctata
Wingspan 2in (50mm)

Nymph emerges after two to three years

Long antennae

Wings fold flat along body

Two long tail filaments

Alderfly
Sialis lutaria
Wingspan 1⅓in (35mm)

Larva pupates before becoming adult

Smoky, veined wings

Long antennae

Mats of eggs laid on reed stem

Like the heroine of a romantic Russian ballet, a mayfly emerges from its larval state on a fine summer morning and mates in the evening; then the female drops her eggs in the water, and both male and female die – all in one day. Before this dramatic climax to its life, the mayfly lives for two years as a nymph in water, breathing through gills on its back. *Ephemera danica* is the largest of the 46 species of mayfly found in Britain.

Clear water throughout British Isles.

Caddisflies are rather like small moths: they have short hairs on their wings and fly mostly at night. There are 190 or so species in Britain, of which *Phryganea grandis* is the biggest. Most are brownish-grey in colour. Adults rarely live much longer than a week. Except for one species, they spend their larval stage in water, usually within a protective case they build from pieces of leaf, shell or other materials. The larvae are part of the diet of many fish.

Still or very slow-moving water.

Stoneflies are secretive creatures, spending most of their two or three weeks of adult life hiding among stones and vegetation beside streams. The 34 British species are rarely found in still water. A stonefly spends two to three years as a nymph, during which time mayfly nymphs provide an important food source. Like the mayfly, the adult is a poor flier, and usually crawls. *Perla bipunctata* is one of the largest species.

Clear, fast-flowing streams mainly in north and west.

The alderfly resembles a caddisfly, but its wings are more strongly veined and free of hairs, and they fold tent-wise over the heavy body. It is a poor flier and is most often seen clinging motionless to a reed frond or stem – rarely, despite its name, on an alder tree – making brief flights mainly at dusk. Eggs are laid in dense mats of up to 2000 each on reeds or other water plants. The larvae spend about two years at the bottom of a lake or river.

Lakes and rivers throughout British Isles.

J F M A M J J A S O N D

J F M A M J J A S O N D

J F M A M J J A S O N D

J F M A M J J A S O N D

These insects fall into two main groups: those, like the mayfly, that live near water as adults and whose nymphs are aquatic – as with dragonflies and damselflies (pp. 315-17) – and the pond skaters and similar bugs and beetles that live on the water surface. For underwater insects, see pp. 376-9.

Giant lacewing

Osmylus fulvicephalus
Wingspan 2in (50mm)

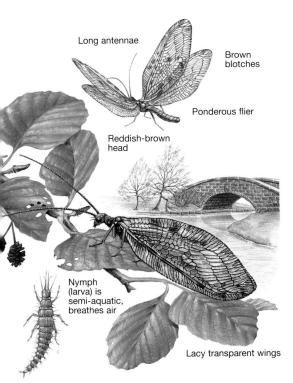

Long antennae

Brown blotches

Ponderous flier

Reddish-brown head

Nymph (larva) is semi-aquatic, breathes air

Lacy transparent wings

Lacewings are aptly named. Their large transparent wings are delicately netted with veins, those of the giant species being distinctively spotted. It is not a good flier, and usually rests by day, often in large groups, on waterside foliage or the underside of bridges, hidden from birds. In the evening it flies along the riverbank in search of mates and egg-laying sites – and small insects to eat. The larvae live in shallow water or wet moss, breathing air through body openings called spiracles. They hibernate, then pupate in spring.

Banks of unpolluted waterways; rare in Scotland.

Pond skater

Gerris lacustris
Length 1/2-5/8in (13-16mm)

Steers with rear legs

Narrow body, brown or black

'Rows' with middle legs

Catches prey with front legs

Hairs and position of claws aid skimming on water

Nymph resembles adult but has no wings

The pond skater is a member of a specialised group of insects – mostly true bugs, but including some beetles – that are adapted to live on the surface of water. Four of its six legs have hairy pads that trap air but repel water, so that it can move about without breaking the surface film. It feeds mainly on dead and dying insects that fall onto the water, using its sharp proboscis to suck their body juices. The female lays eggs under water in jellied clusters in spring. The nymphs (larvae) that emerge moult five times before becoming adults; these, unlike the nymphs, can fly. They hibernate over winter.

Common on still and slow-moving water.

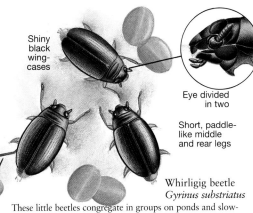

Shiny black wing-cases

Eye divided in two

Short, paddle-like middle and rear legs

Whirligig beetle
Gyrinus substriatus

These little beetles congregate in groups on ponds and slow-flowing rivers, swimming rapidly in small circles at amazing speed. They are scavengers, eating other small insects that fall onto the water. The lower part of the whirligig's eyes is specially adapted for underwater vision. Length 1/4in (6mm). Throughout British Isles, March to October.

Brownish stout body

Legs about same length

Water cricket
Velia caprai

No close relation of the terrestrial cricket despite its name. It is wingless and stouter-bodied than other water-skimming bugs, but like them feeds on small creatures that fall onto the water, and may also devour mosquito larvae. Commonly seen on ponds and streams, August and September. Length to 5/16in (8mm).

Water measurer
Hydrometra stagnorum

A slender bug with an elongated head and long antennae that resemble a fourth pair of legs. Unlike most skimmers, it catches small prey just below the water surface – including water fleas and mosquito larvae. It will jump if disturbed. Length 3/8in (10mm). Still and slow-flowing water, June to September.

Predators and scavengers

Insects fill almost every ecological niche, adapting to all kinds of food sources. Some feed on other insects, on dead and decaying animal matter – often spreading disease in the process – or on plants.

Green lacewing

Chrysopa carnea
Wingspan 1³/₁₆in (30mm)

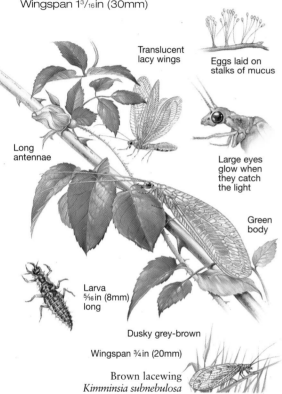

Translucent lacy wings

Eggs laid on stalks of mucus

Long antennae

Large eyes glow when they catch the light

Green body

Larva ⁵/₁₆in (8mm) long

Dusky grey-brown

Wingspan ¾in (20mm)

Brown lacewing
Kimminsia subnebulosa

Scorpionfly

Panorpa communis
Wingspan 1¼in (32mm)

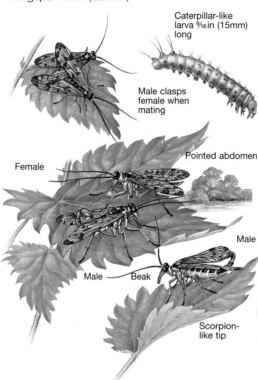

Caterpillar-like larva ⁹/₁₆in (15mm) long

Male clasps female when mating

Female

Pointed abdomen

Male

Male   Beak

Scorpion-like tip

Common cockroach

Blatta orientalis
To 1¹/₁₆in (27mm) long

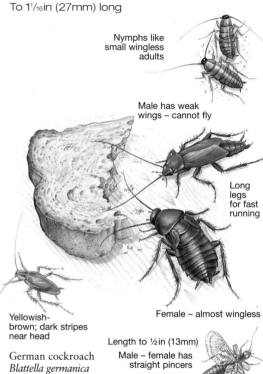

Nymphs like small wingless adults

Male has weak wings – cannot fly

Long legs for fast running

Female – almost wingless

Yellowish-brown; dark stripes near head

German cockroach
Blattella germanica

Length to ½in (13mm)
Male – female has straight pincers

Earwig
Forficula auricularia

The wings of the green lacewing are translucent and delicately veined like leaves. They are rarely stretched in flight during the day, which the insect spends mainly on low vegetation, consuming aphids and the nectar of flowers. The larvae have huge jaws with which they pierce the soft body of aphids and other insects and suck their tissues. Unlike most other lacewing species – there are 14 green and 29 brown in Britain – it hibernates as an adult, often spending the winter inside buildings, where its colour changes to a drab pinkish-brown. The brown lacewing is smaller and also widespread.

Widespread throughout British Isles, especially in south.

Although it is shaped like a scorpion's tail, the distinctive reddish up-curved tip of the male scorpionfly's abdomen is not a sting. At the very tip is a claw which is used to grasp the female while mating. This is a hazardous business for the male. Unless the female is distracted she is likely to attack and kill him. So he usually starts by offering her a diverting drop of his own saliva, placed on a leaf near her beak-like jaws. Scorpionflies are fond of shady hedgerows, nettle beds and bramble thickets. They feed mainly on dead insects, often the left-overs in a spider's web.

Common throughout British Isles.

The commonest cockroaches in the British Isles are not native to these islands at all, but arrived on ships hundreds of years ago from warmer climates. They live in buildings and rubbish tips everywhere, feeding on food scraps and dead animal matter – even dead cockroaches – and spreading bacterial disease. The ½in (13mm) long German cockroach (actually from North Africa) can fly, but the common species (or 'black beetle') from Asia cannot. There are three native cockroaches, inhabitants of woods and heaths in southern England.
Earwigs, common garden pests, like to nibble flowers.

Throughout British Isles, but not in natural habitats.

Flies

Out of some 85,000 known species of flies, a staggering 5200 types live in the British Isles, and a selection representing the main groups is shown. Most are found over a wide area and constitute some of our worst pests.

Black bean aphid
Aphis fabae

Often called blackfly and one of the commonest 'plant lice' that suck the sap of young plant shoots – especially broad beans. They excrete sugary honeydew which is 'milked' by ants. Common, May to July; through winter on spindle. Length to ¹/₈in (3mm).

Females multiply asexually

Cluster thickly in spring

Rose aphid
Macrosiphum rosae

The all-too-familiar greenfly which infests rose bushes in the spring. In summer, it migrates to other plants such as scabious and teasels. Common everywhere, April to October. Length ¹/₈in (3mm).

Green leafhopper
Cicadella viridis

Narrow, green

The green leafhopper is so well camouflaged that it is hard to spot until it jumps from one plant to another. A true bug, it has needle-like mouth-parts for piercing plants to suck the sap. Common throughout British Isles, July to October. Length ¹/₄-⁵/₁₆in (6-8mm).

Striking colours

Black and red froghopper
Cercopis vulnerata

These strikingly coloured tiny insects can be seen clinging to grass and low vegetation. The nymphs live underground, feeding on roots. Commonest in southern and central England, April to August. Length ³/₈in (10mm).

Brilliant green body

Greenbottle
Lucilia caesar

One of a large group of metallic-green flies, related to blowflies, that lay their eggs on meat and dead animals. Common all year. Length ³/₈in (10mm).

Common bluebottle
Calliphora vomitoria

Stout body

The blowfly, best known for the loud buzzing noise the female makes when searching the kitchen looking for a piece of meat on which to lay her eggs. The male normally stays outside. Common everywhere; seen all year. Length ⁷/₁₆in (11mm).

Housefly
Musca domestica

Wherever man goes, the common housefly invariably follows. It is a familiar sight in almost all parts of the world. Eggs are laid in rubbish dumps and manure heaps, and produce adult flies within two weeks, which then join man indoors. Length ⁵/₁₆in (8mm).

Self-grooming spreads dirt

Hoverfly
Syrphus ribesii

One of some 250 species of bee and wasp-like hoverflies, its brilliant markings make this species conspicuous. Its larvae consume huge quantities of aphids. Adults seen April to November, all areas. Length ¹/₂in (13mm).

Mimics wasp

Form dancing swarms, particularly in winter

Winter gnat
Trichocera annulata

A small relative of the daddy long-legs, this is one of ten British species of gnat. Unlike the daddy long-legs it does not readily shed its legs and swarms much of the year. Length ¹/₄in (6mm).

Daddy long-legs
Tipula paludosa

The most common cranefly in Britain, easily recognised with its thin body, narrow wings and very long legs. The legs break off if seized by a predator and do not re-grow. April to September. Length ³/₄in (20mm). Its larvae – leatherjackets – are serious soil pests.

Grey-brown

Mosquito
Culiseta annulata

Flies at night

One of the largest of all mosquitoes, with black and white bands on its legs. It does not carry disease, though its bite can be painful. It is among 30 mosquito species found in Britain. Common in all areas, often indoor October to February. Length ⁵/₁₆in (8mm).

Multicoloured body

Tachinid fly
Echinomyia fera

The larva hatches and lives inside the body of a caterpillar, which it gradually consumes but finally kills only when it is fully grown and ready to pupate. A common fly, most often found on flowers near ponds and streams, August to October. Length ⁵/₈in (16mm).

Long legs; striped body

Spotted cranefly
Nephrotoma appendiculata

The slender body and black and yellow colouring are sufficient to identify this cranefly, which is commonly found in fields and gardens, May to August. It also has long antennae. Length ¹¹/₁₆in (18mm).

Horsefly
Chrysops caecutiens

Although it is especially fond of horses and cattle, this fly will also bite humans. It is recognised by its patterned wings and spotted eyes. May to September. Length ⁷/₁₆in (11mm).

Red and green iridescent eyes

Grasshoppers and crickets

Common field grasshopper

Chorthippus brunneus

Female ¾in (20mm) long; male smaller

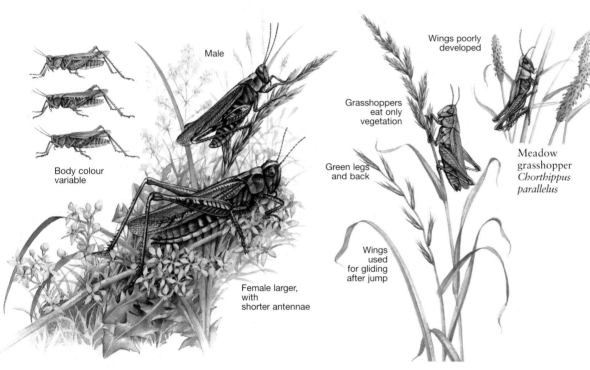

Male

Body colour variable

Female larger, with shorter antennae

Common green grasshopper

Omocestus viridulus

Female ¾in (20mm) long; male smaller

Wings poorly developed

Grasshoppers eat only vegetation

Green legs and back

Meadow grasshopper
Chorthippus parallelus

Wings used for gliding after jump

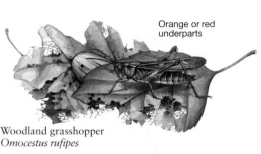

Orange or red underparts

Woodland grasshopper
Omocestus rufipes

Found mainly in woodland clearings, only in southern England, this species is readily distinguished by its orange or red underside. In other respects it resembles the common green grasshopper. Uncommon, June to September. Length of female ¾in (20mm).

Stripe-winged grasshopper
Stenobothrus lineatus

The distinctive feature of this species is the white stripe on each fore-wing. Found on chalk and limestone grassland in southern England, June to September. Female ¾in (20mm) long.

White stripe on fore-wing

Mottled grasshopper
Myrmeleotettix maculatus

This small grasshopper varies greatly in colour, with shades of black, brown and green predominating, but it is always mottled. Found throughout British Isles, June to October. Length of female ⁹⁄₁₆in (15mm).

Colour varies

Likes bog asphodel and bog myrtle

Large marsh grasshopper
Stethophyma grossum

The largest grasshopper in the British Isles can be recognised by its colourful body and short, slow ticking 'song'. Found only in marshland in southern England and western Ireland, July to October. Female 1¼in (32mm) long.

The song of the common field grasshopper is so precise it can be identified with a stopwatch. It produces a series of six to ten half-second chirps, spread evenly over about 12 seconds. The song is produced by a row of tiny pegs on one of the hind legs, which are rubbed against stout veins on the fore-wing. The male chirps to attract a female then, once a potential mate has been found, switches to a mating song. A sunny, grassy hillside comes alive with these mating sounds on a summer day. This species varies in colour according to its habitat. It dies in autumn, leaving eggs in the soil to survive the winter.

Throughout British Isles.

The common green grasshopper is smaller and greener, and therefore even harder to spot on a grassy hillside, than the common field grasshopper. Its call provides the best identification: a continuous chirp for 20 seconds or more, during which the whole body quivers as its legs move up and down up to 20 times a second. If disturbed the insect will usually jump; it should then be possible to follow and watch it.

The meadow grasshopper is the only one of the 11 British species which cannot fly, possessing only vestigial hind-wings. It is nevertheless the most abundant species.

Grassy areas throughout British Isles.

J F M A M J J A S O N D

These relatively large insects are well camouflaged and more often heard than seen. Their chirping is made by a scraping action – wing against wing in the case of crickets, whose males alone chirp, and wing against leg in both grasshopper sexes. Grasshoppers also have shorter antennae.

Dark bush cricket

Pholidoptera griseoaptera
To ³/₄in (20mm)

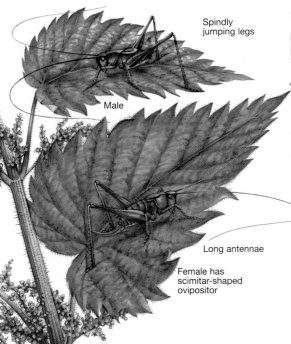

Spindly jumping legs

Male

Long antennae

Female has scimitar-shaped ovipositor

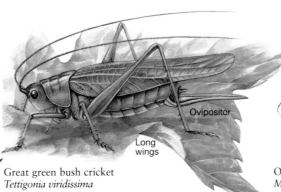

Ovipositor

Long wings

Great green bush cricket
Tettigonia viridissima

Britain's largest bush cricket, with unusually long wings. The male also has a long, almost continuous and very penetrating song. Despite its size it is not easily seen, tending to hide in thick vegetation. Common on south coast of England and Wales, locally inland; July to October. Length to 1⁵/₈in (42mm).

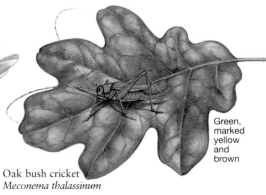

Green, marked yellow and brown

Oak bush cricket
Meconema thalassinum

The only native bush cricket to live in trees, this insect is particularly fond of oaks. It hides under foliage during the day, becoming active after dark. Woodland in England and Wales, July to October. Length to ⁵/₈in (16mm).

Another name for the ten British species of bush cricket is long-horned grasshopper. Bush crickets differ from true grasshoppers in having much longer antennae and more spindly jumping legs. A curious feature is that their ears are situated on their front legs. The dark bush cricket ranges in colour from brown to nearly black and has almost no trace of wings. The female is immediately recognisable by her large egg-laying organ, which is shaped like a scimitar. She lays her eggs in bark or rotting wood. Bush crickets eat both plants and soft-bodied insects, unlike the vegetarian grasshoppers.

Hedgerows and nettlebeds, mainly in southern Britain.

Brown colour

Long antennae and wings

House cricket
Acheta domesticus

Hedgerows hold no attraction for house crickets, which like the warmth and comfort of old buildings. Like some cockroaches they are natives of North Africa, but their musical 'song' makes them welcomed almost as pets. However, they have become less common in recent years. They eat food scraps, and the female lays her eggs in cracks in tiles or planks. Mainly in the south, active indoors all year. Length ³/₄in (20mm).

Mole cricket
Gryllotalpa gryllotalpa

Strong toothed fore-legs enable this brown cricket – one of our largest insects – to tunnel into the ground. Rare; damp areas in the south, mainly May to September. Length 1³/₄in (45mm).

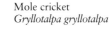

Powerful fore-legs

Stout body

Shiny black and yellow colouring

Large head

Field cricket
Gryllus campestris

The field cricket is seldom seen: not only is it rare, but it also spends much of its life underground in burrows. Eggs are laid near burrows in late spring and summer. The nymphs then retire to the burrows in the autumn to hibernate, emerging in the spring to mature, mate and finally die in late summer. Local in southern England. April to September. Length ⁷/₈in (22mm).

Bees

Bees are among most people's favourite insects – attractive, industrious and long valued for their honey. Yet the honey-bee is but one species among hundreds, most of which do not share its social habits.

Honey-bee

Apis mellifera
Workers ⅝in (16mm), drones ¾in (20mm), queen ⅞in (22mm) long

Queen's cell

Workers' cells

Workers (sterile females) build cells of wax and tend larvae

Drones' cells

Drone (male) has no sting

Queen (fertile female) stays in hive except for one 'nuptial' mating flight, or to lead a swarm to new hive

Workers collect nectar and pollen from flowers, and defend hive

Such is the hectic pace of the worker honey-bee in summer – gathering water, nectar, pollen and resin to meet the colony's constant needs – that few live beyond three or four weeks. Their foraging can extend as far as two miles from the hive, and the honey made from nectar and the protein-rich pollen are used to feed the larvae and the members of the colony that survive over winter. The hive's leader – the queen – lives up to five years; her main job is to lay eggs. Only females (workers and queen) have a sting, a modified ovipositor; it is barbed, and trying to pull it out can rupture the bee's abdomen and kill it.

Throughout British Isles.

Young queen starts colony in mouse's nest or hedgerow bank

Buff-tailed bumble-bee
Bombus terrestris

The largest of Britain's 18 species of bumble-bee. They have much fatter, hairier bodies than honey-bees, and live in much smaller colonies – up to 150 compared with 60,000 honey-bees. Only fertilised young queens survive the winter, as little honey is made. Widespread, March to October. Length ⅝in (16mm).

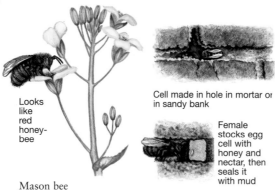

Reddish-brown tail

Red-tailed bumble-bee
Bombus lapidarius

The distinctive reddish-brown 'tail' is mimicked by a 'cuckoo' bee, *Psithyrus rupestris*, which is less hairy. Mainly southern England, March to October. Length ½in (13mm).

Looks like red honey-bee

Cell made in hole in mortar or in sandy bank

Female stocks egg cell with honey and nectar, then seals it with mud

Mason bee
Osmia rufa

One of the solitary bees, of which there are 227 British species. They have no workers, just males and females. Female mason bees clean out holes in mortar or sandy banks to make cells of mud for single eggs. Common throughout British Isles, April to July. Length to ⅝in (16mm).

Wasps

Giant wood wasp

Urocerus gigas
Length ¾in-1½in (20-38mm)

Female has long ovipositor

Male – no narrowing of abdomen

Length ⁵⁄₁₆in (8mm)

Larva ruins fruit

Apple sawfly
Hoplocampa testudinea

Although harmless, the female looks as if she is armed with the most fearsome sting. In fact it is an ovipositor, or egg-laying organ, with which she deposits her eggs in little holes bored in pine and other softwood trees. The larvae spend up to three years in the tree, eating the wood. This species is the largest of the wood wasps. They are black and yellow and without a 'waspie' waist, the female being considerably bigger than the male.
The closely related sawflies have a serrated ovipositor, with which the female lays her eggs in plants, using a sawing action.

Pine woods and plantations throughout Britain.

The black and yellow striped common wasp is very familiar, but it has many relations – including animal and plant parasites – that are hard to recognise as wasps. A wasp's sting has no barb and can be used many times, but in some species it is simply an egg-laying organ.

Common wasp

Vespula vulgaris
Length ⁷/₈in (22mm)

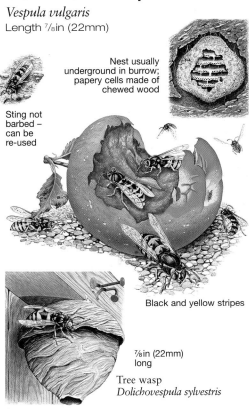

Nest usually underground in burrow; papery cells made of chewed wood

Sting not barbed – can be re-used

Black and yellow stripes

⁷/₈in (22mm) long

Tree wasp
Dolichovespula sylvestris

The common wasp is one of seven species of social wasp found in Britain. It lives in colonies of about 2000 workers, drones and queen. In spring and early summer, workers use their sting to kill insects such as sawfly caterpillars, aphids and flies to feed to their larvae. In return the larvae produce a sweet saliva on which the adult wasps feed. In late summer the queen stops laying eggs; there are no more larvae, and the wasps start to appear in gardens and orchards looking for ripe fruit. The tree wasp makes its nest in enclosed places such as hollow trees, wall cavities and lofts, mostly in the south.

Throughout British Isles but commoner in south.

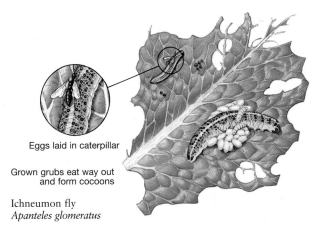

Eggs laid in caterpillar

Grown grubs eat way out and form cocoons

Ichneumon fly
Apanteles glomeratus

Actually a very small parasitic wasp. The female injects her eggs into young butterfly caterpillars, mainly those of the large white; the grubs that hatch proceed to feast inside their helpless host. Very common, killing a great many caterpillars; adults active April to October. Length ¹/₁₀in (2.5mm).

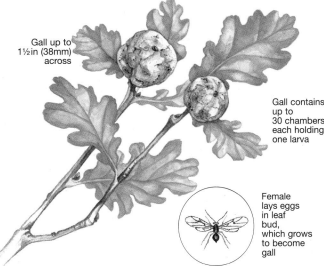

Gall up to 1½in (38mm) across

Gall contains up to 30 chambers, each holding one larva

Female lays eggs in leaf bud, which grows to become gall

Oak apple gall wasp
Biorhiza pallida

The oak apple, a familiar sight on oak trees in early summer, is really a multi-roomed house for the larvae of one of the many species of gall wasp. There are two generations a year, the winter larvae living in galls on the roots of the trees. Adults February to April; June and July. Length ¹/₈in (3mm).

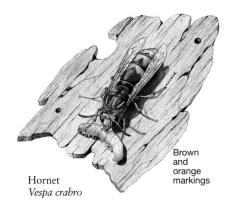

Brown and orange markings

Hornet
Vespa crabro

The biggest wasp in Britain, now also the rarest; it has disappeared from many of its haunts in south-east and central England. The hornet is a social species, nesting in hollow trees, either upright or fallen. The loss of old trees may be a cause of its decline. March to October. Length 1¼in (32mm).

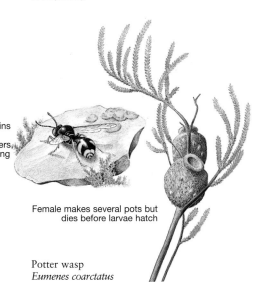

Female makes several pots but dies before larvae hatch

Potter wasp
Eumenes coarctatus

During her two-week life the female makes several tiny pots from sand and mud mixed with her saliva. Into each pot she puts an egg and a paralysed caterpillar. She then seals the pot, leaving the larva to develop in safety. The potter is a solitary wasp living on sandy heathland, June to September. Length ½in (13mm).

Ants

These fascinating insects show their close relationship to bees and wasps in their stinging ability, narrow waist and above all complex social behaviour. This can involve teamwork, mutual aid with different insects such as aphids, and in some cases enslavement of other ant species.

Black garden ant

Lasius niger
Worker in 3/16 (5mm) long; queen 1/2in (13mm); male 1/4in (6mm)

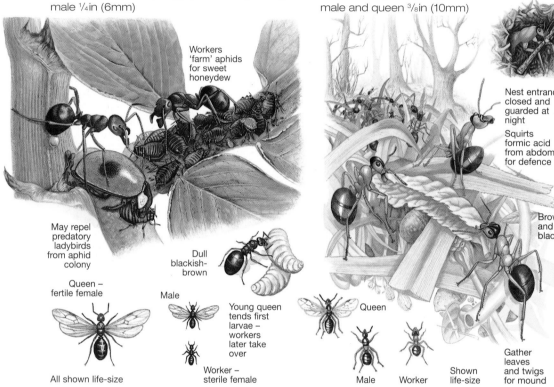

Workers 'farm' aphids for sweet honeydew

May repel predatory ladybirds from aphid colony

Dull blackish-brown

Queen – fertile female

Male

Young queen tends first larvae – workers later take over

Worker – sterile female

All shown life-size

Not only is this the ant most commonly found in gardens, it is the only native British ant which regularly enters houses. It does this in the constant search for sugary foods. Complex nests of many chambers are built out of doors, under plants, old logs and flat stones. Worker ants forage in plants for the sweet honeydew exuded by aphids. In return for honeydew the aphids are protected from predators. In late summer males and queens swarm and mate in mid-air. The male then dies and the queen bites off her wings, founds a new nest and devotes the rest of her life to egg-laying.

Gardens, grasslands, heaths and woods everywhere.

Wood ant

Formica rufa
Worker 1/4in (6mm) long; male and queen 3/8in (10mm)

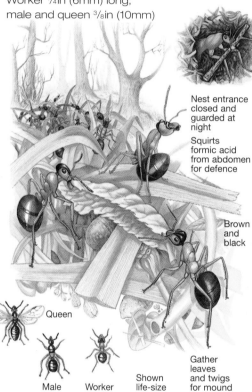

Nest entrance closed and guarded at night

Squirts formic acid from abdomen for defence

Brown and black

Queen

Male   Worker   Shown life-size

Gather leaves and twigs for mound

Wood ants build their nests on a grand scale. They are constructed on the woodland floor under mounds of tiny pieces of leaf and twig that can rise as high as 60in (152cm) – though most are smaller. The nest is a system of underground tunnels which is protected from the weather and predators by the mound and houses up to 300,000 ants. The wood ant workers forage for several yards around the mound for insects, especially the caterpillars of moths and sawflies, which they often kill with a squirt of formic acid. They drag the prey back to the nest whole or in pieces.

Commonest in southern England; related species in Scotland.

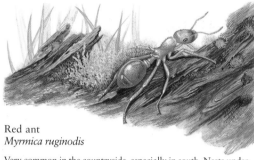

Red ant
Myrmica ruginodis

Very common in the countryside, especially in south. Nests under tree stumps or loose bark and beneath walls. It has a painful sting. April to October. Worker 1/4in (6mm) long.

Turf ant
Tetramorium caespitum

Workers may allow the queen of a different ant species, *Anergates atratulus*, to invade their nest. They then kill their own queen, rear the invader's offspring, and leave their colony to die. Southern England, April to October. Worker 1/8in (3mm) long.

Argentine ant
Iridomyrmex humilis

An immigrant from the tropics, resident here only since 1900, this small, slender ant can survive only in heated buildings in Britain; present all year. Otherwise a vigorous coloniser that eats anything. Worker 1/10in (2.5mm) long.

Blood-red ant
Formica sanguinea

Steals pupae from the nests of other ant species and rears them as slaves. Small areas of south-east England and north-east Scotland, April to October. Worker 1/4in (6mm) long.

Beetles

The habits of many beetles do not endear them, but they often do a useful job as predators of garden pests or as scavengers – disposing of dead organic matter. Some, however, such as the cockchafer, are themselves serious pests. All have hard wing-cases, in fact modified fore-wings, covering the functional hind-wings.

Green tiger beetle

Cicindela campestris
5/8in (16mm) long

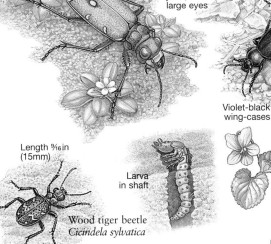

Green, creamy-white spots

Length 9/16in (15mm)

Larva in shaft

Wood tiger beetle
Cicindela sylvatica

Although pretty to look at, the green tiger beetle is a ferocious predator with large, powerful jaws which snap shut on ants and caterpillars with deadly regularity. It is fast: when hunting it can cover the ground at up to 24in (60cm) a second. It can also make short flights. The larvae live in sandy soil, jaws at the ready.
The wood tiger beetle is similar to the green except for its brown colour. It is found in southern England in July and August.

Dry heaths and sandy places in all areas.

J F M A M J J A S O N D

Violet ground beetle

Carabus violaceus
1 1/8in (28mm) long

Wing-cases fused together – cannot fly

Powerful jaws; large eyes

Violet-black wing-cases

Eggs laid on plant leaves

Larvae

After an exhausting night hunting a wide variety of plant-eating insects, the violet ground beetle spends its day resting under moss, bark, or a stone or log. It is one of the largest of the 352 species of ground beetles in Britain, with long legs and a violet, metallic sheen to its wing-cases. Farmers and gardeners should treat it as a friend, for it consumes large numbers of pests. The larvae develop under leaf litter, eating other soft-bodied insects.

Arable fields and gardens in all areas.

J F M A M J J A S O N D

Sexton beetle

Nicrophorus vespilloides
5/8in (16mm) long

Larvae feed on carcass

Black and orange bands

Pair of beetles excavate soil from below carcass and bury it, removing skin

Like vultures, sexton beetles – also called grave-digger or burying beetles – do a good job of disposing of the bodies of other animals, in this case birds and small mammals such as mice. They can smell a carcass up to two miles downwind. Once found, it is usually buried where it lies. The female then lays her eggs nearby and protects them from predators and parasites – eating any competing maggots – until they hatch, with a meal all ready to hand.

Throughout British Isles.

J F M A M J J A S O N D

Stag beetle

Lucanus cervus
Male about 2in (50mm) long; female smaller

Male – antler-shaped jaws

Violet-brown wing-cases

Female – dies after laying eggs

White larva lives 2-3 years

The flight of the male stag beetle on a warm summer night, with its great antler-shaped jaws outstretched, is a marvellous and sometimes alarming sight. While they look capable of inflicting terrible damage, the jaws are mainly ornamental, designed to intimidate other males in the battle to win eligible females. Nevertheless, they help to make it Britain's largest beetle. The female is smaller, with much less impressive jaws, but she can inflict a sharp bite.

Uncommon except in south-east England.

J F M A M J J A S O N D

Beetles

Cockchafer

Melolontha melolontha
1in (25mm) long

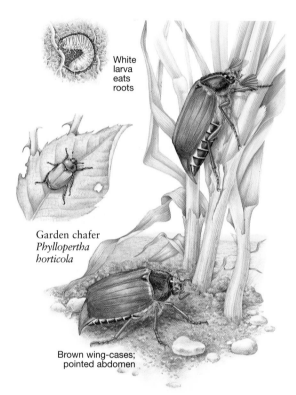

White larva eats roots

Garden chafer
Phyllopertha horticola

Brown wing-cases; pointed abdomen

This beetle – particularly its 2in (50mm) long larva – is a serious pest of farm and garden crops, especially cereals and soft fruits. Commonly called the May bug because it is most often seen in May or early June, it lays large numbers of eggs underground. These then hatch and the larvae proceed to feast on plant roots for the next two or three years. The adult also attacks the leaves of crops and trees. It is a nocturnal creature, normally noticed only by the extent of its depredations.

The garden chafer or June bug is similar in colour, shape and habits to the cockchafer, but only ³/₈in (10mm) long.

Throughout most of British Isles.

Glow-worm

Lampyris noctiluca
⁵/₈in (16mm) long

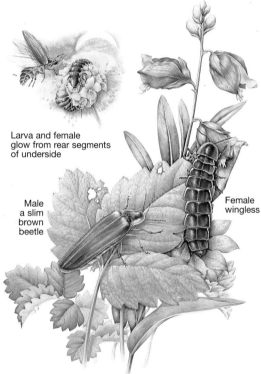

Larva and female glow from rear segments of underside

Male a slim brown beetle

Female wingless

One of the magical experiences of night in the countryside, commonplace 50 years ago but now less so, is to see the light cast by a gathering of female glow-worms to attract males. The glow-worm's light is produced by cells that use oxygen, water and an enzyme to form the substance oxyluciferin, which emits light without generating heat. The light is intense for such a small source, enabling the male to spot the female from up to 10yds away. The male then flies to his wingless, signalling mate. Eggs are laid on low vegetation. The larvae also glow, but from smaller patches.

Widespread but local.

Click beetle

Agriotes lineatus
To ¹/₂in (13mm) long

Sudden release of body's hinge flips beetle into air like spring

Slim, brown beetles; bask on vegetation

Wireworms – 1in (25mm) larvae – eat roots of cereals and root crops

A better name might be flip. When threatened by a bird or other predator, the click beetle or skipjack can jump about a foot in the air at a speed of 8ft (2.5m) per second, sometimes somersaulting half a dozen times in the process. It performs this extraordinary feat with an audible click – hence its name. There are 65 species of click beetle in Britain, the larvae of which – known as wireworms – are serious farm pests. The adult's powerful jump is made possible by a hinged thorax. Lying on its back, the beetle arches its body like a coiled spring and, suddenly releasing the tension, flicks itself into the air.

Fields, woods and gardens throughout British Isles.

Soldier beetle

Rhagonycha fulva
³/₈in (10mm) long

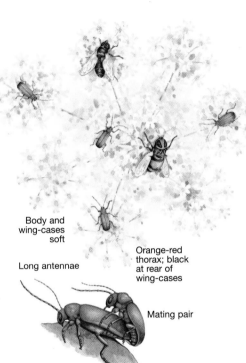

Body and wing-cases soft

Long antennae

Orange-red thorax; black at rear of wing-cases

Mating pair

The brown or black wing-cases and red thorax of this beetle are reminiscent of 19th-century military uniforms. The beetles are often abundant on flower heads such as cow parsley, seeking out small insects or pollen to eat. The colours serve a purpose: they are a warning to birds that these beetles are not pleasant to eat. Soldier beetles have soft wing-cases and body, and long antennae. The flat, velvety larvae have large jaws and feed on small insects such as springtails and silverfish in soil and leaf litter.

Throughout British Isles.

J F M A M J J A S O N D

Spatula-like legs

Bloody-nosed beetle
Timarcha tenebricosa

A large, slow-moving black beetle that ejects a red, bitter fluid from its mouth and joints when attacked. Southern England, April to August, ⁷/₈in (22mm) long.

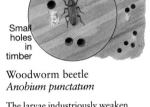

Small holes in timber

Woodworm beetle
Anobium punctatum

The larvae industriously weaken house timbers, furniture and old trees. May to July. ¹/₆in (4mm) long.

Holes slightly larger

Death watch beetle
Xestobium rufovillosum

The mating call – an eerie knocking on wood – is a warning of timbers being eaten. Midlands and south, April to June. Length ¹/₄in (6mm).

Striped wing-cases

Colorado beetle
Leptinotarsa decemlineata

A serious pest of potato crops. An occasional immigrant, May to September. Length ⁷/₁₆in (11mm).

Seven-spot ladybird

Coccinella 7-punctata
¹/₄in (6mm) long

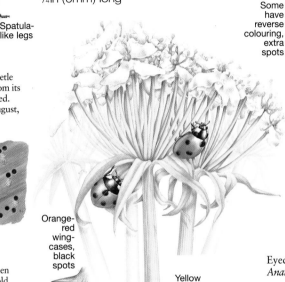

Orange-red wing-cases, black spots

Larva, ½in (13mm) long, devours aphids

Yellow spots on thorax

Although there are 42 different ladybird species in Britain, the seven-spot is the favourite, particularly with children. It is by no means as docile as it looks, being a voracious feeder on aphids. Ladybird eggs are laid on plants infested with aphids and the slate-blue larvae eat hundreds of these pests in the three weeks before they pupate. The striking red, black and yellow colours of the adult are a warning to predatory birds that they are unpleasant to eat.

Throughout British Isles.

J F M A M J J A S O N D

Some have reverse colouring, extra spots

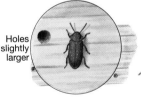

Two-spot ladybird
Adalia 2-punctata

Probably the most common ladybird in Britain and also the one with the widest variations in colouring and number of spots. March to October. Length ¹/₆in (4mm).

Eyed ladybird
Anatis ocellata

Our largest ladybird, with seven or eight black and yellow 'eyes' on each wing-case. Pine woods, June and July. ⁵/₁₆in (8mm) long.

Pale red background

Striking pattern

Fourteen-spot ladybird
Propylea 14-punctata

The bright yellow and black markings make this – one of several yellow species – difficult to overlook. Southern half of Britain. April to September. Length ¹/₆in (4mm).

329

Snails and slugs

Snails and slugs are molluscs related not only to water-snails and many shellfish but also to the octopus and squid. Most species damage crops and garden plants. The need for chalk to make the shell means that few snails live on acid heaths and moors.

Garden snail

Helix aspersa
Shell to 1½in (38mm) wide

Wrinkled fawn shell, up to five dark bands

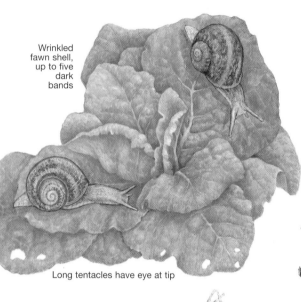

Long tentacles have eye at tip

Pair entwine feet when mating

The garden snail has an unusual sex life. Each snail possesses both male and female sex organs, and mating occurs on a warm, wet night in summer after a short courtship. It culminates in a harpoon-like dart containing sperm being shot from each snail to the other. Then each snail goes away to lay its eggs in a shallow depression in the earth, in batches of up to 40. A month later the eggs hatch and new snails, small replicas of their parents, emerge. The shell enlarges as they grow. Song thrushes eat huge numbers of snails, breaking open the shells on a stone 'anvil', but they can live up to three years.

Throughout British Isles; mainly coastal in Scotland.

J F M A M J J A S O N D

Shell thin, flattened

White-lipped banded snail
Cepaea hortensis

The shell can be all-yellow, but it is often streaked with darker bands, the colour and number of which are variable. The body of the snail itself is dark greenish-grey. Throughout Britain, seen mainly March to October. Shell up to ⅝in (16mm) wide.

Cone-shaped shell

Tree snail
Balea perversa

The other main version of the snail shell – long, narrow and cone-shaped, twirling like a perfectly produced ice-cream cone. Pale brown or greenish. Locally common throughout Britain, seen mainly March to October. Shell to ⅜in (10mm) long.

Creamy shell; yellow-grey body

Roman snail
Helix pomatia

The Romans are reputed to have eaten this snail when in Britain – hence its name. It is rare but unmistakable with its thick, round, cream-coloured shell. Found locally on chalk and limestone in southern England, mostly April to October. Shell to 2in (50mm) wide.

Great black slug

Arion ater
To 6in (15cm) long

Nostril in side of saddle-like mantle

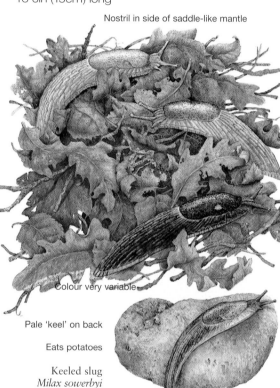

Colour very variable

Pale 'keel' on back

Eats potatoes

Keeled slug
Milax sowerbyi

The name can be misleading, for there are brown, brick-red, orange and grey forms of this slug. But all are big, rounded at the tail and patterned with long tubercles or lumps. The hole in the smooth patch on its back is for breathing. Lacking a shell, slugs solve the problem of preventing their bodies drying out by covering them with a layer of sticky mucus. Much to gardeners' chagrin, they eat plants as well as small dead creatures. They are hermaphrodites like snails.
The keeled slug has a prominent line running down the centre of its back. It lives mostly underground.

Damp places throughout British Isles.

J F M A M J J A S O N D

Soil animals

Decay and renewal go hand in hand on the land, the one following the other. Many small invertebrates help the process of decay, so clearing the way for renewal. The earthworm is perhaps the best known, but there are others – some, such as the millipede, also regarded as pests.

Common centipede

Lithobius forficatus
About 1³/₁₆in (30mm) long

Kills prey by injecting poison from fang-like front legs

Runs with body rigid; most other centipedes wriggle

Although the word centipede means 'hundred feet', there is a surprising variation in the actual number these creatures possess, ranging from 34 to 354 depending on the species. Like insects they have a hard outer skeleton, but their bodies lack the waxy waterproof layer that enables insects to preserve their body moisture. They cannot afford to expose themselves to the sun, and spend their days in dark places under stones, beneath bark and in leaf litter. At night they emerge to catch prey such as woodlice, which they cannot see because they have no eyes. Instead, they rely on vibration, touch and speed.

Woodlands and gardens throughout British Isles.

Curls when attacked

Black snake millipede
Tachypodoiulus niger

It does not help in distinguishing a millipede from a centipede that the black snake millipede has almost a hundred legs – 96 to be precise. But millipedes have two pairs of legs on each body segment – centipedes only one – and millipedes curl up when threatened. Britain's 50 species recycle dead leaves but also attack root crops and young plants. Common throughout Britain, all year. Length 1³/₁₆in (30mm).

Pill millipede
Glomeris marginata

The body is an intriguing combination of black segments divided by yellowish bands. When threatened, it curls up into a tight, defensive, armoured ball. All year. Length ⁹/₁₆in (15mm).

Common woodlouse

Oniscus asellus
To ⁹/₁₆in (15mm) long

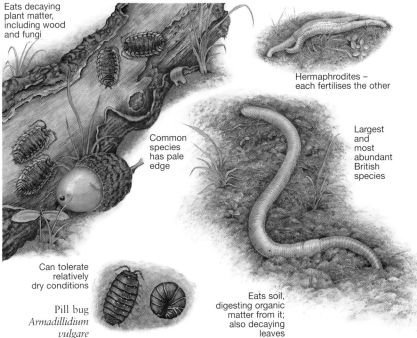

Eats decaying plant matter, including wood and fungi

Common species has pale edge

Can tolerate relatively dry conditions

Pill bug
Armadillidium vulgare

It is difficult, when viewing the common woodlouse, to accept that it is related to the crab and the lobster. But its ancestors crawled from the sea millions of years ago for a drier, but still damp, home on land. It lives under logs and stones, coming out only at night when darkness and moist air protect its body from drying up.
The pill bug, another woodlouse species, rolls up into an armour-plated ball when threatened. It is often found in houses.

Common throughout British Isles.

Common earthworm

Lumbricus terrestris
To 12in (30cm) long

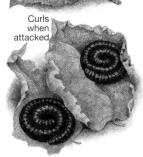

Hermaphrodites – each fertilises the other

Largest and most abundant British species

Eats soil, digesting organic matter from it; also decaying leaves

Earthworms are invaluable to man in helping to maintain soil fertility by aerating it, by recycling minerals to the surface and by increasing its humus content. An acre of land may contain up to 3 million earthworms, whose activities can bring 8 to 10 tons of soil to the surface (in the form of casts) each year. They also take rotting vegetation down into the soil to eat. The common earthworm lives in tunnels as much as 6ft (1.8m) down.

Throughout British Isles.

Spiders

Garden spider

Araneus diadematus
Female's body ½in (13mm) long;
male smaller

When web becomes tattered, female eats it

Small male is scavenger

Trapped insect paralysed with venom, bundled in silk and injected with digestive enzymes

Female has prominent cross on back – pattern varies

Britain has 40 species of orb-web spiders, which spin the disc-shaped webs traditionally associated with spiders. One of the commonest is the garden spider, easily recognised by the white cross on the female's back. Only she builds a web; the male lives by scavenging off it. She constructs the web in two stages. First it is spun with non-sticky silk, with a small inner spiral where the female will await her prey. She then goes back over the outer spiral, eating it and replacing it with sticky silk. She avoids sticking to this by coating her feet in oil. In autumn she lays a mass of eggs which hatch in the spring.

Widespread throughout British Isles.

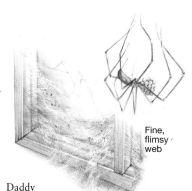

Fine, flimsy web

Daddy long-legs spider
Pholcus phalangioides

A well-named species since its legs are about five times its body length, covered with fine, long hairs. It likes warm conditions and lives indoors in Britain; commonest in south. All year; commonest May to July. Body ⅜in (10mm) long.

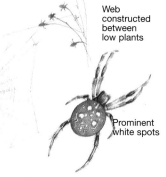

Web constructed between low plants

Prominent white spots

Sprayed silk traps prey

Four-spotted orb spider
Araneus quadratus

The round orange, brown or red female has four large white spots on its back. Found throughout British Isles on low bushes, heather and grass. The web is constructed near the ground. Seen mainly July to October. Body to ¾in (20mm) long.

Spitting spider
Scytodes thoracica

This is no simple expectoration, but a spray squirted over a distance of ⅜in (10mm) which cements the insect prey to the spot, enabling the spider to close in and deliver the fatal bite. Southern England and Wales, mostly April to September. Body length ¼in (6mm).

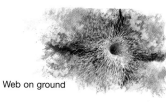

Web on ground

Mesh-web spider
Amaurobius similis

The big, sturdy female builds a web on the ground or other surface to catch beetles and other crawling insects. The male drums on the web to announce he is calling. Throughout British Isles, August to October. Body length ⁷⁄₁₆in (11mm).

Dark brown or grey; hairy

House spider
Tegenaria domestica

Found in all kinds of buildings as well as houses, spinning a dense web, often 6in (15cm) across. The female can live up to four years. Throughout British Isles, all year. Body length ⅜in (10mm).

Veil-web spider
Dictyna arundinacea

The web is spread like a veil over the plant stem, usually on low and especially dry or dead vegetation. Throughout British Isles, March to June. Body ⅛in (3mm) long.

Spiders and harvestmen are arachnids – invertebrates with eight legs and lacking the wings and antennae of insects. They are creatures of extraordinary skill and interest, from web-making to hunting and mating behaviour. The female, larger and longer-lived, dominates the male.

Wolf spider

Pisaura mirabilis

Female's body ⁹/₁₆in (15mm) long; male smaller

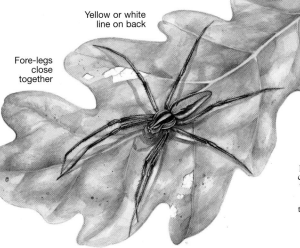

Yellow or white line on back

Fore-legs close together

Female

Male

Male offers courtship 'gift' before mating

Female carries silken egg sac

This large nomadic spider is often seen running across the ground or low vegetation after some unfortunate prey. For the wolf spider is a hunter, relying on its speed rather than a web to catch its victims. They are quickly paralysed by the venom it injects through its fangs, and then sucked dry. The female only spins a web as a nursery in which to place her egg sac when the eggs are ready to hatch. Until then, she carries the sac – a silk ball – around with her. The male always presents her with a gift before mating – often a dead fly wrapped in silk. Without it, she is likely to eat him.

Common in woods and on heaths throughout British Isles.

Striped body

Zebra spider
Salticus scenicus

The black and white zebra spider can jump up to 4in (10cm) – a useful ability in approaching prey. Found on walls and in gardens throughout Britain, May to September. Body length ¹/₄in (6mm).

Buzzing spider
Anyphaena accentuata

The black arrow-head markings on a pale yellow body resemble an escaped convict's uniform of children's comics. The male makes a buzzing sound as it courts the female. Trees and bushes throughout British Isles, April to June. Body ¹/₄in (6mm) long.

One of many similar night-hunting species

Night-hunting spider
Clubiona terrestris

A small, reddish-brown spider mainly active at night; lives in low-lying vegetation. It folds the end of a blade of grass and lines it with silk to make a nest. Common throughout Britain, all year. Body length ¹/₄in (6mm).

Common crab spider
Misumena vatia

The body colour, which can change in the course of a few days, makes good camouflage against pale flowers. The female is usually white, pale green or yellow, with a fat body. The male is much smaller and darker. They sit on flowers, ready to ambush pollinating insects by grasping them with their long crab-like front legs. Common in southern England, May to August. Body ³/₈in (10mm) long.

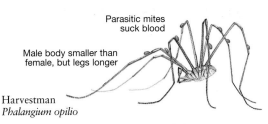

Parasitic mites suck blood

Male body smaller than female, but legs longer

Harvestman
Phalangium opilio

Unlike true spiders, which they closely resemble, harvestmen have a one-piece body and long second pair of legs. They do not make webs. Throughout Britain, June to November. Female's body ³/₈in (10mm) long.

Lives in grassy places

Small wolf spider
Pardosa pullata

When out hunting, the female carries her young on her back. At the egg stage she carries them in a large blue sac. Very common throughout British Isles, April to August. Body length ¹/₄in (6mm).

Crab-like abdomen

Colour matches flower on which it hides

333

mammals, amphibians & reptiles

British species comprise only a small part of the animal kingdom, but they include the most advanced of all creatures – the mammals. Even including feral species, little more than 60 out of a worldwide total of 5000 mammal species live in the British Isles, plus a few marine mammals that enter our waters. Other land vertebrates – the reptiles and amphibians – are even fewer in number. There are six species of British reptile and six native amphibians out of over 3000 species worldwide in each case, though a few more exist here in small colonies following deliberate or accidental introductions.

Our native and feral mammals fall into a small number of groups whose differences are fairly obvious even to the lay observer: the carnivores; deer, ponies and other hoofed animals; the insectivores; the rodents; hares and rabbits; the bats, our only flying mammals; seals; and the whales, including dolphins and porpoises. There is also an introduced marsupial originally from Australia, but very nearly extinct here now.

Amphibians and reptiles are cold-blooded and their body temperature varies with that of the surroundings. As a result, they are active only if it is warm enough, and they all hibernate. As a rule, they lay eggs – amphibians such as frogs in fresh water, reptiles (whose eggs have leathery shells) on land. However, some female reptiles carry eggs inside them, but bear live young. Unlike mammalian mothers, these reptiles do not nourish their young with milk.

Carnivores

Fox

Vulpes vulpes

To 26in (66cm) head and body; 15in (38cm) tail

Pricked-up ears, black behind

Amber eyes

Bushy tail

Tracks narrow; front toes close together

Vixen carries cubs in mouth

Lower legs black

Sparring cubs stand on hind legs

Though basically a creature of the open countryside, the fox is now an increasingly familiar part of the wild life of city and suburbs, often feeding on rats, mice and household scraps. It has a sharp intelligence, acute hearing and keen sense of smell. The fox leads a solitary life for much of the year until late December, when the mating season erupts with short triple barks through the night from competing dog foxes, the vixens responding with eerie screams. Cubs are born about mid-March. Foxes are mostly active at night, when they forage for whatever food is available.

Suburbs and country, except some islands.

Wild cat

Felis sylvestris

To 36in (90cm) head and body; 12in (30cm) tail

Cat rests during day on tree or rock

Black and yellowish-grey striped tabby pattern

Black-ringed tail with rounded tip

Tail more pointed; commonest colours black, tabby and ginger

Feral cat
Felis catus

In prehistoric times, when the country was covered with forest, the wild cat lived all over Britain. But the destruction of woodland and hostility of gamekeepers now confine it to the Scottish Highlands. It is slightly larger than the domestic cat, with longer, softer fur. Rabbits, mountain hares, small rodents and birds are its main prey. Feral cats are domestic cats that have reverted to living wild. They usually live in small colonies of about 15 animals in the grounds of hospitals, factories and even city squares. In Scotland some have interbred with wild cats.

Pure-bred wild cat only in Highlands.

J F M A M J J A S O N D

J F M A M J J A S O N D

Britain's native carnivores – predominantly meat-eaters with sharp canine teeth, powerful jaws and strong claws – belong mostly to the mustelid family, which includes the otter, badger, stoat and weasel. Since the wolf became extinct here only the fox represents the dog family; and the only true wild cat is rare.

Badger

Meles meles

To 30in (76cm) head and body; 6in (15cm) tail

White-tipped ears

Striped head

Fore and hind tracks differ greatly

Hind

Fore

Badgers sharpen claws on trees near sett

Strong forepaws

Emerging from its home (or sett) mainly at night, the badger follows well-trodden paths to and from its feeding grounds and its specially dug latrines on the edge of its territory. A relative of the stoat and weasel, it has a thick-set, barrel-shaped body, short powerful legs with strong claws, and a short, blunt tail. Badgers live in groups of one or two families – up to 15 animals – in a sett made up of underground tunnels and chambers which are kept scrupulously clean and tidy. The badger's diet ranges from earthworms, mice, voles, frogs, snails and wasps to fruit and grass.

Woodland adjoining pastures, especially in south and west; occasionally in suburbs.

Otter

Lutra lutra

To 36in (90cm) head and body; 16in (40cm) tail

Otter runs with hunched, rolling gait

Prey carried ashore with feet or teeth

Tracks show five-toed webbed feet and tail

Fore

Hind

Otters can stay under water for four minutes

Small ears

Often stands upright to look around

Two or three cubs are born in a den, or holt

Thick, tapered tail

Webbed feet

As recently as the 1950s otters could be found throughout most of the country. They are now rarely seen, except on Scotland's west coast, though legal protection and conservation schemes are slowly restoring numbers elsewhere. Streamlined for speed in the water, the otter has a long body with a powerful tail and short, strong legs with webbed feet. Otters eat fish and are as much at home in water as on land. Dog and bitch otters come together only for mating, which may take place at any time of year.

Rivers and coasts, especially in parts of Scotland and Ireland.

Carnivores

Pine marten

Martes martes
To 18in (45cm) head and body; 9in (23cm) tail

May pursue squirrels from tree to tree, but hunts small mammals and birds mainly on ground; also eats eggs and berries

Male

Creamy-yellow throat

Claws grip tree-trunk when climbing

Female

Pine marten leaps from stone to stone across stream

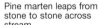

Rich brown fur

Long, bushy tail

Britain's rarest carnivore is about the size of a cat, with long rich brown fur, a bushy tail and unmistakable creamy-yellow throat and ears. It seldom appears in daylight. For centuries the pine marten has been trapped for its fur and persecuted (for no good reason) by gamekeepers, but reafforestation programmes are now making its future brighter. Its legs and feet are well adapted to climbing, and both lair and breeding den are usually a hollow tree, rock fissure or disused bird's nest. Mating is in July or August, but the young are not born until the following March or April.

Remote woods, mainly in north-west Scotland but spreading into new forestry plantations.

Polecat

Mustela putorius
To 15in (38cm) head and body; 5½in (14cm) tail

Dark band over eyes

Dark guard hairs cover creamy-yellow wooly underfur

No distinct mask

Polecat-ferret is crossbreed

Pure ferret is creamy-white, but hybrids occur

Ferret   Pale legs   No mask; pink eyes

The tendency of the polecat to kill for the sake of killing has brought it enemies among poultry farmers and gamekeepers. Trapping has led to the almost total disappearance of the animal from the British countryside, apart from a few areas of Wales. The polecat uses a persistent, foul-smelling scent both defensively and as a territorial boundary marker. A member of the weasel family, it is a mainly nocturnal ground dweller, a good swimmer but a poor climber, preying on rabbits, other small mammals, game birds, poultry, frogs and eggs. The ferret is a domesticated polecat, creamy-white in colour. Escaped ferrets often crossbreed with polecats.

Pure polecats mainly in Wales; polecat-ferrets widespread.

J F M A M J J A S O N D

J F M A M J J A S O N D

Mink

Mustela vison
To 16in (40cm) head and body;
5in (12.5cm) tail

'Pastel' mink are descended from specially bred animals

Chocolate-brown fur

Long body, short legs

Stoat

Mustela erminea
To 10in (25cm) head and body;
3in (76mm) tail

Creamy-white underparts

Black tail-tip

Stoats can kill even relatively large prey, biting deeply into its neck

Weasel

Mustela nivalis
To 8in (20cm) head and body; 2in (50mm) tail

Brown tail

Wavy flank line

Brown throat patch

Weasel often seen streaking across road, its legs a blur

This relative of the weasel was introduced into Britain from North America in the late 1920s to be bred commercially for its fur. So successfully has it managed to establish itself in the wild that it is now officially designated a damaging pest. The mink is mostly active at night, and is rarely seen far from a river or lake. It preys on waterfowl, fish-farm trout and young salmon, which it will kill even when it is not hungry. Its dense, glossy, chocolate-brown fur looks almost black from a distance, especially when wet. Wild mink nest in tree roots or tunnels in steep banks near water; in the spring they produce five or six kittens.

Increasingly common by rivers and lakes in many parts of Britain and Ireland.

The stoat has been relentlessly persecuted for centuries by farmers and gamekeepers, and in the 1950s faced the disappearance of a major food source, the rabbit, through myxomatosis. Yet it not only survived but continued to thrive, and is at last valued for keeping down mice, rats and voles. It is a brilliant hunter, relying mainly on sound and smell. Stoats mate in summer, the young being born the following March in a den in a rock crevice or abandoned rabbit burrow. In winter, stoats in northern Scotland (and sometimes farther south) turn creamy-white, though the tail-tip stays black; they are then called ermine, and are trapped for their valuable fur.

Woods, farms and uplands throughout Britain and Ireland.

The weasel is Britain's smallest carnivore and, with the stoat, one of the most numerous. It looks like a small stoat, but lacks the stoat's distinctive black tail-tip. Its brown fur meets its white underparts in an irregular line along its flanks, and there are small brown patches on its throat. The weasel hunts mainly at night, when it usually patrols a hedgerow or stone wall, investigating every hollow in search of prey; but it is sometimes seen in daylight. Young weasels are usually born in April or May, in a nest of grass or leaves in a shallow hole, tree-stump or haystack. Gamekeepers trap large numbers of weasels, even though they help to keep down rats, mice and voles.

Widespread in Britain, but absent from Ireland.

J F M A M J J A S O N D

J F M A M J J A S O N D

J F M A M J J A S O N D

Deer

Red deer

Cervus elaphus

Stag to 48in (120cm) at shoulder; hind to 45in (114cm)

Largest of our native land animals, the red deer has its main stronghold in the Scottish Highlands, but herds flourish in deer parks all over the country. Stags (males) bear antlers which start to grow in early summer, are hard and fully grown by September, but are shed the following spring. Stags live in separate groups from the hinds, except during the rutting, or mating, season which lasts for about a month in the autumn. Then the stags move into the area where the hinds live and compete with each other to round up a 'harem'. Calves are born – usually singly – in May or June. Red deer feed on grass, heather and the young shoots and bark of trees.

Wild in Scotland, Devon, Somerset, New Forest and Cumbria; common elsewhere in parks.

| J | F | M | A | M | J | J | A | S | O | N | D |

Hoofprint
3½in (83mm) long,
2½in (64mm)
across

Young stag's first antlers are simple spikes

Antlers with increasing number of tines (prongs) each year; can grow to 28in (70cm) long

Hind has no antlers

Short tail, buff rump

Calf brown with white spots

Year-old hind

Stag, summer

Red-brown coat, greyer in winter

Young males chased off by stag with 'harem'

Rival stags clash antlers but seldom fight to death

Stags gather 'harem' of hinds in September

Red and roe deer – the only deer native to Britain – have been joined by several introduced species, some of which have spread after escaping from parks. Most male deer have antlers which grow and are cast each year. Antler shapes and tail and rump markings are important identification aids.

Fallow deer

Dama dama
Buck to 37in (95cm) at shoulder; doe to 30in (76cm)

Doe has no antlers

Broad-bladed antlers

Prominent Adam's apple

Orange-brown coat with white spots; long black and white tail. Some fallow deer are almost black all over

Buck, summer

Fawn are usually pale brown with white spots; this is a dark fawn

Hoofprint 2½in (65mm) long and 1½in (40mm) across

Medieval huntsmen reduced the herds of fallow deer introduced probably by the Romans into our ancient forests. But wild herds still live on in wooded parts of southern England, and protected herds graze the grounds of many a stately home. As in other species, the antlers borne by the buck (male) are covered as they grow with a hairy skin called velvet. In about August this velvet is shed, leaving the antlers clean and hard for the mating season, when rival bucks fight for their does (females). Next spring the antlers are cast (shed) and the cycle starts again. Most fallow deer lose their spots in winter.

Wild herds in south and east; common in parks.

Roe deer

Capreolus capreolus
To 25in (64cm) high at shoulder

Antlers with short branches

Large ears

Black nose; white chin patch

Creamy-buff rump

Buck, summer

Buck, winter

Doe, winter

Very short tail; doe has tuft of hairs on rump

Hoofprint 1¾in (45mm) long and 1⅓in (35mm) across

The roe deer is not much bigger than a large dog. It was widespread in the Middle Ages, but later disappeared and was re-introduced to parts of England more than 100 years ago, since when it has spread to many woodland areas. Roe deer are shy creatures, most active at dawn and dusk, and feed mainly on tree shoots and shrubs. The foxy-red summer coat thickens and darkens to grey-brown in autumn, but the white patch on the rump remains visible; this fluffs out when the deer is alarmed. The buck's antlers are roughened near the base; they are cast by December, and new ones grow during the winter.

Forests in many areas, especially new plantations.

J F M A M J J A S O N D   J F M A M J J A S O N D   341

Deer

Sika deer

Cervus nippon

Stag to 33in (84cm) at shoulder; hind to 28in (70cm)

Stag, summer

Slender antlers; rounded ears

Spotted chestnut-brown coat

White rump and white tail with dark stripe

Hind in grey and brown winter coat

All the sika deer in the wild in Britain are descended from animals that escaped from the deer parks to which they were introduced from east Asia in the 19th century. The sika is smaller than the red deer, to which it is closely related, and its chestnut-brown coat is faintly spotted with white in summer. The coat turns a uniform greyish-brown in winter. Pale hair on the brow gives the sika a frowning look. The stag uses its antlers in combat during the rutting season, and also to mark its territory by thrashing bushes and fraying tree bark in the mixed woodlands which are its usual home.

Isolated patches in many parts of British Isles.

Muntjac

Muntiacus reevesi

To about 19in (48cm) at shoulder

Short antlers; markings in 'V' shape

Buck

Tusk-like teeth

Rounded back, red-brown coat

Dark patch on forehead

Doe

The small and secretive muntjac is difficult to see in the dense woodland where it usually lives. Two distinctive features are the deer's rounded hump-back shape and the buck's short straight antlers, the line of which extends down its face as a V-shaped ridge. The upper jaw has tusk-like teeth which the buck uses as weapons. Muntjac were introduced from China to the Duke of Bedford's estate at Woburn, Bedfordshire, in 1900, and the descendants of escaped animals have become established. Unlike native British deer, muntjac have no fixed breeding season, and females may give birth every seven months.

Woodland and scrub, Midlands and south.

Chinese water deer

Hydropotes inermis

To about 24in (60cm) at shoulder

Large rounded ears

Tusk-like teeth

Short tail

Red-brown summer coat

Coat grey-brown in winter

Unlike any other species of deer in Britain, the male Chinese water deer has no antlers. Instead, tusk-like teeth in its upper jaw protrude about 2¾in (70mm) below its upper lip. The likeliest place to see it is in parks such as Woburn, Bedfordshire, or at Whipsnade. Some wild populations exist in dense reed-beds in East Anglia, but they are shy creatures, only occasionally seen. The sleek red-brown summer coat turns pale grey-brown in winter. Rutting is in November and December, and the spotted fawns are born in June.

Wild in Bedfordshire, Broads and Fens; park herds elsewhere.

Axis deer

Axis axis
To about 37in (95cm) at shoulder

Stag scent-marks territory by thrashing with antlers

Three-pointed antlers to 30in (76cm) long

Rich brown coat

Stag, summer

White bib distinguishes fawn from roe deer

Spotted like the fallow deer, the axis is distinguished by the male's narrower, three-pointed antlers and by its white bib. On the fawn, before the antlers grow, the white bib is already visible against the rich brown coat. The axis is a native of the forests of India and Sri Lanka; in Britain it can be seen in parks such as Woburn and Whipsnade, grazing or browsing on shrubs. The rich brown summer coat darkens slightly in winter. When alarmed, the axis emits a shrill whistle. There is no fixed breeding season, and no fixed season for the stags to shed their antlers.

Woburn, Bedfordshire, and some other deer parks; not in wild.

Père David's deer

Elaphurus davidianus
To 48in (120cm) at shoulder

Backswept points

Tawny-red coat

Long tail

Stag, summer

Spotted calves, born between March and May, are leggy like foals

As large as a red deer, Père David's has a rounded, horse-like rump, a long tail and antlers with backward-sweeping points. Its summer coat is tawny-red, flecked with grey; in winter it is longer and greyish-buff. Père Armand David was a French missionary explorer who saw a herd of the deer in the walled Imperial Hunting Park near Peking in 1865, by which time they were probably extinct in the wild. Exported deer thrived particularly well at Woburn in Bedfordshire, and all the Père David's deer throughout the world are now descended from the Woburn herd. Rutting is in July and August.

Woburn, Bedfordshire, and some other deer parks; not in wild.

Reindeer

Rangifer tarandus
Bull to 48in (120cm) at shoulder; cow to 36in (90cm)

Long, curved antlers in both sexes

Grey-brown coat, paler in winter

Broad hooves spread weight

Calves lack spots, unlike other deer

A small herd from Lapland released in Scotland in 1952 restored reindeer to Britain after an absence of thousands of years. Their long winter coat and soft, dense underfur insulate them against icy temperatures, and broad cloven hooves prevent them sinking too deep into the snow. The long elaborately curved and branched antlers are borne both by bulls and cows – the only female deer to do so. Mature bulls shed their antlers soon after the rut in September and October, but the cows retain theirs through the winter and use them to scrape away the snow to feed on lichens. Calves are born in May or June.

Only in Cairngorm mountains near Aviemore, Highland.

J F M A M J J A S O N D

J F M A M J J A S O N D

J F M A M J J A S O N D

Other hoofed mammals

Exmoor pony

Equus caballus
To 51in (130cm) high at withers

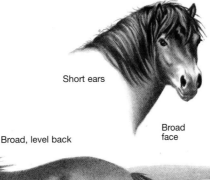

Short ears

Oatmeal-coloured muzzle

Broad face

Broad, level back

Fell pony

Large black or dark brown pony

Up to 55in (140cm) high

New Forest pony

Up to 58in (147cm) high, many smaller; most colours

Shetland pony

Up to 40in (100cm) high; colour varied, with woolly undercoat

Generally dun, bay or brown

Short legs

Strong legs; tough feet; sturdy body

Dartmoor pony
Up to 50in (127cm) high

The herds of ponies that roam Britain's open spaces are not entirely wild; most have owners and are regularly rounded up. Ponies have probably been on Exmoor since about 60,000 BC, and those living there today are the toughest – and truest to the original type – of all British native ponies. Short-legged and broad-chested, they have a thick, wiry coat and a dense undercoat that insulates the body against the hardest winter weather.
Dartmoor ponies are also tough, but few are pure-bred, for any type of stallion can be turned loose on the moor. Other breeds include New Forest ponies, Fell and Dale ponies from the western and eastern sides of the Pennines, and the hardy little Shetland ponies.

J F M A M J J A S O N D

Hoofed mammals account for many of our most important domesticated creatures. Not surprisingly, many have escaped or been allowed to go free and have established colonies in the more isolated parts of the British Isles. In some cases such feral animals can be re-domesticated.

Marsupials

Feral goat

Capra hircus (or domestic)
To 30in (76cm) at shoulder

Spreading horns, to 30in (76cm) long

Shaggy coat

Feral goat shaggy, and smaller than domestic goat

Soay sheep

Ovis aries
To 22in (55cm) at shoulder

Ram has heavy, coiled horns

Lambs born mid-April

Fawn coat

Ewe has small horns or none

Dark brown coat commonest; paler below

Red-necked wallaby

Macropus rufogriseus
To 24in (60cm) head and body; 25in (64cm) tail

Red-brown shoulder patch

Black-tipped feet

Black-tipped tail

Long hind legs and long thick tail

Joey (baby) often grazes from mother's pouch

Small, shaggy, sure-footed and agile, the feral goat has a distinctive appearance but may be hard to spot on the rocky slopes where it roams. These animals are descendants of early domestic goats and have been living wild in mountainous areas for more than 1000 years. Today's herds are small and isolated, and there are many variations of colour and size. Both sexes have horns, which may sweep outwards or backwards. The goats feed on gorse and heather, and on leaves and shoots from trees. For most of the year they keep to high, rocky mountainsides, but in winter they will descend to valleys.

Rocky parts of north and west Britain, and some islands.

Now the only completely wild sheep in Britain, the tiny Soays are descendants of the domestic sheep kept in the Stone Age. For hundreds of years they supplied the inhabitants of the St Kilda islands, in the Outer Hebrides, with wool and meat. Now they roam free and unsupervised, for the last people left the islands in 1930. The sheep, which take their name from one of the islands in the group, are dark brown or fawn, with paler underparts and markings around the eye. During the rutting season in late autumn the rams scuffle for dominance, frequently pushing at each other.

Wild only on St Kilda islands; also zoos and parks.

Many a walker has been surprised to come across a red-necked wallaby, living wild and apparently very much at home thousands of miles from its native Australia. Over the years many wallabies have escaped from zoos and some have managed to survive and establish substantial colonies. The red-necked wallaby is a medium-sized kangaroo, little bigger than a badger, and is named after the patch on its nape and shoulders. A shy creature, it usually hides in thick scrub. It eats heather, grass, bracken and other vegetation.

Very local, restricted to parts of Peak District; large numbers roam freely at Whipsnade in Bedfordshire.

J F M A M J J A S O N D

J F M A M J J A S O N D

J F M A M J J A S O N D

Insectivores

Mole

Talpa europaea
To 6in (15cm) head and body; ½in (13mm) tail

Hedgehog

Erinaceus europaeus
To 10in (25cm) head and body; ½in (13mm) tail

Common shrew

Sorex araneus
To 3in (76mm) head and body; 1½in (40mm) tail

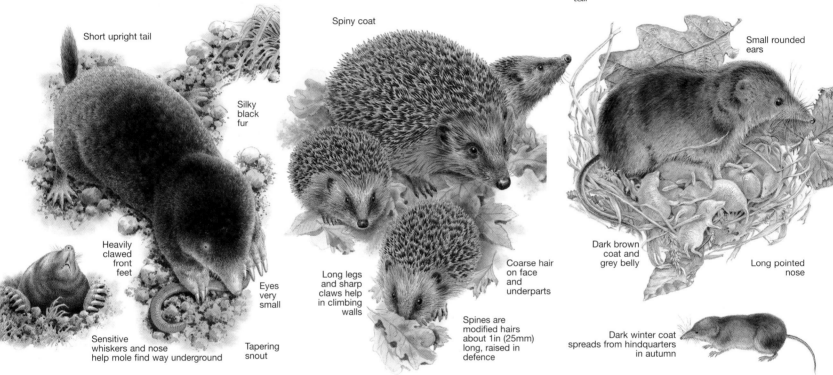

Short upright tail

Silky black fur

Heavily clawed front feet

Sensitive whiskers and nose help mole find way underground

Eyes very small

Tapering snout

Spiny coat

Long legs and sharp claws help in climbing walls

Coarse hair on face and underparts

Spines are modified hairs about 1in (25mm) long, raised in defence

Small rounded ears

Dark brown coat and grey belly

Long pointed nose

Dark winter coat spreads from hindquarters in autumn

Mounds of earth about 6in (15cm) high proclaim the presence of the tunnelling mole; the animal itself is rarely seen. Its front paws can move 8oz (over 200g) of soil – twice the mole's weight – in a minute. Moles spend most of their lives in a burrow system, emerging occasionally, mainly at night, to collect nesting material and food. They are active all year, and feed on earthworms, insect larvae and slugs. They are solitary, and normally make contact only when they mate in spring. The female builds a nest of leaves and grass in which three or four young are born between April and June. Few moles reach the age of three.

Suburbs and country below 3000ft (900m); not Ireland.

Britain's only mammal with a spiny coat is a common sight in the countryside and gardens near dawn and dusk. Sadly, it is also seen as a corpse on the road, because often it does not run from danger but rolls up into a ball. An adult hedgehog has some 5000 white-tipped spines, which deter predators. Hedgehogs are active mostly at night. They eat beetles, caterpillars and earthworms, and also enjoy birds' eggs, slugs and carrion – and a bowl of bread and watered milk in dry seasons. They hibernate fitfully in winter, and are ready to breed in April. Three to five young are born in early summer in a leafy nest.

Widespread in Britain and Ireland.

An angry squeaking in a hedgerow may betray the presence of a couple of shrews fighting over territory, but otherwise they are hard to find. Common shrews are smaller than house mice, and live three-quarters of their lives underground. They are constantly on the move in search of food such as earthworms and woodlice. A shrew can starve to death if it goes without food for more than about three hours. It has poor eyesight and relies mainly on its keen sense of smell and long whiskers to find its way about. Common shrews live about a year and have several litters of six or seven young in that time.

Common in hedges, fields and woods; none in Ireland.

J F M A M J J A S O N D

J F M A M J J A S O N D

J F M A M J J A S O N D

Animals belonging to the order Insectivora – eaters of insects and other small creatures – range from the tiny shrews to the familiar mole and hedgehog. All have a distinctive long pointed nose, small ears and tiny eyes; several live most of the time in underground burrows.

Pygmy shrew

Sorex minutus
2½in (65mm) head and body; 1½in (40mm) tail

Bulbous head; short narrow snout

Grey-brown coat

Thick tail

Tiny size seen against ½in (13mm) yew berry

Even when fully grown, a pygmy shrew is not much bigger than a stag beetle and weighs less than a 10p coin. It is Britain's smallest mammal, and the only shrew found in Ireland. It has to eat its own weight in food every day, and forages busily for insects and soil creatures along tunnels made by other animals or in leaf litter, especially in damp places. Pygmy shrews often live in the same areas as common shrews, but avoid conflict.

Widespread in woods and grassland.

Young born in June in underground nest

JFMAMJJASOND

Water shrew

Neomys fodiens
3⅜in (86mm) head and body; 2¼in (55mm) tail

Black upper coat

Silvery-white underparts

White eyebrows and ear tips

All shrews are attracted to riversides and other damp places by the worms, insect larvae and small spiders that abound there. The water shrew, the largest of Britain's shrews, actually takes to the water, paddling after its prey with the aid of a hairy fringe on each hind toe. A mild poison present in the saliva helps the water shrew to subdue even frogs, which it then hauls ashore to eat. Burrows are often in a river bank, with an underwater entrance, but water shrews also live in drier places. They may live for 18 months, but most die younger or are taken by predators such as owls; pike and mink.

Wetlands, woods and hedgerows; not in Ireland.

JFMAMJJASOND

Lesser white-toothed shrew

Crocidura suaveolens
2⅜in (60mm) head and body; 1½in (40mm) tail

Prominent ears

Feeds on small seashore creatures

Projecting hairs on both species' tails

Neither of Britain's white-toothed shrews is found on the mainland. The lesser species lives on the Isles of Scilly, Jersey and Sark – but not in the other Channel Islands, which are the home of the greater white-toothed shrew, differing from its near relative in its tooth pattern. The overall whiteness of their teeth distinguishes both island shrews from their mainland cousins, which have red tips to their teeth.

Isles of Scilly and Channel Islands.

Greater white-toothed shrew
Crocidura russula

JFMAMJJASOND

Rodents

Brown rat

Rattus norvegicus

To 11in (28cm) head and body; 9in (23cm) tail

Coarse fur

Small, finely haired ears

Thicky, scaly tail, shorter than head and body

Four toes on fore feet, five toes on hind feet; 1⅓in (35mm) long

Entrance to burrow about 1½in (40mm) across, with narrow beaten trail

The big, tough and aggressive brown or common rat is a native of central Asia, and arrived in Britain by boat from Russia at the beginning of the 18th century. It lives wherever there is undisturbed shelter or stored food, from hedgerows and barns to sewers. It is a burrowing animal and can swim well. Elusive and nocturnal, it will eat almost anything, from cereals and fruit to soap and plaster. While enormously prolific, it also has a high mortality rate of about 95 per cent per year, and it rarely lives longer than 18 months. Even so, there are millions of brown rats in Britain.

Common, especially around buildings, farms and hedgerows.

Black rat

Rattus rattus

To 8in (20cm) head and body; 9in (23cm) tail

Long whiskers help rat to find its way in dark buildings

Large pink ears

Thin tail, longer than head and body

Brown forms common, but relative tail length still distinctive

The Black Death of the 14th century, which killed some 25 million people in Europe, is generally attributed (perhaps wrongly) to the black rat – or, more specifically, to bacteria transmitted by the rat flea which it carried. The black rat originated in south-east Asia and first came to Britain in the Middle Ages, possibly in the baggage of a returning Crusader. Since the introduction of the bigger, more aggressive brown rat, the black rat has disappeared from many parts of the country. A few open-air colonies exist on islands such as Lundy, but otherwise they are now virtually extinct in Britain.

Scarce and localised, in ports and towns near coast.

House mouse

Mus musculus

To about 3¼in (83mm) head and body; 3¼in (83mm) tail

Pointed face; big eyes

Greasy, grey-brown fur

Long, scaly tail

Males fight to establish dominance or defend territory

Wherever man is, the house mouse is seldom far away. Indeed, it seems unable to live without him. When the last crofters left the Scottish island of St Kilda in 1930 the house mouse, which had probably lived there since Viking times, became extinct within a few years. Unlike other mice, the house mouse has a strong smell and greasy fur. It can quickly colonise the largest house, since one female produces about 50 young a year. House mice eat anything from soap to cereals. They are extremely adaptable, and have been known to thrive even in frozen-meat stores, growing longer fur to keep out the cold.

In and around buildings throughout Britain and Ireland.

J F M A M J J A S O N D

J F M A M J J A S O N D

J F M A M J J A S O N D

Rats, mice and their relatives outnumber by far all other kinds of mammals in Britain and are of enormous social and economic importance. Their role in spreading disease has declined with improved public health measures, and most species are completely harmless.

Harvest mouse

Micromys minutus
2½in (65mm) head and body; 2½in (65mm) tail

Blunt nose

Small ears

Yellow-brown fur, white underparts

Prehensile tail grasps stalk like fifth foot

Tennis ball with hole, fixed 18in (45cm) off ground, makes nest-box

Wood mouse

Apodemus sylvaticus
3¾in (95mm) head and body; tail longer

Large ears and eyes

Yellow streak on chest

Long tail

Sandy-brown, white underparts

Large hind feet enable wood mouse to leap away like kangaroo

Yellow-necked mouse

Apodemus flavicollis
4in (10cm) head and body; tail longer

Yellow collar

Eats mainly seeds, fruit, small animals

Orange-brown flanks

Long tail

Peeping out from a nest of grass suspended high up among corn stalks, the harvest mouse is part of every child's picture-book image of the countryside. It is one of our smallest mammals, weighing less than a 2p coin, and its unique prehensile tail enables it to climb tall, stiff-stemmed vegetation such as corn stalks with ease. Seeds, grain, grass shoots, soft fruit and insects make up its diet. In winter it seeks cover in low vegetation and barns.

Fields and hedgerows, mainly in south.

The most abundant mouse of the countryside is not often seen, as it is active only at night. Also called the long-tailed field mouse, it lives mostly in woodland, hedgerows, fields and gardens, but will also inhabit outbuildings. It eats seeds, shoots and buds, together with snails and a variety of insects and larvae. Breeding starts in March, the female bearing up to four litters a year, each of about five young, in an underground nest chamber.

Widespread throughout Britain and Ireland.

Slightly bigger than the wood mouse, the much rarer yellow-necked mouse is also brighter-coloured, with a more orange coat, a whiter underside and, on its chest, a distinctive yellow band. It lives in woods, hedgerows and gardens, and only comes out at night. It climbs well, searching the tops of trees for nuts or new buds. In autumn, it often enters outbuildings or houses in search of food and shelter.

Woods and hedgerows in southern and eastern England and Welsh borders.

J F M A M J J A S O N D   J F M A M J J A S O N D   J F M A M J J A S O N D

Rodents

Water vole

Arvicola terrestris
Male to 8in (20cm) head and body; 4³/₄in (12cm) tail

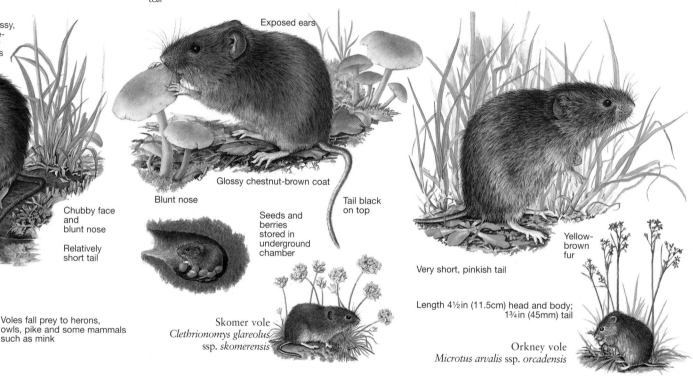

Long, glossy, chocolate-brown fur hides ears

Chubby face and blunt nose

Relatively short tail

Voles fall prey to herons, owls, pike and some mammals such as mink

The first indication of the presence of a water vole is often the plop it makes as it dives for safety under water. It also makes a little V-shaped wave as it swims away. It is the largest British vole, about the size of a rat and sometimes called the water rat, but is differentiated by a blunt snout, tiny ears and shorter tail. Active by day, it feeds almost entirely on waterside plants, and nests in bankside burrows.

Voles burrow in waterside banks

Lowland ponds and slow-running rivers except in Ireland. Now rare.

Bank vole

Clethrionomys glareolus
To 3¹/₂in (90mm) head and body; 2³/₈in (60mm) tail

Exposed ears

Blunt nose

Glossy chestnut-brown coat

Tail black on top

Seeds and berries stored in underground chamber

Skomer vole
Clethrionomys glareolus ssp. *skomerensis*

The bank vole is the smallest of Britain's voles, with a redder coat and more prominent ears than those of the field vole. Bank voles live in deciduous woods and scrub, especially along banks and hedgerows, where they feed on green plants and fruits such as hips and haws. Since they weigh less than 1oz (30g) they can climb delicate shoots to reach for berries. Nests are often built under roots in a chamber reached by tunnels.

The Skomer vole found on Skomer Island, Dyfed, is a 'giant' race twice as heavy as mainland voles and about 4¹/₂in (11.5cm) long in head and body.

Lowland woods, hedges and scrub.

Field vole

Microtus agrestis
To 4in (10cm) head and body; 1¹/₂in (40mm) tail

Yellow-brown fur

Very short, pinkish tail

Length 4¹/₂in (11.5cm) head and body; 1³/₄in (45mm) tail

Orkney vole
Microtus arvalis ssp. *orcadensis*

One of the most numerous of all British mammals, the field or short-tailed vole is an aggressive and noisy little creature, often battling with other field voles in defence of its small territory and emitting loud squeaks and angry chattering noises. It is the main food of barn owls and an important food source for many other predators, including kestrels, foxes, stoats, weasels and snakes. Grass is the main food of the field vole, which nests under logs and other objects in long grass.

The slightly larger Orkney vole is found only in the Orkneys, though a close relative lives on Guernsey.

Widespread in grassland and hedgerows; not in Ireland.

J F M A M J J A S O N D

J F M A M J J A S O N D

Red squirrel

Sciurus vulgaris
To 8in (20cm) head and body; 7in (18cm) tail

The red squirrel was one of the last mammals to colonise Britain before the island became physically separated from mainland Europe. Until the 1940s it was fairly widespread, but it has declined, disappearing altogether from most of southern England, where it survives only in isolated places such as the Isle of Wight. The decline is thought to be due partly to disease, and poor adaptation to deciduous woodland – not just the spread of the grey squirrel. In summer the red squirrel has bright chestnut fur with orange-brown feet and lower legs. As summer progresses, its bushy tail bleaches to a pale cream. Its winter coat is long, silky and chocolate-brown.

Large areas of mature forest, almost entirely in Scotland.

J F M A M J J A S O N D

Most of red squirrel's time is spent in treetops, and courtship involves long chases

Claws grip bark as squirrel runs head first down tree

Squirrels do not hibernate and are often seen in winter

Bushy tail

Ear-tufts much longer in winter

Chestnut fur

Grey squirrel

Sciurus carolinensis
To 10in (25cm) head and body; 8in (20cm) tail

Bushy, grizzled tail

No ear-tufts

Yellowish-brown fur in summer

Winter fur bright silvery-grey on flanks

Unlike the red squirrel, which likes to stay high in the treetops, the grey squirrel – introduced from America at the turn of the century – is an attention-seeker, performing feats of acrobatics in any garden for the prize of a few nuts. Apart from being bigger and stronger than the red squirrel, the grey has another natural advantage in being able to live in hedgerow trees, parkland and gardens, as well as forests. In summer the grey squirrel's fur is yellowish-brown; silvery winter fur begins to grow in the autumn. Neither the grey nor the red squirrel hibernates, but they are less active in winter.

Woods, gardens and hedgerows, particularly in the south.

J F M A M J J A S O N D

Rodents

Common dormouse

Muscardinus avellanarius

3in (75mm) head and body; 2½in (65mm) tail

Orange-yellow fur

Fluffy tail

Domed nest built in a bush, where young grow grey fur

Secretive, nocturnal and rarely seen, dormice are the only mouse-sized rodents to have a fluffy tail. The common dormouse is a chubby-looking creature with orange-yellow fur, creamy-white on the underside, and smaller ears than other mice. Victorian children often kept dormice as pets, but they are now quite rare. They hibernate from October to April, usually below ground. In summer they nest in a loose ball of grass, leaves and shredded honeysuckle bark, often in a bramble thicket. They emerge at dusk to forage, preferably in thick undergrowth, for berries, nuts and insects.

Hedges and woods in the south; uncommon elsewhere.

J F M A M J J A S O N D

Fat dormouse

Glis glis

6in (15cm) head and body; 5in (12.5cm) tail

Dark eye rings

Grey body bur

Bushy grey-brown tail

Fruit and nuts held in paws for eating

The fat dormouse, though eaten as a delicacy at Roman banquets, was not introduced into Britain until 1902, when a few were released at Tring in Hertfordshire. It is now found in many woodland and suburban areas of the Chilterns. It looks like a small grey squirrel, with a bushy tail and fine grey body fur. In summer the fat dormouse eats heartily to prepare for hibernation, often doubling its body weight.

Chiltern woods and houses.

Hibernates October to April

J F M A M J J A S O N D

Golden hamster

Mesocricetus auratus

About 7in (18cm) including tail

A stoutly built burrower from eastern Europe and the Middle East, the soft-furred, nocturnal hamster feeds on fruit, vegetables and grain. It stores food in its large cheek pouches for eating at leisure. An escaped alien, now eliminated, but it may recur.

J F M A M J J A S O N D

Mongolian gerbil

Meriones unguiculatus

About 4in (10cm) head and body; 4in (10cm) tail

The hardy, soft-furred gerbil from the desert regions of Mongolia has been a popular pet in Britain since the 1960s. Many have escaped and established wild colonies, often under sheds and outbuildings. An escaped alien.

J F M A M J J A S O N D

Hares and rabbit

Although they bear some resemblance to rodents, hares and rabbits are classified in a separate group; they can be just as damaging to field crops, however. They are well adapted for life on open grasslands, with strong hind legs for fast movement.

Brown hare

Lepus europaeus
Male to about 22in (55cm) long;
female slightly smaller

Long, black-tipped ears

Big eyes

Long, powerful hind legs enable the hare to run at 35mph (56km/h)

Orange throat and flanks

Mountain hare

Lepus timidus
To 20in (50cm) long

Shorter ears than brown hare

Coat turns wholly or partially white in winter

Spring

Summer (brown all year in Ireland)

Rabbit

Oryctolagus cuniculus
Male to about 19in (48cm) long; female smaller

Prominent eyes set so that rabbit can see all ways at once

Shorter ears than hares, without black tips

Orange nape

Grey-brown fur

Short tail, black on top and white below

Territory is marked by rubbing ground with chin, which has scent-secreting glands

A rare delight, walking across a meadow, is to see two brown hares standing bolt upright on their long hind legs, apparently engaged in boxing. For a few moments these normally alert creatures are totally absorbed in their struggle. But approach too closely and they will disappear in a flash. This spectacular turn of speed enables the brown hare to live successfully in open country, aided by the acute hearing provided by its 4in (10cm) long ears, and its sharp eyesight. Hares breed at almost any time of year, raising litters of two or three leverets, each in a separate form, or depression, in long grass.

Widespread in lowlands, but declining; rare in Ireland.

The mountain hare is smaller than the brown, with shorter ears – about 3in (75mm) long. It is sometimes known as the Irish hare because it is common in all parts of Ireland, not only on upland moors which are its main home in Britain. In winter in Britain its coat turns white, so that it is often impossible to detect as it sits resting in a snowy hollow. Mountain hares feed mostly at night, supplementing their main diet of heather with rushes, cotton grass and bilberry shoots. They breed between February and August, each female having two or three litters with one or two leverets in each.

Common in Ireland and upland Britain (mainly Scotland).

The rabbit was introduced into Britain from France in the 12th century as a source of meat and skins. By the 1950s it had become a serious pest, but it was almost completely wiped out by the flea-borne virus disease myxomatosis. Over 96 per cent of an estimated total population of more than 60 million rabbits died, but virus-resistant strains survived and bred. The average female produces about 20 offspring a year, and as a result rabbits are now almost back to their old numbers in some areas. They live in extensive burrows and eat greenery of many kinds; they eat their own soft droppings to double-digest their food.

Widespread, particularly in farmland and lowland forests.

J F M A M J J A S O N D

J F M A M J J A S O N D

J F M A M J J A S O N D

Bats

Greater horseshoe bat

Rhinolophus ferrumequinum
Wingspan to 13½in (34cm); head and body to 2½in (65mm)

Broad, rounded wings; large body

Fur reddish-brown; young greyer

Pointed ears, no fleshy spike

Horseshoe-shaped snout, used in navigation

The horseshoe-shaped fold of nose skin which gives this species its name plays an important part in its echo-location system. High-pitched notes, emitted through the nostrils, are focused into a narrow beam by a fleshy, cone-shaped trumpet; the bat moves its head from side to side to scan in front. Once common in southern Britain, the greater horseshoe bat is now scarce. For hibernation, it favours spacious, humid places such as old mines, caves or tunnels. Mating is in autumn or winter, but the sperm is stored and does not fertilise the egg until spring.

Largely confined to south-west England and Wales.

Lesser horseshoe bat

Rhinolophus hipposideros
Wingspan to 10in (25cm); head and body 1½in (40mm)

Horseshoe bats hang upside-down by claws and wrap wings around themselves

Broad wings; tiny body

This smaller relative of the greater horseshoe – its body smaller than a man's thumb – is also commoner. It has the same taste for humid hibernating places as its larger cousin, but it can get into narrower tunnels and shafts. The nose construction and echo-location system are similar to those of the greater horseshoe. The annual baby is born in July or August, usually in an attic, a hollow tree or an old building. Tiny insects, caught in flight, are the bat's main source of food. At rest, the bats need roosts where they can hang freely; they do not hide in crevices.

West of England, Wales, Midlands and parts of Ireland.

Pipistrelle

Pipistrellus pipistrellus
Wingspan 8½in (22cm); head and body 1⅜in (35mm)

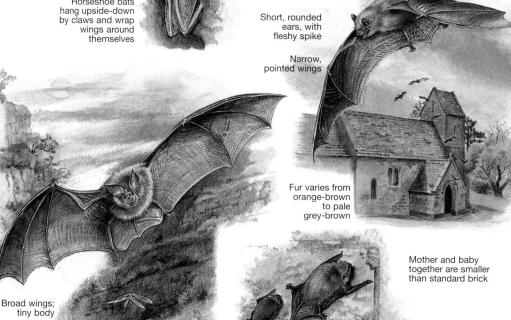

Short, rounded ears, with fleshy spike

Narrow, pointed wings

Fur varies from orange-brown to pale grey-brown

Mother and baby together are smaller than standard brick

Tiniest and commonest of British bats, the pipistrelle is highly gregarious: colonies can number 200 or more. They often congregate in traditional bat haunts such as churches, and can be heard squeaking as they gather to pour out – 15 to 30 minutes before sunset – to forage for insects. Like all British bats except the horseshoes, pipistrelles have a tragus – a fleshy spike – in the ear as part of their echo-location system. They like warm crevices in house roofs as a summer roosting place. The young are born mostly in June.

Widespread in Britain, including cities.

J F M A M J J A S O N D

J F M A M J J A S O N D

Bats are the only flying mammals. In darkness, radar-like echo-location systems help them to navigate and to catch insects. The wings of horseshoe bats are broader and more rounded than others. British bats usually hibernate from October to March. They are protected by law.

Noctule bat

Nyctalus noctula
Wingspan 14in (36cm); head and body 3in (75mm)

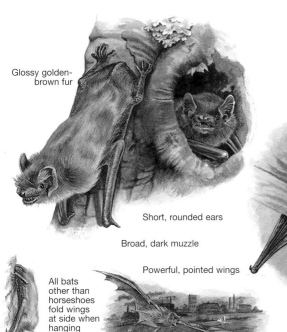

Glossy golden-brown fur

Short, rounded ears

Broad, dark muzzle

Powerful, pointed wings

All bats other than horseshoes fold wings at side when hanging upside-down

A powerful and expert flyer, the noctule bat is often seen on the wing before sunset. It will fly boldly at treetop height, taking steep dives after insects. The noctule has short, rounded ears with a rounded tragus – the central fleshy lobe. Facial glands impart a strong smell to the animal and its roost, usually a hollow or woodpecker hole in a tree.

Common in southern Britain, especially Midlands and south-east.

Single young born in June or July

Serotine bat

Eptesicus serotinus
Wingspan 14in (36cm); head and body 2½in (65mm)

Gable may house colony of 50-100 bats; young born usually in late June

Grizzled fur

Dog-like muzzle

Dark face and ears

Although serotine bats are strong flyers, their distribution is very patchy. They like to raise their young in the attic of an old house; nursing colonies can be noisy, but it is illegal to disturb them. The bats fly out shortly after sunset, catching insects on the wing. Hibernation quarters are in a house roof or hollow tree.

Common in parts of southern and eastern England; rare in north.

Tragus blunt, less than half height of ear; this fleshy spike is typical of all bats other than horseshoes

Common long-eared bat

Plecotus auritus
Wingspan 10in (25cm); head and body without ears 1¾in (45mm)

Huge ears, meeting at base, with long pointed spike (tragus)

Yellow-brown fur, paler beneath

Greyer fur

Grey long-eared bat
Plecotus austriacus

This is one of the few bats fairly easy to identify, if its very long ears can be seen. They are over 1in (25mm) long, nearly three-quarters the length of the rest of the head and body. They enable the bat's echo-location system to locate insect prey not only on the wing but even on leaves. Its wings are broad. Long-eared bats roost mainly in trees, but often breed in attics.

The grey long-eared bat is slightly larger and has greyer hair; it is found only in a few southern counties.

Woods and house roofs in Britain and Ireland, except in far north of Scotland.

Bats

Barbastelle bat

Barbastella barbastellus
Wingspan 10½in (27cm); head and body 1¾in (45mm)

Pug-like face

Black fur, with 'frosty' sheen

Broad, black ears with pointed tragus

Natterer's bat

Myotis nattereri
Wingspan 11in (28cm); head and body 1¾in (45mm)

Light brown above, white below

Long sideways-pointing ears with upturned tips

Whiskered bat

Myotis mystacinus
Wingspan 9½in (24cm); head and body 1½in (40mm)

Narrow, pointed wings

Tiny feet

Ears dark, narrow and upright

Face dark and furry

An odd-looking misfit among British bats, the barbastelle resembles only one other species in the world – a relative found in the Middle East and Asia. It has a pug-like face with a bare, dark brown snout and broad ears that meet at the base between the eyes. Its glossy fur is almost black, and on older bats it is cream-tipped, giving a frosted appearance. The barbastelle lives mostly in trees, occasionally in houses. In summer, females form nursery colonies to raise their young. Barbastelles hibernate with other bat species in caves in the coldest weather, but usually spend the winter in relatively unsheltered places.

Open woodland in southern Britain, but rarely seen.

Natterer's bat was named after the early 19th-century Austrian naturalist who discovered it. It can be recognised by its pure white underparts as it flies – slowly and often at housetop height – soon after sunset, seeking small flying insects. Another distinguishing feature is its tail, whose membranes have a fringe of hairs and often point downwards in flight, in order to catch insects for food. In summer, breeding colonies of females and young form in hollow trees or house roofs. The bats start hibernation comparatively late, often in caves, but are rarely seen after October; they mate during wakeful periods.

Woods, farmland and parkland, except in northern Scotland.

Abundant fur around the eyes and muzzle give this bat its whiskery appearance. It is a small, delicate creature, with narrow, pointed wings and tiny feet. The face is very dark, and so are the upright, pointed ears which give it a rather sprightly look. This bat is common around buildings, hedges or woodland fringes. It can be seen flying from early evening with a slow, fluttering action. Mating takes place during short periods of wakefulness in winter. Pregnancy begins as hibernation ends, and the single young are usually born around June.

Widespread in Britain, except in far north; not recorded in Ireland.

J F M A M J J A S O N D

J F M A M J J A S O N D

Daubenton's bat

Myotis daubentoni
Wingspan 10in (25cm); head and body 1¾in (45mm)

Mouse-eared bat

Myotis myotis
Wingspan 16in (40cm); head and body 2¾in (70mm)

Bechstein's bat

Myotis bechsteini
Wingspan 11in (28cm); head and body 1¾in (45mm)

Short ears; broad, pinkish-brown muzzle

Large feet, splayed toes

Bat crawls backwards into crevice to hibernate

Large wide ears, with long pointed tragus

Long muzzle

Grey-brown back, greyish-white underparts

Wide wingspan

Bare, pinkish face

Ears hang down when bat roosts

Ears do not meet at bases like those of long-eared bat

Narrow muzzle

Long ears, with long narrow tragus

Shallow, fluttering wing-beats enable Daubenton's bat – known also as the water bat – to skim close to the surface of a lake or pond as it hunts. In this way it can catch aquatic insects and even, occasionally, small fish – not preyed upon by other British bats. It hunts mostly by night. It can be confused with the whiskered bat, but at close quarters its large feet, with splayed toes, distinguish it. It sleeps in hollow trees and cracks in walls. Nursery roosts, with adult males excluded, are often in buildings and sometimes contain scores of bats. In winter, the bats tend to hibernate alone in caves or trees.

Widespread near water in most parts of Britain and Ireland.

The biggest British bat is distinguished by its slow, heavy flight, its bare pinkish face and its wide ears. There is a distinct division between its grey-brown back and greyish-white underparts. But sightings of these bats are rare, and they may no longer survive as a native British species. Those seen in the past in Sussex and Dorset could have flown from France, for on the Continent they have been known to travel 125 miles (200km) between summer and winter quarters. In summer the females give birth to their young in a warm dry place, separated from the males. Hibernation is often in a cool, sheltered cave.

Very rare. Some recorded in southern England.

Next to the long-eared bat, Bechstein's has the largest ears of any European bat; they measure about 1in (25mm) – more than half the head and body length. The ears droop to the side when the bat is at rest, and hang down – not folded under the wings – when it roosts upside-down. The muzzle is pinkish-brown, fairly long and narrow. Although the bat is easy to identify, sightings are rare. Most have been in Dorset and Surrey, where it hibernates in old mines. Bechstein's bat was first observed on the Continent and is named after a 19th-century German naturalist. It feeds largely on moths.

Very rare. Woodlands in south of England.

J F M A M J J A S O N D

J F M A M J J A S O N D

J F M A M J J A S O N D

Seals

Clumsy on shore but graceful swimmers, seals are voracious fish-eaters, and there is periodic public argument over whether they should be culled (selectively killed). The two species found around British coasts are both protected by law during their breeding seasons.

Grey seal

Halichoerus grypus
6-7ft (1.8-2.1m) long

Cow seals are paler and lighter-bodied

Young after moulting

Female (cow) has narrower muzzle

Male (bull) has convex profile, heavy muzzle

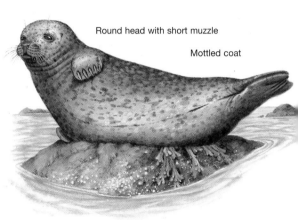

Seals often sleep at sea, floating like upright bottles

Bull looks almost black when wet – largest British mammal

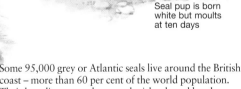

Seal pup is born white but moults at ten days

Common seal

Phoca vitulina
To 5½ft (1.7m) long

Round head with short muzzle

Mottled coat

Head can look like a buoy or fishing float; seal can swim under water for 20-30 minutes

Nostrils in V-shape

Some 95,000 grey or Atlantic seals live around the British coast – more than 60 per cent of the world population. Their breeding grounds are rocky islands and lonely beaches. The grey seal can be recognised by its flattish 'Roman' nose. The cow is smaller than the bull and paler in colour. Grey seals feed largely on cod, whiting and salmon, and they will raid fishing nets to take them. The cow gives birth in the autumn, and the seals are protected by law from September 1 to December 31. Colonies are organised into territories, and the biggest bulls have harems of ten or more cows. Grey seals are noisy animals – barking, hooting, hissing and snarling.

Commonest on rocky shores in the north and west.

The common seal is actually much less common in British waters than the grey, the largest groups gathering around The Wash and on the Norfolk coast. They are considerably smaller than grey seals, the male being bigger than the female. The head is also distinctive; round with a short muzzle. Although common seals may congregate in groups of 100 or more, they do not live in organised herds; they make little noise except for a rather plaintive bark. Fish are caught mainly by diving to search the sea-bed. Common seals are protected by law. Phocine distemper, a viral disease, killed 60 per cent in 1988 and 1989, but the population is now recovering.

Mainly off Ireland, Scotland and eastern England.

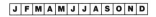

Whales

Toothed whales eat fish, squid and larger sea creatures. They include porpoises and dolphins, which were common in British waters, especially in summer, though they are now generally declining. The huge plankton-feeding baleen whales are rare off our coasts.

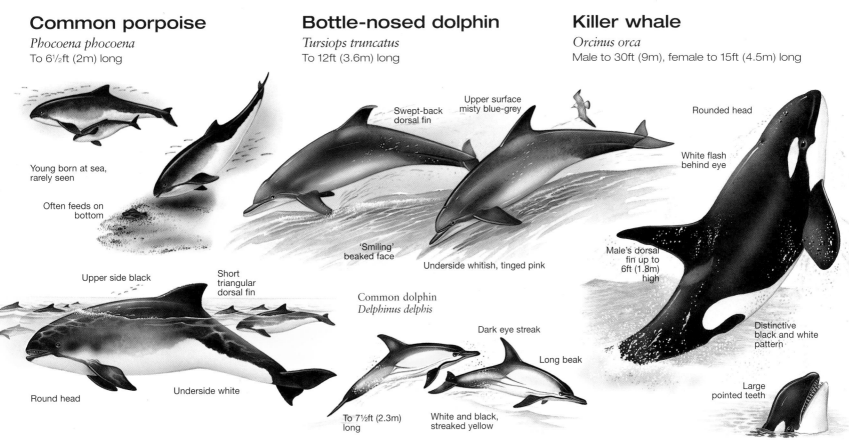

Common porpoise

Phocoena phocoena
To 6½ft (2m) long

Young born at sea, rarely seen

Often feeds on bottom

Upper side black

Short triangular dorsal fin

Round head

Underside white

Bottle-nosed dolphin

Tursiops truncatus
To 12ft (3.6m) long

Swept-back dorsal fin

Upper surface misty blue-grey

'Smiling' beaked face

Underside whitish, tinged pink

Common dolphin
Delphinus delphis

Dark eye streak

Long beak

To 7½ft (2.3m) long

White and black, streaked yellow

Killer whale

Orcinus orca
Male to 30ft (9m), female to 15ft (4.5m) long

Rounded head

White flash behind eye

Male's dorsal fin up to 6ft (1.8m) high

Distinctive black and white pattern

Large pointed teeth

Porpoises are the midgets of the whale world. They migrate seasonally, sometimes in large schools. They have a tremendous appetite, and an adult probably eats about 50 herring-sized fish each day. Porpoises have an extremely accurate echo-location system and acute hearing; even so, some may be found stranded on the shore. Females bear one calf each year – probably far out at sea – after an 11 month pregnancy. The young mature in three to four years and life expectancy is 10-15 years – provided the porpoise does not fall foul of its chief predator, the killer whale.

Seen mostly off north and west coasts, but declining.

After the porpoise, the bottle-nosed dolphin is the whale relative most commonly seen in British waters. It is also the best-known performer in international dolphinarium displays. Bottle-nosed dolphins are playful and highly intelligent, with a fixed 'smile'. Large inshore fish such as cod and salmon are their main food; schools sometimes work together rounding up shoals of fish. These dolphins emit a variety of clicks and whistles used in echo-location. They breed in the summer, when dolphins of both sexes gather in small schools; pregnancy lasts 12 or 13 months. The common dolphin, a frolicsome creature, often travels in large groups, leaping and diving, mainly in deep water.

Seen mainly off south and west coasts.

This fearsome-looking creature with up to 50 large pointed teeth has become familiar to many people as a performer in dolphinariums around the world. Although tame in well-fed captivity, it lives up to its name in the wild. It has an enormous appetite and its prey, in addition to fish, includes seabirds, penguins, seals and even other whales. It will also attack boats. The killer whale is a powerful swimmer, reaching a speed of 30mph (50km/h). Adult males are usually solitary but may be seen with females in small groups. A single baby is born after a year's gestation.

Mostly off Scottish and Irish coasts.

J F M A M J J A S O N D

J F M A M J J A S O N D

J F M A M J J A S O N D

359

Frogs, toads and newts

Common frog

Rana temporaria
Female 3in (75mm); male slightly smaller

The common frog can vary in colour from grey, yellow and brown to orange and red, speckled with black, brown or red. But it always has dark cross-bars on its limbs and a dark patch behind its eyes. Its hibernation can end as early as January in some southern areas.

The tiny tree frog has suction pads on its toes enabling it to climb trees. It was introduced to Britain early in the 20th century and was found mainly in the New Forest and south-east London, but has now died out.

Damp woods and grass near water throughout Britain and Ireland; increasingly scarce with land drainage.

| J | F | M | A | M | J | J | A | S | O | N | D |

Long hind legs enable frog to leap 20in (50cm)

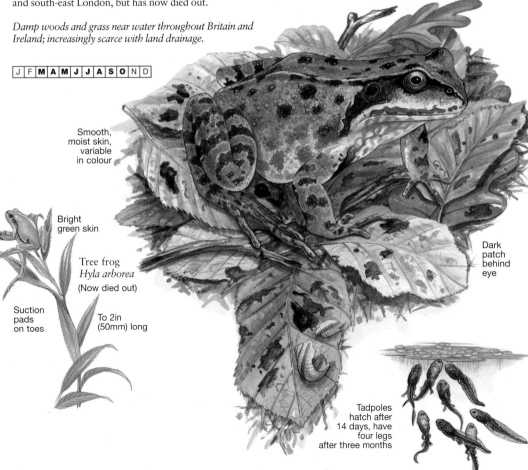

Smooth, moist skin, variable in colour

Bright green skin

Tree frog
Hyla arborea
(Now died out)

Suction pads on toes

To 2in (50mm) long

Dark patch behind eye

Tadpoles hatch after 14 days, have four legs after three months

Marsh frog

Rana ridibunda
Female 5in (12.5cm); male smaller

Pointed snout

Eyes close together, with no dark patch behind them

Vocal sacs inflated

Edible frog
Rana esculenta

Bright green skin

3½in (90mm) long

The biggest frog in Europe is a fairly recent import to Britain. A small number of marsh frogs were released in a garden on the edge of Romney Marsh in 1935. They have since been spread to other parts of England. They are usually light or dark brown, with light green colouring on the back decorated with irregular black spots. From late May the males gather in the water in small groups and croak loudly with the aid of two large, inflated vocal sacs, to attract females.

The closely related bright green edible frog lives in scattered colonies in the south-east and other areas.

Drainage dykes in Romney Marsh and elsewhere.

| J | F | M | A | M | J | J | A | S | O | N | D |

Amphibians live in damp places and lay their eggs in water. They are rather secretive, avoiding exposure to the sun and spending the colder months in hibernation. Frogs, toads and the lizard-like newts all start life as tadpoles, breathing through their gills; later they develop lungs and legs.

Common toad

Bufo bufo
Female to 4in (10cm); male smaller

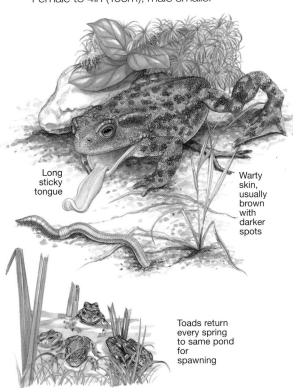

Long sticky tongue

Warty skin, usually brown with darker spots

Toads return every spring to same pond for spawning

The skin of toads is dry and warty compared with that of the frog, which is smooth and wet. Toads also walk rather than leap, and are slow and clumsy in comparison with frogs. Because they are unable to leap out of danger, common toads rely for protection on a bitter poison secreted from glands on their back. They are generally brown in colour, often with darker spots. They prefer to spawn in deep water, the female laying a long jelly-like string up to 10ft (3m) long, containing as many as 7000 black eggs, which becomes entwined with plants in the pond.

Common and widespread in Britain; not in Ireland.

Natterjack toad

Bufo calamita
Head and body 2½in (64mm)

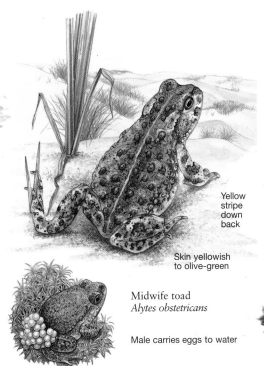

Yellow stripe down back

Skin yellowish to olive-green

Midwife toad
Alytes obstetricans

Male carries eggs to water

A yellow line down the middle of its back distinguishes the natterjack from the common toad. It is also slightly smaller and has a smoother, shinier skin. The toads are fond of sandy places, particularly coastal dunes, where they dig burrows in soft sand, emerging at night to forage on small insects. Winter is spent buried 1-2ft (30-60cm) deep in the sand. Only a very few colonies of midwife toads are known in Britain. They were introduced from France in the 19th century. The male carries the eggs twined round its hind legs, and takes them to water when ready to hatch.

Natterjack: heaths and dunes, spread widely across Britain, but only in about 20 sites.

Smooth newt

Triturus vulgaris
4in (10cm) long overall

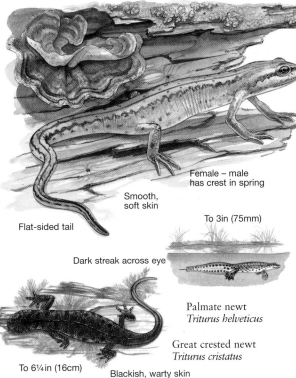

Female – male has crest in spring

Smooth, soft skin

Flat-sided tail

To 3in (75mm)

Dark streak across eye

Palmate newt
Triturus helveticus

Great crested newt
Triturus cristatus

To 6¼in (16cm)

Blackish, warty skin

Although lizard-like in shape, the smooth newt has no scales, moves very slowly and never basks in the sun. It uses its sticky tongue to catch slugs, worms, insects and even other newts. The skin is yellow-olive, spotted on the underside. The smooth newt spends most of the year on land, hiding by day under logs or stones, but breeds in water, wrapping its eggs in water-plant leaves in spring. The great crested newt is the largest native species, with a blackish, warty skin. It is protected by law. The smallest British species is the olive-brown palmate newt found in lowlands and mountainous areas.

Widespread, mostly in lowland areas.

Snakes and lizards

Adder

Vipera berus
Female to 30in (76cm); male to 24in (60cm)

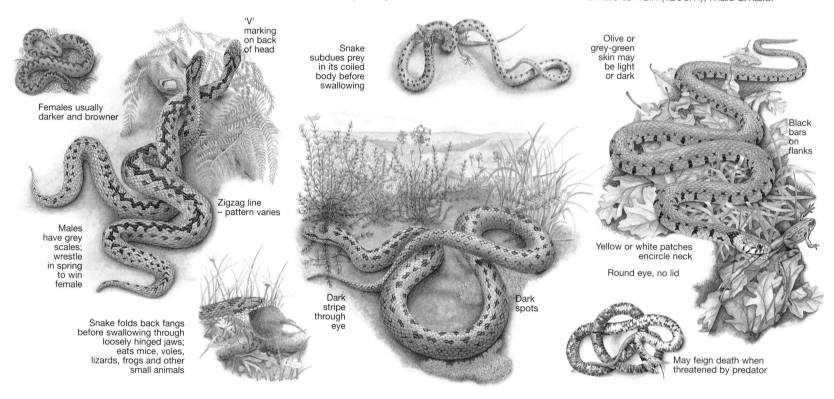

'V' marking on back of head

Females usually darker and browner

Males have grey scales; wrestle in spring to win female

Zigzag line – pattern varies

Snake folds back fangs before swallowing through loosely hinged jaws; eats mice, voles, lizards, frogs and other small animals

Smooth snake

Coronella austriaca
To 24in (60cm)

Snake subdues prey in its coiled body before swallowing

Dark stripe through eye

Dark spots

Grass snake

Natrix natrix
Female to 48in (120cm); male smaller

Olive or grey-green skin may be light or dark

Black bars on flanks

Yellow or white patches encircle neck

Round eye, no lid

May feign death when threatened by predator

Britain's only poisonous snake is shy and nervous, particularly of its greatest enemy, man. Its venom causes pain and swelling followed by diarrhoea and vomiting, but rarely death; only 12 fatalities have been recorded in Britain over the past 60 years. Adders like to bask in open places such as heaths, scrub-covered hillsides and sand-dunes. The female is longer than the male and usually brownish in colour, with a dark brown zigzag stripe on its back, while the male is grey with a black stripe – but colours and patterns vary. They shed their skin from time to time, and hibernate in a dry place underground.

Hedgerows, farmland, moors and woods; not in Ireland.

Although it can bite, the smooth snake is not poisonous. It is one of Britain's rarest animals, seen in only a few southern counties, and protected by law. It is slimmer than an adder, with smoother scales, and has a narrower head with a dark side-stripe through each eye. Colouring varies from grey to brown or red-brown, and there are dark spots down the back, sometimes joined as bars. Prey is mainly other reptiles, such as lizards and slow-worms, together with small mammals and insects. It grips the prey in its mouth, then coils round it, not to crush it but to subdue its struggles.

Rare. Some heaths in southern England.

As the largest and commonest snake in Britain, the grass snake looks threatening but is harmless. Usually it seeks cover when approached – like other snakes it is sensitive to earth vibrations. Two yellow or white patches almost encircle the snake's neck. The flanks may be marked with black bars, and there are black dots along the greenish back. It is usually seen alongside ponds, ditches, streams, field borders and hedgerows. It is an expert swimmer, and hunts fish, frogs, toads and newts, which are swallowed alive. It hibernates under logs or in leaf-litter, dry ditches, or holes in the ground.

Damp heaths, woods or pasture in England and Wales.

J F M A M J J A S O N D

J F M A M J J A S O N D

J F M A M J J A S O N D

Reptiles can live in drier places than amphibians, and bear their young, or lay their eggs, on land. Snakes have dry, scaly skin and move with a gliding motion from side to side. Two lizards have legs and move with a swift, darting wiggle; the third, the slow-worm, is legless and moves like a snake.

Common lizard

Lacerta vivipara
About 6in (15cm)

Sand lizard

Lacerta agilis
About 7in (18cm)

Slow-worm

Anguis fragilis
To 18in (45cm)

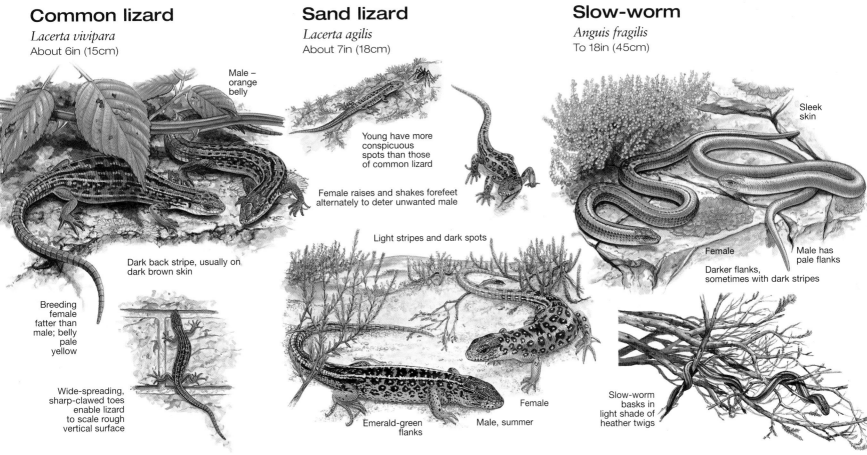

Male – orange belly

Young have more conspicuous spots than those of common lizard

Female raises and shakes forefeet alternately to deter unwanted male

Light stripes and dark spots

Sleek skin

Female

Male has pale flanks

Darker flanks, sometimes with dark stripes

Dark back stripe, usually on dark brown skin

Breeding female fatter than male; belly pale yellow

Wide-spreading, sharp-clawed toes enable lizard to scale rough vertical surface

Emerald-green flanks

Male, summer

Female

Slow-worm basks in light shade of heather twigs

Emerging from hibernation in spring, lizards bask in the sun on heathland, sand-dunes or grassy banks, then court and mate, the young being born in midsummer. As autumn grows colder, they retire into cracks in rocks or under stones to hibernate. They eat a variety of insects and other invertebrates during the warm summer days, spiders being preferred; but they themselves are prey to many animals, including snakes, rats and birds, particularly kestrels. Like other lizards, the common lizard can shed its tail if a predator seizes it – and then grow a new one.

Open spaces throughout British Isles, declining.

During the mating season in May and June the male sand lizard turns bright green on its flanks and underparts. The green gradually fades as the summer passes. More heavily built and solidly marked than the common lizard, the male sand lizard is usually greyish-green and the female brownish. They have two light stripes, sometimes broken, along the back and dark spots between them. They live in dry, open country and hibernate by burrowing deep into the sand. Their main food is large insects such as beetles and grasshoppers. Sand lizards are a rare and declining species, protected by law.

Rare. Sandy heaths in south and dunes in Lancashire.

Though often mistaken for a snake, the slow-worm is, in fact, a legless lizard – and it can move quite fast if disturbed. It is a great slug-eater and is therefore an asset to any garden. Slow-worms like warmth and live on sunny banks and hillsides where there is good cover such as grass, scrub or stones. They hibernate underground from October to March, mate in April and May, and bear their young in August and September. Slow-worms live longer than any other lizards, one in captivity having reached its fifties, but in the wild they are prey to hedgehogs, adders, rats and kestrels.

Widespread throughout Britain; not in Ireland.

| J | F | M | A | M | J | J | A | S | O | N | D |

| J | F | M | A | M | J | J | A | S | O | N | D |

| J | F | M | A | M | J | J | A | S | O | N | D |

363

water life

Under the water's surface lies the most mysterious realm of the natural world, a three-dimensional living space in which water depth, movement and saltiness are just as significant in deciding which species are to be found as geographical location, temperature and other factors.

This section is divided into freshwater and marine life. The latter includes common seaweeds, but plants of rivers and ponds are covered in the wild flowers section at the beginning of the book. Migratory fish such as the salmon and eel are included in the freshwater part, since they are best known as river fish. All the freshwater fish found in the British Isles except a few of the very rarest species are included. They are grouped mainly according to the number of dorsal (back) fins, since they are among the most obvious features by which fish can be identified. The invertebrate life of fresh waters is very diverse, and a representative selection only can be included; distinctive species of ponds and streams are covered separately.

Common fish of salt water are divided similarly. Here it has been possible to cover only a small selection of seashore and rock-pool species and the best known fish caught for eating. Much of the rest of the section is organised on zoological principles, grouping creatures in a way that the amateur as well as the marine biologist will recognise.

Elongated fish

The eel has been observed, caught and eaten for centuries, yet there are still unanswered questions about its behaviour – in particular its migration to its Atlantic spawning grounds. The jawless lampreys are superficially eel-like, but represent remnants of a primitive group.

Eel
Anguilla anguilla
To 40in (100cm)

Brook lamprey
Lampetra planeri
To 10in (25cm)

River lamprey
Lampetra fluviatilis
To 20in (50cm)

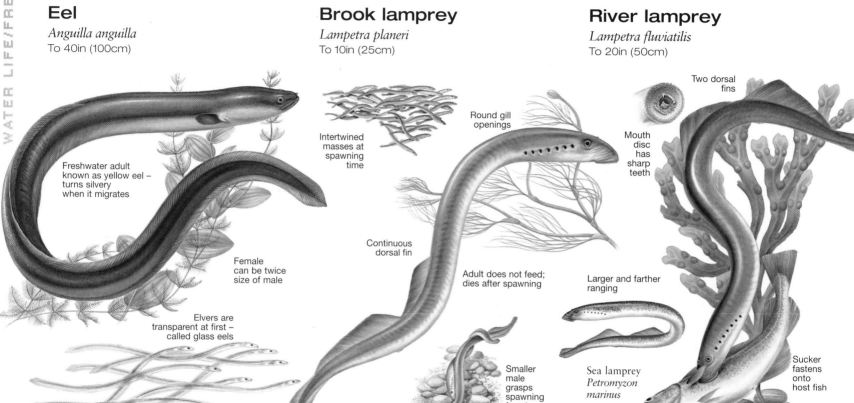

Freshwater adult known as yellow eel – turns silvery when it migrates

Female can be twice size of male

Elvers are transparent at first – called glass eels

Intertwined masses at spawning time

Continuous dorsal fin

Round gill openings

Adult does not feed; dies after spawning

Smaller male grasps spawning female

Two dorsal fins

Mouth disc has sharp teeth

Larger and farther ranging

Sea lamprey
Petromyzon marinus

Sucker fastens onto host fish

Most experts agree that eels spawn in the Sargasso Sea, in the western Atlantic. Mature eels certainly set off from European rivers, apparently to spawn and die, but adults are rarely caught in the ocean. Their transparent, leaf-like larvae are carried by currents to coastal waters where, after three years, they become 2½in (64mm) long elvers. Most find their way into rivers and still waters. Moving upstream in summer, they acquire a brown and yellow colouring. They may live in fresh water for up to 30 years, but eventually return to the coast – now silvery-grey. Eel food includes fish, frogs and snails.

Common in inland waters, estuaries and coastal sea.

A primitive, jawless fish, the lamprey is easily distinguished from an eel by its sucker mouth and seven gill openings on each side of the body. The brook or Planer's lamprey is the smallest and commonest in Britain. It has a dark brown or grey back, with a continuous dorsal fin, and spends all its life in streams. Spawning is between March and June, depending on the water temperature; the male grasps the female, which is larger, with its sucker. After spawning, the adults die. The larvae, or prides, grow slowly for five years, feeding by straining particles from the water.

In many brooks and streams throughout British Isles.

This parasite, also called a lampern, will ride on a swimming fish while sucking its blood. It scrapes a hole in its prey, then clings on by its sucker mouth. After about five years in fresh water it spends a year or two at sea. It returns to fresh water attached to a migrating trout or salmon, drawing nourishment on the way, and spawns the following spring. These lampreys make a tasty dish; King Henry I is said to have died from eating too many. River pollution has reduced their numbers.
The sea lamprey is larger – up to 36in (90cm). It swims farther out to sea, but also enters rivers to spawn.

Many British and Irish streams and rivers, and at sea.

Fish with one dorsal fin

Fish can be grouped for convenience according to the number of dorsal (back) fins they have. A wide variety of fish have only one such fin. The fin's appearance and position can help to identify some species.

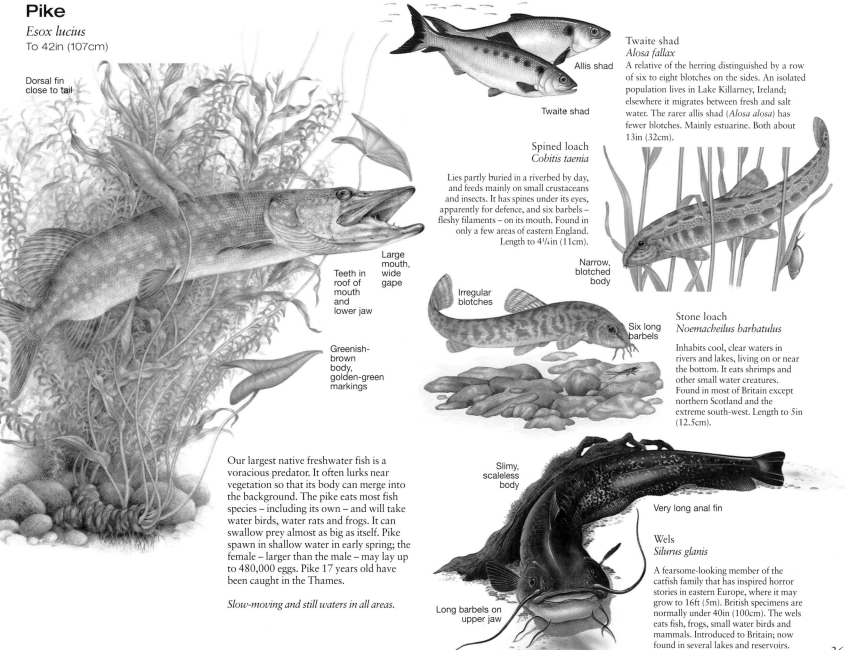

Pike
Esox lucius
To 42in (107cm)

Dorsal fin close to tail

Teeth in roof of mouth and lower jaw

Large mouth, wide gape

Greenish-brown body, golden-green markings

Our largest native freshwater fish is a voracious predator. It often lurks near vegetation so that its body can merge into the background. The pike eats most fish species – including its own – and will take water birds, water rats and frogs. It can swallow prey almost as big as itself. Pike spawn in shallow water in early spring; the female – larger than the male – may lay up to 480,000 eggs. Pike 17 years old have been caught in the Thames.

Slow-moving and still waters in all areas.

Allis shad

Twaite shad

Twaite shad
Alosa fallax
A relative of the herring distinguished by a row of six to eight blotches on the sides. An isolated population lives in Lake Killarney, Ireland; elsewhere it migrates between fresh and salt water. The rarer allis shad (*Alosa alosa*) has fewer blotches. Mainly estuarine. Both about 13in (32cm).

Spined loach
Cobitis taenia

Lies partly buried in a riverbed by day, and feeds mainly on small crustaceans and insects. It has spines under its eyes, apparently for defence, and six barbels – fleshy filaments – on its mouth. Found in only a few areas of eastern England. Length to 4¼in (11cm).

Narrow, blotched body

Irregular blotches

Six long barbels

Stone loach
Noemacheilus barbatulus

Inhabits cool, clear waters in rivers and lakes, living on or near the bottom. It eats shrimps and other small water creatures. Found in most of Britain except northern Scotland and the extreme south-west. Length to 5in (12.5cm).

Slimy, scaleless body

Very long anal fin

Wels
Silurus glanis

A fearsome-looking member of the catfish family that has inspired horror stories in eastern Europe, where it may grow to 16ft (5m). British specimens are normally under 40in (100cm). The wels eats fish, frogs, small water birds and mammals. Introduced to Britain; now found in several lakes and reservoirs.

Long barbels on upper jaw

367

Fish with one dorsal fin/ Carp family

Common carp
Cyprinus carpio
To 23in (58cm)

Crucian carp
Carassius carassius
To 20in (50cm)

Goldfish
Carassius auratus
To 12in (30cm)

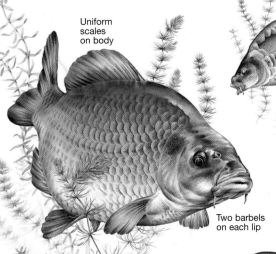

Uniform
scales
on body

Rows of
large scales

Mirror carp

Exceptionally large mirror-like scales
are a distinctive feature of this
variety of common carp. These are
normally present only along the
sides – in the region of the lateral
line – and at the base of the dorsal
fin, but the pattern varies.

Two barbels
on each lip

Scales are
absent

Leather carp

Another variety of common carp,
notable for being wholly or almost
totally scaleless. If there are scales
they will be found just below the
dorsal fin. It was bred for food.

Humped back

No
barbels

No hump on back

To 48in (120cm)

Grass carp
Ctenopharyngodon idella

Colours
vary in
wild
forms

Concave
dorsal fin

Deep body

Originally from central Asia, the common carp
was probably brought to Europe by the Romans,
and is now farmed in many countries. It thrives in
shallow, warm waters. Colour can vary from
muddy slate to brownish-green on the upper part of
the body; the belly is pale. The mouth, which has
four barbels, can be extended like a tube to suck up
food from mud. A carp's diet includes algae, worms,
shellfish and even small fish. It spawns in early
summer, but only in warm water. Carp are very
long-lived – fish over 40 years old are known.

Common in British Isles, especially south-east.

The hardy crucian carp is very variable in
colour, shape and size, depending on its
habitat and available food. Sterile hybrids
sometimes occur due to interbreeding with
common carp. A pure-bred crucian carp
usually has an olive-green or reddish-brown
back, with bronze on the sides and a yellow
or whitish belly. It feeds on plants and small
water creatures. Spawning is from May to
July. The introduced grass carp has a grey-
green back, broad head and large scales.

Eastern and southern England; declining.

The familiar goldfish of the garden pond
has become established in the wild
following many escapes and random
introductions. These fish tend to be olive-
green or brown; gold ones are rare. The
wild goldfish is similar in form to the
larger crucian carp but has a concave-
edged dorsal fin, with a strongly serrated
first spine. Breeding is in June and July if
the water is warm enough. Hardy and
long-lived, it survives well in the wild.

Patchy distribution throughout British Isles.

Carp and their relatives make up the world's largest family of freshwater fishes. The common carp is an important food fish, while others are sought by anglers for sport or kept for ornament. Most species are bottom feeders; some have barbels, fleshy filaments with which they sense food.

Common bream

Abramis brama
To 32in (80cm)

Bitterling

Rhodeus sericeus
To 3½in (90mm)

Tench

Tinca tinca
To 24in (60cm)

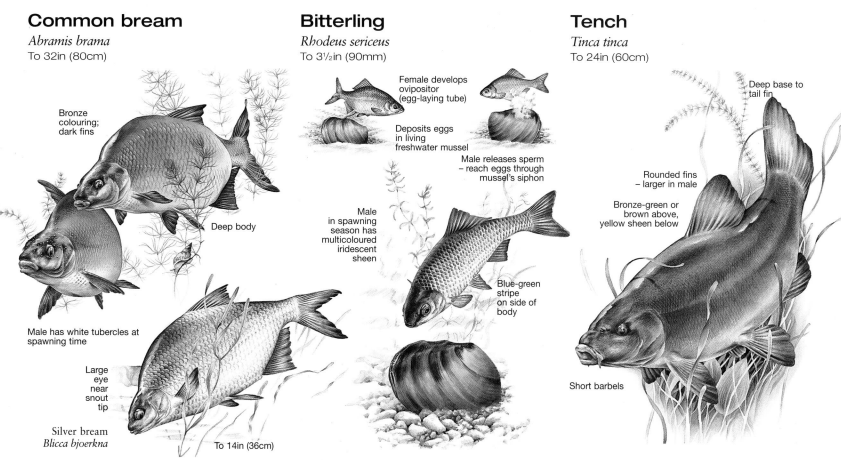

Bronze colouring; dark fins

Deep body

Male has white tubercles at spawning time

Large eye near snout tip

Silver bream
Blicca bjoerkna

To 14in (36cm)

Female develops ovipositor (egg-laying tube)

Deposits eggs in living freshwater mussel

Male releases sperm – reach eggs through mussel's siphon

Male in spawning season has multicoloured iridescent sheen

Blue-green stripe on side of body

Deep base to tail fin

Rounded fins – larger in male

Bronze-green or brown above, yellow sheen below

Short barbels

The deep-bodied common or bronze bream was once a staple fish for poor country folk, but is now valued mainly as a sport fish. It feeds on the bottom of ponds, lakes and slow-moving rivers. Silt deposits with plenty of insect larvae, worms and molluscs suit it best; the mouth is protrusible and can form a tube to suck food from the mud. Bream spawn in shallow water in early summer. The young fish, often called skimmers, are silver at first; their life-span can reach 20 years.

The silver bream is similar in shape, but has larger eyes and lacks the protrusible mouth.

British lowland waters; rarer in the west.

This small fish depends on a living freshwater mussel to reproduce. In spring the male fish, brightly coloured for spawning, selects a mussel and guards it. The female has a long fleshy tube, through which she deposits eggs inside the mussel. The male sperm then reach the eggs through the mussel's siphon system. This process may be repeated by the same pair – or the male may take another partner, using the same mussel. The eggs, protected inside the mussel, hatch two or three weeks later. Bitterling – an introduced species from the Continent – feed on small water creatures, insect larvae and some vegetation.

Only a few places in north-west England.

In the muddy depths of a pond, shallow lake or slow-flowing river the tench is in its element. It can live in water low in oxygen and spends severe winters hibernating in the mud. The tench's food consists mainly of insects, worms, crustaceans, pond snails and plants. Its scales are embedded in a coating of slime which was once credited with healing properties. Tench normally spawn in July and early August, but only if the water is warm enough. An ornamental variety, the golden tench, may be seen in park lakes; its habit of stirring up bottom mud makes it unsuitable for small ponds.

Patchy; commonest in south-east England.

369

Fish with one dorsal fin/ Carp family

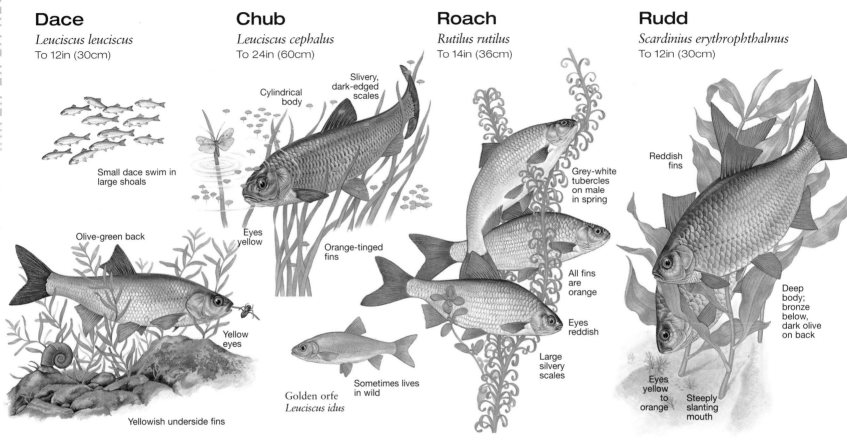

Dace

Leuciscus leuciscus
To 12in (30cm)

Small dace swim in large shoals

Olive-green back

Yellow eyes

Yellowish underside fins

Chub

Leuciscus cephalus
To 24in (60cm)

Cylindrical body

Slivery, dark-edged scales

Eyes yellow

Orange-tinged fins

Golden orfe
Leuciscus idus

Sometimes lives in wild

Roach

Rutilus rutilus
To 14in (36cm)

Grey-white tubercles on male in spring

All fins are orange

Eyes reddish

Large silvery scales

Rudd

Scardinius erythrophthalmus
To 12in (30cm)

Reddish fins

Deep body; bronze below, dark olive on back

Eyes yellow to orange

Steeply slanting mouth

Also called the dart, the dace is similar to the related roach and chub, but can be identified by its yellow eyes and yellowish underside fins. The body is slim and cylindrical, silvery along the flanks and belly, olive-green on the back. The dace's favourite habitat is a clean and fairly fast river. It feeds on insects and their larvae, crustaceans and some vegetation. Spawning time is February to May, and in this period the male develops tubercles.

Common in England, patchy elsewhere; absent from Scotland.

The robust chub is found mainly in slow-flowing rivers; but it also inhabits trout streams and sometimes lakes or gravel pits. It is large-mouthed and omnivorous; bigger chub eat small fish and even frogs and water voles, as well as insects, larvae, freshwater shrimps and plants. Chub spawn in spring, only in flowing waters. The introduced orfe or ide is similar, with reddish fins. Ornamental golden orfe may escape from ponds to survive in the wild.

Common in England and parts of Wales and Scotland.

British anglers catch more roach than any other fish. Roach are abundant and highly adaptable, inhabiting all but the fastest-flowing waters. Interbreeding with rudd and bream can produce hybrids difficult to identify. The true roach can vary from silver to nearly black; orange fins, reddish eyes and large silvery scales distinguish it from the rudd. Its diet includes insect larvae, crustaceans, snails and plants. Spawning is in April and May, the eggs being laid in shallow water attached to plants and stones.

Widespread in British lowland waters.

The 17th-century angler Izaak Walton called the rudd 'a kind of bastard roach', and the two fish are often confused. The rudd has redder fins than the roach, a greener and deeper body and a dorsal fin nearer the tail. The mouth is not suited to bottom feeding, so the rudd eats mainly insects, crustaceans and vegetable matter from the upper and middle depths, although larger rudd will take small fish. They lay transparent yellow eggs among water-weeds from April to June.

In lowland waters of England and Ireland.

Barbel

Barbus barbus
To 36in (90cm)

First ray of dorsal fin is thick and hard

Fleshy lips with four barbels

Reddish tinge to fins

Gudgeon

Gobio gobio
To 8in (20cm)

Purplish iridescent sheen along flanks

Dark spots on back provide camouflage

Two barbels

Flat belly – lives on bottom

Bleak

Alburnus alburnus
To 7in (18cm)

Downward-slanting mouth; large eye

Greenish back

Silvery sides

Narrow body

Minnow

Phoxinus phoxinus
3-4in (75-100mm)

Male's belly and fins redden before spawning in summer

Dark greenish-gold above, dark blotches

Blunt snout; down-turned mouth

Underside white or yellow

The name comes from the four fleshy filaments, called barbels, around this fish's mouth – features shared by a number of other species. It also has thick fleshy lips. The barbel's body, long and round, is of varying colour; it is normally greenish-brown on the back and golden-yellow on the sides. The fish favours flowing water with a stony or sandy bottom. Molluscs, crustaceans, worms and insect larvae are its main diet. Spawning is between May and July.

Clear rivers in many parts of Britain.

This slim fish, once widely caught for the table, is biggest and most plentiful in flooded gravel-pits. It also lives in rivers, canals and ponds. The fish's flat belly enables it to rest on the bottom, where it feeds on plants, small water creatures, and sometimes fish eggs. Gudgeon like shallow water in summer, deeper places in winter, and usually swim in close-packed shoals. Spawning takes place over a long period but peaks in May. The eggs – large for such a small fish – adhere to stones or plants.

England and parts of Ireland; rare elsewhere.

The slim, delicate bleak gives a quicksilver impression as it darts through the water; its lustrous scales were once used to make artificial pearls. Shoals swim near the surface, and occasionally fish will break out of the water to pursue flying insects. Bleak live in slow-flowing lowland rivers and some lakes, feeding on insects and small crustaceans. Larger fish and birds such as kingfishers and grebes eat them in turn. Bleak spawn from April to June, depending on the warmth of the water.

England and parts of Wales.

Small, slender minnows, with blunt snout and down-turned mouth, are often seen darting about in large, compact shoals, relying on their speed to save them from predatory fish and birds. Their presence in a river indicates clean, well-oxygenated water, and they are common in the clear, cool shallows of the upper reaches. When spawning, the male's red belly stands out clearly. Females lay up to 1000 yellow eggs among riverbed stones. The minnow is the smallest British member of the carp family.

In most areas, rarer in Ireland.

Fish with two dorsal fins/ Salmon family

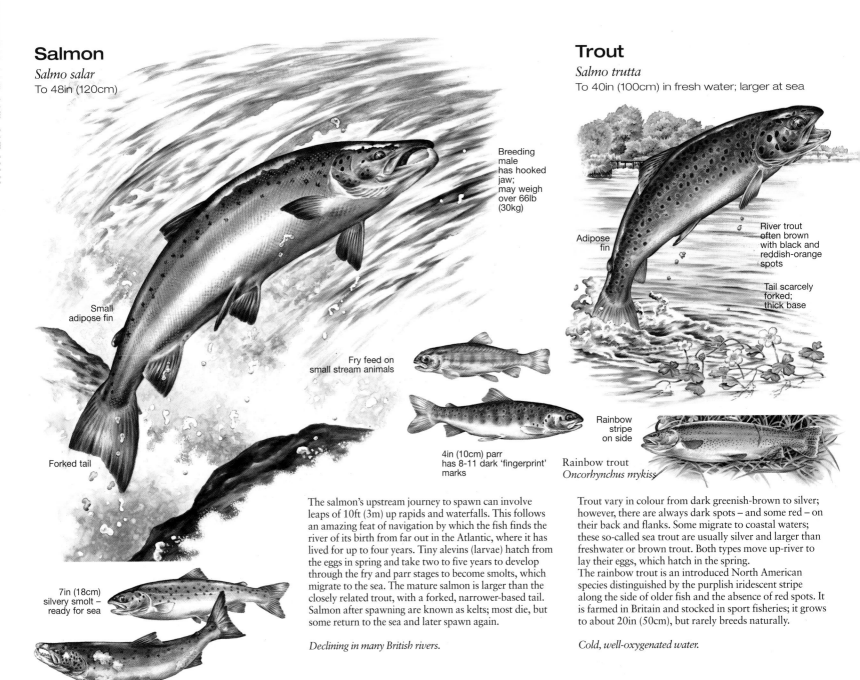

Salmon

Salmo salar
To 48in (120cm)

Breeding male has hooked jaw; may weigh over 66lb (30kg)

Small adipose fin

Fry feed on small stream animals

Forked tail

7in (18cm) silvery smolt – ready for sea

Dark kelt after spawning

4in (10cm) parr has 8-11 dark 'fingerprint' marks

Trout

Salmo trutta
To 40in (100cm) in fresh water; larger at sea

Adipose fin

River trout often brown with black and reddish-orange spots

Tail scarcely forked; thick base

Rainbow stripe on side

Rainbow trout
Oncorhynchus mykiss

The salmon's upstream journey to spawn can involve leaps of 10ft (3m) up rapids and waterfalls. This follows an amazing feat of navigation by which the fish finds the river of its birth from far out in the Atlantic, where it has lived for up to four years. Tiny alevins (larvae) hatch from the eggs in spring and take two to five years to develop through the fry and parr stages to become smolts, which migrate to the sea. The mature salmon is larger than the closely related trout, with a forked, narrower-based tail. Salmon after spawning are known as kelts; most die, but some return to the sea and later spawn again.

Declining in many British rivers.

Trout vary in colour from dark greenish-brown to silver; however, there are always dark spots – and some red – on their back and flanks. Some migrate to coastal waters; these so-called sea trout are usually silver and larger than freshwater or brown trout. Both types move up-river to lay their eggs, which hatch in the spring.
The rainbow trout is an introduced North American species distinguished by the purplish iridescent stripe along the side of older fish and the absence of red spots. It is farmed in Britain and stocked in sport fisheries; it grows to about 20in (50cm), but rarely breeds naturally.

Cold, well-oxygenated water.

Salmon, trout and their relatives form the most important family of freshwater fish with two dorsal fins, and are recognisable by the small, fleshy form of the adipose (fatty) rear fin. In the zander, perch and others the second dorsal fin is spiny like the first.

Char

Salvelinus alpinus
10-12in (25-30cm)

Smelt

Osmerus eperlanus
To 12in (30cm)

Grayling

Thymallus thymallus
To 22in (55cm)

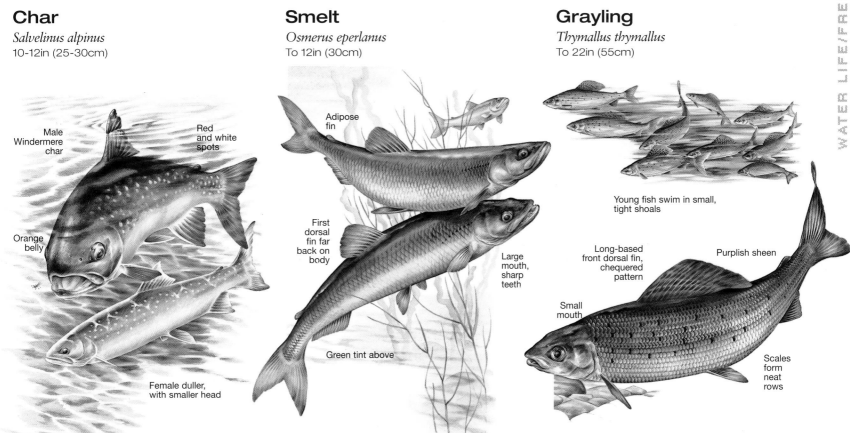

Char
Male Windermere char
Red and white spots
Orange belly
Female duller, with smaller head

Smelt
Adipose fin
First dorsal fin far back on body
Large mouth, sharp teeth
Green tint above

Grayling
Young fish swim in small, tight shoals
Long-based front dorsal fin, chequered pattern
Purplish sheen
Small mouth
Scales form neat rows

The char (or charr) in Britain is a fish of cold, deep lakes. It occurs in some lakes of the Lake District, some Scottish lochs, Irish loughs and in llyns in north Wales. It is a survivor of the Ice Age that is still common in Arctic seas. The char of different waters vary in size and colouring. Among the largest are Windermere char, which can reach 12in (30cm); the male turns orange on the belly at spawning time. Similar colouring gives the name *torgoch*, 'red-belly', to the char of Llyn Padarn in Snowdonia. Char feed mainly on small crustaceans. They reach maturity in four to five years.

Some cold, deep lakes in upland areas of Britain and Ireland.

The slender smelt is a fish of coastal waters and estuaries, moving up-river to spawn. It is silvery, with the front dorsal fin well back on the body. The smelt has a large mouth with many needle-like teeth; its food includes crustaceans and the fry of various fish. The fish reach their river spawning sites in spring but breeding success depends on river conditions between March and July when the eggs hatch and the fry develop. Smelt flesh, when fresh, has an odour of cucumber. The smelt was wiped out by pollution in the 19th century, but there are signs of recovery in some places such as the Thames.

Coastal rivers south of Tay and Clyde; west Ireland.

This fish lives in swift-flowing unpolluted rivers. It can be easily recognised by the long, high, front dorsal fin with its chequered pattern. Grayling feed mainly on small creatures living on the bottom, but may also eat the eggs of trout and salmon. They join in loose shoals for most of the year, pairing off for spawning, which takes place from March to May in gravelly shallows. The male's front dorsal fin – purple-red at this time – curls over the female to keep their bodies close and ensure good fertilisation. A freshly caught grayling has a thyme-like smell – hence the scientific name.

Patchily distributed throughout Britain.

Fish with two dorsal fins

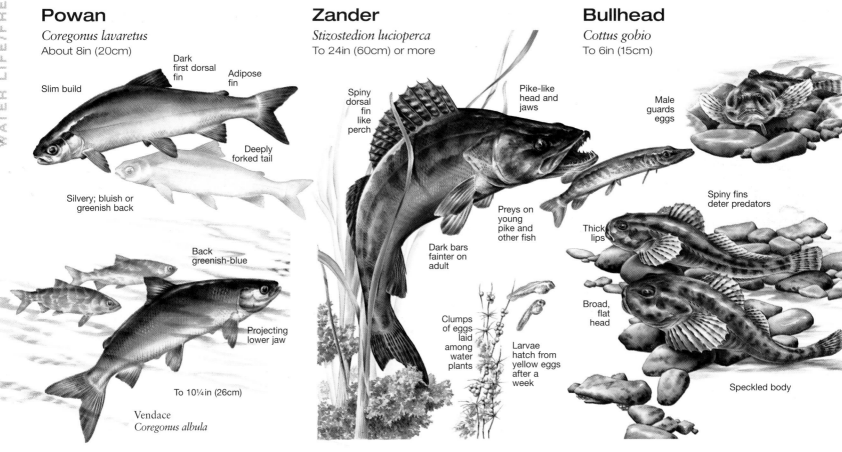

Powan

Coregonus lavaretus
About 8in (20cm)

Slim build

Dark first dorsal fin

Adipose fin

Deeply forked tail

Silvery; bluish or greenish back

Back greenish-blue

Projecting lower jaw

To 10¼in (26cm)

Vendace
Coregonus albula

Zander

Stizostedion lucioperca
To 24in (60cm) or more

Spiny dorsal fin like perch

Pike-like head and jaws

Preys on young pike and other fish

Dark bars fainter on adult

Clumps of eggs laid among water plants

Larvae hatch from yellow eggs after a week

Bullhead

Cottus gobio
To 6in (15cm)

Male guards eggs

Spiny fins deter predators

Thick lips

Broad, flat head

Speckled body

Once the powan was a kind of poor man's herring, widely caught for the market. But it is extremely susceptible to pollution and its numbers have dwindled since the Industrial Revolution. It is still the most widespread of five freshwater species of whitefish, members of the salmon family. It is found in widely separated areas, and has various local names; there is some variation in colour between the populations. The fish are plankton feeders, and spawn from December to March.

The closely related vendace, or pollan, is another species whose numbers have been reduced by pollution.

Very few deep, clean lakes throughout Britain.

This fearsome-looking predator has some of the appearance and characteristics of the pike, but also resembles a perch – hence the alternative name of pike-perch. In fact the zander is a member of the perch family from eastern Europe, introduced to the Woburn Abbey lakes in 1878; it is now quite widespread in East Anglia. Its body is grey-green or brown on the back, paling below; the first dorsal fin is high and spiny. The jaws are powerful and the mouth wide, with fangs and small teeth. The zander lives on a wide variety of smaller fish. A female can lay up to 2 million eggs from April to June.

Lowland rivers, large shallow lakes, fens and canals.

Its broad, heavy head gives this fish both its common names: the second is miller's thumb – millers were supposed to develop broad thumbs by rubbing grain. The bullhead likes shallow rivers with a stony bottom and is also found in the shallows of gravelly lakes and canals. Its body – usually brown or grey, with darker speckles – is flattened underneath. It emerges from hiding places at night to feed on insects, crustaceans, fish eggs or even very small fish. Bullheads spawn in March and April, when heaps of sticky yellow eggs are laid in a hollow scooped out among stones by the male, who then guards them.

Common except in Scotland and Ireland.

Perch

Perca fluviatilis
To 20in (50cm)

Dark spot on front dorsal fin

Bars on side

Lower fins tinged with red

Dorsal fins joined

No bold markings

To 7in (18cm)

Ruffe
Gymnocephalus cernua

The handsome perch shows considerable colour variation; normal colourings are greyish or olive-green on the back, with a paler underside and dark vertical bars on the flanks. The lower fins and sometimes the tail are tinged with red. Perch like clear, slow-moving water. They feed mostly on small fish, lying in wait for them among water plants. Spawning is in April and May, when jelly-like ribbons of eggs are twined around water plants. Once common, many perch were killed by disease in the 1970s. The duller-looking ruffe or pope, a close relative, is quite rare and found in slow-flowing rivers, canals and lakes.

Most of Britain except northern Scotland.

Burbot

Lota lota
To 20in (50cm)

Long rear dorsal fin

Slender body, broad head

Long barbel on chin

The burbot is widely distributed in the northern hemisphere but is probably extinct in Britain. It is the only member of the cod family found in fresh water. It has a slender body, broad head and barbels – two short ones on the nostrils and a long one on the lower jaw. Its colour, generally drab and blotchy, may vary. The burbot lives in lakes and slow-flowing rivers and is largely nocturnal, hiding during the day. The young feed on the larvae of water insects as well as crustaceans, mussels and snails. Older fish are predators, eating perch, roach, gudgeon and other fish. They spawn in sandy or gravel river beds.

Extinct in Britain, due to over-fishing and pollution.

Three-spined stickleback

Gasterosteus aculeatus
To 4in (10cm), usually smaller

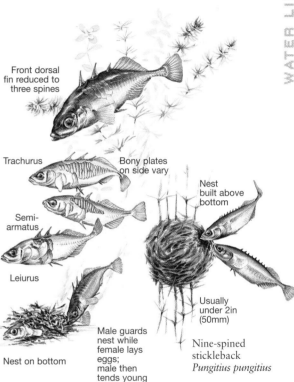

Front dorsal fin reduced to three spines

Trachurus

Bony plates on side vary

Semi-armatus

Nest built above bottom

Leiurus

Nest on bottom

Male guards nest while female lays eggs; male then tends young

Usually under 2in (50mm)

Nine-spined stickleback
Pungitius pungitius

Sticklebacks are our smallest freshwater fish, most being less than 2½in (64mm) long. The commonest is the three-spined, of which there are three forms, with different arrangements of bony plates on the sides. In spring the male develops spawning colours – red throat and belly and bright blue eyes – then performs a courting dance to entice passing females to a nest of vegetation it has built. Sticklebacks eat worms, insects and crustaceans, and in turn are eaten by predatory fish and birds.
The so-called nine-spined stickleback may in fact have 7 to 12 spines. Its breeding behaviour is similar.

Rivers, lakes and ponds; also coastal waters.

375

Invertebrates of still water

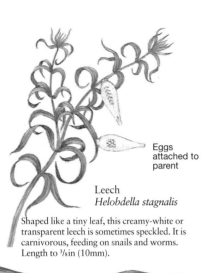

Leech
Helobdella stagnalis

Shaped like a tiny leaf, this creamy-white or transparent leech is sometimes speckled. It is carnivorous, feeding on snails and worms. Length to ³/₈in (10mm).

Eggs attached to parent

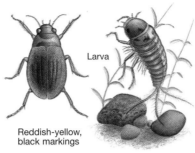

Larva

Reddish-yellow, black markings

Screech beetle
Hygrobia hermanni

Screeches in alarm if picked up. Has prominent eyes and feeds on worms in muddy ponds. Length to ¹/₂in (13mm).

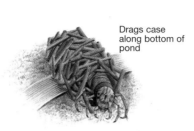

Drags case along bottom of pond

Caddis larva
Limnephilus rhombicus

Caddis larvae build protective cases from a great variety of materials as shelters against predators. This pond-dweller uses plant fragments. Case to 1¹/₂in (38mm) long.

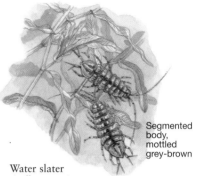

Segmented body, mottled grey-brown

Water slater
Asellus aquaticus

A freshwater crustacean related to the woodlouse; also called the water hog louse. It is common in weedy ponds, canals and moderately polluted rivers where there is plenty of dead organic matter to feed on. Length to ¹/₂in (13mm).

Occasionally floats to surface for air

Great pond snail
Lymnaea stagnalis

Our largest freshwater pulmonate – mollusc with a simple lung – this snail enlarges its shell by adding to the edge of the opening. Height 2in (50mm).

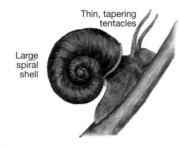

Thin, tapering tentacles

Large spiral shell

Great ramshorn
Planorbarius corneus

Few pond creatures look more elegant on the move than this trumpet snail, the largest ramshorn. Fairly common in ponds, lakes and slow-flowing rivers. Diameter 1³/₄in (30mm).

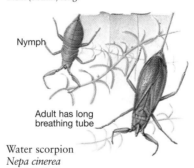

Nymph

Adult has long breathing tube

Water scorpion
Nepa cinerea

The water scorpion has wings but rarely flies. Usually it hangs motionless from the water surface and uses its font limbs to grab prey such as tadpoles, other insects and even small fish. Length 1¹/₃in (35mm).

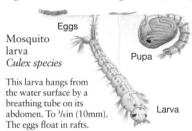

Eggs

Mosquito larva
Culex species

Pupa

This larva hangs from the water surface by a breathing tube on its abdomen. To ³/₈in (10mm). The eggs float in rafts.

Larva

Lesser water boatman
Corixa species

Air trapped among the fine hairs on their bodies makes these insects very buoyant, so they need to use their claws to stay on the bottom. They feed on algae and detritus sucked up through a 'beak' on the rostrum (shell). Length to ¹/₂in (13mm).

Striped wing cases

Sludge worm
Tubifex tubifex

Long thin worms sometimes seen in tangled masses in the mud at the bottom of a pond or stream. The red colouring is caused by the blood pigment haemoglobin, an excellent carrier of oxygen; as a result, it can thrive in water containing very little oxygen. To 3¹/₂in (90mm) long, usually much less.

Lives half-buried in mud

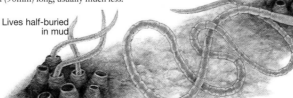

Bristles on all segments

Water fleas
Daphnia and Simocephalus species

Tiny crustaceans found in all kinds of still or slowly flowing water. They feed by filtering small organisms from the water. About ¹/₈in (3mm) long.

Daphnia *Simocephalus*

Flatworm
Polycelis nigra

Minute hair-like cilia covering its body propel this tiny worm. It feeds on living or dead animals which it detects by smell. To ¹/₂in (13mm) long.

Squared head, many eye-spots

An enormous variety of invertebrates – animals without backbones – live in fresh water. Those adapted to the still waters of ponds are becoming scarce as more and more ponds are drained. They include larvae of such flying insects as caddisflies, and beetles, leeches and snails. For surface-dwelling insects, see pp. 318-19.

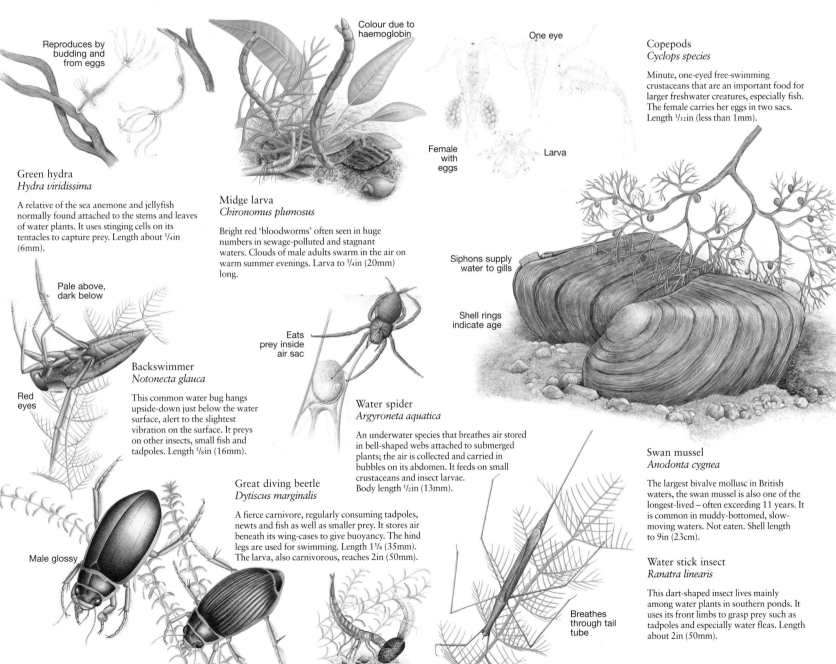

Reproduces by budding and from eggs

Green hydra
Hydra viridissima

A relative of the sea anemone and jellyfish normally found attached to the stems and leaves of water plants. It uses stinging cells on its tentacles to capture prey. Length about $1/4$in (6mm).

Colour due to haemoglobin

Midge larva
Chironomus plumosus

Bright red 'bloodworms' often seen in huge numbers in sewage-polluted and stagnant waters. Clouds of male adults swarm in the air on warm summer evenings. Larva to $3/4$in (20mm) long.

One eye

Female with eggs

Larva

Copepods
Cyclops species

Minute, one-eyed free-swimming crustaceans that are an important food for larger freshwater creatures, especially fish. The female carries her eggs in two sacs. Length $1/32$in (less than 1mm).

Siphons supply water to gills

Shell rings indicate age

Pale above, dark below

Red eyes

Backswimmer
Notonecta glauca

This common water bug hangs upside-down just below the water surface, alert to the slightest vibration on the surface. It preys on other insects, small fish and tadpoles. Length $5/8$in (16mm).

Eats prey inside air sac

Water spider
Argyroneta aquatica

An underwater species that breathes air stored in bell-shaped webs attached to submerged plants; the air is collected and carried in bubbles on its abdomen. It feeds on small crustaceans and insect larvae. Body length $1/2$in (13mm).

Swan mussel
Anodonta cygnea

The largest bivalve mollusc in British waters, the swan mussel is also one of the longest-lived – often exceeding 11 years. It is common in muddy-bottomed, slow-moving waters. Not eaten. Shell length to 9in (23cm).

Male glossy

Great diving beetle
Dytiscus marginalis

A fierce carnivore, regularly consuming tadpoles, newts and fish as well as smaller prey. It stores air beneath its wing-cases to give buoyancy. The hind legs are used for swimming. Length $1^3/8$ (35mm). The larva, also carnivorous, reaches 2in (50mm).

Breathes through tail tube

Water stick insect
Ranatra linearis

This dart-shaped insect lives mainly among water plants in southern ponds. It uses its front limbs to grasp prey such as tadpoles and especially water fleas. Length about 2in (50mm).

Female's wing-cases furrowed

Larva

377

Invertebrates of moving water

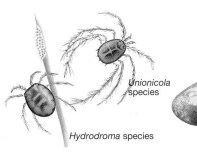

Unionicola species

Hydrodroma species

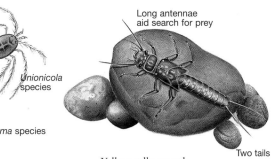

Long antennae aid search for prey

Two tails

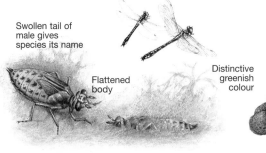

Swollen tail of male gives species its name

Flattened body

Distinctive greenish colour

Water mites
Hydrodroma and Unionicola species

Look like small, headless spiders with round or oval bodies, often brightly coloured, and eight legs. They are predators as adults, but many species have larvae that are parasitic on aquatic insects. Length ⅛in (3mm).

Yellow sally nymph
Isoperla grammatica

A stonefly nymph of fast-flowing rivers and streams, distinguished by its two tails. It is one of the commonest carnivorous stoneflies. The adult is bright yellow. Length ⁷⁄₁₆-⁵⁄₈in (11-16mm).

Club-tailed dragonfly nymph
Gomphus vulgatissimus

Half-buried in river-bottom sand or silt, this broad, yellow-brown and hairy nymph lies in wait for prey such as worms and other small water animals. Length 1¼in (30mm).

Caseless caddis larva
Rhyacophila species

Most caddisfly larvae make protective cases to live in, but some, like this one, are free-living. It is light green and preys on small water animals in stony streams. Length to 1in (25mm).

River limpet
Ancylus fluviatilis

Feeds on minute diatoms and algae

The shape of the thin blackish shell is unmistakable; a cone with a curved top. Like its distant relative the marine limpet, it clings to rocks and water plants with its foot. The soft edge to its shell makes a snug fit. Length of shell ¼-⅓in (6-9mm).

In both still and flowing water

Curved case, almost sealed at rear end

Pea shell cockle
Pisidium species

Common in rivers, canals and lakes, the pea shell cockle has a whitish-yellow or brown shell and a single siphon. It is hermaphrodite. Shell ⅜in (10mm) across.

Welshman's button
Sericostoma personatum

A horn-shaped case, made of grains of sand cemented together, is home for this caddisfly larva, found in fast-flowing streams and some stony lake shores. Case ⅗in (15mm) long. The adult is the model for an angler's fly.

Wandering snail
Lymnaea peregra

A pulmonate, obtaining oxygen via a simple lung. Our commonest freshwater snail, found in rivers, lakes and ponds, even in soft and brackish water. The shape of the shell is very variable; height to ¾in (20mm).

Shell shape varies

Olive upright nymph
Rhithrogena semicolorata

The flat, oval body, edged with rows of paler gills, are striking features of this mayfly nymph. It also has spots on each leg. Length about ½in (13mm). The small angler's curse nymph lives in rivers and on lake shores.

Angler's curse nymph (*Caenis horaria*)

August dun nymph
Ecdyonurus dispar

Another flattened mayfly nymph. Clings to stones and boulders in stony rivers and lake shores. Length ⅝in (16mm).

Three tails distinguish mayfly nymphs

River snail
Viviparus viviparus

A freshwater winkle easily recognised by the three dark bands on its greenish-brown shell. Bears live young. Height 1½in (38mm).

Rivers usually have a more stable temperature than ponds, but the animals that inhabit them have to withstand being washed away by currents. Torrential streams may have few large plants but high oxygen levels. Many river species also survive near the edges of large lakes. For surface-dwelling insects see pp. 318-19.

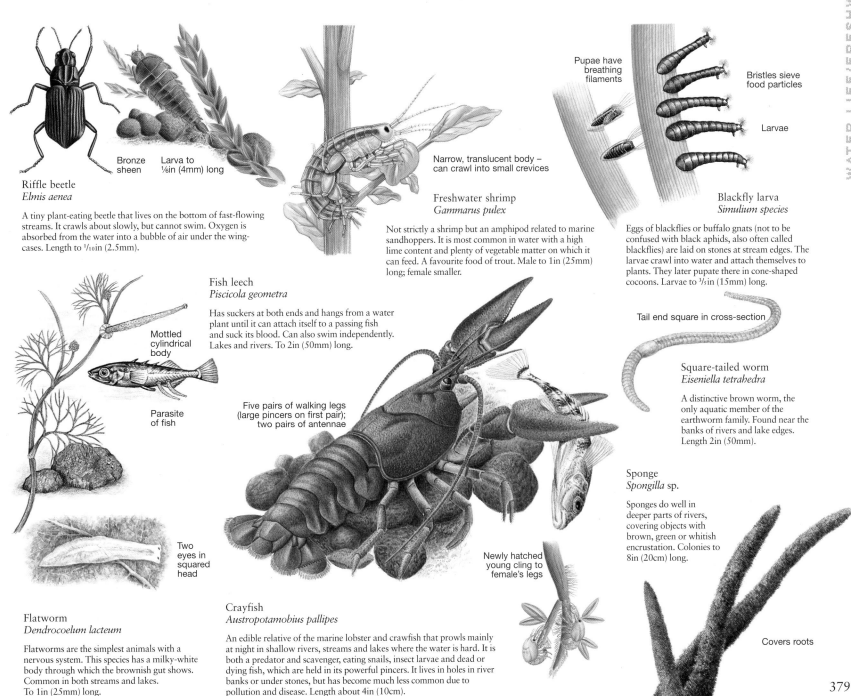

Bronze sheen

Larva to ⅛in (4mm) long

Riffle beetle
Elmis aenea

A tiny plant-eating beetle that lives on the bottom of fast-flowing streams. It crawls about slowly, but cannot swim. Oxygen is absorbed from the water into a bubble of air under the wing-cases. Length to ¹/₁₀in (2.5mm).

Narrow, translucent body – can crawl into small crevices

Freshwater shrimp
Gammarus pulex

Not strictly a shrimp but an amphipod related to marine sandhoppers. It is most common in water with a high lime content and plenty of vegetable matter on which it can feed. A favourite food of trout. Male to 1in (25mm) long; female smaller.

Pupae have breathing filaments

Bristles sieve food particles

Larvae

Blackfly larva
Simulium species

Eggs of blackflies or buffalo gnats (not to be confused with black aphids, also often called blackflies) are laid on stones at stream edges. The larvae crawl into water and attach themselves to plants. They later pupate there in cone-shaped cocoons. Larvae to ³/₅in (15mm) long.

Mottled cylindrical body

Fish leech
Piscicola geometra

Has suckers at both ends and hangs from a water plant until it can attach itself to a passing fish and suck its blood. Can also swim independently. Lakes and rivers. To 2in (50mm) long.

Parasite of fish

Tail end square in cross-section

Square-tailed worm
Eiseniella tetrahedra

A distinctive brown worm, the only aquatic member of the earthworm family. Found near the banks of rivers and lake edges. Length 2in (50mm).

Five pairs of walking legs (large pincers on first pair); two pairs of antennae

Sponge
Spongilla sp.

Sponges do well in deeper parts of rivers, covering objects with brown, green or whitish encrustation. Colonies to 8in (20cm) long.

Two eyes in squared head

Newly hatched young cling to female's legs

Flatworm
Dendrocoelum lacteum

Flatworms are the simplest animals with a nervous system. This species has a milky-white body through which the brownish gut shows. Common in both streams and lakes. To 1in (25mm) long.

Crayfish
Austropotamobius pallipes

An edible relative of the marine lobster and crawfish that prowls mainly at night in shallow rivers, streams and lakes where the water is hard. It is both a predator and scavenger, eating snails, insect larvae and dead or dying fish, which are held in its powerful pincers. It lives in holes in river banks or under stones, but has become much less common due to pollution and disease. Length about 4in (10cm).

Covers roots

379

Flattened fish

Many seabed dwellers have flattened bodies. Some, like the plaice, are compressed sideways, the eyes coming together on one side of the head as the fish grows while the mouth position does not change. But skates and rays are born flattened from top to bottom.

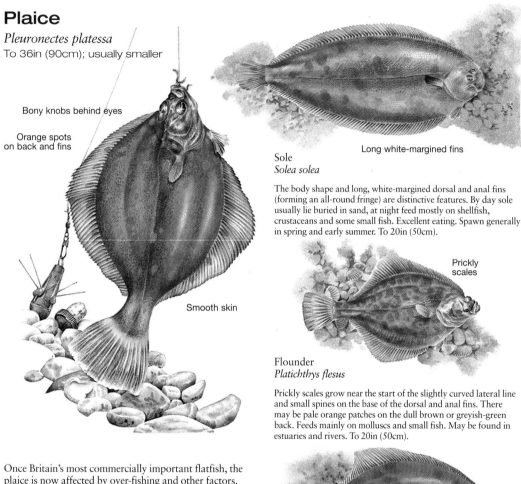

Plaice

Pleuronectes platessa
To 36in (90cm); usually smaller

Bony knobs behind eyes

Orange spots on back and fins

Smooth skin

Long white-margined fins

Sole
Solea solea

The body shape and long, white-margined dorsal and anal fins (forming an all-round fringe) are distinctive features. By day sole usually lie buried in sand, at night feed mostly on shellfish, crustaceans and some small fish. Excellent eating. Spawn generally in spring and early summer. To 20in (50cm).

Prickly scales

Flounder
Platichthys flesus

Prickly scales grow near the start of the slightly curved lateral line and small spines on the base of the dorsal and anal fins. There may be pale orange patches on the dull brown or greyish-green back. Feeds mainly on molluscs and small fish. May be found in estuaries and rivers. To 20in (50cm).

Dark speckles

Dab
Limanda limanda

The eyed side is rough, sandy-brown and usually darkly speckled. Common, especially in the North Sea, and frequently caught by inshore anglers. To 15in (38cm).

Once Britain's most commercially important flatfish, the plaice is now affected by over-fishing and other factors. It is easily distinguished by its brown, orange-spotted top and white underside. Plaice start life as symmetrical fish with an eye on each side; at that stage they feed in surface water. Within two months they start to change shape. The left eye generally moves over the head to join the right, but in some fish the reverse happens. The side with both eyes thus becomes the top and the fish is ready for life as a bottom-dweller. Plaice feed on bristle-worms, shellfish and crabs. They spawn between December and April.

Caught around all coasts.

Thornback ray

Raja clavata
To 34in (86cm)

Egg capsule ('mermaid's purse') 2⅜in (60mm) long; young emerges after 4-5 months

Large spines or 'thorns' on back

Large breathing holes behind eyes

Mainly in shallow water

To about 6ft (1.8m) long

Monkfish
Squatina squatina

Skates and rays, related to sharks, have a skeleton entirely of cartilage. The thornback ray, or roker, is widely sold as skate because the larger 'common' skate is becoming scarce. The thornback's body is mottled grey or fawn, often with brown marbling, and has scattered large-based spines. It feeds mainly on bottom-living crustaceans and small fish at depths of 30-200ft (10-60m). The female lays horned egg capsules – 'mermaid's purses' – from March to August; fertilisation is internal. The related monkfish or angel fish has a curious rounded head. Its flesh is increasingly eaten. It bears live young.

All around coasts, usually over sand and mud.

Elongated fish

The conger eel, one of the great hungry hunters, is entirely a sea creature, but it shares with the freshwater eel - which breeds at sea - some of the mysteries of migration and spawning. But not all elongated fish belong to the eel family; the pipefish, for example, is related to the seahorse.

Conger eel

Conger conger
To 9ft (2.7m)

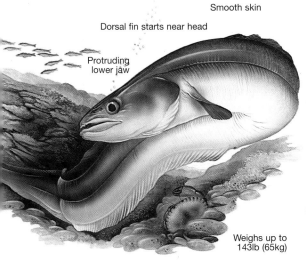

Smooth skin

Dorsal fin starts near head

Protruding lower jaw

Weighs up to 143lb (65kg)

To 6in (15cm)

Lesser sand eel
Ammodytes tobianus

To 8in (20cm)

Black blotch on snout

Greater sand eel
Hyperoplus lanceolatus

Almost anything that moves on the seabed is prey to the voracious conger eel: squid, octopus, crustaceans of all kinds and fish. It has a smooth, apparently scaleless skin and no pelvic fins. Congers like to lurk in rocky crevices and wrecked ships. Like freshwater eels, they go on a long migration – to the east-central Atlantic – before the one spawning of a lifetime, after which they die.
Sand eels, which despite their name and appearance are not true eels, are designed for burrowing in sand. Lesser sand eels can sometimes be seen swimming in shoals near the shore.

Off rocky shores, mostly in north and west.

Butterfish

Pholis gunnellus
To 10in (25cm)

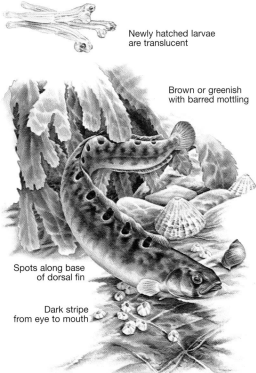

Newly hatched larvae are translucent

Brown or greenish with barred mottling

Spots along base of dorsal fin

Dark stripe from eye to mouth

Its mucus-covered skin and long, spiny dorsal fin make the butterfish difficult to handle – hence its name. Also known as the gunnel, it is familiar in rock pools. The fish bears a distinctive row of 9 to 13 black spots ringed with white along the base of the dorsal fin. It feeds on small crustaceans, molluscs, worms and other creatures. It spawns in January and February, laying eggs in clumps between stones or in empty shells. The female protects them by curling around them, and the male often stands guard. The eggs hatch after about a month, and the young drift out to sea before returning later to inshore waters.

Common on rocky shores.

Greater pipefish

Syngnathus acus
To 18in (45cm)

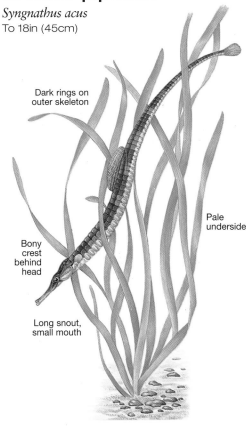

Dark rings on outer skeleton

Pale underside

Bony crest behind head

Long snout, small mouth

Like the seahorse, a close relative, the pipefish has an 'armour-plated' look and swims in an almost vertical position. It is enclosed in bony plates, and its tubular snout, with a small mouth at the end, takes up more than half the length of the head. The fish is well camouflaged in its favourite haunts – the seaweeds and eel grass of inshore waters. It feeds mainly on small crustaceans and fish. Spawning, in spring and summer, starts with a kind of courtship dance. Then the female wraps herself around the male and transfers her eggs to a brood pouch under his tail; there they stay until they hatch.

Shallows around coast and in estuaries.

Fish with one dorsal fin

Like freshwater fish, marine species are conveniently grouped according to the number of dorsal fins. Among those with only one, the appearance and size of the fin vary widely - short and pointed to long and spiny.

Herring
Clupea harengus
To 16in (40cm)

Lumpsucker
Cyclopterus lumpus
Female to 24in (60cm); male smaller

Shanny
Lipophrys pholis
To 6¼in (16cm)

Ballan wrasse
Labrus bergylta
To 24in (60cm)

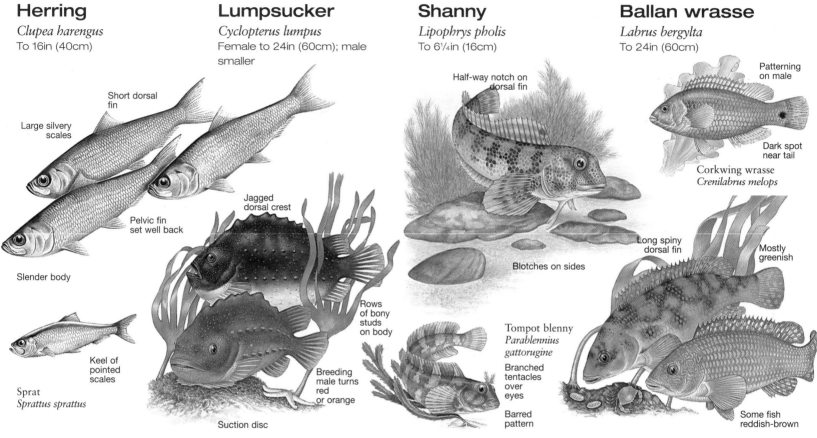

Short dorsal fin

Large silvery scales

Pelvic fin set well back

Slender body

Sprat
Sprattus sprattus

Keel of pointed scales

Jagged dorsal crest

Rows of bony studs on body

Breeding male turns red or orange

Suction disc

Half-way notch on dorsal fin

Blotches on sides

Tompot blenny
Parablennius gattorugine

Branched tentacles over eyes

Barred pattern

Patterning on male

Dark spot near tail

Corkwing wrasse
Crenilabrus melops

Long spiny dorsal fin

Mostly greenish

Some fish reddish-brown

Formerly occurring in vast shoals, the herring is now less common because of over-fishing. The most important stock for the British fishing industry was the Norwegian herring, now sadly reduced. Spawning occurs off the Norwegian coast from February to April; then shoals migrate through the North Sea to spend the summer off Iceland. Herring feed on plankton – tiny floating plants and animals. The sprat is related to the herring but is found in shallower waters. It grows to 6½in (16.5cm).

Less abundant than formerly.

This distinctive shore fish, also called the lumpfish, has a deep, rounded body with rows of bony studs on its sides and, under its belly, a powerful suction disc with which it attaches itself to rocks. In the breeding season the male turns bright red or orange. He devotes himself completely to guarding the eggs, which are laid in a clump among rocks. Lumpsuckers feed mainly on small crustaceans, worms and fish. The flesh is seldom eaten, but the eggs are a common caviar substitute.

All around the coasts of Britain.

Generations of youngsters have tried to catch shannies in rock pools, but their slippery skin and ability to hide in crevices make them elusive. A member of the blenny family, they are usually dull brown or dull green. They feed on a variety of small creatures. Spawning is from April to August. The male stands guard over the eggs – laid in rock crevices – fanning them. The tompot, one of many blennies, lives on south and west coasts. It has a tentacle over each eye, and may grow to 12in (30cm).

Widespread on rocky coasts.

Wrasse – especially breeding males – are among the most colourful fish in British waters. Ballan wrasse are all born female, but some of them later change sex; the mechanism responsible is not fully understood. They feed on crustaceans and molluscs, and in June build a nest from seaweed for their eggs. The bright green young fish are well camouflaged among green seaweed. The corkwing wrasse, greenish to reddish-brown, is 6in (15cm) long. Young ones are common in rock pools.

Rocky coasts; most abundant in the west.

Fish with two dorsal fins

Marine fish with two dorsal fins span the entire size range from the humble gobies, found in many a rock pool, to the mighty and fearsome sharks – of which the harmless dogfish is our commonest example.

Thick-lipped grey mullet

Chelon labrosus
To about 24in (60cm)

Bass

Dicentrarchus labrax
To 40in (100cm)

Dogfish

Scyliorhinus canicula
To 40in (100cm)

Common goby

Pomatoschistus microps
To 2½in (64mm)

Newly hatched fry

Striped appearance

Upper lip appears swollen

Dark patch on gill cover

Spiny front dorsal fin

Body tapers towards tail

Sandy brown, dark spots

Speckled fawn back

Blotches on sides

Length 4¾in (12cm)

Rock goby
Gobius paganellus

A wide, swollen-looking upper lip distinguishes this fish from other mullets around the British coast. It has a torpedo-shaped body with six or seven grey bands running along it. Grey mullet swallow mud and digest the animal and plant life it contains; the thick muscular wall of the stomach grinds the soil and the food to a smooth consistency. Spawning takes place in shallow waters in spring. In the summer, grey mullet may move up estuaries into fresh water.

In coastal waters, estuaries and creeks.

The dorsal fins – the forward one spiny – and the dark patch on its gill covers identify the bass. The adult is a voracious predator, and eats herrings, sprats, pilchards and crabs. It is in turn good to eat and highly regarded by anglers as a sporting fish. Bass spawn in inshore waters during early summer. The younger fish are often seen in estuaries up to the tidal limit. Britain is at the northern extremity of the bass's range, and its growth rate is slower here than in warmer waters.

Most plentiful off southern and western coasts.

This species of dogfish, a harmless creature sold by fish friers as rock salmon, rock fish or flake, is the commonest European shark and is one of three dogfish found in our waters. Its full name is lesser spotted dogfish, but it is also known as the rough hound. Mating takes place in autumn, fertilisation occurring within the female's body, and eggs are laid some weeks later – each in a rectangular capsule, or 'mermaid's purse'. These may take up to 11 months to hatch.

All around coasts; commonest in south.

Tidal pools are the preferred habitat of the common goby, but it is also found in estuaries and drainage channels. A sucker formed from the pelvic fins prevents the fish being swept away by waves. Gobies swim in shoals on or near the bottom, living mostly on small crustaceans. Several spawnings may occur from April to September; the eggs are guarded devotedly by the male. There are 17 goby species in British waters, including the rock goby, which tends to be solitary.

Pools on sandy shores and in salt-marshes.

383

Fish with three or more dorsal fins

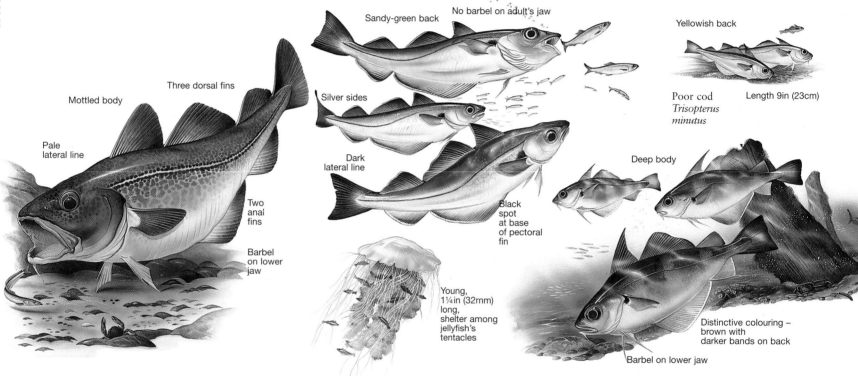

Cod
Gadus morhua
To 48in (120cm), sometimes more

Three dorsal fins

Mottled body

Pale lateral line

Two anal fins

Barbel on lower jaw

Whiting
Merlangius merlangus
To 20in (50cm)

Sandy-green back

No barbel on adult's jaw

Silver sides

Dark lateral line

Black spot at base of pectoral fin

Young, 1¼in (32mm) long, shelter among jellyfish's tentacles

Bib
Trisopterus luscus
To 16in (40cm)

Yellowish back

Poor cod
Trisopterus minutus

Length 9in (23cm)

Deep body

Distinctive colouring – brown with darker bands on back

Barbel on lower jaw

Nine million eggs have been found in a female cod, though two or three million is more normal. Because the cod is so prolific, it has been the mainstay of the British fishing industry. Disputes over cod-fishing rights have gone on for centuries, and the 1960s and 1970s saw the 'cod war' involving Britain, Iceland and Denmark. There are now fears about the cod's decline, due to over-fishing around our shores. Their colour varies from brownish to pale grey, always with dark mottling on the upper body. Cod eat small fish, worms, crustaceans and molluscs. They spawn from February to April, generally in water about 330ft (100m) deep; the fry hatch after about 12 days.

Less abundant around Britain than previously.

Another member of the cod family with a high commercial value, the whiting lives in fairly shallow seas – to around 165ft (50m) – for most of the year and is frequently caught by anglers. It is slimmer and smaller than the cod and has the same distribution of fins, but only young fish have a chin barbel, which disappears with maturity. Food is small fish and crustaceans. Whiting can spawn as early as January and as late as June, according to location. If threatened by larger fish, young whiting often take refuge among the tentacles of a jellyfish. Most predators will not follow for fear of a sting – from which the little whiting seem to be immune.

Common in British inshore waters.

The bib or pout is a common shallow-water fish, and can be distinguished from other members of the cod family by its deeper body and distinctive colouring. The upper body is a striking coppery-brown with four or five darker bands; the sides are yellowish, the underside white. Each pectoral fin has a black spot at the base and the lower jaw carries a long barbel. Bib feed in large shoals on molluscs, small fish and crustaceans. Spawning is normally in March or April; the eggs float in surface waters and hatch in 10-12 days.
The poor cod is similar in appearance but lives mainly offshore. Neither fish has much market value.

All around the British coast.

Some of the most highly prized commercial fish in British waters belong to groups with three or more dorsal fins. Some, like the prolifically breeding cod, are primarily deep-water dwellers, while others, such as the whiting, are commonly found inshore and are often taken by anglers.

Mackerel

Scomber scombrus
To 26in (66cm)

Pollack

Pollachius pollachius
To 51in (130cm)

Haddock

Melanogrammus aeglefinus
To 30in (76cm)

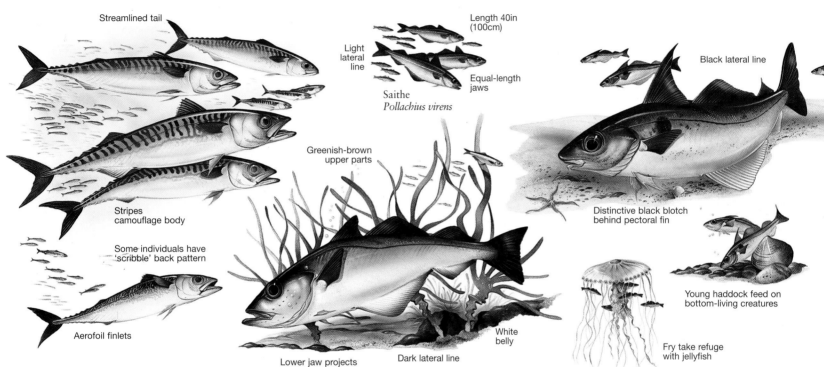

Streamlined tail

Stripes camouflage body

Some individuals have 'scribble' back pattern

Aerofoil finlets

Length 40in (100cm)

Light lateral line

Equal-length jaws

Saithe
Pollachius virens

Greenish-brown upper parts

Lower jaw projects

Dark lateral line

White belly

Black lateral line

Distinctive black blotch behind pectoral fin

Young haddock feed on bottom-living creatures

Fry take refuge with jellyfish

The mackerel is well equipped to elude its hunters and catch its own prey. It is designed for fast swimming: the slender fins can be flattened against the body to reduce drag, aided by a jelly-like covering over the eyes. In addition to the normal fins there are aerofoil-like finlets near the tail. And their silver underside and iridescent blue-green stripes on the upper body make mackerel virtually invisible in the dappled upper waters of the sea. In summer they prey on small fish. Spawning reaches a peak in May and June, when large shoals can be seen off some coasts. In winter they go into deeper waters and eat little.

All around coasts in summer.

The pollack is principally an inshore dweller, but the larger fish tend to spend the winter in deeper waters and move in only for the summer. Rough, rocky waters are preferred. A typical member of the cod family, the pollack has three dorsal fins and two anal fins. Its upper body is dark greenish-brown and its belly white; there is no barbel. The diet consists mainly of crustaceans, small fish and sand eels. Spawning occurs from January to April in depths down to 330ft (100m).
The related saithe, also known as the coley or coalfish, is more important commercially.

All around Britain; commonest in south and west.

Whether eaten fresh or smoked, the haddock is an important commercial fish. It lives close to the seabed in depths of 100-1000ft (30-300m). The upper body is purplish or greenish-grey, the sides silver and the belly white; the lower jaw bears a small barbel. Haddock live on shellfish, crustaceans, sea urchins and marine worms. They spawn in the spring in the northern North Sea and south of Iceland. Eggs drift in the currents for about 15 days before hatching; changing currents could account for a scarcity of young haddock in some years. Inshore migrations of haddock occur in winter.

Chiefly in the North Sea.

385

Crabs

Edible crab

Cancer pagurus
Shell to 8in (20cm) across

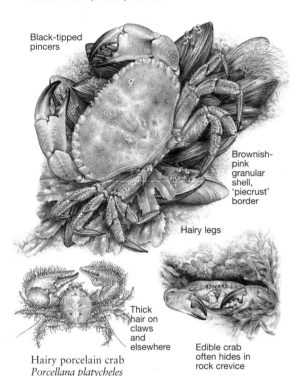

Black-tipped pincers

Brownish-pink granular shell, 'piecrust' border

Hairy legs

Thick hair on claws and elsewhere

Hairy porcelain crab
Porcellana platycheles

Edible crab often hides in rock crevice

Hairy legs and massive black-tipped pincers help to make the edible crab a formidable-looking creature – the largest species of crab commonly found in our waters. Younger and smaller specimens are found under rocks and in seaweed, but the bigger crabs prefer deeper waters. Shellfish are their principal food. A female first mates when she is five or six years old; she may lay 3 million eggs, but few survive. Edible crabs are caught in pots baited with fish.
At the other extreme in size are the porcelain crabs. The hairy species, with a shell no more than 1in (25mm) long, is found under stones, especially on muddy gravel.

Rocky areas around Britain and Ireland.

Shore crab

Carcinus maenas
Shell to 3½in (90mm) across

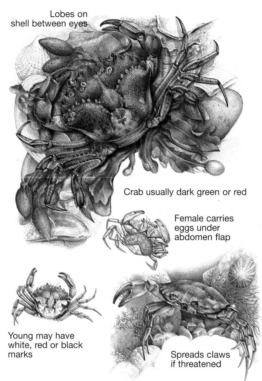

Lobes on shell between eyes

Crab usually dark green or red

Female carries eggs under abdomen flap

Young may have white, red or black marks

Spreads claws if threatened

This is the crab the seashore stroller is most likely to encounter either live or as a discarded shell. It can survive a long time without water, and is often found hiding under stones or in seaweed. The shell, or carapace, has three lobes between the eyes and five sharp points on the edge behind each eye. The female carries an orange-coloured mass of eggs for 12-18 weeks before they hatch. The larvae float in sea currents for several weeks before they descend to the seabed in mature form. Shore crabs are scavengers and will eat almost anything; they themselves are eaten by fish and seabirds. They are edible but are not fished commercially in British waters.

On all types of beach and in some estuaries.

Hermit crab

Pagurus bernhardus
To 4in (10cm) long without shell

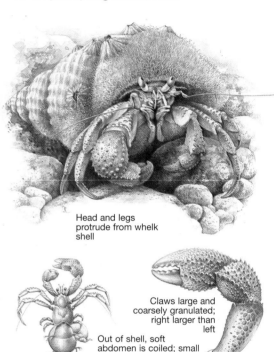

Head and legs protrude from whelk shell

Claws large and coarsely granulated; right larger than left

Out of shell, soft abdomen is coiled; small limbs grasp shell

The hermit crab lives its life in another creature's armour. It hides its soft abdomen from attackers in a discarded shell – often a whelk's – and walks on its two front pairs of legs, dragging the shell behind. A great scavenger, the hermit crab eats almost anything organic, and in turn is preyed upon by seabirds – if they can break its shell. In normal crab fashion, it sheds the shell covering of its claws and front legs as it grows. Periodically it will also change its borrowed shell. This is often shared with marine worms, and parasitic anemones attach themselves to the outside. A female hermit crab, after mating, carries her eggs until they hatch.

Most kinds of beach, including sand-flats.

Crabs are shore or seabed dwellers with two pincers and four pairs of walking legs. As a crab grows the shell is moulted at intervals and replaced by a new growth beneath. Mating generally occurs only when the female has recently moulted. Damaged limbs can be shed and regrown.

Velvet crab

Necora puber
Shell to 3½in (90mm) across

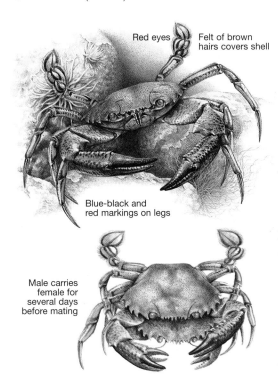

Red eyes

Felt of brown hairs covers shell

Blue-black and red markings on legs

Male carries female for several days before mating

Masked crab

Corystes cassivelaunus
Shell to 1½in (40mm) across

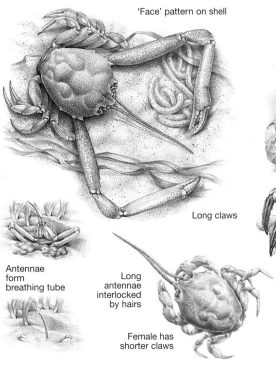

'Face' pattern on shell

Long claws

Antennae form breathing tube

Long antennae interlocked by hairs

Female has shorter claws

Spiny spider crab

Maja squinado
Shell to 8in (20cm) across

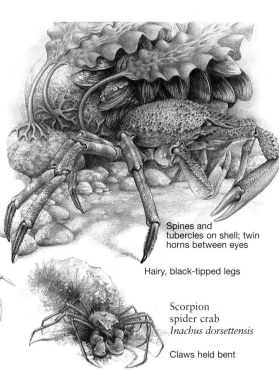

Spines and tubercles on shell; twin horns between eyes

Hairy, black-tipped legs

Scorpion spider crab
Inachus dorsettensis

Claws held bent

This handsome, aggressive crab gets its name from the velvety covering of fine hairs on its shell; it is also called the fiddler crab. Other distinctive features are the bold blue-black and red leg markings and the bright red eyes, which can help to frighten possible attackers. It uses its flattened, paddle-like hind legs to swim rapidly sideways to avoid predators such as cuttlefish. If cornered, it tries to repel an enemy by sitting back and holding its claws wide apart, thus looking as large as possible. Velvet crabs will eat just about anything they can find as they hunt through the kelp forest and under seaweed on the shore.

In rocky places, on and offshore to most coasts.

Furrows on the upper shell resembling a human face give this crab its name. But its most remarkable feature is a pair of extremely long antennae which are linked throughout their length by a network of interlocking hairs. In daylight the crab lies buried in several inches of sand, with only the antennae tips extending into the water above. It breathes by taking in water through the tube formed by the hairs and passing it over the gills. In darkness the crab emerges to hunt for shrimps, worms and other small sea creatures. Large numbers of masked crabs come ashore to mate in early summer; unlike other species, the females do not have to moult before mating.

Clean sand and shallow water all around British Isles.

Long legs and a more or less triangular body characterise the spider crabs. All adopt a form of camouflage, draping themselves with seaweed, sponges and other debris. The spiny spider crab, or thornback crab, has sharp spines on its body shell, and is sometimes eaten. Its food is chiefly small sea animals and seaweed. In summer it moves from deeper to shallower inshore waters to breed, forming large mating mounds of up to 100 crabs. The fertilised female carries up to 150,000 eggs for nine months. The reddish-brown scorpion spider crab has short, stout claws. The shell bears five tubercles and grows to a length of 1in (25mm).

Mainly rocks and sand in south and south-west.

Lobsters, prawns and shrimps

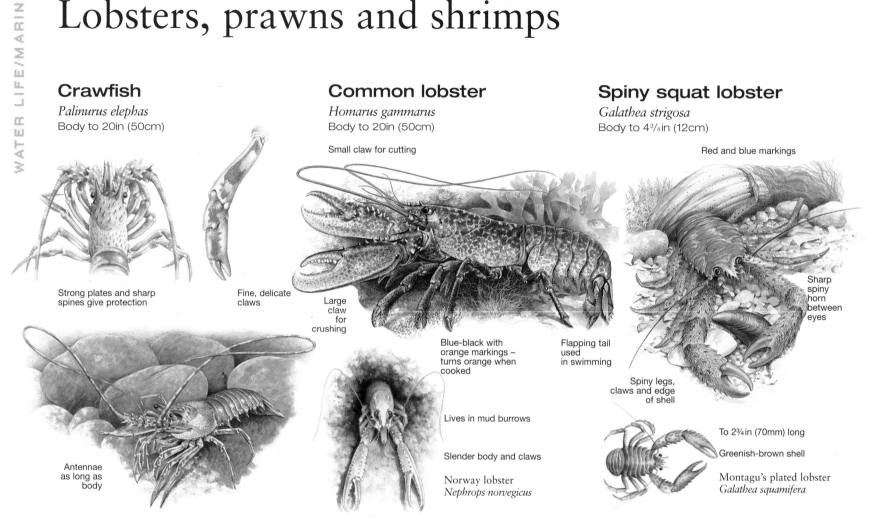

Crawfish
Palinurus elephas
Body to 20in (50cm)

Strong plates and sharp spines give protection

Fine, delicate claws

Antennae as long as body

Common lobster
Homarus gammarus
Body to 20in (50cm)

Small claw for cutting

Large claw for crushing

Blue-black with orange markings – turns orange when cooked

Flapping tail used in swimming

Lives in mud burrows

Slender body and claws

Norway lobster
Nephrops norvegicus

Spiny squat lobster
Galathea strigosa
Body to 4¾in (12cm)

Red and blue markings

Sharp spiny horn between eyes

Spiny legs, claws and edge of shell

To 2¾in (70mm) long

Greenish-brown shell

Montagu's plated lobster
Galathea squamifera

Lacking the lobster's powerful pincers, the crawfish or spiny lobster cannot crush hard-shelled creatures for food, so it lives on small, soft-bodied animals and sometimes dead fish. For protection it has a heavy, very spiny shell which can inflict a serious wound if it is mishandled. It is reddish-brown, with antennae twice as long as the body. In winter, crawfish mate in deep water; females carry the eggs for several months until they hatch, and the young may drift many miles before dropping to the seabed. Crawfish are not widely eaten in Britain but are popular in France as *langouste*.

Warm waters off south-west, western Scotland and Ireland.

The common lobster's massive claws cut and crush its food and protect it against predators, such as seals. It feeds mainly at night on crabs, shellfish and other sea creatures. Like the crab, it grows by periodically shedding its shell. It can live for 30 years. A female mates at about seven years old and lays an orange mass of eggs in late summer. She carries them under the abdomen for up to ten months until the shrimp-like larvae hatch.
The Norway lobster is probably better known by the restaurant name of Dublin Bay prawn. Orange-coloured and like a slender lobster, it is about 6in (15cm) long.

Rocky areas all around coasts, mostly below low tide mark.

Although its shell is vividly marked with red and blue, this creature – actually a close relative of the hermit crab – is not easy to find. It can hide its flattened body under stones, in crevices or under overhanging rocks, where it clings upside-down. The tail and last rudimentary pair of legs curl under the body. The squat lobster feeds at night, mainly on particles collected from the sea bottom. After mating, the eggs are carried by the female.
Montagu's plated lobster has a greenish-brown shell sometimes flecked with red. It is fairly common on the lower shore, especially under stones.

On all rocky British and Irish coasts.

Lobsters are classified, like crabs, in the decapod division of the crustaceans – meaning that they have a hard external skeleton and ten legs. The same applies to the squat lobsters, but these are more closely related to hermit crabs than to true lobsters, crawfish, prawns and shrimps.

Long-clawed squat lobster

Munida rugosa
Shell to 2½in (65mm)

Long antennae

Flattened body

Claws twice as long as body

Three spines between eyes

With claws at least twice as long as its body and antennae almost as long as its claws, this creature can afford to bury the rest of itself. It likes areas where sediment accumulates between rocks. By digging out a hole under a boulder it can stay hidden and protected by day with only the antennae and claws protruding. At night it emerges to hunt for small crustaceans and worms. It uses its delicate claws to probe crevices and sift through sediment, but they are not as effective for defence as the strong claws of its spiny relative. Both are sometimes eaten locally.

Below low-water mark off western coasts, especially Scottish sea-lochs.

Common prawn

Palaemon serratus
Body to 4in (10cm)

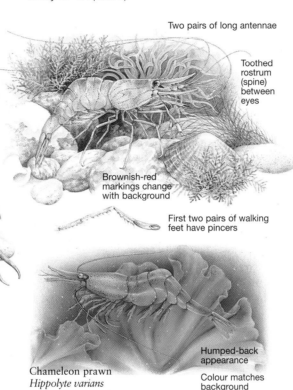

Two pairs of long antennae

Toothed rostrum (spine) between eyes

Brownish-red markings change with background

First two pairs of walking feet have pincers

Humped-back appearance

Colour matches background

Chameleon prawn
Hippolyte varians

An almost transparent body makes the prawn difficult to spot; it turns red only when boiled. When feeding, it walks delicately on its last three pairs of legs; the first two pairs have pincers for picking up small food particles. Two pairs of antennae and eyes on movable stalks can sense food or danger in any direction – very necessary, for the prawn is prey to many sea dwellers. In summer, prawns mate after the females have moulted.
The chameleon prawn, as its name indicates, can change colour to match its background – green, red or brown by day, transparent blue by night. It grows to 1in (25mm).

Rock pools in south and west; also among eel grass on sand.

Common shrimp

Crangon crangon
Body to 3½in (90mm), usually smaller

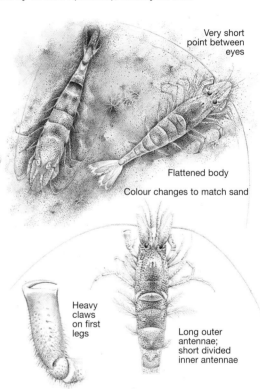

Very short point between eyes

Flattened body

Colour changes to match sand

Heavy claws on first legs

Long outer antennae; short divided inner antennae

The shrimp's flat body can change colour from yellow to almost black to match the sand over which it moves and in which it hides during the day. At night it comes out to feed, rarely swimming but walking slowly on its last two pairs of legs while using the long third pair to investigate hard objects. The shrimp's stout claws enable it to tackle almost anything – including worms, young fish, small crustaceans and plants. The female carries fertilised eggs on her swimmerets – paddle-like limbs beneath the tail. Shrimps are hardy and tolerate a wide temperature range. They are caught commercially in several areas.

Abundant in sandy areas, including bays and estuaries.

389

Small seashore creatures

The seashore supports a fascinating variety of small animals – primitive insects, crustaceans and even spiders – all of which have established a successful way of living with the tide's ebb and flow

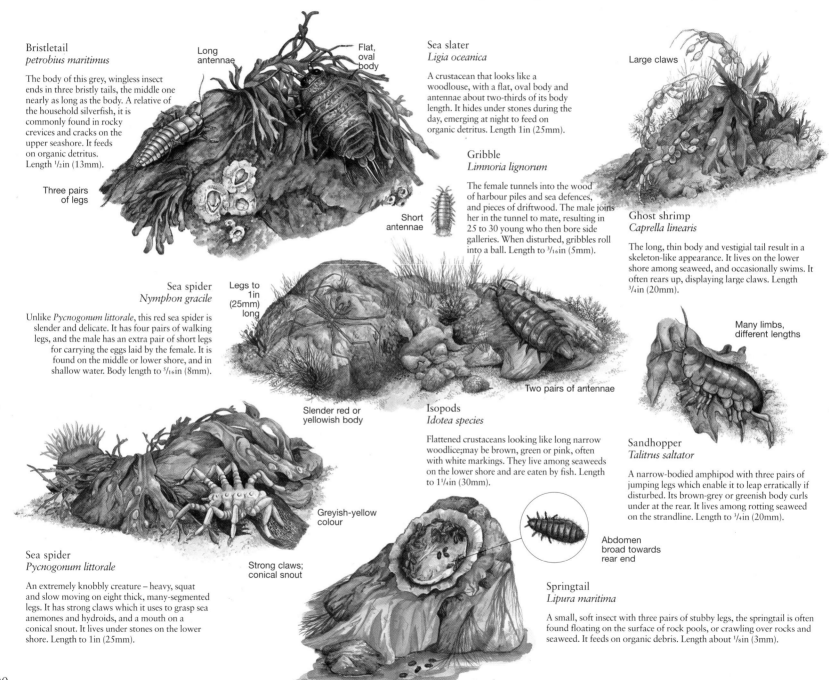

Bristletail
petrobius maritimus

The body of this grey, wingless insect ends in three bristly tails, the middle one nearly as long as the body. A relative of the household silverfish, it is commonly found in rocky crevices and cracks on the upper seashore. It feeds on organic detritus. Length 1/2in (13mm).

Long antennae

Flat, oval body

Three pairs of legs

Sea slater
Ligia oceanica

A crustacean that looks like a woodlouse, with a flat, oval body and antennae about two-thirds of its body length. It hides under stones during the day, emerging at night to feed on organic detritus. Length 1in (25mm).

Short antennae

Large claws

Gribble
Limnoria lignorum

The female tunnels into the wood of harbour piles and sea defences, and pieces of driftwood. The male joins her in the tunnel to mate, resulting in 25 to 30 young who then bore side galleries. When disturbed, gribbles roll into a ball. Length to 3/16in (5mm).

Ghost shrimp
Caprella linearis

The long, thin body and vestigial tail result in a skeleton-like appearance. It lives on the lower shore among seaweed, and occasionally swims. It often rears up, displaying large claws. Length 3/4in (20mm).

Sea spider
Nymphon gracile

Unlike *Pycnogonum littorale*, this red sea spider is slender and delicate. It has four pairs of walking legs, and the male has an extra pair of short legs for carrying the eggs laid by the female. It is found on the middle or lower shore, and in shallow water. Body length to 5/16in (8mm).

Legs to 1in (25mm) long

Slender red or yellowish body

Two pairs of antennae

Many limbs, different lengths

Isopods
Idotea species

Flattened crustaceans looking like long narrow woodlice;may be brown, green or pink, often with white markings. They live among seaweeds on the lower shore and are eaten by fish. Length to 1 1/4in (30mm).

Sandhopper
Talitrus saltator

A narrow-bodied amphipod with three pairs of jumping legs which enable it to leap erratically if disturbed. Its brown-grey or greenish body curls under at the rear. It lives among rotting seaweed on the strandline. Length to 3/4in (20mm).

Sea spider
Pycnogonum littorale

An extremely knobbly creature – heavy, squat and slow moving on eight thick, many-segmented legs. It has strong claws which it uses to grasp sea anemones and hydroids, and a mouth on a conical snout. It lives under stones on the lower shore. Length to 1in (25mm).

Greyish-yellow colour

Strong claws; conical snout

Abdomen broad towards rear end

Springtail
Lipura maritima

A small, soft insect with three pairs of stubby legs, the springtail is often found floating on the surface of rock pools, or crawling over rocks and seaweed. It feeds on organic debris. Length about 1/8in (3mm).

Marine worms

Many species of worms live among the rocks, sand and mud of shores and estuaries, showing a wide range of adaptations to their environment. Some build solid, if rather rough and rustic, structures, some live in tunnels in the sand and some, like the ragworm, are highly mobile carnivorous hunters.

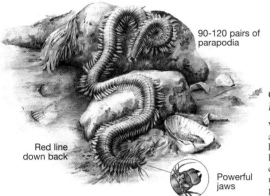

90-120 pairs of parapodia

Red line down back

Powerful jaws

Common ragworm
Hediste diversicolor

With four black eyes, several pairs of feelers and large, black, pincer-like jaws, this worm hunts small creatures. Large specimens can bite humans. The ragworm swims or crawls over sediment with the aid of numerous leg-like outgrowths called parapodia. Length to 4³/₄in (12cm).

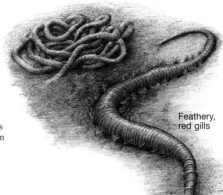

Lugworm
Arenicola marina

The familiar casts seen on sandy beaches indicate the presence of the lugworm. The worm swallows sand, digests the food in it, and ejects the rest onto the surface. It has 13 pairs of feathery red gills on the middle part of the body. Length to 8in (20cm).

Feathery, red gills

Lives in U-shaped burrow

Coiled tube worm
Spirorbis spirorbis

The chalky-white protective tube of this tiny worm coils clockwise in a spiral. It is found in large numbers on seaweeds and rocks. Tiny green feeding tentacles, like a florist's decorative spray, sprout from the mouth of the tube. Tube ³/₁₆in (5mm) diameter.

Worm inhabits chalky coiled tube

Peacock worm
Sabella pavonina

With its tentacles extended, the peacock worm looks like an underwater fly-whisk. The tentacles vary in colour from brown to red and violet, and emerge from a mud-coloured tube. The worm itself remains permanently hidden inside the tube, which is made of mud and fine sand, bound together by hardened mucus. Tube about 10in (25cm) long.

Tentacles absorb oxygen and trap food particles

Sea mouse
Aphrodite aculeata

A broad-backed, oval worm often cast up on the shore after storms. Coarse, iridescent green and golden hairs and brown bristles run along the sides; its back is covered in a dense felt of fine grey hairs. Length about 4in (10cm).

Scales are hidden under grey hair

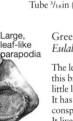

Large, leaf-like parapodia

Green leaf worm
Eulalia viridis

The leg-like outgrowths (parapodia) of this brilliant green worm are shaped like little leaves, and are used for swimming. It has a small, round head, with two conspicuous eyes and a long proboscis. It lives in rock crevices in shallow water and preys on small animals. Length 4-6in (10-15cm).

Sand mason worm
Lanice conchilega

The worm lives in a rustic-looking tube made of sand grains and shell fragments, only the top of which projects from the sand. Its finger-like extensions support the worm's fine, sticky feeding tentacles; below them are three pairs of red tufted gills. Found on the lower shore. Tube to 10in (25cm) long.

Red ribbon worm
Lineus ruber

A very common red-brown seashore worm. It is carnivorous, catching tiny sea animals by a tubular feeding proboscis which extends from an opening in front of the mouth. It has a flattish head with up to eight pairs of eyes, and is found under stones and boulders on muddy sand. Length to 6in (15cm).

Proboscis extends to catch prey

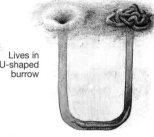

Keel worm
Pomatoceros triqueter

The triangular-section tapering tubes in which this worm lives are often found in great profusion on pebbles, rocks and shells. The tube 'stopper' is shaped like a trumpet mute, and the crown of feeding tentacles is banded. Length about 1¹/₄in (30mm).

Head has crown of feeding tentacles

Fat pink, red or brown body

Only part of tube is above sand

391

Barnacles and chitons

Despite appearances, barnacles are crustaceans more closely related to crabs than to the superficially similar limpets. Chitons – reminiscent of woodlice – are, on the other hand, vegetarian molluscs with eight-part articulating armoured shells.

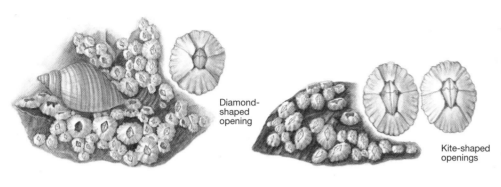

Diamond-shaped opening

Kite-shaped openings

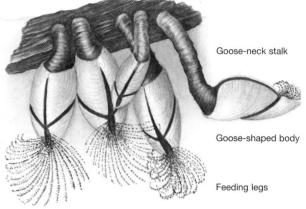

Goose-neck stalk

Goose-shaped body

Feeding legs

Common acorn barnacle
Semibalanus balanoides

Barnacles are identified by the size, number and arrangement of their outer plates and the shape of the opening through which their feathery legs extend when feeding. The very common acorn barnacle forms a greyish-white cone with six plates; the aperture is diamond-shaped and the joint lines meet at an oblique angle. It is most abundant on northern coasts. To ³/₅in (15mm) across.

Star acorn barnacle
Chthamalus stellatus and C. montagui

Two almost identical species often living in great numbers on the upper shores of exposed west and south-west coasts. The conical shell has six grooved, whitish plates – the front ones overlapped by the edges of their neighbours. The opening is oval or kite-shaped, with joint lines meeting at right-angles. To ³/₅in (15mm) across.

Goose barnacle
Lepas anatifera

The name refers to this barnacle's remarkable appearance: the five shiny white plates supposedly resemble the bird's body and the tough grey-brown stalk – which attaches the barnacle to a rock or other surface – its neck. Once it was believed that the crustacean grew feathers and turned into a goose – and one species of goose (p. 211) bears its name. It is large for a barnacle; the shell may be 2in (50mm) long and the stalk 8in (20cm). Frequently clings to ships' hulls and driftwood.

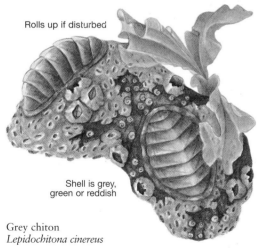

Rolls up if disturbed

Shell is grey, green or reddish

Grey chiton
Lepidochitona cinereus

Chitons are plentiful on or under rocks on the lower shore. A shell of eight plates protects the body; when moved by any outside force the chiton curls up like a woodlouse. The grey species, which can also be green or reddish, grows to ³/₄in (20mm) long. The shell is three-quarters of the total body width – a way of distinguishing similar species.

Granules on shell

Coat-of-mail chiton
Lepidopleurus asellus

A grey or brown chiton whose plates bear lines of granules rather like a corn-cob. There is a narrow, fleshy margin covered by delicate, overlapping rectangular scales. The species grows to ³/₄in (20mm) long, the shell width being four-fifths the total.

Glossy plates

Red chiton
Tonicella rubra

Glossy red or brownish-red 'armour' distinguishes this species. The shell plates have a central keel and the fleshy edge has a granular appearance. One of the smaller chitons, growing to only about ¹/₂in (13mm); shell width three-quarters the total.

Shelled molluscs

Most sea shells are the hard outer skeletons of molluscs. Apart from the soft-bodied octopuses, cuttlefish and sea slugs, molluscs are broadly divided into gastropods – snail-like creatures with one shell – and bivalves, with a pair of hinged shells.

Common limpet

Patella vulgata
Shell to 2½in (65mm) across; usually less

Common whelk

Buccinum undatum
Shell to 4¾in (12cm) long

Common mussel

Mytilus edulis
To 4in (10cm) long

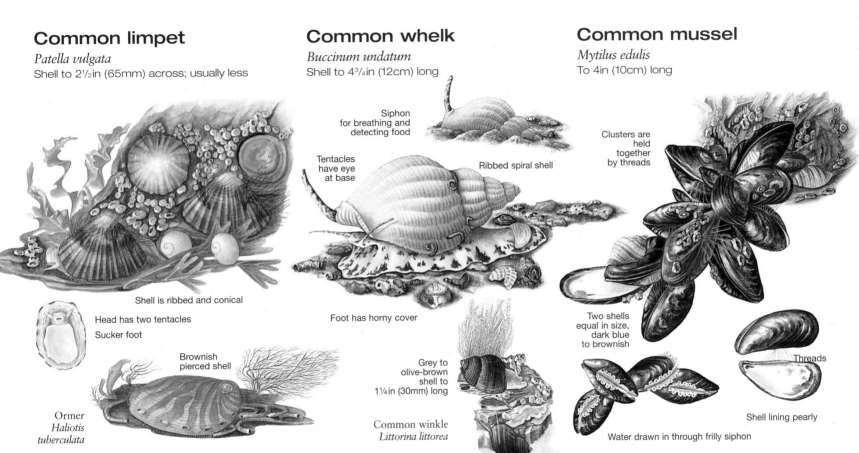

Shell is ribbed and conical

Head has two tentacles

Sucker foot

Brownish pierced shell

Ormer
Haliotis tuberculata

Siphon for breathing and detecting food

Tentacles have eye at base

Ribbed spiral shell

Foot has horny cover

Grey to olive-brown shell to 1¼in (30mm) long

Common winkle
Littorina littorea

Clusters are held together by threads

Two shells equal in size, dark blue to brownish

Threads

Shell lining pearly

Water drawn in through frilly siphon

The sticking power of the limpet's sucker-like foot is celebrated: it is almost impossible to dislodge one except by a sharp, sudden blow. Every limpet has a spot that it treats as home. If this is on soft rock, the edge of the shell carves out a depression into which it fits; on hard rock, the shell itself is worn down to a perfect fit to retain water at low tide. A limpet will crawl under water up to 36in (90cm) from its base to scrape algae off rocks for food. Most limpets start as males but become female later in life. The ormer is a Mediterranean species found as far north as the Channel Islands. Holes in the 4in (10cm) shell allow water to escape after respiration.

On rocky shores all around British Isles.

The common whelk – whose shell countless children have held to their ear to 'hear the sound of the sea' – lives mainly below low-tide mark, but is sometimes found in low-water rock pools. It can extend a long siphon like a submarine's periscope to breathe clean water and detect food – mainly marine carrion – even when half submerged in mud. Whelks are caught commercially in baited pots. After mating, the females collect for mass egg-laying. Long after whelks die, their tough shells remain intact and often become homes for hermit crabs. The common winkle, or periwinkle, has no siphon. It is abundant on most coasts and eats seaweeds.

Common and widespread, lower shore and below.

Mussels, a food for man since ancient times, are among the commonest bivalves around Britain's coasts. They often grow in tightly packed clusters, attached to rocks and to each other by threads extending from the shell openings. Mussels spawn in early spring; the larvae float in sea currents before settling down to colonise new areas. Wild mussels should be collected for food only from clean water; they extract not only food particles from the water, but also pollutants. Some mussels contain minute pearls produced by coating irritating particles with mother-of-pearl – which coats the inside of the shell.

All round British coasts and in estuaries.

Sea shells/ Gastropods

Shells are the hard outer skeletons of molluscs. Those of gastropods are coiled like a snail's or bowl-shaped like a limpet's. The illustrations are about life-size; the labels indicate the maximum for the species.

Common winkle
Littorina littorea

On rocks and seaweeds, middle and lower shore. Widespread. 1¼in (30mm).

Rough winkle
Littorina saxatilis

In crevices and on stones, upper and middle shore. Widespread. 5/16in (8mm).

Small winkle
Littorina neritoides

In crevices, upper shore. Widespread. 3/16in (5mm).

Flat winkle
Littorina obtusata

On seaweeds, middle and lower shore. Widespread. 3/8in (10mm).

Dog whelk
Nucella lapillus

Crevices and rocks, among barnacles and mussels, middle and lower shore. Widespread. 1¼in (30mm).

Common whelk
Buccinum undatum

Sand, mud and rocks, low water and below. Widespread. 4¼in (12cm).

Netted dog whelk
Hinia reticulatus

Under stones and in crevices, lower shore and shallows. Widespread. 1½in (40mm).

Grey top shell
Gibbula cineraria

On and under stones and seaweeds, lower shore and below. Widespread. ½in 13mm).

Painted top shell
Calliostoma zizyphinum

On rocks and kelp, lower shore and below. Widespread. 1in (25mm).

Oyster drill (sting winkle)
Ocenebra erinacea

On rocks, lower shore and below. Widespread. 2³/8in (60mm).

Blue-rayed limpet
Helicon pellucidum

On kelp plants, lower shore and below. Widespread. 1in (25mm).

Purple top shell
Gibbula umbilicalis

On rocks, upper and middle shore. Widespread except east coasts. ½in (13mm).

Common necklace shell
Polinices polianus

In sand, lower shore and below. Widespread. 3/5in (15mm).

Thick-lipped dog whelk
Hinia incrassata

Under stones and in crevices, lower shore and shallows. Widespread. 3/5in (15mm).

Common limpet
Patella vulgata

On rocks, stones and in pools throughout the shore. Widespread. 2½in (60mm).

Inside view

Side view

Slipper limpet
Crepidula fornicata

Shallow water, in chains attached to stones and shells. Widespread except northern Scotland. 2in (50mm).

Tower shell
Turritella communis

In sand and mud, offshore. Widespread. 2³/8in (60mm).

Needle shell
Bittium reticulatum

Among rocks and stones, lower shore and shallows. Widespread. ½in (13mm).

Northern cowrie
Trivia arctica

Among rocks and sea squirts, lower shore (rarely) and below. Widespread. 3/8in (10mm).

European cowrie
Trivia monacha

Among rocks and sea squirts, lower shore and below. Widespread. ½in (13mm).

Sea shells/ Bivalves

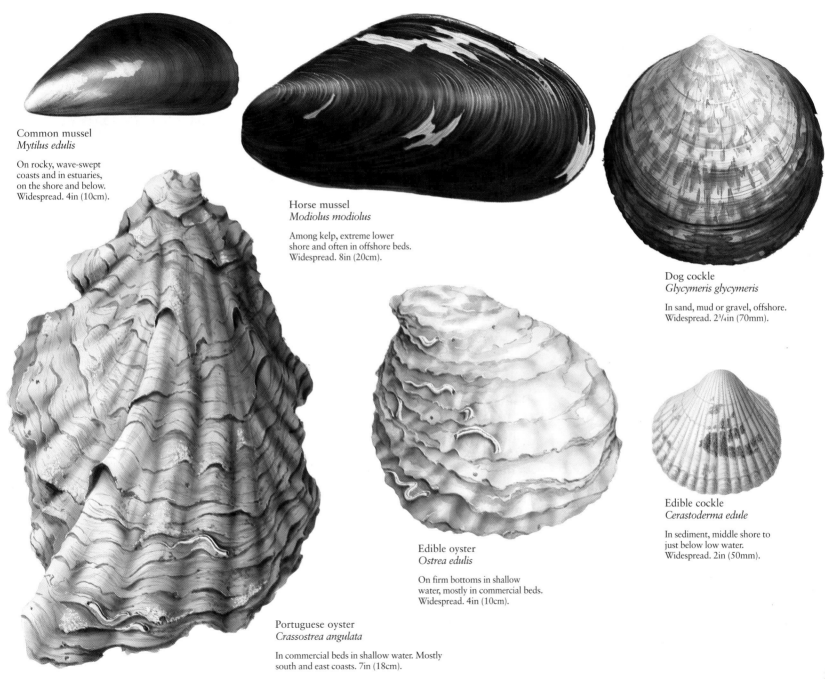

Common mussel
Mytilus edulis

On rocky, wave-swept
coasts and in estuaries,
on the shore and below.
Widespread. 4in (10cm).

Horse mussel
Modiolus modiolus

Among kelp, extreme lower
shore and often in offshore beds.
Widespread. 8in (20cm).

Dog cockle
Glycymeris glycymeris

In sand, mud or gravel, offshore.
Widespread. 2³⁄₄in (70mm).

Edible oyster
Ostrea edulis

On firm bottoms in shallow
water, mostly in commercial beds.
Widespread. 4in (10cm).

Edible cockle
Cerastoderma edule

In sediment, middle shore to
just below low water.
Widespread. 2in (50mm).

Portuguese oyster
Crassostrea angulata

In commercial beds in shallow water. Mostly
south and east coasts. 7in (18cm).

395

Sea shells/Bivalves

Common otter shell
Lutraria lutraria

In sand and sandy mud, extreme lower shore and below. Widespread. 4³/₄in (12cm).

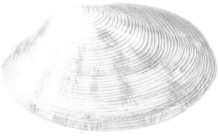

Banded carpet shell
Venerupis rhomboides

In sand and gravel, extreme lower shore and below. Widespread. 2¹/₂in (60mm).

Common nut shell
Nucula nucleus

In gravel and coarse sand, offshore. Widespread. ¹/₂in (13mm).

Thin tellin
Angulus tenuis

In fine sand, middle shore down to shallows. Widespread. ³/₄in (20mm).

Baltic tellin
Macoma balthica

In mud and muddy sand, especially on estuary shores. Widespread. 1in (25mm).

Banded wedge shell
Donax vittatus

In clean sand, from middle shore downwards. Widespread. 1¹/₂in (40mm).

Common piddock
Pholas dactylus

Bores into soft rock, wood or firm sand, lower shore and shallows. South and south-west coasts. 6in (15cm).

Sand gaper
Mya arenaria

In mud and sand, lower shore and below. Widespread. 6in (15cm).

Oval piddock
Zirfaea crispata

Bores into soft rock, lower shore and shallows. Widespread. 3¹/₂in (90mm).

Oval venus
Timoclea ovata

In sand and gravel, offshore. Widespread. ³/₄in (20mm).

Striped venus
Chamelea gallina

In sand, lower shore downwards. Widespread. 1³/₄in (45mm).

Bivalve molluscs such as oysters have two hinged shells, but those washed up on the beach have often broken apart. The best time to collect shells is at low water during spring tides – when the moon is full or new. Then the greatest expanse of beach and the lowest rocks are exposed.

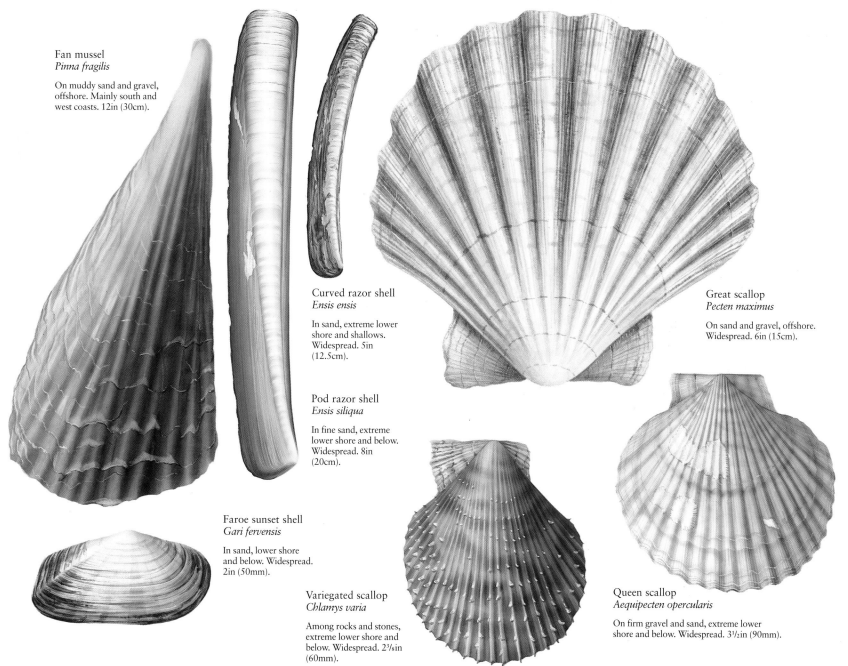

Fan mussel
Pinna fragilis

On muddy sand and gravel, offshore. Mainly south and west coasts. 12in (30cm).

Curved razor shell
Ensis ensis

In sand, extreme lower shore and shallows. Widespread. 5in (12.5cm).

Pod razor shell
Ensis siliqua

In fine sand, extreme lower shore and below. Widespread. 8in (20cm).

Great scallop
Pecten maximus

On sand and gravel, offshore. Widespread. 6in (15cm).

Faroe sunset shell
Gari fervensis

In sand, lower shore and below. Widespread. 2in (50mm).

Variegated scallop
Chlamys varia

Among rocks and stones, extreme lower shore and below. Widespread. 2³/₈in (60mm).

Queen scallop
Aequipecten opercularis

On firm gravel and sand, extreme lower shore and below. Widespread. 3¹/₂in (90mm).

397

Octopuses and their relatives

Soft-bodied marine molluscs range from the highly intelligent octopus, squid and cuttlefish – jet-propelled tentacled creatures – to the crawling sea hare and sea slugs.

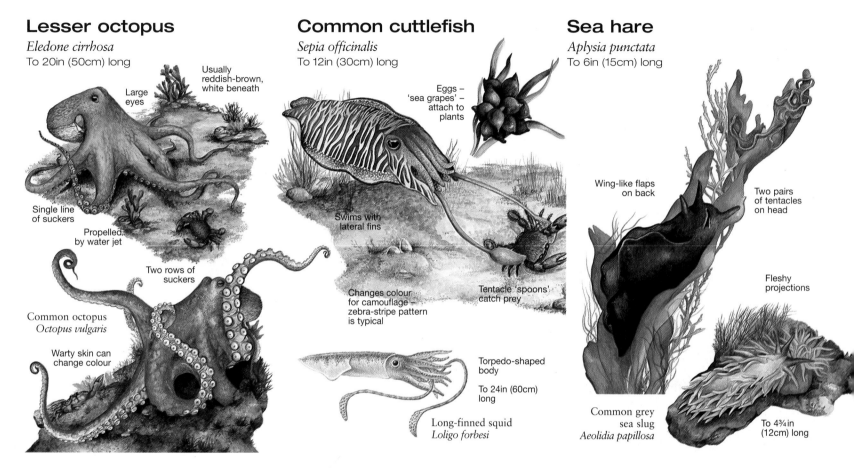

Lesser octopus
Eledone cirrhosa
To 20in (50cm) long

Large eyes

Usually reddish-brown, white beneath

Single line of suckers

Propelled by water jet

Two rows of suckers

Common octopus
Octopus vulgaris

Warty skin can change colour

Common cuttlefish
Sepia officinalis
To 12in (30cm) long

Eggs – 'sea grapes' – attach to plants

Swims with lateral fins

Changes colour for camouflage – zebra-stripe pattern is typical

Tentacle 'spoons' catch prey

Torpedo-shaped body

To 24in (60cm) long

Long-finned squid
Loligo forbesi

Sea hare
Aplysia punctata
To 6in (15cm) long

Wing-like flaps on back

Two pairs of tentacles on head

Fleshy projections

Common grey sea slug
Aeolidia papillosa

To 4¾in (12cm) long

In proportion to its size, the octopus has a larger brain than any other invertebrate, and it can carry out complex activities. It feeds mainly on crabs, lobsters and fish, biting and killing them with a venomous saliva and eating them with its parrot-like beak. The octopus can change colour rapidly to merge with its background and produces a screen of inky fluid when in danger. It is identified by the single line of suckers on each tentacle.

The common octopus, despite its name, is less commonly seen around Britain. It reaches a length of 36in (90cm) and has a double row of suckers on its tentacles.

Rocky coasts; common octopus confined mainly to south.

Large numbers of cuttlefish live off British shores, and the porous white internal shells of dead ones – cuttlebones – are familiar to beachcombers. By day the cuttlebone's chambers are flooded with water and the cuttlefish sinks to the seabed. At night, when it comes out to hunt, the water is replaced by gas, giving buoyancy. Two of the ten tentacles are long, with spoon-like ends; it uses these to seize crabs, prawns and other prey. The cuttlefish can camouflage itself and eject inky fluid to confuse attackers. The long-finned squid, long-bodied, reddish and speckled, lives mainly in deeper waters.

Close inshore all around the coast.

When its large head tentacles are prominent, this creature looks something like a crouching hare. But there is nothing hare-like about its progress; it crawls steadily over the seabed on a single large foot, and feeds on seaweed. It is a link between shelled snails, such as the whelk, and the true sea slugs. Sea hares are hermaphrodites and mate in chains. Up to 26 million eggs are laid in strings as much as 65ft (20m) long.

The common grey sea slug's back is covered with long, fleshy projections called cerata, and there are four tentacles on the head. This slug is found among rocks.

Shallow waters around all coasts.

Sea urchins and sea cucumbers

These starfish relatives crawl on suckered tube feet. Sea urchins have a shell-like skeleton and movable spines; chalky plates protect sea cucumbers.

Common sea urchin

Echinus esculentus
To 8in (20cm) across

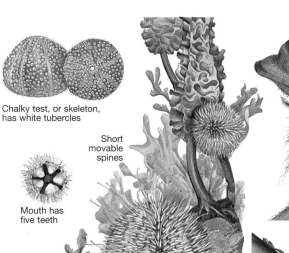

Chalky test, or skeleton, has white tubercles

Short movable spines

Mouth has five teeth

Suckered tube feet enable urchin to climb rocks

This large, almost spherical species is deep reddish in colour, sometimes tinged with blue or purple. It is also called the edible sea urchin; inside the brittle skeleton, or test, is the edible part – its roe. Moving by means of tube feet, the creature uses the five powerful teeth on its underside to scrape seaweed and tiny animals off rocks. These cannot regenerate where large numbers of urchins feed, and in some areas the rocks are scraped completely bare. Young urchins are prey to fish and seals, but the formidable movable spines protect mature ones.

Rocky coasts, especially off Scotland.

Heart urchin

Echinocardium cordatum
2in (50mm) long

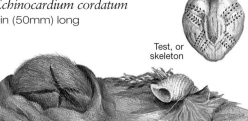

Test, or skeleton

Grooves in test roofed by spines

Green sea urchin
Psammechinus miliaris

Purple-tipped green spines

Short, densely set brownish spines point backwards on this burrowing urchin, also known as the sea potato. They serve for digging in sand. When the creature has burrowed vertically down, it ploughs slowly forward, eating small particles of plant and animal matter which the tube feet pass into its toothless mouth. As with other urchins, eggs and sperm are shed freely into the water at breeding time. Heart urchins live in groups.

The green sea urchin's flattened globe shape is covered with short, purple-tipped green spines. It is 2in (50mm) across, and is widely distributed.

Offshore and near low-water mark all around coasts.

Cotton spinner

Holothuria forskali
To 10in (25cm) long

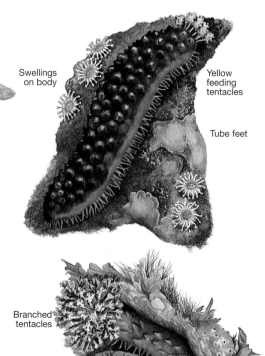

Swellings on body

Yellow feeding tentacles

Tube feet

Branched tentacles

Sea gherkin
Cucumaria normani

Swellings cover the black, cucumber-shaped body of the cotton spinner, largest of the sea cucumbers living around Britain. It has a strange system of defence, squirting out sticky white threads from its rear. These threads – which give the creature its name – swell up and entangle or at least distract the enemy. The cotton spinner, also known as the black sea cucumber, moves on sucker-like tube feet. At its mouth end are yellow tentacles which scoop up food from the seabed.

Sea gherkins – with branched tentacles at the front end – live in rock crevices or cling to seaweed.

Mostly Atlantic and Channel coasts.

399

Starfish

Starfish and their close relatives the brittlestars and featherstars are bottom-dwellers belonging to the echinoderms – a group of spiny-skinned animals with no head but usually five-rayed symmetry (though the number of arms can vary). The spines arise from a skeleton of bony plates.

Common starfish

Asterias rubens
To 20in (50cm) across; usually 2-4in (5-10cm)

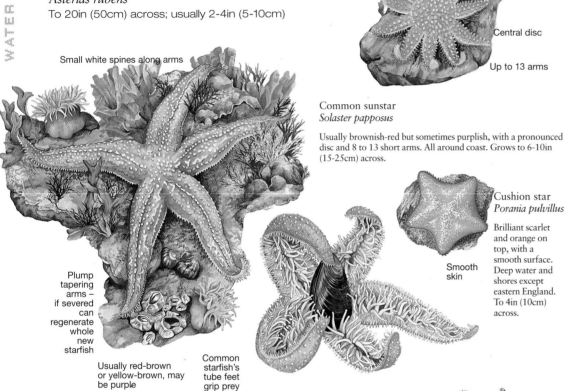

Small white spines along arms

Plump tapering arms – if severed can regenerate whole new starfish

Usually red-brown or yellow-brown, may be purple

Common starfish's tube feet grip prey

Central disc

Up to 13 arms

Common sunstar
Solaster papposus

Usually brownish-red but sometimes purplish, with a pronounced disc and 8 to 13 short arms. All around coast. Grows to 6-10in (15-25cm) across.

Cushion star
Porania pulvillus

Brilliant scarlet and orange on top, with a smooth surface. Deep water and shores except eastern England. To 4in (10cm) across.

Smooth skin

Spiny starfish
Marthasterias glacialis

Bulbous grey-green spines grow along the brownish-yellow arms, which are often tipped red or purple; underside is yellowish-white. Rocky areas, especially in west and south-west. To 32in (80cm) across; usually less.

Bulbous spines

Any arm can take the lead when this slow-moving predator walks across the rocky or sandy seabed on its tube feet. These are connected to an internal water system and work by hydraulic pressure. When the starfish comes upon its shellfish prey – mussels and oysters are favourites – they grip each shell and pull them apart. Then the starfish pushes its stomach around the prey's soft flesh and digests it. In spring and summer, the female sheds millions of eggs into the sea while the male releases sperm to fertilise them. The larvae drift for some time before settling on the seabed to develop.

All around British Isles, both inshore and in deeper water.

Common brittlestar

Ophiothrix fragilis
Arms to 4in (10cm) long; disc ³/₄in (20mm) across

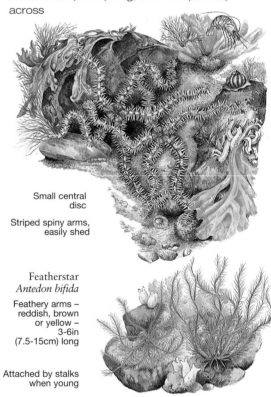

Small central disc

Striped spiny arms, easily shed

Featherstar
Antedon bifida

Feathery arms – reddish, brown or yellow – 3-6in (7.5-15cm) long

Attached by stalks when young

The arms of a brittlestar, arranged symmetrically around a central disc, are much thinner than those of a starfish. As the name suggests, they are easily shed but, as with true starfish, a missing one is soon regenerated. The common brittlestar's five spiny arms are often banded with bright colours. Spectacular beds of many millions can occur where currents bring in large supplies of dead organic matter and plankton for them to eat. The raised arms trap food particles and move them to the mouth. The less common featherstar has ten mobile, feathery arms projecting from a cup-shaped body.

On shore or in deeper water on most coasts.

Jellyfish and comb jellies

These superficially similar, more or less transparent creatures belong to several distinct zoological groups. All drift in currents or propel themselves only slowly through the water.

Common jellyfish

Aurelia aurita
Body to 12in (30cm) across

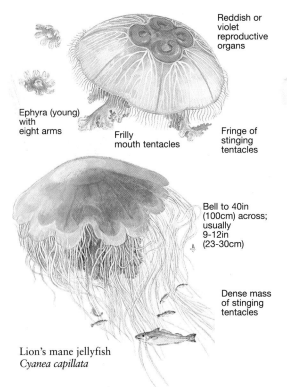

Reddish or violet reproductive organs

Ephyra (young) with eight arms

Frilly mouth tentacles

Fringe of stinging tentacles

Bell to 40in (100cm) across; usually 9-12in (23-30cm)

Dense mass of stinging tentacles

Lion's mane jellyfish
Cyanea capillata

Four circular or horseshoe-shaped reproductive organs adorn the common jellyfish's transparent body, or bell. By pulsating the bell it progresses slowly through the water. Stinging tentacles at the edge rarely hurt humans, and some fish also appear immune. Small fish and crustaceans that are stunned by the sting are pushed into the creature's mouth by the four frilly mouth tentacles. Larval jellyfish develop through an anemone-like stage and then a small star-shaped form before becoming adults.

The lion's mane is one of the largest jellyfish seen around Britain – mainly in the south-west. Its sting is painful.

All around British Isles, often in swarms.

Portuguese man-of-war

Physalia physalis
Float to 12in (30cm) long

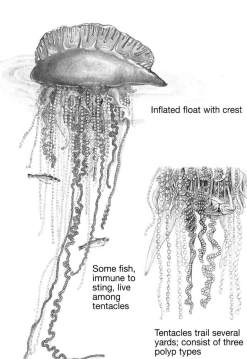

Inflated float with crest

Some fish, immune to sting, live among tentacles

Tentacles trail several yards; consist of three polyp types

A blue gas-filled float surmounted by a sail-like crest gives this strange creature its name. Although commonly regarded as a jellyfish, it is a siphonophore – a complex floating colony of many individual animals, or polyps, performing different functions. It cannot swim like a true jellyfish, but drifts with wind and current. Its trailing tentacles, each attached to a stinging polyp, are so long that a bather can be badly injured without even seeing the animal. Small fish and shrimps paralysed by the sting are passed to feeding polyps for digestion. Other polyps specialise in reproduction.

Ocean-dwelling, driven ashore by south-west winds.

Comb jelly

Beroë cucumis
To 5in (12.5cm) long

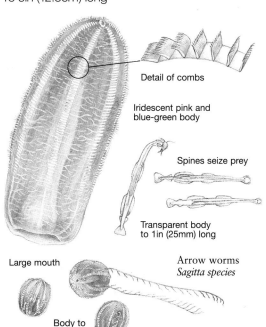

Detail of combs

Iridescent pink and blue-green body

Spines seize prey

Transparent body to 1in (25mm) long

Large mouth

Arrow worms
Sagitta species

Body to 1¼in (30mm) long

Branched tentacles retractable

Sea gooseberry
Pleurobrachia pileus

Comb jellies were once classed with jellyfish but now form a separate group, the ctenophores. Their name comes from the rows of tiny comb-like plates along their bodies; these beat rhythmically to move the animal along. Comb jellies are hermaphrodites and predators. The elongated *Beroë cucumis* has a wide mouth able to engulf prey almost its own size. The sea gooseberry is round, with long sticky tentacles, which can be drawn in like fishing lines after ensnaring prey such as fish larvae. Arrow worms form a unique group. They have torpedo-shaped bodies and curved spines on the head to catch prey.

Mainly in the open sea, but often driven inshore.

Plant-like sea animals

Sea anemones are the best known examples of marine animals that superficially resemble plants and spend their lives attached to rocks. Others include corals – soft and hard – hydroids, sponges and moss animals.

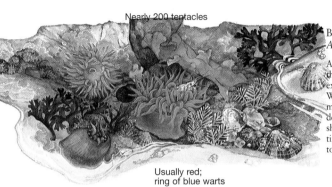

Nearly 200 tentacles

Usually red; ring of blue warts

Beadlet anemone
Actinia equina

A mere blob of red, brown or green jelly when the tide is out, but in the sea stinging tentacles expand – to trap shrimps or small fish for food. When beadlet anemones touch tentacles, one may attack the other by stinging its column; the defeated anemone slowly retreats. Common in shallow water on all coasts, on rocks and timbers. Column to 3in (75mm) tall; tentacles to 2¾in (70mm) across.

Each 'finger' 1¼in (30mm) wide

Dead man's fingers
Alcyonium digitatum

A soft coral, living as a colony of polyps strengthened by a skeleton of limy splinters. Dislodged pieces of the colony, when washed ashore, can resemble bloated white fingers – hence the macabre name. They feed on planktonic animals. Colonies can grow to 10in (25cm) high. All coasts.

Two types of tentacles

Tube anemone
Cerianthus lloydii

A burrowing anemone, living partly buried in muddy sand. It has two rings of different tentacles – the inner short and erect, the outer spanning about 2¾in (70mm). May be brown, reddish or white. If touched, the anemone draws back into its tube. All coasts.

Inhabits slimy tube

Yellow in shady places

Forms green patches 8in (20cm) across

Purse sponge
Scypha compressa

A white, vase-shaped stalked sponge found hanging under rocks in shallow water. Collapses to purse shape out of water. All coasts. Length ¾-2in (20-50mm).

Lives among seaweeds

Zooid inhabits compartment 1/12 × 1/25 in (2 × 1mm)

Forms patches to 8in (20cm) wide, often on kelp

Breadcrumb sponge
Halichondria panicea

Sponges are among the simplest animals. They draw in water with oxygen and food particles through pores. The breadcrumb sponge – named because of the way it crumbles – encrusts rocks on all coasts.

To 6in (15cm) tall

Oaten pipes hydroid
Tubularia indivisa

Another colonial creature, related to the Portuguese man-of-war (p. 401), with a horny 'stem' crowned by a reddish-pink head of polyps with white tentacles. These surround the mouth opening and are responsible for trapping food particles. The hydroid grows freely where currents are strong. It reproduces itself by budding.

Sea mat
Membranipora membranacea

Like the corals, a colony of many individual small animals, or zooids. It belongs to the bryozoan (moss animal) group and encrusts brown seaweeds and other surfaces like fine white lace. Each tiny rectangular compartment contains one zooid, which extends minute tentacles under water to trap food particles.

To 5in (12.5cm) tall

Common sea squirt
Ciona intestinalis

Little more than a translucent bag with two openings, or siphons, through which water is passed to extract food particles and oxygen. Yet its larvae have a primitive 'backbone', and it may represent an evolutionary link between invertebrates and vertebrate animals such as fish. Clusters on sheltered rocks, walls and piles on all coasts.

Eel grass and green seaweeds

Eel grasses are marine flowering plants, but seaweeds are more primitive non-flowering algae – the green ones mostly thin and delicate.

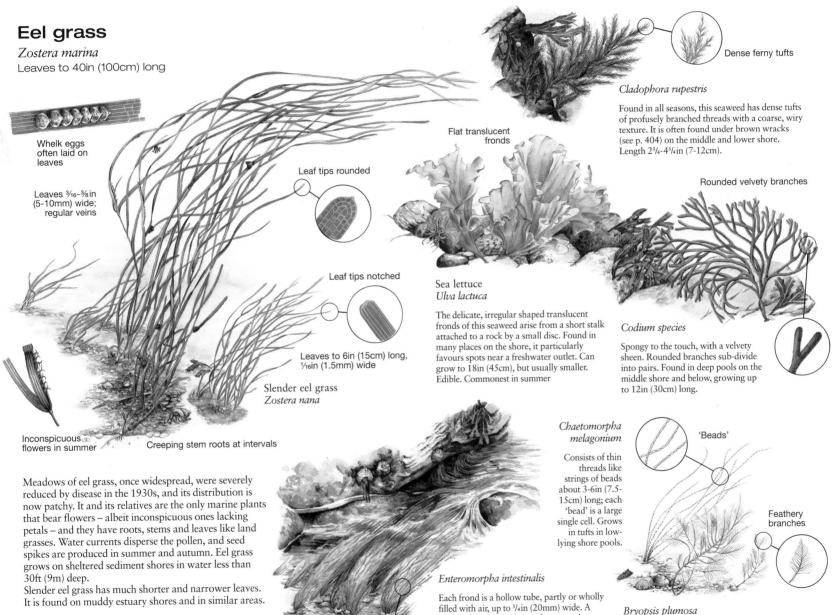

Eel grass

Zostera marina
Leaves to 40in (100cm) long

Whelk eggs often laid on leaves

Leaves ³⁄₁₆-³⁄₈in (5-10mm) wide; regular veins

Leaf tips rounded

Leaf tips notched

Leaves to 6in (15cm) long, ¹⁄₁₆in (1.5mm) wide

Slender eel grass
Zostera nana

Inconspicuous flowers in summer

Creeping stem roots at intervals

Meadows of eel grass, once widespread, were severely reduced by disease in the 1930s, and its distribution is now patchy. It and its relatives are the only marine plants that bear flowers – albeit inconspicuous ones lacking petals – and they have roots, stems and leaves like land grasses. Water currents disperse the pollen, and seed spikes are produced in summer and autumn. Eel grass grows on sheltered sediment shores in water less than 30ft (9m) deep.
Slender eel grass has much shorter and narrower leaves. It is found on muddy estuary shores and in similar areas.

Sheltered creeks, estuaries and sea-lochs.

Flat translucent fronds

Sea lettuce
Ulva lactuca

The delicate, irregular shaped translucent fronds of this seaweed arise from a short stalk attached to a rock by a small disc. Found in many places on the shore, it particularly favours spots near a freshwater outlet. Can grow to 18in (45cm), but usually smaller. Edible. Commonest in summer

Dense ferny tufts

Cladophora rupestris

Found in all seasons, this seaweed has dense tufts of profusely branched threads with a coarse, wiry texture. It is often found under brown wracks (see p. 404) on the middle and lower shore. Length 2³⁄₄-4³⁄₄in (7-12cm).

Rounded velvety branches

Codium species

Spongy to the touch, with a velvety sheen. Rounded branches sub-divide into pairs. Found in deep pools on the middle shore and below, growing up to 12in (30cm) long.

Chaetomorpha melagonium

Consists of thin threads like strings of beads about 3-6in (7.5-15cm) long; each 'bead' is a large single cell. Grows in tufts in low-lying shore pools.

'Beads'

Feathery branches

Enteromorpha intestinalis

Each frond is a hollow tube, partly or wholly filled with air, up to ³⁄₄in (20mm) wide. A minute disc attaches it in tufts to a rock or the back of a limpet. Found mainly on the upper shore where there is fresh water, and in summer on estuary mud-flats. Length 2-12in (5-30cm).

Thin hollow tubes

Bryopsis plumosa

A feathery seaweed with paired branches found near and below low-water mark. By growing in pools and steep-sided gullies, it avoids drying out. Length about 4in (10cm).

403

Brown seaweeds

These include many of the toughest and most familiar seaweeds. The large kelps have thick stalks and big fronds; a structure called the holdfast attaches them to rocks, mostly at or below low-water mark and often in a dense 'forest'. Wracks, leathery and slippery, grow at various levels.

Holdfast

Crinkled edges

Sugar kelp
Laminaria saccharina

Also known as sea belt. Has a small holdfast and short stalk from which single fronds with crinkled edges emerge. It tends to flop over even under water. Sugar kelp is used as cattle fodder and is also eaten by humans; it yields a kind of sugar that is used pharmaceutically. May grow longer than 13ft (4m).

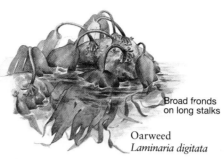

Broad fronds on long stalks

Oarweed
Laminaria digitata

The long, bending stalk is too smooth for many other seaweeds to attach themselves to it, as they do to most kelps. It is oval in cross-section, and expands gradually into a broad frond with finger-like blades. May be found covering large areas of the lower shore, reaching up to 6½ft (2m) in length.

Air bladders

Bladder wrack
Fucus vesiculosus

Each leathery, ribbon-like frond has a distinct midrib with pairs of air bladders on either side. These float the fronds up towards the light. Found on the middle shore; to 40in (100cm) long

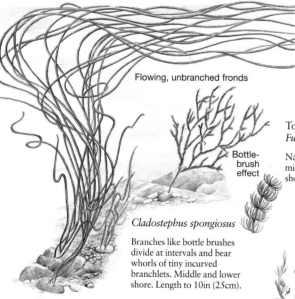

Flowing, unbranched fronds

Bottle-brush effect

Cladostephus spongiosus

Branches like bottle brushes divide at intervals and bear whorls of tiny incurved branchlets. Middle and lower shore. Length to 10in (25cm).

Bootlace weed
Chorda filum

More romantically known as mermaid's tresses, the long, unbranched fronds ¼in (6mm) thick grow in clumps on rocks in pools and gullies of the lower shore. Fine hairs covering the fronds are small species of brown seaweed. Length to 18ft (5.5m).

Serrated frond edges

Toothed wrack
Fucus serratus

Named after the serrated edges of the fronds; there is a distinct midrib but no air bladders. Found in rock pools and on the lower shore. To 24in (60cm) long.

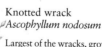

Knotted wrack
Ascophyllum nodosum

Largest of the wracks, growing profusely on sheltered shores. Its fronds are held up towards the light by oval bladders. Side branches bear raisin-shaped fruiting bodies. To 10ft (3m) long.

Raisin-shaped reproductive parts

Pod-like reproductive parts

Spiral wrack
Fucus spiralis

The fronds are smooth-edged with no air bladders; they often grow with a slight twist. The tips of the branches bear rounded pod-like fruiting (reproductive) bodies. Grows on the upper shore; may reach a length of 16in (40cm).

Swollen, granular reproductive parts

Fronds curled

Channelled wrack
Pelvetia canaliculata

Grows high on the shore, and to conserve water the frond edges curl inwards to form a channel. Brittle-looking, tufted branches hang from rocks with the channel side inwards. Fronds about 6in (15cm) long, tipped by fruiting bodies.

Red seaweeds

Many red seaweeds grow in deep water or a sheltered position because they cannot survive drying out. They need little light, and some can be found in kelp forests or under large wracks. A few species are tough enough to survive on rocks above low-water mark, or in rock pools. Most are fairly small.

Gigartina stellata

Similar to carragheen, distinguished by inrolled edges to the fronds, which are also less divided and fan-like. It is dark red and is common on rocks on the middle and lower shore, often among other algae. 2³/₄-8in (7-20cm) tall.

Sharp angle between branches

Fronds triply branched

Pepper dulse
Laurencia pinnatifida

Tough, fleshy and flattened fronds branch alternately from the main stem. Subsequent sub-branches are shorter and bunched like stubby fingers. May be yellowish-red or dark green. To 8in (20cm) tall in pools or lower shore; shorter if exposed on middle shore.

Dulse
Rhodymenia palmetta

One of the commonest red seaweeds, growing singly or in groups. It is abundant on the lower and middle shore, often attached to larger, brown seaweeds such as kelp stalks. Fronds may be undivided or have lobes and side branches. Eaten in Iceland. To 12in (30cm) long.

Fronds may be forked

Coral weed
Corallina officinalis

Abundant in rock pools, growing in extensive tufts up to 4³/₄in (12cm) high. It has a chalky 'skeleton' giving it a stiff, brittle appearance. The colour varies according to exposure to light; the plant may be bleached to complete whiteness.

Segmented stems

Purple laver
Porphyra umbilicalis

When exposed to air this seaweed turns black and flattens against the rock face. But it can survive in a dried state, even high on the shore. Its rose-purple fronds show their true beauty only under water. Found on stones and timbers at most beach levels. To 8in (20cm) long.

Thin, delicate fronds; blacken in air

Carragheen
Chondrus crispus

A low, bushy plant, also called Irish moss. Its tough flat fronds are divided regularly and often form a broad fan shape that tapers to a narrow base and small, disc-shaped holdfast. Can be eaten cooked with sugar. Found on many types of shore, but not on mud. 3-6in (7.5-15cm) tall.

Tough, flat fronds

***Lithothamnium* and *Lithophyllum* species**

A widespread group of species scarcely recognisable as seaweed. The plants form a hard, chalky crust, ranging from pink to purple, over rocks and shells. It may be smooth or knobbly, or may spread as thin lobes with concentric growth rings.

Encrust rocks

index

The common English names of our wild plants and animals are often picturesque and appealing, but they can lead to confusion. For example, the Scotsman's bluebell is called the harebell south of the border. So this book – and thus this index – gives both the common and the more precise (if less romantic-sounding) scientific name. The latter is printed in *italics* and is in Latin or Latinised Greek, with two parts. The first name indicates the genus (plural genera) – a group of closely related plants or animals, distinct but usually with obvious similarities. The second indicates the species, a group of more or less identical organisms. For example, the genus *Viola*, the violets, includes such species as *Viola canina* (the dog violet) and *Viola odorata* (sweet violet). A group of related genera together form a family; an example is Rosaceae, the rose family. Family names are not usually printed in italics.

Occasionally, members of a species living in different places show slight but distinctive variations – as with the sea radish, a coastal variant of the wild radish, *Raphanus raphanistrum*. Such a variant is called a subspecies (abbreviated ssp.); the sea radish is classified as *Raphanus raphanistrum* ssp. *maritimus*. A true-breeding form of a plant with an abnormal flower colour or leaf shape is called a variety (abbreviated var.) and is named similarly. Such a variant arising in cultivation is called a cultivar, and its name is given in quotation marks – as with the Lombardy poplar, *Populus nigra* 'Italica'. On rare occasions two species (or even members of different genera) cross-breed to produce a hybrid; this is indicated by a x ('cross') sign, as with the London plane, *Platanus* x *hispanica*, and the Leyland cypress, x *Cupressocyparis leylandii*. The x sign is ignored when listing the names alphabetically.

409

410

411

413

acknowledgments

The illustrations in *Britain's Wildlife, Plants and Flowers* were created by the following artists. Where more than one artist contributed to a page the positions of the illustrations are indicated by letters: t (top), b (bottom), c (centre), l (left) and r (right).

1–9: Jim Russell
10: l–cr, Brenda Katté; r, Shirley Hooper
11: l, Helen Cowcher; cl, Leonora Box; cr, Roger Hughes; r, Helen Senior
12: l–cr, Wendy Bramall; r, Shirley Hooper
13: l & r, Wendy Bramall; c, Shirley Hooper
14: l & r, Wendy Bramall; c, Shirley Hooper
15: l & cl, John Rignall; cr, Helen Cowcher; r, Brenda Katté
16: l, Helen Cowcher; c & r, Shirley Hooper
17: Shirley Hooper
18: l & r, Wendy Bramall; c, Stuart Lafford
19: Victoria Goaman
20: l & c, Victoria Goaman; r. Wendy Bramall
21: l & r, Wendy Bramall; c, Leonora Box
22: l–cr, Paul Wrigley; r, John Rignall
23: l & cl, Paul Wrigley; cr, Leonora Box; r, Colin Emberson
24: l, Barbara Walker; c, Josiane Campan; r, John Rignall
25: l, Roger Hughes; c, Brenda Katté; r, Line Mailhé
26: l & cl, Line Mailhé; cr, Colin Emberson; r, Helen Senior
27: l, Leonora Box; c, Colin Emberson; r, Victoria Goaman
28–29: Stuart Lafford
30: l & cl, Wendy Bramall; cr, Frankie Coventry; r, Marie-Claude Guyetand
31: l, Sarah Fox-Davies; c & r, Frankie Coventry
32: John Rignall
33: l & c, Line Mailhé; r, Shirley Hooper
34: l & c, John Rignall; r, Shirley Hooper
35: l, Shirley Hooper; c, Brenda Katté; r, Guy Michel
36: l & c, Colin Emberson; r, Stuart Lafford
37: l, Leonora Box; r, Barbara Walker
38: Victoria Goaman
39: l, Sarah Fox-Davies; c, Wendy Bramall; r, Brenda Katté
40: l, cl & r, Wendy Bramall; cr, Guy Michel
41: l, Victoria Goaman; cl–r, Helen Cowcher
42: l & cl, Brenda Katté; cr, John Rignall; r, Shirley Hooper
43: l, cr & r, Colin Emberson; cl, Norman Lacey
44: Colin Emberson
45: l, Norman Lacey; c, Stuart Lafford; r, Colin Emberson
46: l & c, Brenda Katté; r, Guy Michel
47: l, Guy Michel; c, Leonora Box; r, Rosemary Wise
48: Shirley Hooper

49: l, Leonora Box; r, Shirley Hooper
50: l, Line Mailhé; c & r, Helen Cowcher
51: Line Mailhé; c, Stuart Lafford; r, Helen Senior
52: l, Victoria Goaman; r, Paul Wrigley
53: Paul Wrigley
54: l, Leonora Box; c & r, John Rignall
55: Colin Emberson
56: l–cr, Stuart Lafford; r, Colin Emberson
57: l, Shirley Hooper; c, Brenda Katté; r, Leonora Box
58: Barbara Walker
59: l, Josiane Campan; c & r, Leonora Box
60: l & r, Victoria Goaman; c, Stephanie Harrison
61: l, Peter Wrigley; c, Victoria Goaman; r, Stuart Lafford
62: l, Colin Emberson; c, Guy Michel; r, Roger Hughes
63: l, Victoria Goaman; c, Philippe Couté; r, Maurice Espérance
64: l & cl, Sarah Fox-Davies; cr, Brenda Katté; r, Guy Michel
65: l & cl, Leonora Box; cr & r, Derek Rodgers
66: l, Colin Emberson; cl, Line Mailhé; cr, Marie-Claire Nivoix; r, Leonora Box
67: l & cl, Colin Emberson; cr, Line Mailhé; r, Marie-Claire Nivoix
68: l & c, Wendy Bramall; r, Helen Senior
69: l, Philippe Couté; c & r, Colin Emberson
70: l, Philippe Couté; c, Colin Emberson; r, Shirley Hooper
71: l & c, Sarah Fox-Davies; r, John Rignall
72: l & cl, Marie-Claire Nivoix; r, Stuart Lafford
73: Stuart Lafford
74: Brenda Katté
75: l & c, Brenda Katté; cr, Stuart Lafford; r, Helen Cowcher
76: l & c, Barbara Walker; r, Leonora Box
77: l, Leonora Box; c, Frankie Coventry; r, Line Mailhé
78: John Rignall
79: Josiane Campan
80: l, Josiane Campan; c, Sarah Fox-Davies; r, Stuart Lafford
81: l & c, Helen Cowcher; r, Sarah Fox-Davies
82: l, Helen Senior; r, Colin Emberson
83: Colin Emberson
84: l, Roger Hughes; r, Helen Cowcher
85: l & cl, Delyth Jones; cr, Line Mailhé; r, Wendy Bramall
86: Marjory Saynor
87: l & c, Marie-Claire Nivoix, r, Shirley Hooper
88: Brenda Katté
89: l, Brenda Katté; c, Roger Hughes; r, Sarah Fox-Davies
90: l–cr, Colin Emberson; r, Helen Cowcher
91: Helen Cowcher
92: l, Frankie Coventry; cl, Wendy Bramall; cr, Colin Emberson; r, Leonora Box
93: l, John Rignall; cl & c, Stuart Lafford; r, Roger Hughes
94–95: Brenda Katté
96: l & c, Brenda Katté; r, Sarah Fox-Davies
97: l, Helen Cowcher; c, Rosemary Wise; r, Shirley Hooper
98: l, Shirley Hooper; c, Line Mailhé; r, Helen Cowcher

99: l & c, Helen Cowcher; r, Leonora Box
100: l, Helen Cowcher; c, Roger Hughes; r, Stuart Lafford
101: l & r, Stuart Lafford; c, Victoria Goaman
102: l, Stuart Lafford; c, Victoria Goaman; r, Wendy Bramall
103: l, John Rignall; c & r, Leonora Box
104: Colin Emberson
105: l, Shirley Hooper; c, Barbara Walker; r, Colin Emberson
106: l & c, Barbara Walker; r, Roger Hughes
107: l, Josiane Campan; cl & r, Colin Emberson; cr, Guy Michel
108: l, Helen Senior; c & r, Brenda Katté
109: Leonora Box
110: l & c, Leonora Box; r, Stuart Lafford
111: l, Stuart Lafford; r, John Rignall
112: l, Colin Emberson; c, Guy Michel; r, Stuart Lafford
113: l, Leonora Box; cl, Line Mailhé; cr, Brenda Katté; r, Colin Emberson
114: l, Brenda Katté; c, Colin Emberson
115: l, Wendy Bramall; c, Maurice Espérance; r, Helen Senior
116: l, Guy Michel; cl & cr, Helen Senior; r, Line Mailhé
117: Helen Senior
118: Paul Wrigley; cl, Wendy Bramall; cr & r, Victoria Goaman
119: Colin Emberson
120: l & c, Guy Michel; r, Barbara Walker
121: l, Guy Michel; c, Marjory Saynor; r, Barbara Walker
122: l, Brenda Katté; cl, Sarah Fox-Davies; cr, Stuart Lafford; r, Leonora Box
123: l, Leonora Box; c, Peter Wrigley; r, Stuart Lafford
124: l, Frankie Coventry; c, Leonora Box; r, Colin Emberson
125: l, Stuart Lafford; c, Leonora Box; r, Philippe Couté
126–7: Wendy Bramall
128: l, Helen Cowcher; cl, Paul Wrigley; cr, Josiane Campan; r, Colin Emberson
129: l, Marie-Claude Guyetand; cl, Colin Emberson; cr & r, Guy Michel
130: l–cr, Colin Emberson; r, Philippe Couté
131: l, Philippe Couté; c & r, Stuart Lafford
132: l, Line Mailhé; c, Paul Wrigley; r, Stephanie Harrison
133: l, Sarah Fox-Davies; c, John Rignall; r, Guy Michel
134: l & cr, Colin Emberson; cl, Brenda Katté; r, Stuart Lafford
135: Stuart Lafford
136: Stuart Lafford; cl, Leonora Box; cr & r, Wendy Bramall
137: l–cr, Brenda Katté; r, Leonora Box
138: l & cl, Brenda Katté; cr & r, Leonora Box
139: Leonora Box
140: l, Brenda Katté; r, Leonora Box
141: Guy Michel; c, Wendy Bramall; r, Sarah Fox-Davies
142–3: Jim Russell
144: l & c, Richard Bonson; r, Brian Delf
145: l & r, Brian Delf; c, Shirley Felts
146: Brian Delf

147: l & c, Ann Savage; r, Nicholas Hall
148: Ann Savage
149: l & c, Ann Savage; r, Brian Delf
150–1: Brian Delf
152: l, Brian Delf; c & r, Ian Garrard
153: Ian Garrard
154: l & c, Ian Garrard; r, Ann Savage
155: Ann Savage
156: l, Nicholas Hall; c, Ann Savage; r, Ian Garrard
157: l, Ian Garrard; c, Nicholas Hall; r, Ann Savage
158: l & c, Ian Garrard; r, Ann Savage
159–61: Ann Savage
162: l & c, Derek Rodgers; r, Shirley Felts
163: Shirley Felts
164–5: Derek Rodgers
166: Brian Delf
167: l, Brian Delf; r, Ann Savage
168–9: Brian Delf
170: l & c, Ann Savage; r, Brian Delf
171: l & r, Ann Savage; c, Brian Delf
172–7: David Salariya
178–83: Richard Bonson
184–5: Jim Russell
186: l, Guy Michel; cl & r, Erhard Ludwig; cr, Brenda Katté
187: l, Pierre Brochard; cl, Guy Michel; cr, François Vitalis; r, Erhard Ludwig
188: lb & clb, Richard Bonson; lt, clt & r, Brenda Katté; cr, Erhard Ludwig
189: l, Helga Marxmüller; cl & cr, Guy Michel; r, Erhard Ludwig
190: l, Erhard Ludwig; cl, Line Mailhé; cr, Brenda Katté; r, Guy Michel
191: l, Erhard Ludwig; cl, Guy Michel; crt, Brenda Katté; crb & r, Line Mailhé
192: ltr & cr, Brenda Katté; lb, Josiane Campan; cl. Helga Marxmüller; r, Line Mailhé
193: l, cr, tr & r, Brenda Katté; cl, Helga Marxmüller; crb, Line Mailhé
194: l & cl, Guy Michel; cr, Brenda Katté; r, Line Mailhé
195: l & rb; Guy Michel; clt, Josiane Campan; clb & rt, Brenda Katté; cr, Erhard Ludwig
196: Brenda Katté
197: Liz Pepperell
198: Wendy Bramall
199–201: Ann Winterbotham
202–3: Jim Russell
204–5: John Francis
206: l, Trevor Boyer; c & r, Tim Hayward
207: Trevor Boyer
208: l, Robert Morton; c & r, Robert Gillmor
209: Robert Morton
210: l & c, Sean Milne; r, Stephen Adams
211: Stephen Adams
212: l, Stephen Adams, cl–r, Robert Morton
213–17: Robert Morton
218–22: Ken Wood
223–6: Tim Hayward
227: John Francis
228: l, John Francis; c & r, Trevor Boyer
229–30: Trevor Boyer
231: l & c, Tim Hayward; r, Trevor Boyer
232: l, Tim Hayward; c, Robert Morton; r, Trevor Boyer

233: Robert Morton
234: l & c, John Francis; r, Trevor Boyer
235: l, John Francis; c & r, Trevor Boyer
236–41: Trevor Boyer
242: Tim Hayward
243: l, Tim Hayward; c & r, Peter Barrett
244–5: Trevor Boyer
246: l, Peter Barrett; r, Robert Morton
247–9: Robert Morton
250: Norman Arlott
251–2: Denys Ovenden
253: l & c, Denys Ovenden; r, Norman Arlott
254–6: Peter Barrett
257–66: Norman Arlott
267: l & c, Ken Wood; r, Norman Arlott
268: l, Norman Arlott; c & r, Ken Wood
269–70: Ken Wood
271: l, Robert Morton; c & r, Ken Wood
272: Ken Wood
273: l & r, Robert Morton; c, Ken Wood
274–5: Robert Morton
276–7: Jim Russell
278–81: Leonora Box
282: l & c, Leonora Box; r, Colin Emberson
283–8: Colin Emberson
289: l, Colin Emberson; c & r, Pat Flavel
290–4: Helen Senior
295: Barbara Walker
296–7: Liz Pepperell
298–300: Rachel Birkett
301: Guy Michel
302: l & c, Brenda Katté; r, Guy Michel
303: l & c, Line Mailhé; r, Richard Lewington
304: l, Richard Lewington; c, Barbara Walker;
r, Josiane Campan
305: l, Barbara Walker; cl, Line Mailhé,
cr, Guy Michel; r, Brenda Katté
306: l, Josiane Campan; cl & rb, Brenda Katté;
cr & rc, Guy Michel; rt, Pat Flavel
307: l–cr, Line Mailhé; r, Josiane Campan
308: l & r, Guy Michel; cr, Josiane Campan
309: l, Guy Michel; cl–r, Line Mailhé
310: br, Line Mailhé;
remainder, Guy Michel/Colin Emberson
311: l & cl, Line Mailhé; cr & r, Brenda Katté
312: l & cl, Josiane Campan;
cr & r, Brenda Katté
313: l & c, Guy Michel; r, Josiane Campan
314: Brenda Katté
315–17: Richard Bonson
318: Norman Lacey
319: l, Norman Lacey; c & r, Andrew Robinson
320: Norman Lacey
321: lt, Jeane Colville; lb, Stephen Adams;
clb & rt, Barbara Walker;
crb & rb, Norman Lacey;
remainder, Sally Smith
322–3: Richard Bonson
324: l & c, Ann Savage; r, David Baird
325: cb, Barbara Walker; remainder, David Baird
326: Ann Savage
327: l & cl, Elizabeth Rice; cr, Sandra Pond;
r, Richard Lewington
328: Sandra Pond
329: l, Adrian Williams; clt & r, Sandra Pond;
clc & clb, Richard Lewington
330–1: Tricia Newell

332–3: Richard Bonson
334–5: Jim Russell
336: l, Peter Barrett; r, Eric Robson
337–9: Peter Barrett
340–3: John Francis
344: br, Eric Robson; remainder, Libby Turner
345: l & c, Eric Robson; r, Peter Barrett
346: l & c, Brian Delf; r, Gill Tomblin
347: Gill Tomblin
348–50: Sarah Fox-Davies
351: Brian Delf
352: l & c, Sarah Fox-Davies; r, Peter Barrett
353: l & c, Jim Channell; r, Robert Morton
354–7: Gill Tomblin
358: Peter Barrett
359: Colin Newman
360–1: Phil Weare
362: Rosalind Hewitt
363: Phil Weare
364–5: Jim Russell
366: Denys Ovenden
367: l, Richard Bonson; ct & cb, Colin Newman;
cc & rt, Stuart Lafford
368–9: Mick Loates
370: Stuart Lafford
371: l & cl, Mick Loates; cr & r, Stuart Lafford
372–5: Mick Loates
376: lb, clt, crt, crc, Andrew Robinson;
rt, Phil Weare; remainder, Tricia Newell
377: lt, clt, crt & r, Tricia Newell,
remainder, Andrew Robinson
378: lt, clt & rc, Andrew Robinson;
crt, Richard Bonson; br, Norman Lacey;
remainder, Tricia Newell
379: lt & cb, Andrew Robinson; ct, Phil Weare;
remainder, Tricia Newell
380: l & c, Mick Loates; r, Colin Newman
381: l, Colin Newman; c, Mick Loates,
r, Stuart Lafford
382: l, Robin Armstrong;
remainder, Stuart Lafford
383: l & cl Mick Loates, cr, Colin Newman;
r, Stuart Lafford
384–5: Colin Newman
386–9: Richard Bonson
390: Wendy Bramall
391: Sue Stitt
392–3: Sue Wickison
394–7: Jim Channell
398–400: Wendy Bramall
401–2: Ann Winterbotham
403–5: Sue Stitt

The publishers acknowledge their indebtedness to the following books, which were consulted for reference during the creation of this volume:

An Angler's Entomology by J.R. Harris (Collins)
The Atlas of Breeding Birds in Britain and Ireland by J.T.R. Sharrock (Poyser)
Atlas of the British Flora edited by F.H. Perring and S.M. Walters (Thomas Nelson)
The Audubon Society Field Guide to North American Birds (Alfred Knopf)

The Birdlife of Britain by Peter Hayman and Philip Burton (Mitchell Beazley)
Birds of Britain and Europe by Nicholas Hammond and Michael Everett (Pan)
The Birds of Britain and Europe by Hermann Heinzel, Richard Fitter and John Parslow (Collins)
The Birdwatcher's Key by Bob Scott and Don Forrest (Warne)
Birdwatcher's Pocket Guide by Peter Hayman (Mitchell Beazley)
British and European Birds by Walter Thiede (Chatto & Windus)
British and European Fishes by Fritz Terofal (Chatto & Windus)
British Ferns and Mosses by P.G. Taylor (Eyre and Spottiswoode)
British Trees and Shrubs by R.D. Meikle (Eyre and Spottiswoode)
Butterflies and Moths in Britain and Europe by David Carter (Pan)
The Butterflies of Britain and Europe by Lionel Higgins and Brian Hargreaves (Collins)

Collecting and Studying Mushrooms, Toadstools and Fungi by Alan Major (Bartholomew)
The Complete Guide to British Wildlife by N. Arlott, R. Fitter and A. Fitter (Collins)
The Concise British Flora in Colour by W. Keeble Martin (Edbury Press and Michael Joseph)

Docks and Knotweeds of the British Isles by J.E. Lousley and D.H. Kent (Botanical Society of the British Isles)
The Dragonflies of Great Britain and Ireland by Cyril O. Hammond (Harley Books)
Drawings of British Plants by Stella Ross-Craig (Bell)

Evergreen Garden Trees and Shrubs by Anthony Huxley (Blandford)

A Field Guide to the British Countryside by Alfred Leutscher (New English Library)
A Field Guide to the Trees of Britain and Northern Europe by Alan Mitchell (Collins)
Flora of the British Isles by A.R. Clapham, T.G. Tutin and E.F. Warburg (Cambridge University Press)

Freshwater Fish of Britain, Ireland and Europe by Roger Phillips and Martyn Rix (Pan)
The Freshwater Life of the British Isles and *The Observer's Book of Pond Life* by John Clegg (Warne)
Fungi of Northern Europe by Sven Nilsson and Olle Persson (Penguin)

Grasses, Ferns, Mosses and Lichens of Great Britain and Ireland by Roger Phillips (Pan)
Grasses, Sedges and Rushes in Colour by M. Skytte Christiansen (Blandford)
Guide to Birds of Britain and Europe by Bertel Bruun (Hamlyn)
Guide to Mushrooms and Toadstools by Morten Lange and F. Bayard Hora (Collins)
Guide to Spiders of Britain and Northern Europe by Dick Jones (Country Life Books)
Guide to the Countryside of Britain and Northern Europe edited by Pat Morris (Hamlyn)
Guide to the Ferns, Mosses and Lichens of Britain and Northern and Central Europe by Hans Martin Jahns (Collins)
Guide to the Freshwater Fishes of Britain and North-Western Europe and *Guide to the Sea Fishes of Britain and Europe* by Bent J. Muus and Preben Dahlstrom (Collins)
Guide to the Insects of Britain and Western Europe by Michael Chinery (Collins)
Guide to the Seashore and Shallow Seas of Britain and Europe by A.C. Campbell (Hamlyn)
Guide to Wild Flowers of Britain and Europe by Barry Tebbs (Usborne)

Handbook of the Birds of Europe, the Middle East and North Africa edited by Stanley Cramp (Oxford University Press)
Handguide to the Butterflies and Moths of Britain and Europe by John Wilkinson and Michael Tweedie (Collins)
Handguide to the Sea Coast by Denys Ovenden and John Barrett (Collins)
Hillier's Manual of Trees and Shrubs (David and Charles)
A History of Britain's Trees by Gerald Wilkinson (Hutchinson)

The Insects by Peter Farb (Time-Life)

The Kingfisher Nature Handbook by Jeanette Harris (Ward Lock)

Life in Lakes and Rivers by T.T. Macan and E.B. Worthington (Collins)

Moths by E.B. Ford (Collins)
The Moths of the British Isles by Richard South (Warne)
Mushrooms and Other Fungi of Great Britain and Europe by Roger Phillips (Pan)
Mushrooms and Toadstools by Derek Reid (Kingfisher)

The Natural History of the Garden by Michael Chinery (Fontana/Collins)

The New Field Guide to Fungi by Eric Soothill and Alan Fairhurst (Michael Joseph)
The Oxford Book of Flowerless Plants by F.H. Brightman and B.E. Nicholson (Oxford University Press)
The Oxford Book of Insects by John Burton (Oxford University Press)
The Oxford Book of Trees by B.E. Nicholson and A.R. Clapham (Oxford University Press)
The Oxford Book of Wild Flowers by B.E. Nicholson, S. Ary and M. Gregory (Oxford University Press)

Pocket Guide to Butterflies by Paul Whalley (Mitchell Beazley)
Pocket Guide to Mushrooms and Toadstools by David N. Pegler (Mitchell Beazley)

RNSC Guide to British Wild Flowers by Franklyn Perring (Country Life Books)
RSPB Guide to British Birds by David Saunders (Hamlyn)

The Tree Key by Herbert L. Edlin (Warne)
Trees by Elizabeth Martin (Ward Lock)
Trees and Bushes by Helga Vedel and Johan Lange (Methuen)

Trees and Bushes of Europe by Oleg Polunin and Barbara Everard (Oxford University Press)
Trees and Shrubs by Kurt Harz (Chatto & Windus)
Trees in Britain, Europe and North America by Roger Phillips (Pan)
Trees of the British Isles edited by Barry Tebbs (Orbis)

The Wildflower Key by Francis Rose (Warne)
Wildflowers of Britain by Roger Phillips (Pan)
The Wild Flowers of Britain and Northern Europe by Richard Fitter, Alastair Fitter and Marjorie Blamey (Collins)
The Wild Flowers of the British Isles by Ian Garrard and David Streeter (Macmillan)
Willows and Poplars of Great Britain and Ireland by R.D. Meikle (Botanical Society of the British Isles)

The Young Specialist Looks at Marine Life by W. de Haas and F. Knorr (Burke Books)
The Young Specialist Looks at Pond Life by W. Engelhardt (Burke Books)
The Young Specialist Looks at Seashore by A. Kosch, H. Frieling and H. Janus (Burke Books)